Patty

Laverty

WAVELET THEORY

WAVELET THEORY

An Elementary Approach With Applications

David K. Ruch
Metropolitan State College of Denver

Patrick J. Van Fleet
University of St. Thomas

WILEY

A JOHN WILEY & SONS, INC., PUBLICATION

For general information on our other products and services or for technical support, please contact our Customer Care Department within the United States at (800) 762-2974, outside the United States at (317) 572-3993 or fax (317) 572-4002.

Wiley also publishes its books in a variety of electronic formats. Some content that appears in print may not be available in electronic format. For information about Wiley products, visit our web site at www.wiley.com.

Library of Congress Cataloging-in-Publication Data:

Ruch, David K., 1959–
  Wavelet theory: an elementary approach with applications / David K. Ruch, Patrick J. Van Fleet
      p. cm.
  Includes bibliographical references and index.
  ISBN 978-0-470-38840-2 (cloth)
  1. Wavelets (Mathematics) 2. Transformations (Mathematics) 3. Digital Images—Mathematics. I. Van Fleet, Patrick J., 1962– II. Title.
  QA403.3.V375 2009
  515'.2433—dc22   2009017249

Printed in the United States of America.

10 9 8 7 6 5 4 3 2 1

To Pete and Laurel
for a lifetime of encouragement
(DKR)

To Verena, Sam, Matt, and Rachel
for your unfailing support
(PVF)

CONTENTS

Preface

This book presents some of the most current ideas in mathematics. Most of the theory was developed in the past twenty years, and even more recently, wavelets have found an important niche in a variety of applications. The filter pair we present in Chapter 8 is used by JPEG2000 [59] and the Federal Bureau of Investigation [8] to perform image and fingerprint compression, respectively. Wavelets are also used in many other areas of image processing as well as in applications such as signal denoising, detection of the onset of epileptic seizures [2], modeling of distant galaxies [3], and seismic data analysis [34, 35].

The development and advancement of the theory of wavelets came through the efforts of mathematicians with a variety of backgrounds and specialties, and of engineers and scientists with an eye for better solutions and models in their applications. For this reason, our goal was to write a book that provides an introduction to the essential ideas of wavelet theory at a level accessible to undergraduates and at the same time to provide a detailed look at how wavelets are used in "real-world" applications. Too often, books are heavy on theory and pay little attention to the details of application. For example, the discrete wavelet transform is but one piece of an image compression algorithm, and to understand this application, some attention must be given to quantization and coding methods. Alternatively, books might provide a detailed description of an application that leaves the reader curious about the theoretical foundations of some of the mathematical concepts used in the model. With this book, we have attempted to balance these two competing yet related tenets, and it is ultimately up to the reader to determine if we have succeeded in this endeavor.

To the Student

If you are reading this book, then you are probably either taking a course on wavelets or are working on your own to understand wavelets. Very often students are naturally curious about a topic and wish to understand quickly their use in applications. Wavelets provide this opportunity — the discrete Haar wavelet transformation is easy to understand and use in applications such as image compression. Unfortunately, the discrete Haar wavelet transformation is *not* the best transformation to use in many applications. But it does provide us with a concrete example to which we can refer as we learn about more sophisticated wavelets and their uses in applications. For this reason, you should study carefully the ideas in Chapters 3 and 4. They provide a framework for all that follows. It is also imperative that you develop a good working knowledge of the Fourier series and transformations introduced in Chapter 2. These ideas are very important in many areas of mathematics and are the basic tools we use to construct the wavelet filters used in many applications.

If you are a mathematics major, you will learn to write proofs. This is quite a change from lower-level mathematics courses where computation was the main objective. Proof-writing is sometimes a formidable task and the best way to learn is to practice. An indirect benefit of a course based on this book is the opportunity to hone your proof-writing skills. The proofs of most of the ideas in this book are straightforward and constructive. You will learn about proof by induction and contraposition. We have provided numerous problems that ask you to complete the details of a portion of a proof or mimic the ideas of one case in a proof to complete another. We strongly encourage you to tackle as many of these problems as possible. This course should provide a good transition from the proofs you see in a sophomore linear algebra course to the more technical proofs you might see in a real analysis course.

Of course, the book also contains many computational problems as well as problems that require the use of a computer algebra system (CAS). It is important that you learn how to use a CAS — both to solve problems and to investigate new concepts. It is amazing what you can learn by taking examples from the book and using a CAS to understand them or even change them somewhat to see the effects. We strongly encourage you to install the software packages described below and to visit the course Web site and work through the many labs and projects that we have provided.

To the Instructor

In this book we focus on bridging the gap often left between discrete wavelet transformations and the traditional multiresolution analysis-based development of wavelet theory. We provide the instructor with an opportunity to balance and integrate these ideas, but one should be wary of getting bogged down in the finer details of either topic. For example, the material on Fourier series and transforms is a place where instructors should use caution. These topics can be explored for entire semesters, and deservedly so, but in this course they need to be treated as tools rather than the thrust of the course.

The heart of wavelet theory is covered in Chapters 3, 5, and 6 in a comprehensive approach. Extensive details and examples are given or outlined via problems, so students should be able to gain a full understanding of the theory without hand-waving at difficult material. Having said that, some proofs are omitted to keep a nice flow to the book. This is not an introductory analysis book, nor is the level of rigor up to that of a graduate text. For example, the technical proofs of the completeness and separation properties of multiresolution analyses are left to future courses. The order of infinite series and integration are occasionally swapped with comment but not rigorous justification. We choose not to develop fully the theory of Riesz bases and how they lead to true dual multiresolutions of $L^2(\mathbb{R})$, for this would leave too little time for the very real applications of biorthogonal filters. We hope students will whet their appetites for future courses from the taste of theory they are given here!

We feel that the discrete wavelet transform material is essential to the spirit of the book, and based on our experience, students will find the applications quite gratifying. It may be tempting to expand on our introduction to these ideas after the Haar spaces are built, but we hope sufficient time is left for the development of multiresolution analyses and the Daubechies wavelets, which can take considerable time.

We also hope that instructors will take the time for thorough treatment of the connections between standard wavelet theory and discrete wavelet transforms. Our experience, both personally and with teaching other faculty at workshops, is that these connections are very rewarding but are not obvious to most beginners in the field. Some interesing problems crop up as we move between $L^2(\mathbb{R})$ and finite-dimensional approximations.

Text Topics

In Chapter 1, we provide a quick introduction to the complex plane and $L^2(\mathbb{R})$, with no prior experience assumed, emphasizing only the properties that will be needed for wavelet development. We believe that the fundamentals of wavelets can be studied in depth without getting into the intricacies of measure theory or the Lesbesgue integral, so we discuss briefly the measure of a set and convergence in norm versus pointwise convergence, but we do not dwell heavily on these ideas.

In Chapter 2 we present Fourier series and the Fourier transform in a limited and focused fashion. These ideas and their properties are developed only as tools to the extent that we need them for wavelet analysis. Our goal here is to prepare quickly for the study of wavelets in the transform domain. For example, the transform rules on translation and dilation are given, since these are critical for manipulating scaling function symbols in the transform domain. B-splines are introduced in this chapter as an important family of functions that will be used throughout the book, especially in Chapter 8.

In Chapter 3 we begin our study of wavelets in earnest with a comprehensive examination of Haar spaces. All the major ideas of multiresolution analysis are here, cast in the accessible Haar setting. The properties and standard notations of approximation spaces V_j and detail spaces W_j are developed in detail with numerous examples.

Students may be ready for some applications after the long Haar space analysis, and we present some classics in Chapter 4. The ideas behind filters and the discrete Haar wavelet transform are introduced first. The basics of processing signals and images are developed in Sections 4.1 and 4.2, with sufficient detail so that students can carry out the calculations and fully understand what software is doing while processing large images. The attractive and very accessible topics of image compression and edge detection are introduced as applications in Section 4.3.

In Chapter 5 we generalize the Haar space concepts to a general multiresolution analysis, beginning with the main properties in the time domain. Section 5.2 begins the development of critical multiresolution properties in the transform domain. In Section 5.3 we present some concrete examples of functions satisfying multiresolution properties. In addition to Haar, the Shannon wavelet and B-splines are discussed, each of which has some desirable properties but is missing others. This also provides some motivation for the formidable challenge of developing Daubechies wavelets. We return to B-splines in Chapter 8.

Chapter 6 centers on the Daubechies construction of continuous, compactly supported scaling functions. After a detailed development of the ideas, a clear algorithm is given for the construction. The next two sections are devoted to the cascade algorithm, which we delay presenting until after the Daubechies construction, with the motivation of plotting these amazing scaling functions with only a dilation equation to guide us. The cascade algorithm is introduced in the time domain, where examples make it intuitively clear, and is then discussed in the transform domain. Finally, we study the practical issue of coding the algorithm with discrete vectors.

After the rather heavy theory of Chapters 5 and 6, an investigation of the discrete Daubechies wavelet transform and applications in Chapter 7 provides a nice change of pace. An important concept in this chapter is that of handling the difficulties encountered when the decomposition and reconstruction formula are truncated, which are investigated in Section 7.2. Our efforts are rewarded with applications to image compression, noise reduction and image segmentation in Section 7.3.

In Chapter 8 we introduce scaling functions and wavelets in the biorthogonal setting. This is a generalization of an orthogonal multiresolution analysis with a single scaling function to a dual multiresolution analysis with a pair of biorthogonal scaling functions. We begin by introducing several new ideas via an example from B-splines, with an eye toward creating symmetric filters to be used in later applications. The main structural framework for dual multiresolution analyses and biorthogonal wavelets is developed in Section 8.2. We then move to constructing a family of biorthogonal filters based on B-splines using the methods due to Ingrid Daubechies in Section 8.3. The Cohen–Daubechies–Feauveau CDF97 filter pair is used in the JPEG2000 and FBI fingerprint compression standards, so it is natural to include them in the book. The method of building biorthogonal spline filters can be adjusted fairly easily to create the CDF97 filter pair, and this construction is part of Section 8.3. The pyramid algorithm can be generalized for the biorthogonal setting and is presented in Section 8.4. The discrete biorthogonal wavelet transform is discussed in Section 8.5. An advantage of biorthogonal filter pairs is that they can be made symmetric, and this desirable property affords a method, also presented in Section 8.5, of dealing with edge conditions in signals or digital images. A fundamental theoretical underpinning of dual multiresolution analyses is the concept of a Riesz basis, which is a generalization of orthogonal bases. The very formidable specifics of Riesz bases have been suppressed throughout most of this chapter in an effort to provide a balance between theory and applications. As a final and optional topic in this chapter, a brief examination of Riesz bases is provided in Section 8.6.

Wavelet packets, the topic of Chapter 9, provide an alternative wavelet decomposition method but are more computationally complex since the decomposition includes splitting the detail vectors as well as the approximations. We introduce wavelet packet functions in Section 9.1 and wavelet packet spaces in Section 9.2. The discrete wavelet packet transform is presented in Section 9.3 along with the best basis algorithm. The wavelet packet decomposition allows for redundant representations of the input vector or matrix, and the best basis algorithm chooses the "best" representation. This is a desirable feature of the transformation as this algorithm can be made application-dependent. The FBI fingerprint compression standard uses the CDF97 biorthogonal filter pair in conjunction with a wavelet packet transformation, and we outline this standard in Section 9.4.

Prerequisites

The minimal requirements for students taking this course are two semesters of calculus and a course in sophomore linear algebra. We use the ideas of bases, linear independence, and projection throughout the book so students need to be comfortable with these ideas before proceeding. The linear algebra prerequisite also provides the necessary background on matrix manipulations that appear primarily in sections dealing with discrete transformations. Students with additional background in Fourier series or proof-oriented courses will be able to move through the material at a much faster pace than will students with the minimum requirements. Most proofs in the book are of a direct and constructive nature, and some utilize the concept of mathematical induction. The level of sophistication assumed increases steadily, consistent with how students should be growing in the course. We feel that reading and writing proofs should be a theme throughout the undergraduate curriculum, and we suggest that the level of rigor in the book is accessible by advanced juniors or senior mathematics students. The constant connection to concrete applications that appears throughout the book should give students a good understanding of why the theory is important and how it is implemented. Some algorithms are given and experience with CAS software is very helpful in the course, but significant programming experience is not required.

Possible Courses for this Book

The book can serve as a stand-alone introduction to wavelet theory and applications for students with no previous exposure to wavelets. If a brisk pace is kept in line with the prerequisites discussed above, the course could include the first six chapters plus the discrete Daubechies transform and a sample of its applications. While considerable time can be spent on applied projects, we strongly recommend that any course syllabus include Chapter 6, on Daubechies wavelets. The construction of these wavelets is a remarkable mathematical achievement accomplished during our lifetime (if not those of our students) and should be covered if at all possible.

Some instructors may prefer to first cover Chapters 3 and 4 on Haar spaces before introducing the Fourier material of Chapter 2. This approach will work well since aside from a small discussion of the Fourier series associated with the Haar filter, no ideas from Fourier analysis are used in Chapters 3 and 4.

A very different course can be taught if students have already completed a course using Van Fleet's book *Discrete Wavelet Transformations: An Elementary Approach with Applications* [60]. Our book can be viewed as a companion text, with consistent notation, themes, and software packages. Students with this experience can move quickly through the applications, focusing on the traditional theory and its connections to discrete transformations. Students completing the discrete course should have a good sense of where the material is headed, as well as motivation to see the theoretical development of the various discrete transform filters. In this case, some sections of the text can be omitted and the entire book could be covered in one semester.

A third option exists for students who have a strong background in Fourier analysis. In this case, the instructor could concentrate heavily on the theoretical ideas in Chapters 5, 6, 8, and 9 and develop a real appreciation for how Fourier methods can be used to drive the theory of multiresolution analysis and filter design.

Problem Sets, Software Package, and Web Site

Problem solving is an essential part of learning mathematics, and we have tried to provide ample opportunities for the student to do so. After each section there are problem sets with a variety of exercises. Many allow students to fill in gaps in proofs from the text narrative, as well as to provide proofs similar to those given in the text. Others are fairly routine paper–pencil exercises to ensure that students understand examples, theorem statements, or algorithms. Many require computer work, as discussed in the next paragraph. We have provided 430 problems in the book to facilitate student comprehension material covered. Problems marked with a ★ should be assigned and address ideas that are used later in the text.

Many concepts in the book are better understood with the aid of computer visualization and computation. For these reasons, we have built the software package ContinuousWavelets to enhance student learning. This package is modeled after the DiscreteWavelets package that accompanies Van Fleet's book [60]. These packages are available for use with the computer algebra systems (CAS)

Mathematica®, Matlab®, and Maple™. This new package is used in the text to investigate a number of topics and to explore applications. Both packages contain modules for producing all the filters introduced in the course as well as discrete transformations and their inverses for use in applications. Visualization tools are also provided to help the reader better understand the results of transformations. Modules are provided for applications such as data compression, signal/image denoising, and image segmentation. The ContinuousWavelets package includes routines for constructing scaling functions (via the cascade algorithm) and wavelet functions. Finally, there are routines to easily implement the ideas from Chapter 3 — students can easily construct piecewise constant functions and produce nice graphs of projections into the various V_j and W_j spaces.

The course Web site is

http://www.stthomas.edu/wavelets

On this site, visitors will find the software packages described above, several computer labs and projects of varying difficulty, instructor notes on teaching from the text, and some solutions to problems.

DAVID K. RUCH

PATRICK J. VAN FLEET

Denver, Colorado USA
St. Paul, Minnesota USA
March 2009

Acknowledgments

We are grateful to several people who helped us with the manuscript. Caroline Haddad from SUNY Geneseo, Laurel Rogers, and University of St. Thomas mathematicians Doug Dokken, Eric Rawdon, Melissa Shepart–Loe, and Magdalena Stolarska read versions of the first four chapters. They caught several errors and made many suggestions for improving the presentation.

We gratefully acknowledge the National Science Foundation for their support through a grant (DUE–0717662) for the development of the book and the computer software. We wish to thank our editor Susanne Steitz–Filler, for her help on the project. Radka Tezaur and David Kubes provided digital images that were essential in the presentation. We also wish to salute our colleagues Peter Massopust, Wasin So, and Jianzhang Wang, with whom we began our journey in this field in the 1990s.

Dave Ruch would like to thank his partner, Tia, for her support and continual reminders of the importance of applications, and their son, Alex, who worked through some of the Haar material for a school project.

Patrick Van Fleet would like to express his deep gratitude to his wife, Verena, and their three children, Sam, Matt, and Rachel. They allowed him time to work on the book and provided support in a multitude of ways. The project would not have been possible without their support and sacrifice.

D.K.R. and P.V.F.

CHAPTER 1

THE COMPLEX PLANE AND THE SPACE $L^2(\mathbb{R})$

We make extensive use of complex numbers throughout the book. Thus for the purposes of making the book self-contained, this chapter begins with a review of the complex plane and basic operations with complex numbers. To build wavelet functions, we need to define the proper space of functions in which to perform our constructions. The space $L^2(\mathbb{R})$ lends itself well to this task, and we introduce this space in Section 1.2.

We discuss the inner product in $L^2(\mathbb{R})$ in Section 1.3, as well as vector spaces and subspaces. In Section 1.4 we talk about bases for $L^2(\mathbb{R})$. The construction of wavelet functions requires the decomposition of $L^2(\mathbb{R})$ into nested subspaces. We frequently need to approximate a function $f(t) \in L^2(\mathbb{R})$ in these subspaces. The tool we use to form the approximation is the *projection* operator. We discuss (orthogonal) projections in Section 1.4.

1.1 COMPLEX NUMBERS AND BASIC OPERATIONS

Any discussion of the complex plane starts with the definition of the *imaginary unit*:

$$i = \sqrt{-1}$$

Wavelet Theory: An Elementary Approach with Applications. By D. K. Ruch and P. J. Van Fleet
Copyright © 2009 John Wiley & Sons, Inc.

We immediately see that

$$i^2 = (\sqrt{-1})^2 = -1, \quad i^3 = i^2 \cdot i = -i, \quad i^4 = (-1) \cdot (-1) = 1$$

In Problem 1.1 you will compute i^n for any integer n.

A *complex number* is any number of the form $z = a + bi$ where $a, b \in \mathbb{R}$. The number a is called the *real part* of z and b is called the *imaginary part* of z. The set of complex numbers will be denoted by $\mathbb{C}$. It is easy to see that $\mathbb{R} \subset \mathbb{C}$ since real numbers are those complex numbers with the imaginary part equal zero.

We can use the *complex plane* to envision complex numbers. The complex plane is a two-dimensional plane where the horizontal axis is used for the real part of complex numbers and the vertical axis is used for the imaginary part of complex numbers. To plot the number $z = a + bi$, we simply plot the ordered pair (a, b). In Figure 1.1 we plot some complex numbers.

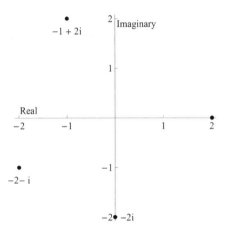

Figure 1.1   Some complex numbers in the complex plane.

Complex Addition and Multiplication

Addition and subtraction of complex numbers is a straightforward process. Addition of two complex numbers $u = a + bi$ and $v = c + di$ is defined as $y = u + v = (a + c) + (b + d)i$. Subtraction is similar: $z = u - v = (a - c) + (b - d)i$.

To multiply the complex numbers $u = a + bi$ and $v = c + di$, we proceed just as we would if $a + bi$ and $c + di$ were binomials:

$$u \cdot v = (a + bi)(c + di) = ac + adi + bci + bdi^2 = (ac - bd) + (ad + bc)i$$

Example 1.1 (Complex Arithmetic) *Let $u = 2 + i$, $v = -1 - i$, $y = 2i$, and $z = 3 + 2i$. Compute $u + v$, $z - v$, $u \cdot y$, and $v \cdot z$.*

Solution

$$u + v = (2 - 1) + (1 - 1)i = 1$$
$$z - v = (3 - (-1)) + (2 - (-1))i = 4 + 3i$$
$$u \cdot y = (2 + i) \cdot 2i = 4i + 2i^2 = -2 + 4i$$
$$v \cdot z = (-1 - i) \cdot (3 + 2i) = (3(-1) - (-1)2) + (3(-1) + 2(-1))i = -1 - 5i$$

■

Complex Conjugation

One of the most important operations used to work with complex numbers is *conjugation*.

Definition 1.1 (Conjugate of a Complex Numbers) *Let* $z = a + bi \in \mathbb{C}$. *The* conjugate *of* z, *denoted by* $\overline{z}$, *is defined by*

$$\overline{z} = a - bi$$

■

Conjugation is used to divide two complex numbers and also has a natural relation to the length of a complex number.

To plot $z = a + bi$, we plot the ordered pair (a, b) in the complex plane. For the conjugate $\overline{z} = a - bi$, we plot the ordered pair $(a, -b)$. So geometrically speaking, the conjugate $\overline{z}$ of z is simply the reflection of z over the real axis. In Figure 1.2 we have plotted several complex numbers and their conjugates.

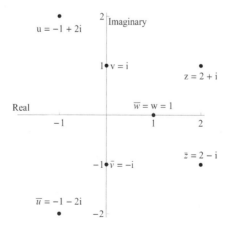

Figure 1.2   Complex numbers and their conjugates in the complex plane.

A couple of properties of the conjugation operator are immediate and we state them in the proposition below. The proof is left as Problem 1.3.

Proposition 1.1 (Properties of the Conjugation Operator) *Let* $z = a + bi$ *be a complex number. Then*

(a) $\overline{\overline{z}} = z$

(b) $z \in \mathbb{R}$ *if and only if* $\overline{z} = z$

■

Proof: Problem 1.3.

■

Note that if we graph the points $z = \cos\theta + i\sin\theta$ as θ ranges from 0 to 2π, we trace a circle with center $(0,0)$ with radius 1 in a counterclockwise manner. Note that if we produce the graph of $\overline{z} = \cos\theta - i\sin\theta$ as θ ranges from 0 to 2π, we get the same picture, but the points are drawn in a clockwise manner. Figure 1.3 illustrates this geometric interpretation of the conjugation operator.

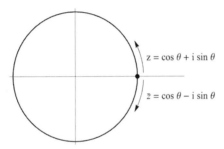

Figure 1.3   A circle is traced in two ways. Both start at $\theta = 0$. As θ ranges from 0 to 2π, the points z trace the circle in a counterclockwise manner while the points $\overline{z}$ trace the circle in a clockwise manner.

Modulus of a Complex Number

We can use the distance formula to determine how far the point $z = a + bi$ is away from $0 = 0 + 0i$ in the complex plane. The distance is $\sqrt{(a-0)^2 + (b-0)^2} = \sqrt{a^2 + b^2}$. This computation gives rise to the following definition.

Definition 1.2 (Modulus of a Complex Number) *The modulus of the complex number* $z = a + bi$ *is denoted by* $|z|$ *and is defined as*

$$|z| = \sqrt{a^2 + b^2}$$

■

Other names for the value $|z|$ are *length*, *absolute value*, and *norm* of z.

There is a natural relationship between $|z|$ and $\overline{z}$. If we compute the product $z \cdot \overline{z}$ where $z = a + bi$, we obtain

$$z \cdot \overline{z} = (a + bi)(a - bi) = a^2 - b^2 i^2 = a^2 + b^2$$

The right side of the equation above is simply $|z|^2$ so we have the following useful identity:

$$\boxed{|z|^2 = z \cdot \overline{z}} \tag{1.1}$$

In Problem 1.5 you are asked to compute the norms of some complex numbers.

Division of Complex Numbers

We next consider division of complex numbers. That is, given $z = a + bi$ and $y = c + di \neq 0$, how do we express the quotient z/y as a complex number? We proceed by multiplying both the numerator and denominator of the quotient by $\overline{y}$:

$$\frac{z}{y} = \frac{a + bi}{c + di} = \frac{a + bi}{c + di} \cdot \frac{c - di}{c - di} = \frac{(ac + bd) + (bc - ad)i}{c^2 + d^2} = \frac{ac + bd}{c^2 + d^2} + \frac{bc - ad}{c^2 + d^2} i$$

PROBLEMS

1.1 Let n be any integer. Find a closed formula for i^n.

1.2 Plot the numbers $3 - i, 5i, -1$, and $\cos\theta + i\sin\theta$ for $\theta = 0, \pi/4, \pi/2, 5\pi/6, \pi$ in the complex plane.

1.3 Prove Proposition 1.1.

1.4 Compute the following values.

(a) $(3 - i) + (2 + i)$

(b) $(1 + i) - \overline{(3 + i)}$

(c) $-i^3 \cdot (-2 + 3i)$

(d) $\overline{(2 + 5i)} \cdot (4 - i)$

(e) $\overline{(2 + 5i) \cdot (4 - i)}$

(f) $(2 - i) \div i$

(g) $(1 + i) \div (1 - i)$

1.5 For each complex number z, compute $|z|$.

(a) $z = 2 + 3i$

(b) $z = 5$

(c) $z = -4i$

(d) $z = \tan\theta + i$ where $\theta \in \left(-\frac{\pi}{2}, \frac{\pi}{2}\right)$

(e) z satisfies $z \cdot \overline{z} = 6$

1.6   Let $z = a + bi$ and $y = c + di$. For parts (a) – (d) show that:

(a) $\overline{y \cdot z} = \overline{y} \cdot \overline{z}$

(b) $|z| = |\overline{z}|$

(c) $|y \cdot z| = |y| \cdot |z|$

(d) $\overline{y + z} = \overline{y} + \overline{z}$

(e) Find the real and imaginary parts of $z^{-1} = \dfrac{1}{z}$.

★**1.7**   Suppose $z = a + bi$ with $|z| = 1$. Show that $\overline{z} = z^{-1}$.

★**1.8**   We can generalize Problem 1.6(d). Suppose that $z_k = a_k + b_k i$, for $k = 1, \ldots, n$. Show that

$$\overline{\sum_{k=1}^{n} z_k} = \sum_{k=1}^{n} \overline{z_k} = \sum_{k=1}^{n} a_k - i\sum_{k=1}^{n} b_k$$

★**1.9**   Suppose that $\sum\limits_{k \in \mathbb{Z}} a_k$ and $\sum\limits_{k \in \mathbb{Z}} b_k$ are convergent series where $a_k, b_k \in \mathbb{R}$. For $z_k = a_k + ib_k$, $k \in \mathbb{Z}$, show that

$$\overline{\sum_{k=1}^{\infty} z_k} = \sum_{k=1}^{\infty} \overline{z_k} = \sum_{k=1}^{\infty} a_k - i\sum_{k=1}^{\infty} b_k$$

★**1.10**   The identity in this problem is key to the development of the material in Section 6.1. Suppose that $z, w \in \mathbb{C}$ with $|z| = 1$. Show that

$$|(z - w)(z - 1/\overline{w})| = |w|^{-1}|z - w|^2$$

The following steps will help you organize your work:

(a) Using the fact that $|z| = 1$, expand $|z - w|^2 = (z - w)\overline{(z - w)}$ to obtain

$$|z - w|^2 = 1 + |w|^2 - w\overline{z} - \overline{w}z$$

(b) Factor $-\overline{w}z^{-1}$ from the right side of the identity in part (a) and use Problem 1.7 to show that

$$|z - w|^2 = -\overline{w}z^{-1}\left(z^2 - \frac{1 + |w|^2}{\overline{w}}z + \frac{w}{\overline{w}}\right)$$

(c) Show that the quadratic on the right-hand side of part (b) can be factored as $(z - w)(z - 1/\overline{w})$.

(d) Take norms of both sides of the identity obtained in part (c) and simplify the result to complete the proof.

1.2   THE SPACE $L^2(\mathbb{R})$

In order to create a mathematical model with which to build wavelet transforms, it is important that we work in a vector space that lends itself to applications in digital imaging and signal processing. Unlike $\mathbb{R}^N$, where elements of the space are N-tuples $\mathbf{v} = (v_1, \ldots, v_N)^T$, elements of our space will be functions. We can view a digital image as a function of two variables where the function value is the gray-level intensity, and we can view audio signals as functions of time where the function values are the frequencies of the signal. Since audio signals and digital images can have abrupt changes, we will not require functions in our space to necessarily be continuous. Since audio signals are constructed of sines and cosines and these functions are defined over all real numbers, we want to allow our space to hold functions that are supported (the notion of support is formally provided in Definition 1.5) on $\mathbb{R}$. Since rows or columns of digital images usually are of finite dimension and audio signals taper off, we want to make sure that the functions $f(t)$ in our space decay sufficiently fast as $t \to \pm\infty$. The rate of decay must be fast enough to ensure that the energy of the signal is finite. (We will soon make precise what we mean by the *energy* of a function.) Finally, it is desirable from a mathematical standpoint to use a space where the inner product of a function with itself is related to the size (norm) of the function. For this reason, we will work in the space $L^2(\mathbb{R})$. We define it now.

$L^2(\mathbb{R})$ Defined

Definition 1.3 (The Space L²(ℝ)) *We define the space $L^2(\mathbb{R})$ to be the set*

$$L^2(\mathbb{R}) = \left\{ f : \mathbb{R} \to \mathbb{C} \mid \int_{\mathbb{R}} |f(t)|^2 \, dt < \infty \right\} \qquad (1.2)$$

∎

Note: A reader with some background in analysis will understand that a rigorous definition of $L^2(\mathbb{R})$ requires knowledge of the Lebesgue integral and sets of measure zero. If the reader is willing to accept some basic properties obeyed by Lebesgue integrals, then Definition 1.3 will suffice.

We define the *norm* of a function in $L^2(\mathbb{R})$ as follows:

Definition 1.4 (The $\mathbf{L^2(\mathbb{R})}$ Norm) *Let* $f(t) \in L^2(\mathbb{R})$. *Then the norm of* $f(t)$ *is*

$$
\boxed{\|f(t)\| = \left(\int_{\mathbb{R}} |f(t)|^2 \, dt \right)^{\frac{1}{2}}} \tag{1.3}
$$

∎

The norm of the function is also referred to as the *energy* of the function. There are several properties that the norm should satisfy. Since it is a measure of energy or size, it should be nonnegative. Moreover, it is natural to expect that the only function for which $\|f(t)\| = 0$ is $f(t) = 0$. Some clarification of this property is in order before we proceed.

If $f(t) = 0$ for all $t \in \mathbb{R}$, then certainly $|f(t)|^2 = 0$, so that $\|f(t)\| = 0$. But what about the function that is 0 everywhere except, say, for a finite number of values? It is certainly possible that a signal might have such abrupt changes at a finite set of points. We learned in calculus that such a finite set of points has no bearing on the integral. That is, for $a < c < b$, $f(c)$ might not even be defined, but

$$
\int_a^b f(t) \, dt = \lim_{L \to c^-} \int_a^L f(t) \, dt + \lim_{L \to c^+} \int_L^b f(t) \, dt
$$

could very well exist. This is certainly the case when $f(t) = 0$ except at a finite number of values.

This idea is generalized using the notion of *measurable sets*. Intervals (a, b) are measured by their length $b - a$, and in general, sets are measured by writing them as a limit of the union of nonintersecting intervals. The measure of a single point a is 0, since for an arbitrarily small positive measure $\epsilon > 0$, we can find an interval that contains a and has measure less than ϵ (the interval $(a - \epsilon/4, a + \epsilon/4)$ with measure $\epsilon/2$ works). We can generalize this argument to claim that a finite set of points has measure 0 as well. The general definition of sets of measure 0 is typically covered in an analysis text (see Rudin [48], for example).

The previous discussion leads us to the notion of *equivalent functions*. Two functions $f(t)$ and $g(t)$ are said to be equivalent if $f(t) = g(t)$ except on a set of measure 0.

We state the following proposition without proof.

Proposition 1.2 (Functions for Which $\|\mathbf{f(t)}\| = \mathbf{0}$) *Suppose that* $f(t) \in L^2(\mathbb{R})$. *Then* $\|f(t)\| = 0$ *if and only if* $f(t) = 0$ *except on a set of measure 0.* ∎

Examples of Functions in $L^2(\mathbb{R})$

Our first example of elements of $L^2(\mathbf{R})$ introduces functions that are used throughout the book.

Example 1.2 (The Box ⊓(t), Triangle ∧(t), and Sinc Functions) *We define the* box function

$$\sqcap(t) = \begin{cases} 1, & 0 \le t < 1 \\ 0, & \textit{otherwise} \end{cases} \tag{1.4}$$

the triangle function

$$\wedge(t) = \begin{cases} t, & 0 \le t < 1 \\ 2 - t, & 1 \le t < 2 \\ 0, & \textit{otherwise} \end{cases} \tag{1.5}$$

and the sinc function

$$\text{sinc}(t) = \begin{cases} 1, & t = 0 \\ \dfrac{\sin(t)}{t}, & \textit{otherwise} \end{cases} \tag{1.6}$$

These functions are plotted in Figure 1.4.

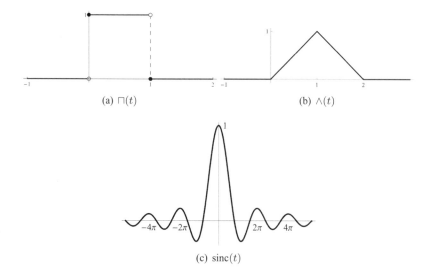

(a) ⊓(t)   (b) ∧(t)

(c) sinc(t)

Figure 1.4   The functions $\sqcap(t)$, $\wedge(t)$, and $\text{sinc}(t)$.

The box function is an element of $L^2(\mathbb{R})$. Since $\sqcap^2(t) = \sqcap(t)$, we have

$$\int_{\mathbb{R}} \sqcap^2(t)\,dt = \int_{\mathbb{R}} \sqcap(t)\,dt = \int_0^1 1\cdot dt = 1$$

To show that $\wedge(t) \in L^2(\mathbb{R})$, *we first note that*

$$\wedge(t)^2 = \begin{cases} t^2, & 0 \le t < 1 \\ (2-t)^2, & 1 \le t < 2 \\ 0, & \text{otherwise} \end{cases}$$

so that

$$\int_{\mathbb{R}} \wedge(t)^2\,dt = \int_0^1 t^2\,dt + \int_1^2 (2-t)^2\,dt = \frac{1}{3} + \frac{1}{3} = \frac{2}{3}$$

To see that $\text{sinc}(t) \in L^2(\mathbb{R})$, *we consider the integral*

$$\int_{\mathbb{R}} \text{sinc}^2(t)\,dt = \int_{-\infty}^0 \frac{\sin^2(t)}{t^2}\,dt + \int_0^\infty \frac{\sin^2(t)}{t^2}\,dt \qquad (1.7)$$

We split the second integral in (1.7) as follows:

$$\int_0^\infty \frac{\sin^2(t)}{t^2}\,dt = \int_0^1 \frac{\sin^2(t)}{t^2}\,dt + \int_1^\infty \frac{\sin^2(t)}{t^2}\,dt \qquad (1.8)$$

Note that the second integral in (1.8) is certainly nonnegative and we can bound the integral above by 1:

$$0 \le \int_1^\infty \frac{\sin^2(t)}{t^2}\,dt = \lim_{L\to\infty} \int_1^L \frac{\sin^2(t)}{t^2}\,dt$$

$$\le \lim_{L\to\infty} \int_1^L \frac{1}{t^2}\,dt$$

$$= \lim_{L\to\infty} -\frac{1}{t}\Big|_1^L$$

$$= \lim_{L\to\infty} -\frac{1}{L} + 1 = 1 \qquad (1.9)$$

Now we analyze the first integral on the right side of (1.8). By L'Hôpital's rule, we know that $\lim_{t\to 0} \frac{\sin^2(t)}{t^2} = 1$ *so* $\text{sinc}^2(t)$ *is a continuous function on* $[0,1]$. *From calculus, we recall that a continuous function on a closed interval achieves a maximum value, and it can be shown that the maximum value of* $\text{sinc}^2(t)$ *on* $[0,1]$ *is* 1 *(see Problem 1.12). Thus*

$$0 \le \int_0^1 \frac{\sin^2(t)}{t^2}\,dt \le \int_0^1 1\cdot dt = 1 \qquad (1.10)$$

Combining (1.9) and (1.10) gives

$$\int_0^\infty \text{sinc}^2(t)\,dt = \int_0^\infty \frac{\sin^2(t)}{t^2}\,dt \le 2$$

In a similar manner we can bound the first integral in (1.7) by 2 as well so that $\text{sinc}(t) \in L^2(\mathbb{R})$.   ∎

Let's look at some other functions in $L^2(\mathbb{R})$.

Example 1.3 (Functions in $L^2(\mathbb{R})$) *Determine whether or not the following functions are in $L^2(\mathbb{R})$. For those functions in $L^2(\mathbb{R})$, compute their norm.*

(a) $f_1(t) = t^n$, *where* $n = 0, 1, 2, \ldots$

(b) $f_2(t) = t^2 \sqcap (t/4)$, *where* $\sqcap(t)$ *is the box function defined in Example 1.2*

(c) $f_3(t) = \begin{cases} \dfrac{i}{\sqrt{t}}, & t \geq 1 \\ 0, & \text{otherwise} \end{cases}$

(d) $f_4(t) = \begin{cases} \dfrac{1}{t}, & t \geq 1 \\ 0, & \text{otherwise} \end{cases}$

Solution
   For $f_1(t)$ we have $|f_1(t)|^2 = t^{2n}$, so that

$$\int_{\mathbb{R}} t^{2n} \, dt = \lim_{L \to \infty} \int_{-L}^{0} t^{2n} \, dt + \lim_{L \to \infty} \int_{0}^{L} t^{2n} \, dt$$

Both of these integrals diverge — in particular

$$\lim_{L \to \infty} \int_{0}^{L} t^{2n} \, dt = \lim_{L \to \infty} \frac{1}{2n+1} L^{2n+1} \to \infty$$

Thus we see that no monomials are elements of $L^2(\mathbb{R})$. We could generalize part (a) to easily show that polynomials are not elements of $L^2(\mathbb{R})$.
   Since $\sqcap(t/4)$ *is 0 whenever* $t \notin [0,4)$, *we can write* $f_2(t)$ *as*

$$f_2(t) = \begin{cases} t^2, & 0 \leq t < 4 \\ 0, & \text{otherwise} \end{cases}$$

so that

$$\int_{\mathbb{R}} |f_2(t)|^2 \, dt = \int_{0}^{4} (t^2)^2 \, dt = \int_{0}^{4} t^4 \, dt = \frac{1024}{5}$$

Thus $f_2(t) \in L^2(\mathbb{R})$ and $\|f_2(t)\| = \frac{32\sqrt{5}}{5}$. Computing the modulus of $f_3(t)$ gives

$$|f_3(t)|^2 = \begin{cases} \dfrac{1}{t}, & t \geq 1 \\ 0, & \text{otherwise} \end{cases}$$

$$\int_{\mathbb{R}} |f_3(t)|^2 \, | \, dt = \int_{1}^{\infty} \frac{1}{t} \, dt = \lim_{L \to \infty} \int_{1}^{L} t^{-1} \, dt = \lim_{L \to \infty} \ln(L) \to \infty$$

Thus $f_3 \notin L^2(\mathbb{R})$. Finally, $|f_4(t)|^2 = t^{-2}$ so integrating over $t \geq 1$ gives

$$\int_{\mathbb{R}} |f_4(t)|^2 \, dt = \int_1^{\infty} t^{-2} \, dt = \lim_{L \to \infty} \int_1^L t^{-2} \, dt = \lim_{L \to \infty} 1 - 1/L = 1$$

So we see that $f_4(t) \in L^2(\mathbb{R})$ and $\|f_4(t)\| = 1$.   ∎

The Support of a Function

As we will learn in subsequent chapters, the support of a function plays an important role in the theory we develop. We define it now:

Definition 1.5 (Support of a Function) *Suppose that $f(t) \in L^2(\mathbb{R})$. We define the support of f, denoted $\text{supp}(f)$, to be the set*

$$\text{supp}(f) = \{t \in \mathbb{R} \mid f(t) \neq 0\} \tag{1.11}$$

∎

Here are some examples to better illustrate Definition 1.5.

Example 1.4 (Examples of Function Support) *Find the support of each of the following functions:*

(a) $f(t) = \dfrac{1}{1 + |t|}$

(b) $\sqcap(t)$

(c) $\wedge(t)$

(d) $g(t) = \sum_{k=0}^{\infty} c_k \wedge (t - 2k)$, where $c_k \neq 0$, $k \in \mathbb{Z}$, and $\sum_{k=0}^{\infty} c_k^2 < \infty$

Solution
  We observe that $f(t) > 0$, so $\text{supp}(f) = \mathbb{R}$. In Problem 1.15 you will show that $f(t) \in L^2(\mathbb{R})$. The box function $\sqcap(t)$ is supported on the interval $[0,1)$. The triangle function $\wedge(t)$ is supported on the interval $(0,2)$.
  The final function is a linear combination of even-integer translates of triangle functions. It is zero on the interval $(-\infty, 0]$ and $g(2k) = 0$, $k = 1, 2, \ldots$ So

$$\text{supp}(g) = (0,2) \cup (2,4) \cup (4,6) \cup \cdots = \bigcup_{k=0}^{\infty} (2k, 2k + 2)$$

In Problem 1.17 you will show that $g(t) \in L^2(\mathbb{R})$.   ∎

The support of the functions in Example 1.4(b) and (c) were finite-length intervals. We are often interested in functions whose supports are contained in a finite–length interval and we define this type of support below.

Definition 1.6 (Functions of Compact Support) *Let $f(t) \in L^2(\mathbb{R})$. We say that f is compactly supported if* supp(f) *is contained in a closed interval of finite length. In this case we say that the compact support of f is the smallest closed interval $[a, b]$ such that* supp(f) $\subseteq [a, b]$. *This interval is denoted by* $\overline{\text{supp}(f)}$.   ∎

We know from Example 1.4 that supp($\wedge$) $= (0,2)$. The compact support of $\wedge(t)$ is $\overline{\text{supp}(\wedge)} = [0,2]$. In a similar manner, supp($\sqcap$) $= [0,1]$.

$L^2(\mathbb{R})$ Functions at $\pm\infty$

When motivating the definition of $L^2(\mathbb{R})$, we stated that we want functions that tend to 0 as $t \to \pm\infty$ in such a way that $\|f\|^2 < \infty$. The following proposal shows a connection between the rate of decay and the finite energy of a function in $L^2(\mathbb{R})$.

Proposition 1.3 (Integrating the "Tails" of an $L^2(\mathbb{R})$ Function) *Suppose that $f(t) \in L^2(\mathbb{R})$ and let $\epsilon > 0$. Then there exists a real number $L > 0$ such that*

$$\int_{-\infty}^{-L} |f(t)|^2 \, dt + \int_{L}^{\infty} |f(t)|^2 \, dt = \int_{|t|>L} |f(t)|^2 \, dt < \epsilon$$

∎

Before we prove Proposition 1.3, let's understand what it is saying. Consider the $L^2(\mathbb{R})$ function $f_4(t)$ from Example 1.3(d). Let's pick $\epsilon = 10^{-16}$. Then Proposition 1.3 says that for some $L > 0$, the "tail" $\int_{L}^{\infty} |f_4(t)|^2 \, dt$ of the integral $\int_{\mathbb{R}} |f_4(t)|^2 \, dt$ will satisfy

$$\int_{L}^{\infty} \frac{1}{t^2} \, dt < 10^{-16} = .0000000000000001$$

Proof: This proof requires some ideas from infinite series in calculus. Suppose that $\epsilon > 0$ and let's first consider the integral $\int_{0}^{\infty} |f(t)|^2 \, dt$. Since $f(t) \in L^2(\mathbb{R})$, we know that this integral converges to some value s. Now write the integral as a limit. For $N \in \mathbb{N}$ we write

$$\int_{0}^{\infty} |f(t)|^2 \, dt = \lim_{N \to \infty} \int_{0}^{N} |f(t)|^2 \, dt$$

$$= \lim_{N \to \infty} \int_{0}^{1} |f(t)|^2 \, dt + \int_{1}^{2} |f(t)|^2 \, dt + \cdots + \int_{N-1}^{N} |f(t)|^2 \, dt$$

$$= \lim_{N \to \infty} \sum_{k=0}^{N} \int_{k}^{k+1} |f(t)|^2 \, dt$$

$$= \lim_{N \to \infty} \sum_{k=0}^{N} a_k$$

where $a_k = \int\limits_{k}^{k+1} |f(t)|^2\, dt \geq 0$. Thus we can view the integral as the sum of the series $\sum\limits_{k=0}^{\infty} a_k$ and we know that the series converges to s.

Recall from calculus that a series converges to s if its sequence of partial sums $s_N = \sum\limits_{k=0}^{N} a_k$ converges to s. The formal definition of a convergent sequence says that for all $\epsilon > 0$, there exists an $N_0 \in \mathbb{N}$ such that whenever $N \geq N_0$, we have $|s_N - s| < \epsilon$. We use this formal definition with $\epsilon/2$. That is, for $\frac{\epsilon}{2} > 0$, there exists an $N_0 \in \mathbb{N}$ such that $N \geq N_0$ implies that $|s_N - s| < \frac{\epsilon}{2}$.

In particular, for $N = N_0$ we have

$$|s_{N_0} - s| < \frac{\epsilon}{2} \tag{1.12}$$

But

$$
\begin{aligned}
|s_{N_0} - s| &= \left| \sum_{k=0}^{N_0} a_k - \int_0^{\infty} |f(t)|^2\, dt \right| \\
&= \left| \sum_{k=0}^{N_0} \int_k^{k+1} |f(t)|^2\, dt - \int_0^{\infty} |f(t)|^2\, dt \right| \\
&= \left| \int_0^{N_0} |f(t)|^2\, dt - \int_0^{\infty} |f(t)|^2\, dt \right|
\end{aligned}
$$

Now $\int\limits_{0}^{\infty} |f(t)|^2\, dt \geq \int\limits_{0}^{N_0} |f(t)|^2\, dt$, so that we can drop the absolute-value symbols and rewrite the last identity as

$$
\begin{aligned}
|s_{N_0} - s| &= \left| \int_0^{N_0} |f(t)|^2\, dt - \int_0^{\infty} |f(t)|^2\, dt \right| \\
&= \int_0^{\infty} |f(t)|^2\, dt - \int_0^{N_0} |f(t)|^2\, dt \\
&= \int_{N_0}^{\infty} |f(t)|^2\, dt \tag{1.13}
\end{aligned}
$$

Combining (1.12) and (1.13) gives

$$\int_{N_0}^{\infty} |f(t)|^2\, dt < \frac{\epsilon}{2}$$

In a similar manner (see Problem 1.19), we can find $N_1 > 0$ such that

$$\int_{-\infty}^{-N_1} |f(t)|^2\, dt < \frac{\epsilon}{2}$$

Now choose L to be the bigger of N_0 and N_1. Then $L \geq N_0$, $-L \leq -N_1$, and from this we can write

$$\int_L^\infty |f(t)|^2 \, dt \leq \int_{N_0}^\infty |f(t)|^2 \, dt < \frac{\epsilon}{2} \tag{1.14}$$

and

$$\int_{-\infty}^{-L} |f(t)|^2 \, dt \leq \int_{-\infty}^{-N_1} |f(t)|^2 \, dt < \frac{\epsilon}{2} \tag{1.15}$$

Combining (1.14) and (1.15) gives

$$\int_L^\infty |f(t)|^2 \, dt + \int_{-\infty}^L |f(t)|^2 \, dt < \frac{\epsilon}{2} + \frac{\epsilon}{2} = \epsilon$$

and the proof is complete.   ∎

Convergence in $L^2(\mathbb{R})$

In an elementary calculus class, you undoubtedly talked about sequences and limits. We are also interested in looking at sequences in $L^2(\mathbb{R})$. We need to be a little careful when describing convergence. It does not make sense to measure convergence at a point since equivalent functions might disagree on some set of measure 0. Since we are measuring everything using the norm, it is natural to view convergence in this light as well.

Definition 1.7 (Convergence in $\mathbf{L^2}(\mathbb{R})$) *Suppose that $f_1(t), f_2(t), \ldots$ is a sequence of functions in $L^2(\mathbb{R})$. We say that $\{f_n(t)\}_{n \in \mathbb{N}}$ converges to $f(t) \in L^2(\mathbb{R})$ if for all $\epsilon > 0$, there exists $L \geq 0$ such that whenever $n > L$, we have $\|f_n(t) - f(t)\| < \epsilon$.* ∎

If a sequence of functions $f_n(t)$, $n = 1, 2, \ldots$ converges in $L^2(\mathbb{R})$ to $f(t)$, we also say that the sequence of functions *converges in norm* to $f(t)$.

Other than the use of the norm, the definition should look similar to that of the formal definition of the limit of a sequence, which is covered in many calculus books (see Stewart [55], for example). The idea is that no matter how small a distance ϵ we pick, we are guaranteed that we can find an $N > 0$ so that $n > N$ ensures that the distance between $f_n(t)$ and $f(t)$ is smaller than ϵ.

Example 1.5 (Convergence in $\mathbf{L^2}(\mathbb{R})$) *Let $f_n(t) = t^n \sqcap (t)$. Show that $f_n(t)$ converges (in an $L^2(\mathbb{R})$ sense) to 0.*
Solution
Let $\epsilon > 0$. We note that $\sqcap(t)^2 = \sqcap(t)$ and compute

$$\|f_n(t) - 0\| = \left(\int_{\mathbb{R}} t^{2n} \sqcap (t) \, dt \right)^{1/2}$$

$$= \left(\int_0^1 t^{2n} \, dt \right)^{1/2}$$

$$= 1/\sqrt{2n+1}$$

Now, if we can say how to choose L so that $1/\sqrt{2n+1} < \epsilon$ whenever $n > L$, we are done. The natural thing to do is to solve the inequality $1/\sqrt{2n+1} < \epsilon$ for n. We obtain

$$n > \frac{1}{2}\left(\frac{1}{\epsilon^2} - 1\right)$$

Now the right-hand side is negative when $\epsilon > 1$, and in this case we can choose any $L \geq 0$. Then $n > L$ gives $\frac{1}{\sqrt{2n+1}} < 1 < \epsilon$.

When $\epsilon \leq 1$, we take $L = \frac{1}{2}\left(\frac{1}{\epsilon^2} - 1\right)$ to complete the proof.   ∎

In a real analysis class, we study *pointwise* convergence of sequences of functions. Convergence in norm is quite different, and in Problem 1.20 you will investigate the pointwise convergence properties of the sequence of functions from Example 1.5.

PROBLEMS

1.11   Suppose that $\int_{\mathbb{R}} f(t)\,dt < \infty$. Show that:

(a) $\int_{\mathbb{R}} f(t-a)\,dt = \int_{\mathbb{R}} f(t)\,dt$ for any $a \in \mathbb{R}$

(b) $\int_{\mathbb{R}} f(mt+b)\,dt = \dfrac{1}{m}\int_{\mathbb{R}} f(t)\,dt$ for $m, b \in \mathbb{R}$, with $m \neq 0$

1.12   Use the definition (1.6) of $\operatorname{sinc}(t)$ in Example 1.2 and L'Hôpital's rule to show that $f(t) = \operatorname{sinc}^2(t)$ is continuous for $t \in \mathbb{R}$ and then show that the maximum value of $f(t)$ is 1.

1.13   Determine whether or not the following functions are elements of $L^2(\mathbb{R})$. For those functions that are in $L^2(\mathbb{R})$, compute their norms.

(a) $f(t) = e^{-|t|}$

(b) $r(t) = \begin{cases} 1, & t \geq 1 \\ 0, & t < 1 \end{cases}$

(c) $g(t) = (1+t^2)^{-1/2}$

(d) $h(t) = \sqcap(t)\left(\cos(2\pi t) + i\sin(2\pi t)\right)$

1.14   Give examples of $L^2(\mathbb{R})$ functions f such that $\|f(t)\| = 0$ and $f(t) \neq 0$ at

(a) a single point

(b) five points

(c) an infinite number of points

★**1.15**   Show that the function $f(t) = \dfrac{1}{1+|t|} \in L^2(\mathbb{R})$.

★**1.16**   Consider the function

$$f(t) = \begin{cases} \frac{1}{2}t^3 - \frac{1}{2}t + 3, & -2 < t < 1 \\ 2t - 4, & 1 \leq t < 2 \\ 0, & \text{otherwise} \end{cases}$$

Show that $f(t) \in L^2(\mathbb{R})$.

1.17   Show that the function $g(t)$ from Example 1.4(d) is in $L^2(\mathbb{R})$.

1.18   Find the support of each of the following functions. For those functions f that are compactly supported, identify $\text{supp}(f)$.

(a) $f_1(t) = e^{-t^2}$

(b) $f_2(t) = \sqcap(2t - k), k \in \mathbb{Z}$

(c) $f_3(t) = \sqcap(2^j t - k), j, k \in \mathbb{Z}$

(d) $f_4(t) = \text{sinc}(t)$

(e) $f_5(t) = \sqcap\left(\frac{t}{n\pi}\right) \text{sinc}(\pi t)$, where n is a positive integer

(f) $f_6(t) = \sum\limits_{k=0}^{n} c_k v(t - 2k)$, where $c_k \neq 0, k \in \mathbb{Z}$,

$$v(t) = \begin{cases} |t|, & -1 < t < 1 \\ 0, & \text{otherwise} \end{cases}$$

and n is a positive integer

1.19   Complete the proof of Proposition 1.3. That is, show that if $f(t) \in L^2(\mathbb{R})$ and $\epsilon > 0$, there exists an $N_1 > 0$ such that

$$\int_{-\infty}^{-N_1} |f(t)|^2 \, dt < \epsilon/2$$

1.20   In Example 1.5 we showed that $f_n(t) = t^n \sqcap(t)$ converges in norm to $f(t) = 0$. Does $f_n(t)$ converge pointwise to 0 for all $t \in \mathbb{R}$?

1.21   Consider the sequence of functions

$$g_n(t) = \begin{cases} t^n, & -1 \leq t < 1 \\ 0, & \text{otherwise} \end{cases}$$

(a) Show that $g_n(t)$ converges in norm to 0.

(b) Show that $\lim\limits_{n \to \infty} g_n(a) = 0$ for $a \neq \pm 1$.

(c) Compute $\lim\limits_{n \to \infty} g_n(1)$ and $\lim\limits_{n \to \infty} g_n(-1)$.

1.3   INNER PRODUCTS

Recall that for vectors $\mathbf{u}, \mathbf{v} \in \mathbb{R}^N$, with $\mathbf{u} = (u_1, u_2, \ldots, u_N)^T$ and $\mathbf{v} = (v_1, v_2, \ldots, v_N)^T$, we define the inner product as

$$\mathbf{u} \cdot \mathbf{v} = \sum_{k=1}^{N} u_k v_k$$

Inner Products Defined

We can also define an inner product for functions in $L^2(\mathbb{R})$. In some sense it can be viewed as an extension of the definition above (see Van Fleet [60]). We state the formal definition of the inner product of two functions $f(t), g(t) \in L^2(\mathbb{R})$ at this time:

Definition 1.8 (The Inner Product in $\mathbf{L^2(\mathbb{R})}$) *Let $f(t), g(t) \in L^2(\mathbb{R})$. Then we define the* inner product *of $f(t)$ and $g(t)$ as*

$$\langle f(t), g(t) \rangle = \int_{\mathbb{R}} f(t)\overline{g(t)}\, dt \tag{1.16}$$

■

Here are some examples of the inner products:

Example 1.6 (Computing $\mathbf{L^2(\mathbb{R})}$ Inner Products) *Compute the following inner products:*

(a) *The triangle function $f(t) = \wedge(t)$ and the box function $g(t) = \sqcap(t)$ (see Example 1.2)*

(b) $f(t) = \begin{cases} t^{-1}, & t \geq 1 \\ 0, & t < 1 \end{cases}$ *and* $g(t) = \begin{cases} it^{-2}, & t \geq 1 \\ 0, & t < 1 \end{cases}$

(c) $f(t) = \wedge(t+1)$ *and* $g(t) = \begin{cases} \sin(2\pi t), & -1 \leq t \leq 1 \\ 0, & \text{otherwise} \end{cases}$

Solution
It is easy to verify that both functions in part (a) are in $L^2(\mathbb{R})$ and we also know that $\overline{g(t)} = g(t)$ since $g(t)$ is real-valued. Thus we can compute

$$\langle \wedge(t), \sqcap(t) \rangle = \int_{\mathbb{R}} \wedge(t) \sqcap(t)\, dt = \int_0^1 \wedge(t)\, dt = \int_0^1 t\, dt = \frac{1}{2}$$

For part (b), $\overline{g(t)} = -g(t)$. We have

$$\langle f(t), g(t) \rangle = -i \int_1^\infty t^{-1} t^{-2}\, dt = -i \int_1^\infty t^{-3}\, dt = -i \lim_{L \to \infty} \left. -\frac{1}{2t^2} \right|_1^L = -\frac{i}{2}$$

For the final inner product, both functions are real-valued and compactly supported on $[-1,1]$. *Note that* $\wedge(t+1)$ *is even and* $g(t)$ *is odd, so that the product* $\wedge(t+1)g(t)$ *is odd. Recall from calculus (see Stewart [55]) that the integral of an odd function over any finite interval symmetric about zero is zero. Thus* $\langle \wedge(t+1), g(t) \rangle = 0$. ∎

Properties of Inner Products

Example 1.6(b) illustrates an important point about the inner product that we have defined. You should verify that if we had reversed the order of $f(t)$ and $g(t)$, the resulting inner product would be $\frac{i}{2}$. On the other hand, in Example 1.6(a) it is easy to verify that $\langle \wedge(t), \sqcap(t) \rangle = \langle \sqcap(t), \wedge(t) \rangle = \frac{1}{2}$. Thus we see that the inner product is not necessarily a commutative operation. The following proposition describes exactly what happens if we commute functions and then compute their inner product.

Proposition 1.4 (The $L^2(\mathbb{R})$ Inner Product Is Not Commutative) *If* $f(t), g(t)$ *are functions in* $L^2(\mathbb{R})$, *then their inner product (1.16) satisfies*

$$\boxed{\langle g(t), f(t) \rangle = \overline{\langle f(t), g(t) \rangle}}$$

∎

Proof: The proof of this proposition is left as Problem 1.24. ∎

Since many of the inner products we will compute involve only real-valued functions. We have the following simple corollary:

Corollary 1.1 (Inner Products of Real-Valued $L^2(\mathbb{R})$ Functions Commute) *Suppose* $f(t), g(t) \in L^2(\mathbb{R})$ *and further assume* $f(t)$ *and* $g(t)$ *are real-valued functions. Then*

$$\boxed{\langle f(t), g(t) \rangle = \langle g(t), f(t) \rangle}$$

∎

Proof: This simple proof is left as Problem 1.25. ∎

One of the nice properties of $L^2(\mathbb{R})$ is the relationship between the inner product and the norm. We have the following proposition.

Proposition 1.5 (Relationship Between the $L^2(\mathbb{R})$ Norm and Inner Product) *For* $f(t) \in L^2(\mathbb{R})$, *we have*

$$\boxed{\|f\|^2 = \langle f(t), f(t) \rangle} \tag{1.17}$$

∎

Proof: Let $f(t) \in L^2(\mathbb{R})$. We use (1.1) to write $f(t)\overline{f(t)} = |f(t)|^2$. Integrating both sides over $\mathbb{R}$ gives

$$\int_{\mathbb{R}} f(t)\overline{f(t)}\, \mathrm{d}t = \int_{\mathbb{R}} |f(t)|^2 \, \mathrm{d}t$$

The left-hand side of this identity is $\langle f(t), f(t) \rangle$ and the right-hand side is $\| f(t) \|^2$. ∎

The next result describes how the inner product is affected by scalar multiplication.

Proposition 1.6 (Scalar Multiplication and the Inner Product) *Suppose that $f(t)$ and $g(t)$ are functions in $L^2(\mathbb{R})$, and assume that $c \in \mathbb{C}$. Then*

$$\langle cf(t), g(t) \rangle = c\langle f(t), g(t) \rangle \tag{1.18}$$

and

$$\langle f(t), cg(t) \rangle = \overline{c}\,\langle f(t), g(t) \rangle \tag{1.19}$$

∎

Proof: For (1.18), we have

$$\langle cf(t), g(t) \rangle = \int_{\mathbb{R}} cf(t)\overline{g(t)}\, dt = c \int_{\mathbb{R}} f(t)\overline{g(t)}\, dt = c\langle f(t), g(t) \rangle$$

and for (1.19), we have

$$\langle f(t), c\,g(t) \rangle = \int_{\mathbb{R}} f(t)\overline{c\,g(t)}\, dt = \overline{c} \int_{\mathbb{R}} f(t)\overline{g(t)}\, dt = \overline{c}\,\langle f(t), g(t) \rangle$$

∎

In the sequel we frequently compute inner products of the form $\langle f(t-k), g(t-\ell) \rangle$ or $\langle f(2t-k), g(2t-\ell) \rangle$, where $k, \ell \in \mathbb{Z}$ and $f(t), g(t) \in L^2(\mathbb{R})$. The following proposition gives reformulations of these inner products.

Proposition 1.7 (Translates and Dilates in Inner Products) *Suppose that $f(t)$ and $g(t)$ are functions in $L^2(\mathbb{R})$ and $k, \ell, m \in \mathbb{Z}$. Then*

$$\langle f(t-k), g(t-\ell) \rangle = \langle f(t), g(t-(\ell-k)) \rangle \tag{1.20}$$

and

$$\langle f(2^m t - k), g(2^m t - \ell) \rangle = 2^{-m}\,\langle f(t), g(t-(\ell-k)) \rangle \tag{1.21}$$

∎

Proof: The proof of this proposition is straightforward and is left as Problem 1.29. ∎

The Cauchy–Schwarz Inequality

Recall for vectors $\mathbf{u}, \mathbf{v} \in \mathbb{R}^N$, the *Cauchy–Schwarz* inequality (see Strang [56]) states that

$$\mathbf{u} \cdot \mathbf{v} \le \| \mathbf{u} \|\, \| \mathbf{v} \|$$

Functions in $L^2(\mathbb{R})$ also satisfy the *Cauchy–Schwarz inequality*. Before stating and proving the Cauchy–Schwarz inequality, we need to return once again to the concept of equivalent functions and their role in computing inner products. If $f(t), g(t) \in L^2(\mathbb{R})$ are equivalent, we would expect their inner products with any other function $h(t) \in L^2(\mathbb{R})$ to be the same. The following proposition, stated without proof, confirms this fact.[1]

Proposition 1.8 (Integrals of Equivalent Functions) *Suppose that $f(t)$, $g_1(t)$, and $g_2(t)$ are functions in $L^2(\mathbb{R})$ with $g_1(t) = g_2(t)$ except on a set of measure zero. Then $\langle f(t), g_1(t) \rangle = \langle f(t), g_2(t) \rangle$.* ∎

We are now ready to state the Cauchy–Schwarz inequality for functions in $L^2(\mathbb{R})$.

Proposition 1.9 (Cauchy–Schwarz Inequality) *Suppose that $f(t), g(t) \in L^2(\mathbb{R})$. Then*

$$\boxed{|\langle f(t), g(t) \rangle| \leq \|f(t)\| \cdot \|g(t)\|} \tag{1.22}$$

∎

Proof: First, suppose that

$$\|g(t)\|^2 = \int_{\mathbb{R}} |g(t)|^2 \, \mathrm{d}t = \langle g(t), g(t) \rangle = 0$$

Then by Proposition 1.2 (with $g_1(t) = g(t)$ and $g_2(t) = 0$), $g(t) = 0$ except on a set of measure 0. We next employ Proposition 1.8 to see that $\langle f(t), g(t) \rangle = 0$ and the result holds.

Now assume that $\|g(t)\|^2 > 0$ and for any $z \in \mathbb{C}$, consider

$$0 \leq \|f(t) + zg(t)\|^2 = \langle f(t) + zg(t), f(t) + zg(t) \rangle$$

We can expand the right-hand side of this identity to write

$$0 \leq \langle f(t), f(t) \rangle + \langle f(t), zg(t) \rangle + \langle zg(t), f(t) \rangle + \langle zg(t), zg(t) \rangle$$

Using Proposition 1.1, Proposition 1.5, and (1.19), we can rewrite the preceding equation as

$$0 \leq \|f(t)\|^2 + \overline{z} \langle f(t), g(t) \rangle + z\overline{\langle f(t), g(t) \rangle} + |z|^2 \|g(t)\|^2 \tag{1.23}$$

We now make a judicious choice for z. Since $\|g(t)\| \neq 0$, we take

$$z = -\frac{\langle f(t), g(t) \rangle}{\|g(t)\|^2}$$

The second term in (1.23) becomes

$$\overline{z} \langle f(t), g(t) \rangle = -\frac{\overline{\langle f(t), g(t) \rangle}}{\|g(t)\|^2} \langle f(t), g(t) \rangle = -\frac{|\langle f(t), g(t) \rangle|^2}{\|g(t)\|^2} \tag{1.24}$$

---

[1]The reader interested in the proof of Proposition 1.8 should consult an analysis text such as Rudin [48].

and the third term in (1.23) is

$$z\overline{\langle f(t), g(t)\rangle} = -\frac{\langle f(t), g(t)\rangle}{\|g(t)\|^2}\overline{\langle f(t), g(t)\rangle} = -\frac{|\langle f(t), g(t)\rangle|^2}{\|g(t)\|^2} \tag{1.25}$$

The last term in (1.23) can be written as

$$|z|^2\|g(t)\|^2 = \frac{|\langle f(t), g(t)\rangle|^2}{\|g(t)\|^4}\|g(t)\|^2 = \frac{|\langle f(t), g(t)\rangle|^2}{\|g(t)\|^2} \tag{1.26}$$

Inserting (1.24), (1.25), and (1.26) into (1.23) gives

$$0 \leq \|f(t)\|^2 - 2\frac{|\langle f(t), g(t)\rangle|^2}{\|g(t)\|^2} + \frac{|\langle f(t), g(t)\rangle|^2}{\|g(t)\|^2}$$

$$= \|f(t)\|^2 - \frac{|\langle f(t), g(t)\rangle|^2}{\|g(t)\|^2}$$

Adding the second term on the right side to both sides of the identity above gives

$$\frac{|\langle f(t), g(t)\rangle|^2}{\|g(t)\|^2} \leq \|f(t)\|^2$$

Finally, we multiply both sides of this inequality by $\|g(t)\|^2$ to obtain

$$|\langle f(t), g(t)\rangle|^2 \leq \|f(t)\|^2 \cdot \|g(t)\|^2$$

Taking square roots of this last inequality gives the desired result.   ∎

We have referred to the space $\mathbb{R}^N$ several times in this chapter. The space $\mathbb{R}^N$ is a standard example of a *vector space*. Basically, a vector space is a space where addition and scalar multiplication obey fundamental properties. We require the sum of any two vectors, or the product of a scalar and a vector, to remain in the space. We also want addition to be commutative, associative, and distributive over scalar multiplication. We want the space to contain a zero element and additive inverses for all elements in the space. Scalar multiplication should be associative as well as distributive over vectors, and we also require a multiplicative identity. Certainly, $\mathbb{R}^N$ is an example of a vector space (see Strang [56] for more details on vector spaces) with standard vector addition and real numbers as scalars.

Vector Spaces

We summarize these properties in the following formal definition of a vector space.

Definition 1.9 (Vector Space) *Let* $\mathcal{V}$ *be any space such that addition of elements of* $\mathcal{V}$ *and multiplication of elements by scalars from a set* $\mathcal{F}$ *are well defined.*[2] *Then* $\mathcal{V}$ *is called a* vector space *or* linear space *over* $\mathcal{F}$ *if*

---

[2]The set of scalars $\mathcal{F}$ for $\mathcal{V}$ is called a *field*. For our purposes, our set of scalars will be either the real or the complex numbers.

(a) $\mathcal{V}$ is closed under addition. That is, for all $\mathbf{u}, \mathbf{v} \in \mathcal{V}$, we have $\mathbf{u} + \mathbf{v} \in \mathcal{V}$.

(b) Addition is commutative and associative. That is, for all $\mathbf{u}, \mathbf{v} \in \mathcal{V}$, we have

$$\mathbf{u} + \mathbf{v} = \mathbf{v} + \mathbf{u}$$

and for all $\mathbf{u}, \mathbf{v}, \mathbf{w} \in \mathcal{V}$, we have

$$\mathbf{u} + (\mathbf{v} + \mathbf{w}) = (\mathbf{u} + \mathbf{v}) + \mathbf{w}$$

(c) There exists an additive identity $\mathbf{0} \in \mathcal{V}$ so that for all $\mathbf{v} \in \mathcal{V}$, we have

$$\mathbf{v} + \mathbf{0} = \mathbf{v}$$

(d) For each $\mathbf{v} \in \mathcal{V}$, there exists an additive inverse $-\mathbf{v} \in \mathcal{V}$ so that for all $\mathbf{v} \in \mathcal{V}$, we have

$$\mathbf{v} + -\mathbf{v} = \mathbf{0}$$

(e) $\mathcal{V}$ is closed under scalar multiplication. That is, for $c \in \mathcal{F}$ and $\mathbf{v} \in \mathcal{V}$, we have $c\mathbf{v} \in \mathcal{V}$.

(f) Scalar multiplication is associative. That is, for $c, d \in \mathcal{F}$ and $\mathbf{v} \in \mathcal{V}$, we have

$$(cd)\mathbf{v} = c(d\mathbf{v})$$

(g) For all $\mathbf{v} \in \mathcal{V}$, we have $1 \cdot \mathbf{v} = \mathbf{v}$.

(h) Addition distributes over scalar multiplication and scalar multiplication distributes over addition. That is, for $\mathbf{u}, \mathbf{v} \in \mathcal{V}$ and $c, d \in \mathcal{F}$, we have

$$c(\mathbf{u} + \mathbf{v}) = c\mathbf{u} + c\mathbf{v}$$
$$(c + d)\mathbf{v} = c\mathbf{v} + d\mathbf{v}$$

■

If we review the properties listed in Definition 1.9, it is easy to verify that $L^2(\mathbb{R})$ satisfies properties (b)–(h). You are asked to do so in Problem 1.30. The most difficult property to verify is the fact that $L^2(\mathbb{R})$ is closed under addition. The triangle inequality, stated and proved below, ensures that $L^2(\mathbb{R})$ is closed under addition and is thus a vector space.

Proposition 1.10 (The Triangle Inequality) *Suppose that $f(t), g(t) \in L^2(\mathbb{R})$. Then*

$$\boxed{\|f(t) + g(t)\| \leq \|f(t)\| + \|g(t)\|} \tag{1.27}$$

■

Proof:

Let $f(t), g(t) \in L^2(\mathbb{R})$. We begin by computing

$$
\begin{aligned}
\|f(t) + g(t)\|^2 &= \langle f(t) + g(t), f(t) + g(t) \rangle \\
&= \|f(t)\|^2 + \langle f(t), g(t) \rangle + \langle g(t), f(t) \rangle + \|g(t)\|^2 \\
&\le \|f(t)\|^2 + |\langle f(t), g(t) \rangle| + |\langle g(t), f(t) \rangle| + \|g(t)\|^2
\end{aligned}
$$

Now we use the Cauchy–Schwarz inequality on the two inner products in the previous line. We have

$$
\begin{aligned}
\|f(t) + g(t)\|^2 &\le \|f(t)\|^2 + \|f(t)\| \cdot \|g(t)\| + \|g(t)\| \cdot \|f(t)\| + \|g(t)\|^2 \\
&= \|f(t)\|^2 + 2\|f(t)\| \cdot \|g(t)\| + \|g(t)\|^2 \\
&= (\|f(t)\| + \|g(t)\|)^2
\end{aligned}
$$

Taking square roots of both sides of the previous inequality gives the desired result. ∎

Subspaces

In many applications we are interested in special subsets of a vector space $\mathcal{V}$. We will insist that these subsets carry all the properties of a vector space. Such subsets are known as *subspaces*.

Definition 1.10 (Subspace) *Suppose that $\mathcal{W}$ is any nonempty subset of a vector space $\mathcal{V}$. $\mathcal{W}$ is called a* subspace *of $\mathcal{V}$ if whenever $\mathbf{u}, \mathbf{v} \in \mathcal{W}$ and c, d are any scalars, we have $c\mathbf{u} + d\mathbf{v} \in \mathcal{W}$.* ∎

In Problem 1.31 you will show that if $\mathcal{W}$ is a subspace of vector space $\mathcal{V}$, then $\mathcal{W}$ is a vector space. Below we give an example of a subspace.

Example 1.7 (An Example of a Subspace) *Consider the vector space $L^2(\mathbb{R})$ and let $\mathcal{W}$ be the set of all piecewise constant functions with possible breakpoints at the integers with the added condition that each function in $\mathcal{W}$ is zero outside the interval $[-N, N)$ for N some positive integer. Show that $\mathcal{W}$ is a subspace of $\mathcal{V}$.*
Solution
Suppose that $f(t), g(t) \in \mathcal{W}$. Then f and g must be linear combinations of some integer translates of the box function $\sqcap(t)$. That is, there exist scalars (complex numbers) $a_k, b_k, k = -N, \ldots, N-1$ such that

$$
f(t) = \sum_{k=-N}^{N-1} a_k \sqcap(t - k) \quad and \quad g(t) = \sum_{k=-N}^{N-1} b_k \sqcap(t - k)
$$

Now for $c, d \in \mathbb{C}$, we form the function

$$h(t) = cf(t) + dg(t)$$

$$= c \sum_{k=-N}^{N-1} a_k \sqcap (t-k) + d \sum_{k=-N}^{N-1} b_k \sqcap (t-k)$$

$$= \sum_{k=-N}^{N-1} (ca_k + db_k) \sqcap (t-k)$$

and note that $h(t)$ is a piecewise constant function with possible breakpoints at the integers and $h(t) = 0$ for $t \notin [-N, N)$. Since $h(t) \in \mathcal{W}$ we have $\mathcal{W} \neq \emptyset$ and by Definition 1.10, we see that $\mathcal{W}$ is a subspace of $L^2(\mathbb{R})$. ■

PROBLEMS

1.22 Compute the following inner products $\langle f(t), g(t) \rangle$:

(a) $f(t) = \wedge(t)$ and $g(t) = (1 + t^2)^{-1/2}$

(b) $f(t) = g(t) = e^{-|t|}$ (Compare your answer with Problem 1.13(a))

(c) $f(t) = \sqcap(t-k)$ and $g(t) = \wedge(t)$ where $k \in \mathbb{Z}$

1.23 In this problem we illustrate the result of Proposition 1.8. Let $h(t) = \sqcap\left(\frac{t}{2}\right)$. Define the functions

$$f(t) = \begin{cases} e^{-t}, & t > 0 \\ 0, & t \leq 0 \end{cases}$$

and

$$g(t) = \begin{cases} f(t), & t \notin \mathbb{Z} \\ 0, & t \in \mathbb{Z} \end{cases}$$

(a) Verify that $f(t), g(t)$, and $h(t)$ are in $L^2(\mathbb{R})$ and that $f(t)$ and $g(t)$ are equivalent functions.

(b) Show that $\langle f(t), h(t) \rangle = \langle g(t), h(t) \rangle$.

1.24 Prove Proposition 1.4.

1.25 Prove Corollary 1.1.

1.26 Let $f, g \in L^2(\mathbb{R})$. Show that $\|f(t) + g(t)\|^2 = \|f(t)\|^2 + \|g(t)\|^2$ if and only if $\langle f(t), g(t) \rangle = 0$.

1.27 Verify the Cauchy–Schwarz inequality for $f(t)$ and $h(t)$ in Problem 1.23.

1.28 Find two functions $f(t)$ and $g(t)$ that satisfy the equality part of the Cauchy–Schwarz inequality. That is, find $f(t)$ and $g(t)$ so that $|\langle f(t), g(t) \rangle| = \|f(t)\| \cdot \|g(t)\|$.

Can you state conditions in general that guarantee equality in the Cauchy–Schwarz inequality?

1.29   Prove Proposition 1.7. (*Hint:* Substitute for $t - k$ in the first identity and $2^m t - k$ in the second identity.)

1.30   Show that $L^2(\mathbb{R})$ satisfies properties (b)–(h) of Definition 1.9. Since $L^2(\mathbb{R})$ also satisfies the triangle inequality (1.27), we see that $L^2(\mathbb{R})$ is a vector space.

1.31   Suppose that $\mathcal{W}$ is a subspace of a vector space $\mathcal{V}$ over a set of scalars $\mathcal{F} = \mathbb{R}$ or $\mathcal{F} = \mathbb{C}$. Show that $\mathcal{W}$ is a vector space over $\mathcal{F}$.

1.4 BASES AND PROJECTIONS

Our derivation of wavelets depends heavily on the construction of orthonormal bases for subspaces of $L^2(\mathbb{R})$. Recall (see Strang [56]) that a *basis* for the vector space $\mathbb{R}^N$ is a set of N linearly independent vectors $\{\mathbf{u}^1, \ldots, \mathbf{u}^N\}$ that span $\mathbb{R}^N$. This basis is called *orthonormal* if $\mathbf{u}^j \cdot \mathbf{u}^k = 0$ whenever $j \neq k$, and $\mathbf{u}^j \cdot \mathbf{u}^j = 1$.

Bases

We can easily extend the ideas of basis and orthornormal basis to $L^2(\mathbb{R})$.

Definition 1.11 (Basis and Orthonormal Basis in $\mathbf{L^2(\mathbb{R})}$) *Suppose that $\mathcal{W}$ is a subspace of $L^2(\mathbb{R})$ and suppose that $\{e_k(t)\}_{k \in \mathbb{Z}}$ is a set of functions in $\mathcal{W}$. We say that $\{e_k(t)\}_{k \in \mathbb{Z}}$ is a basis for $\mathcal{W}$ if the functions span $\mathcal{W}$ and are linearly independent. We say that $\{e_k(t)\}_{k \in \mathbb{Z}}$ is an orthonormal basis for $\mathcal{W}$ if*

$$\langle e_j(t), e_k(t) \rangle = \begin{cases} 1, & j = k \\ 0, & j \neq k \end{cases} \tag{1.28}$$

∎

Suppose that $\mathcal{W}$ is a subspace of $L^2(\mathbb{R})$ and $\{e_k(t)\}_{k \in \mathbb{Z}}$ is an orthonormal basis for $\mathcal{W}$. Since $\{e_k(t)\}_{k \in \mathbb{Z}}$ is a basis for $\mathcal{W}$, we can write any $f(t) \in \mathcal{W}$ as

$$f(t) = \sum_{k \in \mathbb{Z}} a_k e_k(t) \tag{1.29}$$

where $a_k \in \mathbb{C}$. It would be quite useful to know more about the coefficients a_k, $k \in \mathbb{Z}$. We compute the inner product of both sides of (1.29) with $e_j(t)$ to obtain

$$\langle f(t), e_j(t) \rangle = \left\langle \sum_{k \in \mathbb{Z}} a_k e_k(t), e_j(t) \right\rangle$$

$$= \sum_{k \in \mathbb{Z}} a_k \langle e_k(t), e_j(t) \rangle$$

$$= \sum_{k \in \mathbb{Z}} a_k \int_{\mathbb{R}} e_k(t) \overline{e_j(t)} \, \mathrm{d}t$$

Since $\{e_k(t)\}_{k\in\mathbb{Z}}$ is an orthonormal basis, the integral in each term of the last identity is 0 unless $j = k$. In the case where $j = k$, the integral is 1 and the right-hand side of the last identity reduces to the single term:

$$a_j = \langle f(t), e_j(t)\rangle = \int_{\mathbb{R}} f(t)\overline{e_j(t)}\,\mathrm{d}t \qquad (1.30)$$

Projections

An orthonormal basis gives us a nice representation of the coefficients a_k, $k \in \mathbb{Z}$. Now suppose that $g(t)$ is an arbitrary function in $L^2(\mathbb{R})$. The representation (1.30) suggests a way that we can approximate $g(t)$ in the subspace $\mathcal{W}$. We begin by defining a *projection*.

Definition 1.12 (Projection) *Let $\mathcal{W}$ be a subspace of $L^2(\mathbb{R})$. Then $P\colon L^2(\mathbb{R}) \to \mathcal{W}$ is a* projection *from $L^2(\mathbb{R})$ into $\mathcal{W}$ if for all $f(t) \in \mathcal{W}$, we have $P(f(t)) = f(t)$.* ∎

Thus a projection is any linear transformation from $L^2(\mathbb{R})$ to subspace $\mathcal{W}$ that is an identity operator for $f(t) \in \mathcal{W}$. If you have taken a multivariable calculus class, you probably learned how to project vectors from $\mathbb{R}^2$ into the subspace $\mathcal{W} = \{c\mathbf{a} \mid c \in \mathbb{R}\}$ where $\mathbf{a}$ is some nonzero vector in $\mathbb{R}^2$ (see Stewart [55]). This is an example of a projection (using the vector space $\mathbb{R}^2$ instead of $L^2(\mathbb{R})$) with $P(\mathbf{v}) = \left(\frac{\mathbf{a}^T\mathbf{v}}{\|\mathbf{a}\|^2}\right)\mathbf{a}$.

A useful way to project $g(t) \in L^2(\mathbb{R})$ into a subspace $\mathcal{W}$ is to take an orthonormal basis $\{e_k(t)\}_{k\in\mathbb{Z}}$ and write

$$P(g(t)) = \sum_{k\in\mathbb{Z}} \langle g(t), e_k(t)\rangle e_k(t) \qquad (1.31)$$

We need to show that (1.31) is a projection from $L^2(\mathbb{R})$ into $\mathcal{W}$. To do so, we need the following auxiliary results. The proofs of both results are outlined as exercises.

Proposition 1.11 (The Norm of $P(g(t))$) *Let $\mathcal{W}$ be a subspace of $L^2(\mathbb{R})$ with orthonormal basis $\{e_k(t)\}_{k\in\mathbb{Z}}$. For $g(t) \in L^2(\mathbb{R})$, the function $P(g(t))$ defined in (1.31) satisfies*

$$\|P(g(t))\|^2 = \sum_{k\in\mathbb{Z}} |\langle g(t), e_k(t)\rangle|^2 \qquad (1.32)$$

∎

Proof: Problem 1.34. ∎

Proposition 1.12 (An Upper Bound for $\|P(g(t))\|$) *Let $\mathcal{W}$ be a subspace of $L^2(\mathbb{R})$ with orthonormal basis $\{e_k(t)\}_{k\in\mathbb{Z}}$. Then $\|P(g(t))\|$, where $P(g(t))$ is defined in (1.31), satisfies*

$$\|P(g(t))\| \leq \|g(t)\| \qquad (1.33)$$

$\blacksquare$

Proof: Problem 1.35. $\blacksquare$

Proposition 1.12 tells us that if $g(t) \in L^2(\mathbb{R})$, then so is $P(g(t))$. This fact is required to establish the following result.

Proposition 1.13 (A Projection from $\mathbf{L^2}(\mathbb{R})$ into $\mathcal{W}$) *Let $\mathcal{W}$ be a subspace of $L^2(\mathbb{R})$ with orthonormal basis $\{e_k(t)\}_{k \in \mathbb{Z}}$. Then the function $P(g(t))$ defined by (1.31) is a projection from $L^2(\mathbb{R})$ into $\mathcal{W}$.* $\blacksquare$

Proof: From Proposition 1.12, we know that $P(g(t)) \in L^2(\mathbb{R})$ whenever $g(t) \in L^2(\mathbb{R})$. We need to show that for any $f(t) \in \mathcal{W}$, we have $P(f(t)) = f(t)$.

Since $f(t) \in \mathcal{W}$, we can write it as a linear combination of basis functions:

$$f(t) = \sum_{j \in \mathbb{Z}} a_j e_j(t)$$

Then

$$P(f(t)) = \sum_{k \in \mathbb{Z}} \langle f(t), e_k(t) \rangle \, e_k(t)$$

$$= \sum_{k \in \mathbb{Z}} \left\langle \sum_{j \in \mathbb{Z}} a_j e_j(t), e_k(t) \right\rangle e_k(t)$$

$$= \sum_{k \in \mathbb{Z}} \sum_{j \in \mathbb{Z}} a_j \, \langle e_j(t), e_k(t) \rangle \, e_k(t)$$

Since $\{e_k(t)\}_{k \in \mathbb{Z}}$ is an orthonormal basis for $\mathcal{W}$, the inner product $\langle e_j(t), e_k(t) \rangle$ is nonzero only when $j = k$. In this case the inner product is 1. Thus the double sum in the last identity reduces to a single sum with j replaced by k. We have

$$P(f(t)) = \sum_{k \in \mathbb{Z}} a_k \, e_k(t) = f(t)$$

and the proof is complete. $\blacksquare$

PROBLEMS

1.32   Suppose that $\{e_k(t)\}$ is an orthonormal basis for $L^2(\mathbb{R})$. For $L \in \mathbb{Z}, L > 0$, let $f_L(t) = \sum\limits_{k=-L}^{L} a_k e_k(t)$. Show that

$$\|f_L(t)\|^2 = \sum_{k=-L}^{L} a_k^2$$

1.33   Suppose that P is a projection from vector space $\mathcal{V}$ into subspace $\mathcal{W}$. Show that $P^2 = P$. That is, show that for all $\mathbf{v} \in \mathcal{V}$, we have $P^2 \mathbf{v} = P\mathbf{v}$.

1.34   Expand the right-hand side of

$$\|P(g(t))\|^2 = \langle P(g(t)), P(g(t)) \rangle$$

$$= \left\langle \sum_{k \in \mathbb{Z}} \langle g(t), e_k(t) \rangle \, e_k(t), \sum_{k \in \mathbb{Z}} \langle g(t), e_k(t) \rangle \, e_k(t) \right\rangle$$

to provide a proof of Proposition 1.11.

1.35   In this problem you will prove Proposition 1.12. The following steps will help you organize the proof. Let $g(t) \in L^2(\mathbb{R})$ and suppose that $\{e_k(t)\}_{k \in \mathbb{Z}}$ is an orthonormal basis for subspace $\mathcal{W}$.

(a) Let $g_L(t) = \sum_{k=-L}^{L} \langle g(t), e_k(t) \rangle \, e_k(t)$. Show that

$$\langle g(t), g_L(t) \rangle = \langle g_L(t), g(t) \rangle = \sum_{k=-L}^{L} |\langle g(t), e_k(t) \rangle|^2$$

(b) Use part (a) to show that

$$\|g(t) - g_L(t)\|^2 = \|g(t)\|^2 - 2 \sum_{k=-L}^{L} |\langle g(t), e_k(t) \rangle|^2 + \|g_L(t)\|^2$$

(c) Show that $\|g_L(t)\|^2 = \sum_{k=-L}^{L} |\langle g(t), e_k(t) \rangle|^2$. This is a special case of Problem 1.32.

(d) Use parts (b) and (c) to write

$$\|g(t) - g_L(t)\|^2 + \sum_{k=-L}^{L} |\langle g(t), e_k(t) \rangle|^2 = \|g(t)\|^2$$

and thus infer that

$$\sum_{k=-L}^{L} |\langle g(t), e_k(t) \rangle|^2 \leq \|g(t)\|^2$$

(e) Use Proposition 1.11 and let $L \to \infty$ in part (d) to complete the proof.

CHAPTER 2

FOURIER SERIES AND FOURIER TRANSFORMATIONS

Fourier series and transformations are very popular and powerful tools in applied mathematics. For our purposes, Fourier series and transformations will be used to design *scaling filters* and *wavelet filters* which in turn allow us to construct *scaling functions* and *wavelet functions*. The ability to "work in the transform domain" is not only useful in wavelet theory — it is a key method of solution in many areas of mathematics. We give only a brief introduction to Fourier series and transformations in this chapter. For a more thorough introduction to Fourier analysis and its wide assortment of applications, the reader is encouraged to see the book by Kammler [38].

In Section 2.1 we introduce the complex exponential function. This is the essential construct for the Fourier series that are introduced in Section 2.2. In this section we compute Fourier series and learn some calculus necessary to construct one series from another. Another important idea that is covered in Section 2.2 is the notion of expanding functions built from $\sin(\omega)$ and $\cos(\omega)$ into Fourier series. We use this idea repeatedly throughout the book and it is important that you master it.

Section 2.3 introduces the Fourier transformation. For our purposes, this extension of Fourier series to the real line allows us to better analyze scaling and wavelet functions. The final section of the chapter begins with an introduction to convolution.

Wavelet Theory: An Elementary Approach with Applications. By D. K. Ruch and P. J. Van Fleet   **31**
Copyright © 2009 John Wiley & Sons, Inc.

This important mathematical operation, which can be viewed as a "moving inner product", has applications in many areas of mathematics. In particular, when we process a discrete signal or image, we typically convolve it with some prescribed filter. For functions in $L^2(\mathbb{R})$, convolution is defined as an integral and we use this definition to develop *B-spline functions*. This construction makes use of an important theorem that connects the convolution of two functions with the product of their Fourier transformations.

2.1 EULER'S FORMULA AND THE COMPLEX EXPONENTIAL FUNCTION

One of the most famous formulas for complex numbers is *Euler's formula*. To derive this formula, we recall the Maclaurin series for the functions $\cos(t)$, $\sin(t)$, and e^t. We have:

$$\cos(t) = 1 - \frac{t^2}{2!} + \frac{t^4}{4!} - \frac{t^6}{6!} + \cdots \tag{2.1}$$

$$\sin(t) = t - \frac{t^3}{3!} + \frac{t^5}{5!} - \frac{t^7}{7!} + \cdots \tag{2.2}$$

$$e^t = 1 + t + \frac{t^2}{2!} + \frac{t^3}{3!} + \frac{t^4}{4!} + \frac{t^5}{5!} + \frac{t^6}{6!} + \cdots \tag{2.3}$$

These series are quite similar. Indeed, if we replace t with it in (2.3), we obtain

$$e^{it} = 1 + it + \frac{(it)^2}{2!} + \frac{(it)^3}{3!} + \frac{(it)^4}{4!} + \frac{(it)^5}{5!} + \frac{(it)^6}{6!} + \cdots$$

$$= 1 + it - \frac{t^2}{2!} - i\frac{t^3}{3!} + \frac{t^4}{4!} + i\frac{t^5}{5!} - \frac{t^6}{6!} + \cdots$$

$$= \left(1 - \frac{t^2}{2!} + \frac{t^4}{4!} - \frac{t^6}{6!} + \cdots\right) + i\left(t - \frac{t^3}{3!} + \frac{t^5}{5!} - \frac{t^7}{7!} + \cdots\right)$$

Since the Maclaurin series for $\cos t$ and $\sin t$ converge for all real numbers t, we arrive at Euler's formula:

$$\boxed{e^{it} = \cos(t) + i\sin(t)} \tag{2.4}$$

Properties of e^{it}

At first glance it seems odd to see a connection between an exponential function and trigonometric functions. But as we will see throughout the book, representing numbers in terms of a *complex exponential* is quite common. We have already seen the right-hand side of (2.4). Indeed, Figure 1.3 shows us that $\cos(t) + i\sin(t)$ is a circle with center $(0,0)$ and radius 1 for $0 \leq t < 2\pi$, so we see that the graph of e^{it} is the same circle.

If we compute the conjugate of e^{it}, we obtain

$$\overline{e^{it}} = \overline{\cos(t) + i\sin(t)} = \cos(t) - i\sin(t)$$

On the other hand,

$$e^{-it} = \cos(-t) + i\sin(-t) = \cos(t) - i\sin(t)$$

so that

$$\boxed{\overline{e^{it}} = e^{-it} = \cos(t) - i\sin(t)} \tag{2.5}$$

In Problem 2.2 you are asked to evaluate e^{it} for various values of t. We will often have cause to evaluate e^{it} at the values $k\pi$ or $2k\pi$ where $k \in \mathbb{Z}$. In the former case we have

$$\boxed{e^{ik\pi} = \cos(k\pi) + i\sin(k\pi) = \cos(k\pi) = (-1)^k}$$

and in the case of the latter we compute

$$e^{2\pi i k} = \cos(2\pi k) + i\sin(2\pi k) = 1$$

Using (2.4), we can immediately compute the modulus of e^{it}. Using (1.1), we have

$$|e^{it}|^2 = \cos^2(t) + \sin^2(t) = 1$$

so that

$$\boxed{|e^{it}| = 1}$$

Trigonometric Functions in Terms of e^{it}

From (2.4) we know that the function $f(\omega) = e^{i\omega}$ is defined in terms of $\cos(\omega)$ and $\sin(\omega)$. Conversely, it is possible to express these trigonometric functions in terms of complex exponentials. We have

$$e^{i\omega} = \cos(\omega) + i\sin(\omega) \tag{2.6}$$
$$e^{-i\omega} = \cos(\omega) - i\sin(\omega) \tag{2.7}$$

Note: From this point on we use ω as our independent variable when working with complex exponentials and Fourier series.

If we add (2.6) and (2.7), we obtain

$$e^{i\omega} + e^{-i\omega} = \cos(\omega) + i\sin(\omega) + \cos(\omega) - i\sin(\omega)$$
$$= 2\cos(\omega)$$

Solving for $\cos(\omega)$ gives

$$\cos(\omega) = \frac{e^{i\omega} + e^{-i\omega}}{2} \tag{2.8}$$

In a similar manner (see Problem 2.5) we can write

$$\sin(\omega) = \frac{e^{i\omega} - e^{-i\omega}}{2i} \tag{2.9}$$

Complex Exponential Function

Euler's formula allows us to define one of the most important functions used in wavelet analysis.

Definition 2.1 (Complex Exponential Function) *Let* $k \in \mathbb{Z}$*. We define the* complex exponential function *as*

$$e_k(\omega) = e^{ik\omega} = \cos(k\omega) + i\sin(k\omega) \tag{2.10}$$

■

Since $e_k(\omega)$ is constructed using $\cos(k\omega)$ and $\sin(k\omega)$, it is easy to see that $e_k(\omega)$ is a $\frac{2\pi}{k}$–periodic function. In some cases, it is better to view k as the number of copies of $\cos(\omega)$ ($\sin(\omega)$) that appears in the real (imaginary) part of $e_k(\omega)$. Figure 2.1 illustrates this point.

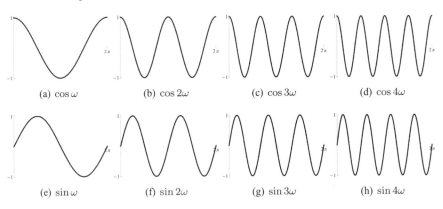

(a) $\cos \omega$ (b) $\cos 2\omega$ (c) $\cos 3\omega$ (d) $\cos 4\omega$

(e) $\sin \omega$ (f) $\sin 2\omega$ (g) $\sin 3\omega$ (h) $\sin 4\omega$

Figure 2.1 A graph of the real (top) and imaginary (bottom) parts of $e_k(\omega)$ for $k = 1, 2, 3, 4$.

Inner Product of Complex Exponential Functions

An important characteristic of the complex exponential function $e_k(\omega)$ is the value of the inner product of $e_k(\omega)$ and $e_j(\omega)$ over any interval of length 2π. We perform the computation over the interval $[-\pi, \pi)$, but in Problem 2.7 you will show that the computation is valid over any interval $[a, a + 2\pi)$, $a \in \mathbb{R}$.

We consider two cases for the integral

$$\langle e_k(\omega), e_j(\omega) \rangle = \int_{-\pi}^{\pi} e_k(\omega)\overline{e_j(\omega)} \, d\omega \tag{2.11}$$

Where $j = k$, the integrand is

$$e_k(\omega)\overline{e_k(\omega)} = |e_k(\omega)|^2$$

But $|e_k(\omega)|^2 = \cos^2(k\omega) + \sin^2(k\omega) = 1$, so that (2.11) becomes

$$\int_{-\pi}^{\pi} 1 \, d\omega = 2\pi \tag{2.12}$$

When $j \neq k$, we write

$$\int_{-\pi}^{\pi} e_k(\omega)\overline{e_j(\omega)} \, d\omega = \int_{-\pi}^{\pi} e^{ik\omega} e^{-ij\omega} \, d\omega = \int_{-\pi}^{\pi} e^{i(k-j)\omega} \, d\omega$$

$$= \int_{-\pi}^{\pi} \cos\big((k-j)\omega\big) \, d\omega + i \int_{-\pi}^{\pi} \sin\big((k-j)\omega\big) \, d\omega$$

We next make the substitution $u = (k-j)\omega$ so that $\frac{1}{k-j} \, du = d\omega$. We have

$$\int_{-\pi}^{\pi} e_k(\omega)\overline{e_j(\omega)} \, d\omega = \int_{-\pi}^{\pi} \cos\big((k-j)\omega\big) \, d\omega + i \int_{-\pi}^{\pi} \sin\big((k-j)\omega\big) \, d\omega$$

$$= \frac{1}{k-j} \int_{-(k-j)\pi}^{(k-j)\pi} \cos u \, du + \frac{i}{k-j} \int_{-(k-j)\pi}^{(k-j)\pi} \sin u \, du$$

In the second integral, the integrand is an odd function and the interval of integration is symmetric about 0 so that the integral is 0. Since the integrand in the first integral is an even function, we write

$$\frac{1}{k-j} \int_{-(k-j)\pi}^{(k-j)\pi} \cos u \, du = \frac{2}{k-j} \int_{0}^{(k-j)\pi} \cos u \, du = \frac{2}{k-j} \sin u \Big|_{0}^{(k-j)\pi} = 0 \tag{2.13}$$

Combining (2.12) and (2.13), we conclude that

$$\boxed{\langle e_k(\omega), e_j(\omega) \rangle = \int_{-\pi}^{\pi} e_k(\omega)\overline{e_j(\omega)} \, d\omega = \begin{cases} 2\pi, & j = k \\ 0, & j \neq k \end{cases}} \tag{2.14}$$

PROBLEMS

2.1   Suppose that $z \in \mathbb{C}$. Write down a formula (in terms of z) for all complex numbers w where $|w| = |z|$.

2.2   Compute e^{it} for $t = 0, \dfrac{\pi}{4}, \dfrac{\pi}{2}, \pi, 2\pi$, and $u + 2\pi$, where $u \in \mathbb{R}$.

2.3   Show that $f(t) = e^{it}$ is a 2π-periodic function. That is, show that $f(t + 2\pi) = f(t)$.

2.4   Prove *DeMoivre's theorem*. That is, show that for any nonnegative integer n, we have
$$(\cos\omega + i\sin\omega)^n = \cos n\omega + i\sin n\omega$$

2.5   Use (2.6) and (2.7) to establish (2.9).

2.6   Use (2.8) and (2.9) to establish the following trigonometric identities:

(a) $\cos^2(\omega) + \sin^2(\omega) = 1$

(b) $\sin(2\omega) = 2\sin(\omega)\cos(\omega)$

(c) $\cos(2\omega) = \cos^2(\omega) - \sin^2(\omega)$

(d) $\sin^2(\omega) = \dfrac{1 - \cos(2\omega)}{2}$

(e) $\cos^2(\omega) = \dfrac{1 + \cos(2\omega)}{2}$

★2.7   Show that for $a \in \mathbb{R}$ and $f(\omega)$ any integrable 2π-periodic function, we have
$$\int_a^{2\pi+a} f(\omega)\,d\omega = \int_0^{2\pi} f(\omega)\,d\omega$$

(*Hint:* Rewrite the integral as
$$\int_a^{2\pi+a} f(\omega)\,d\omega = \int_a^0 f(\omega)\,d\omega + \int_0^{2\pi} f(\omega)\,d\omega + \int_{2\pi}^{2\pi+a} f(\omega)\,d\omega$$

and then show that
$$\int_a^0 f(\omega)\,d\omega = -\int_{2\pi}^{2\pi+a} f(\omega)\,d\omega$$
The substitution $u = \omega + 2\pi$ will be helpful here.)

★2.8   Use identity (2.9) to prove the following useful identity:
$$1 - e^{-i\omega} = 2ie^{-i\omega/2}\sin\left(\frac{\omega}{2}\right)$$

2.2 FOURIER SERIES

In Chapter 1 we defined the space $L^2(\mathbb{R})$, the inner product of functions in $L^2(\mathbb{R})$, and the norm of $f \in L^2(\mathbb{R})$. We also discussed vector spaces, subspaces, and projects of $L^2(\mathbb{R})$ functions.

In some cases we are not interested in square-integrable functions on the entire real line; applications often consider only square-integrable functions defined on some interval $[a, b]$. In these cases, we typically work with the space $L^2([a, b])$. Here is the definition of this space:

Definition 2.2 (The Space $L^2([a, b])$) *Let $a, b \in \mathbb{R}$ with $a < b$. We define the space $L^2([a, b])$ to be the set*

$$L^2([a,b]) = \left\{ f : [a,b] \to \mathbb{C} \mid \int_a^b |f(t)|^2 \, \mathrm{d}t < \infty \right\} \tag{2.15}$$

∎

We can define the norm of $f \in L^2([a, b])$:

$$\|f(\omega)\| = \left(\int_a^b f(\omega)^2 \, \mathrm{d}\omega \right)^{1/2}$$

and the inner product of two functions $f, g \in L^2([a, b])$:

$$\langle f(\omega), g(\omega) \rangle = \int_a^b f(\omega)\overline{g(\omega)} \, \mathrm{d}\omega$$

and note that $\|f(\omega)\|^2 = \langle f(\omega), f(\omega) \rangle$.

We are particularly interested in the space $L^2([-\pi, \pi])$; the length of this interval is 2π and it is easy to verify that the 2π-periodic functions $\cos(k\omega)$ and $\sin(k\omega)$, $k \in \mathbb{Z}$, are in $L^2([-\pi, \pi])$ (see Problem 2.9).

We can state and prove the Cauchy–Schwarz inequality and the triangle inequality for functions $f, g \in L^2([-\pi, \pi])$ and thus see that $L^2([-\pi, \pi])$ is a vector space. Since $\cos(k\omega), \sin(k\omega) \in L^2([-\pi, \pi])$ and $e_k(\omega) = \cos(k\omega) + i \sin(k\omega)$, we see that $e_k(\omega) \in L^2([-\pi, \pi])$ as well.

In Section 2.1 we learned that the set $\{e_k(\omega)\}_{k \in \mathbb{Z}}$ forms an *orthogonal set*. That is, if we define

$$\delta_{j,k} = \begin{cases} 1, & j = k \\ 0, & j \neq k \end{cases} \tag{2.16}$$

we see that

$$\langle e_k(\omega), e_j(\omega) \rangle = 2\pi \delta_{j,k}$$

Moreover, it can be shown[3] (see e. g., Saxe [50]) that the set of complex exponential functions $\{e_k(\omega)\}_{k\in\mathbb{Z}}$ forms a basis for the space $L^2([-\pi,\pi])$. One way to view this statement is as follows:

> *Any 2π-periodic function, square integrable on the interval $[-\pi,\pi]$, can be expressed as a linear combination of complex exponential functions.*

Fourier Series

This linear combination is typically called a *Fourier series* and gives rise to the following definition:

Definition 2.3 (Fourier Series) *Suppose that $f \in L^2([-\pi,\pi])$. Then the* Fourier series *for f is given by*

$$f(\omega) = \sum_{k=-\infty}^{\infty} c_k e^{ik\omega} \tag{2.17}$$

∎

Some explanation of the "=" in (2.17) is in order. If we let

$$s_n(\omega) = \sum_{k=-n}^{n} c_k e^{ik\omega}$$

then $f(\omega)$ in (2.17) is the limit, in the L^2 sense, $\lim_{n\to\infty} s_n(\omega) = f(\omega)$. That is,

$$\|s_n(\omega) - f(\omega)\| \to 0 \quad \text{as} \quad n \to \infty$$

The definition says nothing about the *pointwise* convergence of the sequence $\{s_n(\omega)\}$. As we learned in Example 1.5 and Problem 1.20, it is entirely possibly for $s_n(\omega)$ to converge in norm to $f(\omega)$, but not converge pointwise to $f(\omega)$ for certain values of ω. Of course, there are conditions that we can impose on $f(\omega)$ to guarantee that $s_n(\omega)$ converges pointwise to $f(\omega)$. The interested reader is referred to Kammler [38] or Boggess and Narcowich [4]. The reader with an advanced background in analysis should see Körner [40].

Fourier Coefficients

The c_k for $k \in \mathbb{Z}$ are called the *Fourier coefficients* for $f(\omega)$. For applications we will need a formula for the c_k. The following proposition tells us how to write the Fourier coefficients in terms of $f(\omega)$ and complex exponential functions.

---

[3]The proof of this result is beyond the scope of this book.

Proposition 2.1 (Fourier Coefficients) *Suppose that* $f(\omega) \in L^2([-\pi, \pi])$ *has the Fourier series given by (2.17). Then the* Fourier coefficients c_k *can be expressed as*

$$c_k = \frac{1}{2\pi} \int_{-\pi}^{\pi} f(\omega)\overline{e_k(\omega)}\, d\omega \qquad (2.18)$$

■

Proof: We rewrite (2.17) using a counter index j instead of k and then multiply both sides of (2.17) by $\overline{e_k(\omega)}$ to obtain

$$f(\omega)\overline{e_k(\omega)} = \sum_{j=-\infty}^{\infty} c_j e^{ij\omega}\overline{e_k(\omega)} = \sum_{j=-\infty}^{\infty} c_j e_j(\omega)\overline{e_k(\omega)}$$

Next we integrate the above identity over the interval $-[\pi, \pi]$. Some analysis is required to justify passing the integral through the infinite summation, but the step is valid (see Rudin [48]) and we obtain

$$\int_{-\pi}^{\pi} f(\omega)\overline{e_k(\omega)}\, d\omega = \int_{-\pi}^{\pi} \sum_{j=-\infty}^{\infty} c_j e_j(\omega)\overline{e_k(\omega)}\, d\omega$$

$$= \sum_{j=-\infty}^{\infty} c_j \int_{-\pi}^{\pi} e_j(\omega)\overline{e_k(\omega)}\, d\omega$$

$$= 2\pi c_k$$

The $2\pi c_k$ appears in the last line by virtue of the fact that the inner product $\langle e_j(\omega), e_k(\omega)\rangle$ is 0 for $j \neq k$. When $j = k$, which happens exactly once in the entire sum, the inner product is 2π. Solving the last equality for c_k gives the desired result.   ■

Examples of Fourier Series

Let's look at some examples of Fourier series for functions in $L^2([-\pi, \pi])$.

Example 2.1 (Fourier Series for a Piecewise Constant Function) *Consider the function*

$$f(\omega) = \begin{cases} 1, & -\dfrac{\pi}{2} \leq \omega < \dfrac{\pi}{2} \\ 0, & otherwise \end{cases}$$

The function is plotted in Figure 2.2. Find the Fourier series for $f(\omega)$.
Solution
   We need to find the Fourier coefficients c_k *for* $f(\omega)$. *Thus we consider the integral*

Figure 2.2   A piecewise constant function on $[-\pi, \pi)$.

$$c_k = \frac{1}{2\pi} \int_{-\pi}^{\pi} f(\omega)\overline{e_k(\omega)}\,d\omega$$

$$= \frac{1}{2\pi} \int_{-\pi/2}^{\pi/2} e^{-ik\omega}\,d\omega$$

$$= \frac{1}{2\pi} \left(\int_{-\pi/2}^{\pi/2} \cos(k\omega)\,d\omega - i \int_{-\pi/2}^{\pi/2} \sin(k\omega)\,d\omega \right)$$

Now the integrand of the second integral above is an odd function and since we are integrating it over an interval symmetric about 0, we see that the integral value is 0. The integrand of the first integral is an even function so we can rewrite this integral as

$$c_k = \frac{1}{2\pi} \cdot 2 \int_{0}^{\pi/2} \cos(k\omega)\,d\omega = \frac{1}{\pi} \int_{0}^{\pi/2} \cos(k\omega)\,d\omega$$

If $k = 0$, we obtain $c_0 = \frac{1}{\pi} \cdot \frac{\pi}{2} = \frac{1}{2}$. If $k \neq 0$, we see that

$$c_k = \frac{1}{k\pi} \sin(k\omega)\Big|_0^{\pi/2}$$

$$= \frac{1}{k\pi} \sin\left(\frac{k\pi}{2}\right)$$

The value of c_k is 0 for $k \neq 0$ even, so the Fourier series for $f(\omega)$ is

$$f(\omega) = \frac{1}{2} + \frac{1}{\pi} \sum_{k \text{ odd}} \frac{1}{k} \sin\left(\frac{k\pi}{2}\right) e^{ik\omega} \qquad (2.19)$$

We can further simplify the series above. We split the summation into two sums:

$$f(\omega) = \frac{1}{2} + \frac{1}{\pi} \sum_{\substack{k \text{ odd} \\ k<0}} \frac{1}{k} \sin\left(\frac{k\pi}{2}\right) e^{ik\omega} + \frac{1}{\pi} \sum_{\substack{k \text{ odd} \\ k>0}} \frac{1}{k} \sin\left(\frac{k\pi}{2}\right) e^{ik\omega}$$

In the sum over the negative odd integers, replace k with $-k$ and use the fact that $\sin\left(-\frac{k\pi}{2}\right) = -\sin\left(\frac{k\pi}{2}\right)$ *to write*

$$f(\omega) = \frac{1}{2} + \frac{1}{\pi}\sum_{\substack{k \text{ odd} \\ k>0}}\left(-\frac{1}{k}\right)\sin\left(\frac{-k\pi}{2}\right)e^{-ik\omega} + \frac{1}{\pi}\sum_{\substack{k \text{ odd} \\ k>0}}\frac{1}{k}\sin\left(\frac{k\pi}{2}\right)e^{ik\omega}$$

$$= \frac{1}{2} + \frac{1}{\pi}\sum_{\substack{k \text{ odd} \\ k>0}}\frac{1}{k}\sin\left(\frac{k\pi}{2}\right)e^{-ik\omega} + \frac{1}{\pi}\sum_{\substack{k \text{ odd} \\ k>0}}\frac{1}{k}\sin\left(\frac{k\pi}{2}\right)e^{ik\omega}$$

$$= \frac{1}{2} + \frac{1}{\pi}\sum_{\substack{k \text{ odd} \\ k>0}}\frac{1}{k}\sin\left(\frac{k\pi}{2}\right)\left(e^{ik\omega} + e^{-ik\omega}\right)$$

$$= \frac{1}{2} + \frac{2}{\pi}\sum_{\substack{k \text{ odd} \\ k>0}}\frac{1}{k}\sin\left(\frac{k\pi}{2}\right)\cos(k\omega)$$

$$= \frac{1}{2} + \frac{2}{\pi}\sum_{k=1}^{\infty}\frac{1}{2k-1}\sin\left(\frac{(2k-1)\pi}{2}\right)\cos\big((2k-1)\omega\big)$$

Note that we used (2.8) to write $2\cos(k\omega) = e^{ik\omega} + e^{-ik\omega}$. *Now* $\sin\left(\frac{(2k-1)\pi}{2}\right) = (-1)^{k+1}$ *for* $k = 1,2,\ldots$, *so the series becomes*

$$f(\omega) = \frac{1}{2} + \frac{2}{\pi}\sum_{k=1}^{\infty}\frac{1}{2k-1}(-1)^{k+1}\cos\big((2k-1)\omega\big)$$

$$= \frac{1}{2} - \frac{2}{\pi}\sum_{k=1}^{\infty}\frac{(-1)^k}{2k-1}\cos\big((2k-1)\omega\big) \qquad (2.20)$$

Figure 2.3 contains plots of the partial sum

$$f_n(\omega) = \frac{1}{2} - \frac{2}{\pi}\sum_{k=1}^{n}\frac{(-1)^k}{2k-1}\cos\big((2k-1)\omega\big)$$

for various values of n.   ∎

It should not be a surprise that the Fourier series for $f(\omega)$ is reduced to a series in terms of cosine. Since $f(\omega)$ is an even function, we do not expect any contributions to the series from the "odd" parts $\sin(k\omega)$ of the complex exponential functions $e_k(\omega)$. In Problems 2.11 and 2.12 you will investigate further the symmetry properties of Fourier series and coefficients.

Example 2.1 allows us to better understand the convergence properties of Fourier series. The sequence of functions $f_n(\omega)$ converges in norm to $f(\omega)$ and converges pointwise to $f(\omega)$ everywhere but $\omega = \pm\frac{\pi}{2}$. Indeed, if you look at the plots of $f_n(\omega)$ for $n = 10, 20, 100$ in Figure 2.3, you will notice "overshoots" and "undershoots" forming at $\omega = \pm\frac{\pi}{2}$. This is the famous *Gibbs phenomenon* [29, 30]. For a nice

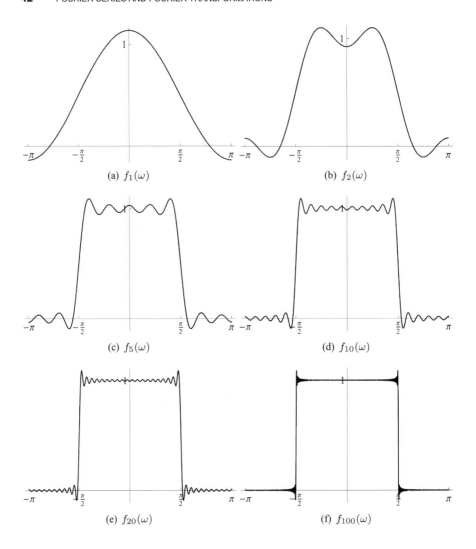

(a) $f_1(\omega)$ (b) $f_2(\omega)$

(c) $f_5(\omega)$ (d) $f_{10}(\omega)$

(e) $f_{20}(\omega)$ (f) $f_{100}(\omega)$

Figure 2.3   Plots of the partial sum $f_n(\omega)$ from Example 2.1 for $n = 1, 2, 5, 10, 20, 100$.

treatment of the Gibbs phenomenon, see Kammler [38]. Our next example considers the Fourier series for a continuous function in $L^2\big([-\pi, \pi]\big)$.

Example 2.2 (Fourier Series for the Triangle Function) *Consider the triangle function on* $[-\pi, \pi)$:

$$
g(\omega) = \begin{cases} 1 + \dfrac{\omega}{\pi}, & -\pi \le \omega < 0 \\[2mm] 1 - \dfrac{\omega}{\pi}, & 0 \le \omega < \pi \\[2mm] 0, & \text{otherwise} \end{cases}
$$

$$
= \begin{cases} 1 - \dfrac{|\omega|}{\pi}, & |\omega| \le \pi \\[2mm] 0, & \text{otherwise} \end{cases}
$$

The triangle function is plotted in Figure 2.4. Find the Fourier series for $g(\omega)$.

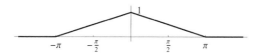

Figure 2.4 A piecewise linear function on $[-\pi, \pi)$.

Solution

   We use Problem 2.11(c) to realize that since $g(\omega)$ *is even, the Fourier series reduces to the cosine series:*

$$
g(\omega) = c_0 + 2 \sum_{k=1}^{\infty} c_k \cos(k\omega)
$$

We can compute c_0 *using the formula*

$$
c_0 = \frac{1}{2\pi} \int_{-\pi}^{\pi} g(\omega)\, \mathrm{d}\omega
$$

The integral above represents the area of the triangle in Figure 2.4, which is π, *so we see that*

$$
c_0 = \frac{1}{2\pi} \cdot \pi = \frac{1}{2}
$$

We use Problem 2.11(a) to find c_k for $k > 0$:

$$c_k = \frac{1}{\pi} \int_0^\pi g(\omega) \cos(k\omega)\, d\omega$$

$$= \frac{1}{\pi} \int_0^\pi \left(1 - \frac{\omega}{\pi}\right) \cos(k\omega)\, d\omega$$

$$= \frac{1}{\pi} \left(\int_0^\pi \cos(k\omega)\, d\omega - \frac{1}{\pi} \int_0^\pi \omega \cos(k\omega)\, d\omega \right)$$

$$= -\frac{1}{\pi^2} \int_0^\pi \omega \cos(k\omega)\, d\omega$$

since $\int_0^\pi \cos(k\omega)\, d\omega = 0$. We integrate the above integral by parts to obtain

$$c_k = -\frac{1}{\pi^2} \left(\frac{\omega \sin(k\omega)}{k} \Big|_0^\pi - \frac{1}{k} \int_0^\pi \sin(k\omega)\, d\omega \right)$$

$$= \frac{1}{k\pi^2} \int_0^\pi \sin(k\omega)\, d\omega$$

$$= -\frac{1}{k^2 \pi^2} \cos(k\omega) \Big|_0^\pi$$

$$= \frac{1}{k^2 \pi^2} \left(1 - \cos(k\pi)\right)$$

$$= \begin{cases} \dfrac{2}{k^2 \pi^2}, & k \text{ odd} \\[2mm] 0, & k \text{ even} \end{cases}$$

So the series for $g(\omega)$ is

$$g(\omega) = \frac{1}{2} + 2 \sum_{\substack{k \text{ odd} \\ k > 0}} \frac{2}{k^2 \pi^2} \cos(k\omega)$$

$$= \frac{1}{2} + \frac{4}{\pi^2} \sum_{k=1}^\infty \frac{1}{(2k-1)^2} \cos\big((2k-1)\omega\big) \tag{2.21}$$

Figure 2.5 contains plots of the partial sum

$$g_n(\omega) = \frac{1}{2} - \frac{2}{\pi} \sum_{k=1}^n \frac{(-1)^k}{2k-1} \cos\big((2k-1)\omega\big)$$

for various values of n. ∎

Examples 2.1 and 2.2 showed how to construct Fourier series by using Proposition 2.1 to compute the Fourier coefficients. In the next example we see that this direct computation is sometimes unnecessary.

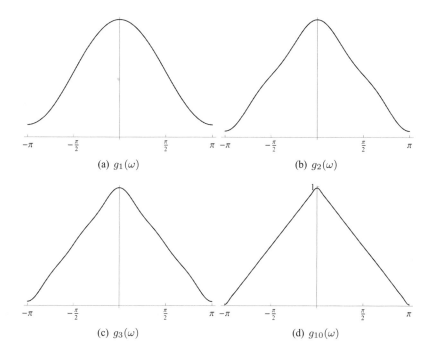

(a) $g_1(\omega)$

(b) $g_2(\omega)$

(c) $g_3(\omega)$

(d) $g_{10}(\omega)$

Figure 2.5 Plots of the partial sum $g_n(\omega)$ from Example 2.2 for $n = 1, 2, 3, 10$.

Example 2.3 (Expanding to Find the Fourier Series) *Find the Fourier series for the function $h(\omega) = \sin^3(\omega)\,(1 + \cos\omega)$.*
Solution
We observe that $h(\omega)$ is an odd function so we could use Problem 2.12(a) to realize that the Fourier coefficients are purely imaginary and can be computed as

$$c_k = -\frac{i}{\pi}\int_0^{\pi} h(\omega)\sin(k\omega)\,d\omega$$

$$= -\frac{i}{\pi}\int_0^{\pi} \sin^3(\omega)\,(1 + \cos\omega)\sin(k\omega)\,d\omega$$

The above integral is quite tedious to compute, so we use an alternative means of finding the Fourier series for $h(\omega)$. We use (2.8) and (2.9) to expand $\cos\omega$ and $\sin\omega$, respectively, in terms of complex exponentials. We have

$$\sin^3(\omega)\,(1 + \cos\omega) = \left(\frac{e^{i\omega} - e^{-i\omega}}{2i}\right)^3 \cdot \left(1 + \frac{e^{i\omega} + e^{-i\omega}}{2}\right)$$

Cubing the first factor gives

$$\left(\frac{e^{i\omega} - e^{-i\omega}}{2i}\right)^3 = -\frac{i}{8}e^{-3i\omega} + \frac{3i}{8}e^{-i\omega} - \frac{3i}{8}e^{i\omega} + \frac{i}{8}e^{3i\omega} \qquad (2.22)$$

Multiplying (2.22) by $\left(1 + \frac{e^{i\omega} + e^{-i\omega}}{2}\right)$ *and expanding gives*

$$h(\omega) = -\frac{i}{16} e^{-4i\omega} - \frac{i}{8} e^{-3i\omega} + \frac{i}{8} e^{-2i\omega} + \frac{3i}{8} e^{-i\omega} - \frac{3i}{8} e^{i\omega} - \frac{i}{8} e^{2i\omega}$$

$$+ \frac{i}{8} e^{3i\omega} + \frac{i}{16} e^{4i\omega} \qquad (2.23)$$

We can read the Fourier coefficients from (2.23). They are listed in Table 2.1.

Table 2.1 The Fourier coefficients for the function from Example 2.3.

k	h_k	k	h_k
-4	$-\dfrac{i}{16}$	4	$\dfrac{i}{16}$
-3	$-\dfrac{i}{8}$	3	$\dfrac{i}{8}$
-2	$\dfrac{i}{8}$	2	$-\dfrac{i}{8}$
-1	$\dfrac{3i}{8}$	1	$-\dfrac{3i}{16}$
$|k| > 4,\ k = 0$	0		

We can easily verify that as indicated by Problem 2.12(a), the Fourier coefficients are pure imaginary numbers and satisfy $c_k = -c_{-k}$ *(Problem 2.12(b)). We can also write (2.23) as a series in terms of* $\sin(k\omega)$ *as suggested by Problem 2.12(c). Grouping conjugate terms in (2.23) together and using (2.9) gives*

$$h(\omega) = \frac{i}{16}\left(e^{4i\omega} - e^{-4i\omega}\right) + \frac{i}{8}\left(e^{3i\omega} - e^{-3i\omega}\right) - \frac{i}{8}\left(e^{2i\omega} - e^{-2i\omega}\right)$$

$$+ \frac{3i}{8}\left(e^{i\omega} - e^{-i\omega}\right)$$

$$= \frac{i}{16} \cdot 2i \sin(4\omega) - \frac{i}{8} \cdot 2i \sin(3\omega) - \frac{i}{8} \cdot 2i \sin(2\omega) + \frac{3i}{8} \cdot 2i \sin(\omega)$$

$$= -\frac{1}{8} \sin(4\omega) + \frac{1}{4} \sin(3\omega) + \frac{1}{4} \sin(2\omega) - \frac{3}{4} \sin(\omega)$$

■

Computing One Fourier Series from Another

The computation of Fourier coefficients can be quite difficult so, as was the case in Example 2.3, we look for alternative methods for computing Fourier coefficients. In many cases, we can use a known Fourier series for one function to help us construct the Fourier series for another function. Here is an example.

Example 2.4 (Fourier Series of a Piecewise Constant Function) *Consider the expanded box function $p(\omega) \in L^2\left([-\pi, \pi]\right)$ given by*

$$p(\omega) = \sqcap\left(\frac{\omega}{\pi}\right) = \begin{cases} 1, & 0 \le \omega < \pi \\ 0, & otherwise \end{cases}$$

The function $p(\omega)$ is plotted in Figure 2.6.

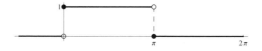

Figure 2.6   A piecewise constant function on $[-\pi, \pi)$.

Find the Fourier series for $p(\omega)$.

Solution

Rather than use Proposition 2.1 to compute the Fourier coefficients for $p(\omega)$ directly, we observe that $p(\omega)$ can be obtained from $f(\omega)$ in Example 2.1 by translation. That is,

$$p(\omega) = f\left(\omega - \frac{\pi}{2}\right)$$

We can use (2.19) to find the Fourier series for $p(\omega)$:

$$
\begin{aligned}
p(\omega) = f\left(\omega - \frac{\pi}{2}\right) \\
= \frac{1}{2} + \frac{1}{\pi} \sum_{k \text{ odd}} \frac{1}{k} \sin\left(\frac{k\pi}{2}\right) e^{ik\left(\omega - \frac{\pi}{2}\right)} \\
= \frac{1}{2} + \frac{1}{\pi} \sum_{k \text{ odd}} \frac{1}{k} \sin\left(\frac{k\pi}{2}\right) e^{-ik\pi/2} e^{ik\omega} \\
= \frac{1}{2} - \frac{1}{\pi} \sum_{k \text{ odd}} \frac{1}{k} \sin\left(\frac{k\pi}{2}\right) i^k e^{ik\omega}
\end{aligned}
$$

We can read the Fourier coefficients d_k for $p(\omega)$ from the identity above:

$$
d_k = \begin{cases}
0, & k \ne 0, k \text{ even} \\
\dfrac{1}{2}, & k = 0 \\
-\dfrac{\sin\left(\frac{k\pi}{2}\right) i^k}{k\pi}, & k \text{ odd}
\end{cases}
$$

■

Rules for Computing Fourier Series

Example 2.4 is a good example of how to construct the Fourier series of one function from the Fourier series of another function. It turns out that there are several rules we can develop for computing Fourier series in this way.

Proposition 2.2 (Translation Rule) *Suppose that $f(\omega) \in L^2([-\pi, \pi])$ is a 2π-periodic function that has the Fourier series representation*

$$f(\omega) = \sum_{k=-\infty}^{\infty} c_k e^{ik\omega}$$

and for $a \in \mathbb{R}$, $g(\omega) = f(\omega - a)$. If $g(\omega)$ has the Fourier series representation

$$g(\omega) = \sum_{k=-\infty}^{\infty} d_k e^{ik\omega}$$

then $d_k = e^{-ika} c_k$.   ∎

Proof: Using Proposition 2.1, we have

$$d_k = \frac{1}{2\pi} \int_{-\pi}^{\pi} g(\omega) e^{-ik\omega} \, d\omega = \frac{1}{2\pi} \int_{-\pi}^{\pi} f(\omega - a) e^{-ik\omega} \, d\omega$$

The u-substitution $u = \omega - a$ gives

$$
\begin{aligned}
d_k &= \frac{1}{2\pi} \int_{-\pi-a}^{\pi-a} f(u) e^{-ik(u+a)} \, du \\
&= \frac{1}{2\pi} e^{-ika} \int_{-\pi-a}^{\pi-a} f(u) e^{-iku} \, du \\
&= \frac{1}{2\pi} e^{-ika} \int_{-\pi}^{\pi} f(u) e^{-iku} \, du \\
&= e^{-ika} c_k
\end{aligned}
$$

Note that in the computation above, we used Problem 2.7.   ∎

There are several other rules for obtaining one Fourier series from another. We state the *modulation rule* below and leave its proof as Problem 2.20. Rules involving conjugation, reflection, and differentiation are discussed in Problem 2.21.

Proposition 2.3 (Modulation Rule) *Suppose that $f(\omega) \in L^2([-\pi, \pi])$ is a 2π-periodic function that has Fourier series representation*

$$f(\omega) = \sum_{k=-\infty}^{\infty} c_k e^{ik\omega}$$

and for $m \in \mathbb{Z}$, $g(\omega) = e^{im\omega} f(\omega)$. *If* $g(\omega)$ *has the Fourier series representation*

$$g(\omega) = \sum_{k=-\infty}^{\infty} d_k e^{ik\omega}$$

then $d_k = c_{k-m}$.   ∎

Proof: This proof is left as Problem 2.20.   ∎

PROBLEMS

2.9   Show that for $k \in \mathbb{Z}$, the functions $\cos(k\omega)$ and $\sin(k\omega)$ are members of $L^2\big([-\pi, \pi]\big)$.

2.10   Suppose that $f(\omega) = e^{-|\omega|}$. Show that the Fourier series coefficients for $f(\omega)$ are $c_k = \dfrac{1 - (-1)^k e^{-\pi}}{1 + k^2}$.

2.11   Suppose that $f(\omega) \in L^2\big([-\pi, \pi]\big)$ is an even function (i.e., $f(\omega) = f(-\omega)$).

(a) Show that the Fourier coefficients c_k for $f(\omega)$ are real numbers and can be expressed as

$$c_k = \frac{1}{\pi} \int_0^\pi f(\omega) \cos(k\omega)\, d\omega$$

(b) Using part (a), show that $c_k = c_{-k}$.

(c) Using part (b), show that the Fourier series for $f(\omega)$ can be written as

$$f(\omega) = c_0 + 2 \sum_{k=1}^{\infty} c_k \cos(k\omega)$$

(*Hint:* Use the ideas from Example 2.1 and (2.8) to express the Fourier series as a series in cosine.)

2.12   Suppose that $f(\omega) \in L^2\big([-\pi, \pi]\big)$ is an odd function (i.e., $f(-\omega) = -f(\omega)$).

(a) Show that the Fourier coefficients c_k for $f(\omega)$ are imaginary numbers and can be expressed as

$$c_k = -\frac{i}{\pi} \int_0^\pi f(\omega) \sin(k\omega)\, d\omega$$

(b) Using part (a), show that $c_k = -c_{-k}$.

(c) Using part (b), show that the Fourier series for $f(\omega)$ can be written as

$$f(\omega) = 2i \sum_{k=1}^{\infty} c_k \sin(k\omega)$$

(*Hint:* Use the ideas from Problem 2.11, Example 2.1, and (2.9) to express the Fourier series as a series in sine.)

2.13   Let $f(\omega) = \sum_{k \in \mathbb{Z}} c_k e^{ik\omega}$ be a function in $L^2([-\pi, \pi])$ with $c_k = c_{-k}$. Show that $f(\omega)$ is an even function.

2.14   Let $f(\omega) = \sum_{k \in \mathbb{Z}} c_k e^{ik\omega}$ be a function in $L^2([-\pi, \pi])$ with $c_k = -c_{-k}$. Show that $f(\omega)$ is an odd function.

⋆**2.15**   Suppose that $f(\omega) = \sum_{k \in \mathbb{Z}} c_k e^{ik\omega}$ is a function in $L^2([-\pi, \pi])$. Let $g(\omega) = |f(\omega)|$. Show that:

(a) $g(\omega)$ is a 2π-periodic function.

(b) $g(\omega)$ is an even function. That is, show $g(-\omega) = g(\omega)$. (*Hint:* $f(-\omega) = \overline{f(\omega)}$ and for $z \in \mathbb{C}$, $|z| = |\overline{z}|$.)

2.16   Use (2.21) to show that

$$\frac{\pi^2}{8} = \sum_{k=1}^{\infty} \frac{1}{(2k-1)^2}$$

⋆**2.17**   Suppose that $f(\omega) \in L^2([-\pi, \pi])$ with $\langle f(\omega), e_k(\omega) \rangle = 0$ for all integers $k \neq 0$. Show that $f(\omega)$ is a constant function.

2.18   Find Fourier series for

(a) $f_1(\omega) = \cos^2\left(\frac{\omega}{2}\right)$.

(b) $f_2(\omega) = \sin^2\left(\frac{\omega}{2}\right)$.

⋆**2.19**   Let $f(\omega) = \sum_{k=-\infty}^{\infty} c_k e^{ik\omega}$ be a function in $L^2([-\pi, \pi])$ with $c_k \in \mathbb{R}$. Show that $f(-\omega) = \overline{f(\omega)}$. (*Hint:* Problem 1.9 will be useful.)

2.20   Prove Proposition 2.3.

2.21   Suppose $f(\omega), g(\omega) \in L^2([-\pi, \pi])$ have the Fourier series representation

$$f(\omega) = \sum_{k=-\infty}^{\infty} c_k e^{ik\omega} \qquad g(\omega) = \sum_{k=-\infty}^{\infty} d_k e^{ik\omega}$$

Prove the following:

(a) If $g(\omega) = f(-\omega)$, then $d_k = c_{-k}$.

(b) If $g(\omega) = \overline{f(-\omega)}$, then $d_k = \overline{c_k}$.

(c) If $g(\omega) = f'(\omega)$, then $d_k = ikc_k$.

★**2.22**   An important operation involving bi-infinite sequences is that of *convolution*. In this problem, we will define the convolution product of two bi-infinite sequences and then prove a rule for computing the Fourier series of a function whose Fourier coefficients are obtained via convolution.

Let $\mathbf{h} = (\ldots, h_{-2}, h_{-1}, h_0, h_1, h_2, \ldots)$ and $\mathbf{v} = (\ldots, v_{-2}, v_{-1}, v_0, v_1, v_2, \ldots)$ be two bi-infinite sequences. Then the *convolution product* of $\mathbf{h}$ and $\mathbf{v}$, denoted by $\mathbf{h} * \mathbf{v}$, is the bi-infinite sequence $\mathbf{y}$ whose elements are given by

$$y_n = \sum_{k \in \mathbb{Z}} h_k v_{n-k}, \quad n \in \mathbb{Z} \tag{2.24}$$

We assume that there is sufficient decay in the tails $\mathbf{h}$ and $\mathbf{v}$ so that each value y_n exists.

Let $\mathbf{y} = \mathbf{h} * \mathbf{v}$ and

$$Y(\omega) = \sum_{n \in \mathbb{Z}} y_n e^{in\omega} \tag{2.25}$$

$$H(\omega) = \sum_{k \in \mathbb{Z}} h_k e^{ik\omega} \tag{2.26}$$

$$V(\omega) = \sum_{k \in \mathbb{Z}} v_k e^{ik\omega} \tag{2.27}$$

Show that $Y(\omega) = H(\omega)V(\omega)$. The following steps will help you organize your work.

(i) Replace y_n in (2.25) by the series given in (2.24) and then change the order of summations to obtain

$$Y(\omega) = \sum_{k \in \mathbb{Z}} h_k \left(\sum_{n \in \mathbb{Z}} v_{n-k} e^{in\omega} \right)$$

(ii) Make the substitution $j = n - k$ on the index of the inner sum in part (i).

(iii) Factor an $e^{ik\omega}$ out of the inner summation. The inner summation should now be $V(\omega)$.

(iv) Factor out $V(\omega)$ from the remaining sum and recognize this sum as $H(\omega)$.

2.23   Use Problem 2.22 to show that convolution of bi-infinite sequences is a commutative operation. That is, show that $\mathbf{h} * \mathbf{v} = \mathbf{v} * \mathbf{h}$.

2.24   Let $h(\omega)$ be the function from Example 2.3. Use Table 2.1, Proposition 2.2, Proposition 2.3, and Problem 2.21 to find the Fourier coefficients of the following functions.

(a) $f_1(\omega) = h(\omega - \pi)$

(b) $f_2(\omega) = h(-\omega)$

(c) $f_3(\omega) = e^{2i\omega} h(\omega)$

(d) $f_4(\omega) = h'(\omega)$

2.25   Let $f(\omega)$ be the function plotted in Figure 2.7 and further assume that $f(\omega)$ has the Fourier series representation $f(\omega) = \sum\limits_{k=-\infty}^{\infty} c_k e^{ik\omega}$.

Figure 2.7   The function $f(\omega)$ from Problem 2.25.

Use Proposition 2.2, Proposition 2.3, and Problem 2.21 to sketch the following functions:

(a) $g_1(\omega) = \sum\limits_{k=-\infty}^{\infty} (-1)^k c_k e^{ik\omega}$

(b) $g_2(\omega) = \sum\limits_{k=-\infty}^{\infty} c_{-k} e^{ik\omega}$

(c) $g_3(\omega) = i \sum\limits_{k=-\infty}^{\infty} k c_k e^{ik\omega}$

★**2.26**   Let L be a positive integer and suppose that $H(\omega) = \sum\limits_{k=0}^{L} h_k e^{ik\omega}$. Here $h_k = 0$ for $k < 0, k > L$. If $G(\omega) = \sum\limits_{k=0}^{L} g_k e^{ik\omega}$, where $g_k = (-1)^k h_{L-k}$, write $G(\omega)$ in terms of $H(\omega)$.

★**2.27**   Suppose that $N = 2M$ is a positive integer and let $H(\omega) = \cos^N\left(\frac{\omega}{2}\right)$. Find the Fourier series for $H(\omega)$. (*Hint:* Use (2.8) to write $\cos\left(\frac{\omega}{2}\right)$ in terms of complex exponentials and then show that $H(\omega) = \dfrac{1}{2^N} e^{-iM\omega} \left(e^{i\omega} + 1\right)^N$. Use the binomial theorem.)

★**2.28**   We have developed the notion of Fourier series for functions $f(\omega) \in L^2\left([-\pi, \pi]\right)$, but we could easily develop the theory for $f(\omega) \in L^2\left([-L, L]\right)$, where $L > 0$ is a real number.

(a) Show that the complex exponential function $e_k^L(\omega) = e^{2\pi i k \omega / 2L}$ is a $2L$-periodic function and that $e_k^L(\omega) \in L^2\left([-L, L]\right)$.

(b) Show that for $j, k \in \mathbb{Z}$, we have

$$\langle e_j^L(\omega), e_k^L(\omega) \rangle = \begin{cases} 2L, & j = k \\ 0, & j \neq k \end{cases}$$

(c) It can be shown (see, e.g., Saxe [50]) that the set $\{e_k^L(\omega)\}_{k \in \mathbb{Z}}$ forms a basis for $L^2([-L, L])$. Thus we can express $f(\omega) \in L^2([-L, L])$ as

$$f(\omega) = \sum_{k=-\infty}^{\infty} c_k e^{2k\pi i \omega / 2L}$$

Using an argument similar to the proof of Proposition 2.1, show that

$$c_k = \frac{1}{2L} \int_{-L}^{L} f(\omega) e^{-2\pi i k \omega / 2L} \, d\omega$$

2.3   THE FOURIER TRANSFORM

One of the most useful tools for analysis of functions in $L^2(\mathbb{R})$ is the *Fourier transformation*. In this section we define the Fourier transform of a function $f(t) \in L^2(\mathbb{R})$, look at some transform pairs that will be useful in later chapters, and derive some properties obeyed by Fourier transforms.

Motivating the Fourier Transform from Fourier Series

In Section 2.2 we defined the Fourier series[4]

$$f(t) = \sum_{k \in \mathbb{Z}} c_k e^{ikt}$$

where

$$c_k = \frac{1}{2\pi} \int_{-\pi}^{\pi} f(t) e^{-ikt} \, dt$$

and $f(t) \in L^2([-\pi, \pi])$.

A natural question to ask is whether we can formulate a similar representation for functions $f(t)$ defined on the real line and not necessarily periodic.

In Problem 2.28 we learned that we could create Fourier series for functions of arbitrary period $2L$ ($L > 0$). That is,

$$f(t) = \sum_{k \in \mathbb{Z}} c_k e^{2\pi i k t / (2L)} \tag{2.28}$$

---

[4]To motivate the Fourier transformation for functions in $L^2(\mathbb{R})$, we will use t as our independent variable in Fourier series representations instead of the ω that was used as the independent variable in Section 2.2.

where

$$c_k = \frac{1}{2L} \int_{-L}^{L} f(t)e^{-2\pi i k t/(2L)} \, dt \qquad (2.29)$$

and $f(t) \in L^2\left([-L, L]\right)$.

It is quite inviting to let $L \to \infty$ and see what happens to the series representation of $f(t)$. There are several steps in what follows that need justification, but formally inserting (2.29) into (2.28) and letting $L \to \infty$ gives

$$
\begin{aligned}
f(t) &= \lim_{L \to \infty} \sum_{k \in \mathbb{Z}} \left(\frac{1}{2L} \int_{-L}^{L} f(u)e^{-2\pi i k u/(2L)} \, du \right) e^{2\pi i k t/(2L)} \\
&= \lim_{L \to \infty} \sum_{k \in \mathbb{Z}} \frac{1}{2L} \int_{-L}^{L} f(u)e^{\pi i k(t-u)/L} \, du \\
&= \lim_{L \to \infty} \sum_{k \in \mathbb{Z}} \left(\frac{1}{2\pi} \int_{-L}^{L} f(u)e^{i(t-u)\left(\frac{\pi k}{L}\right)} \, du \right) \frac{\pi}{L} \qquad (2.30)
\end{aligned}
$$

We want to view (2.30) as a Riemann sum. To do so, we need to define sample points ω_k that we can use to partition the real line. We also need the width $\Delta \omega$ of each interval. We let $\omega_k = \frac{\pi k}{L}$. We set $\Delta \omega$ is the width of each subinterval $[\omega_k, \omega_{k+1})$ so that $\Delta \omega = \omega_{k+1} - \omega_k = \frac{\pi(k+1)}{L} - \frac{\pi k}{L} = \frac{\pi}{L}$.

We also note that the integral in parentheses in (2.30) is a function of ω_k. We let

$$F_L(\omega) = \frac{1}{2\pi} \int_{-L}^{L} f(u)e^{i(t-u)\omega} \, du \qquad (2.31)$$

so that (2.30) becomes

$$f(t) = \lim_{L \to \infty} \sum_{k \in \mathbb{Z}} F_L(\omega_k) \Delta \omega \qquad (2.32)$$

Note that as $L \to \infty$, the right-hand side of (2.31) formally becomes

$$\frac{1}{2\pi} \int_{u \in \mathbb{R}} f(u)e^{i(t-u)\omega} \, du \qquad (2.33)$$

The right-hand side of (2.32) is a Riemann sum for the integral

$$\int_{\omega \in \mathbb{R}} F_L(\omega) \, d\omega$$

and as $L \to \infty$ we use the limit value (2.33) of (2.31) to write

$$
\begin{aligned}
f(t) &= \int_{\omega \in \mathbb{R}} \left(\frac{1}{2\pi} \int_{u \in \mathbb{R}} f(u)e^{i(t-u)\omega} \, du \right) d\omega \\
&= \frac{1}{\sqrt{2\pi}} \int_{\omega \in \mathbb{R}} \left(\frac{1}{\sqrt{2\pi}} \int_{u \in \mathbb{R}} f(u)e^{-iu\omega} \, du \right) e^{it\omega} \, d\omega \qquad (2.34)
\end{aligned}
$$

Definition of the Fourier Transform

We define the Fourier transform[5] $\hat{f}(\omega)$ of $f(t) \in L^2(\mathbb{R})$ as the inner integral in (2.34). We have the following definition:

Definition 2.4 (The Fourier Transform of f(t) ∈ L²(ℝ)) *Let $f(t) \in L^2(\mathbb{R})$. Then we define the* Fourier transformation $\hat{f}(\omega)$ *of* $f(t)$ *as*

$$\boxed{\hat{f}(\omega) = \frac{1}{\sqrt{2\pi}} \int_{\mathbb{R}} f(t)e^{-it\omega}\,dt}$$ (2.35)

∎

We can insert (2.35) into (2.34) to obtain an inversion formula that allows us to recover $f(t)$ from its Fourier transform. In the proposition that follows, we have listed the conditions[6] that $f(t)$ must satisfy to make the discussion leading to (2.34) rigorous.

Proposition 2.4 (The Inverse Fourier Transform) *Suppose that* $f(t) \in L^2(\mathbb{R})$ *is piecewise continuous and satisfies the following conditions:*

(i) $\int_{\mathbb{R}} |f(t)|\,dt < \infty.$

(ii) *On any bounded interval* $(a, b) \subset \mathbb{R}$, $f(t)$ *possesses only a finite number of extreme values.*

(iii) *On any bounded interval* $(a, b) \subset \mathbb{R}$, $f(t)$ *can only possess a finite number of discontinuities. Moreover, the left-hand and right-hand limits at these points must be finite.*

Then we can recover $f(t)$ *from its Fourier transform (2.35)* $\hat{f}(\omega)$ *using the formula*

$$\boxed{f(t) = \frac{1}{\sqrt{2\pi}} \int_{\mathbb{R}} \hat{f}(\omega)e^{i\omega t}\,d\omega}$$ (2.36)

∎

Proof: The discussion preceding Definition 2.4 provides an outline for the proof but is by no means rigorous. We will not provide a rigorous proof of this result. The interested reader is referred to either Boggess and Narcowich [4] or Walnut [62]. A nice derivation of the inversion formula also appears in Kammler [38]. ∎

---

[5] In (2.34) we have "split" the $\frac{1}{2\pi}$ between the Fourier transform $\hat{f}(\omega)$ and the outer integral. Some authors prefer to define the Fourier transform using $\frac{1}{2\pi}$ so that there is no factor multiplying the outer integral in (2.34). Other authors introduce a 2π in the complex exponential that appears in (2.35). This eliminates the multiplier $\frac{1}{2\pi}$ altogether (see the definition of the Fourier transform $F(s)$ in Kammler [38] as well as Problem 1.4).

[6] These conditions are known as the *Dirichlet conditions*. For more information, please see Kammler [38].

Examples of Fourier Transforms

We are primarily interested in computing Fourier transforms. To see how this is done, let's look at a couple of examples.

Example 2.5 (The Fourier Transforms of $\sqcap(t)$ and $f(t) = e^{-|t|}$) *Compute the Fourier transforms of the box function $\sqcap(t)$ and the function $f(t) = e^{-|t|}$.*
Solution
  Recall from (1.4) that the box function $\sqcap(t)$ is given by

$$\sqcap(t) = \begin{cases} 1, & 0 \le t < 1 \\ 0, & otherwise \end{cases}$$

Thus the Fourier transform $\widehat{\sqcap}(\omega)$ becomes

$$\widehat{\sqcap}(\omega) = \frac{1}{\sqrt{2\pi}} \int_{\mathbb{R}} \sqcap(t) e^{-it\omega} \, dt = \frac{1}{\sqrt{2\pi}} \int_0^1 e^{-it\omega} \, dt$$

Now if $\omega = 0$, the value of the integral above is 1 and we have

$$\widehat{\sqcap}(0) = \frac{1}{\sqrt{2\pi}} \tag{2.37}$$

  When $\omega \ne 0$, we make the u-substitution $u = -it\omega$, so that $du = -i\omega \, dt$ or $dt = \frac{du}{-i\omega} = \frac{i}{\omega} \, du$. We can find an antiderivative

$$\int e^{-it\omega} \, dt = \frac{i}{\omega} \int e^u \, du = \frac{i}{\omega} e^u = \frac{i}{\omega} e^{-it\omega}$$

so that the Fourier transform becomes

$$\begin{aligned}
\widehat{\sqcap}(\omega) &= \frac{1}{\sqrt{2\pi}} \int_0^1 e^{-it\omega} \, dt \\
&= \frac{1}{\sqrt{2\pi}} \frac{i}{\omega} e^{-it\omega} \Big|_0^1 \\
&= \frac{1}{\sqrt{2\pi}} \frac{i}{\omega} \left(e^{-i\omega} - 1 \right)
\end{aligned}$$

  We will now do a little algebra to put the Fourier transform $\widehat{\sqcap}(\omega)$ in a form that typically appears in other books. We begin by factoring an $e^{-i\omega/2}$ from the last term. We have

$$\widehat{\sqcap}(\omega) = \frac{1}{\sqrt{2\pi}} \frac{i}{\omega} \left(e^{-i\omega} - 1 \right)$$

$$= \frac{1}{\sqrt{2\pi}} \frac{i}{\omega} e^{-i\omega/2} \left(e^{-i\omega/2} - e^{i\omega/2} \right)$$

$$= -\frac{1}{\sqrt{2\pi}} \frac{i}{\omega} e^{-i\omega/2} \left(e^{i\omega/2} - e^{-i\omega/2} \right)$$

$$= -\frac{1}{\sqrt{2\pi}} \frac{i}{\omega} e^{-i\omega/2} \, 2i \, \sin\left(\frac{\omega}{2}\right)$$

$$= \frac{1}{\sqrt{2\pi}} e^{-i\omega/2} \left(\frac{\sin(\omega/2)}{\omega/2} \right)$$

We can use the definition (1.6) for the sinc *function from Example 1.2 to express the Fourier transformation of the box function* $\sqcap(t)$ *as*

$$\boxed{\widehat{\sqcap}(\omega) = \frac{1}{\sqrt{2\pi}} e^{-i\omega/2} \, \mathrm{sinc}\left(\frac{\omega}{2}\right)} \tag{2.38}$$

The real and imaginary parts of $\widehat{\sqcap}(\omega)$ *are plotted in Figure 2.8.*

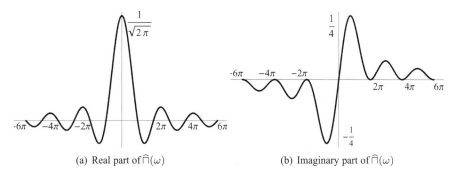

(a) Real part of $\widehat{\sqcap}(\omega)$ (b) Imaginary part of $\widehat{\sqcap}(\omega)$

Figure 2.8   The real and imaginary parts of $\widehat{\sqcap}(\omega)$.

Note that the definition of $\mathrm{sinc}(\omega)$ *allows us to include the* $\omega = 0$ *case (2.37) in (2.38). Recall from Problem 1.12 that* $\mathrm{sinc}(\omega)$ *is continuous at* $\omega = 0$, *so* $\widehat{\sqcap}(\omega)$ *is a continuous function.*

We now consider the Fourier transform of $f(t) = e^{-|t|}$. *This function is plotted in Figure 2.9. We first write down the Fourier transform integral and then, using the fact that* $e^{-i\omega t} = \cos \omega t - i \sin \omega t$, *split it into real and imaginary parts.*

$$\hat{f}(\omega) = \frac{1}{\sqrt{2\pi}} \int_{\mathbb{R}} f(t) e^{-i\omega t} \, dt$$

$$= \frac{1}{\sqrt{2\pi}} \int_{\mathbb{R}} f(t) \cos(\omega t) \, dt - i \frac{1}{\sqrt{2\pi}} \int_{\mathbb{R}} f(t) \sin(\omega t) \, dt$$

Note that the integrand of the imaginary part of $\hat{f}(\omega)$ is odd. This means that the integral that represents the imaginary part of $\hat{f}(\omega)$, provided that it converges, is 0. In Problem 2.29 you will show that this integral converges.

Now the integrand for the real part of $\hat{f}(\omega)$ is an even function, so we can write the Fourier transform as

$$\hat{f}(\omega) = \frac{2}{\sqrt{2\pi}} \int_0^\infty f(t)\cos(\omega t)\,dt = \frac{2}{\sqrt{2\pi}} \int_0^\infty e^{-t}\cos(\omega t)\,dt \qquad (2.39)$$

To simplify the integral above, we need an antiderivative of $I = \int e^{-t}\cos\omega t\,dt$. This integral is very similar those found in a standard calculus text (see, e. g., Stewart [55] for example) and requires integration by parts twice. We let $u = \cos(\omega t)$, so that $dv = e^{-t}\,dt$. These substitutions give $du = -\omega\sin\omega t\,dt$ and $v = -e^{-t}$. Thus the integral becomes

$$I = -\cos(\omega t)\,e^{-t} - \omega\int e^{-t}\sin(\omega t)\,dt \qquad (2.40)$$

We now integrate by parts again. We let $u = \sin(\omega t)$ and $dv = e^{-t}\,dt$ so that $du = \omega\cos(\omega t)\,dt$ and $v = -e^{-t}$. The integral in (2.40) becomes

$$\int e^{-t}\sin(\omega t)\,dt = -\sin(\omega t)e^{-t} + \omega\int\cos(\omega t)e^{-t}\,dt = -\sin(\omega t)e^{-t} + \omega I$$

Inserting this result into (2.40) gives

$$I = -\cos(\omega t)\,e^{-t} - \omega\left(-\sin(\omega t)\,e^{-t} + \omega I\right)$$

Solving for I gives

$$I = \frac{e^{-t}\left(\omega\sin(\omega t) - \cos(\omega t)\right)}{1 + \omega^2} \qquad (2.41)$$

We use the antiderivative (2.41) to simplify (2.39) further. We have

$$\begin{aligned}
\hat{f}(\omega) &= \frac{2}{\sqrt{2\pi}} \int_0^\infty e^{-t}\cos(\omega t)\,dt = \frac{2}{\sqrt{2\pi}} I\Big|_0^\infty \\
&= \frac{2}{\sqrt{2\pi}} \frac{e^{-t}\left(\omega\sin(\omega t) - \cos(\omega t)\right)}{1 + \omega^2}\Big|_0^\infty
\end{aligned} \qquad (2.42)$$

It is straightforward (see Problem 2.30) to show that

$$\lim_{t\to\infty} \frac{e^{-t}\left(\omega\sin(\omega t) - \cos(\omega t)\right)}{1 + \omega^2} = 0$$

and if we plug $t = 0$ into (2.41), we obtain $-1/(1 + \omega^2)$.
Thus (2.42) becomes

$$\hat{f}(\omega) = \frac{2}{\sqrt{2\pi}} \frac{1}{1 + \omega^2} = \sqrt{\frac{2}{\pi}} \cdot \frac{1}{1 + \omega^2} \qquad (2.43)$$

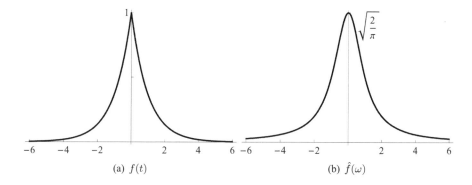

(a) $f(t)$ (b) $\hat{f}(\omega)$

Figure 2.9   The function $f(t) = e^{-|t|}$ and its Fourier transform $\hat{f}(\omega) = \sqrt{\dfrac{2}{\pi}} \cdot \dfrac{1}{1+\omega^2}$.

We have plotted $\hat{f}(\omega)$ in Figure 2.9.   ∎

As you can see by Example 2.5, direct computation of Fourier transforms can be quite tedious. The example did reveal one shortcut however. The even function $f(t) = e^{-|t|}$ reduced the transformation integral to an integral involving $f(t)$ and $\cos(\omega t)$. That is, if $f(t)$ is an even function, then

$$\hat{f}(\omega) = \sqrt{\frac{2}{\pi}} \int_0^\infty f(t) \cos(\omega t)\, dt$$

In Problem 2.31 you will investigate the impact that an odd function has on the Fourier transform.

Rules for Computing Fourier Transforms

In calculus, we typically learn the definition of the derivative, compute a few derivatives directly by definition, and then learn several rules that make it possible for us to compute derivatives of more complicated functions. In a similar fashion, we can develop rules that often afford us a more efficient way to compute Fourier transforms.

We now state eight rules that can be used to compute Fourier transforms. Each of these rules assumes that you have in hand a *transform pair* (i.e., $f(t)$ and $\hat{f}(\omega)$). A new function $g(t)$ is created from $f(t)$ and the rule shows how to construct $\hat{g}(\omega)$ from $\hat{f}(\omega)$. The proofs of each rule are straightforward. In most cases we can start with the definition of the Fourier transform (2.35) and manipulate it to obtain the desired result.

Our first rule says that the Fourier transform is linear.

Proposition 2.5 (Linearity Rule for Fourier Transforms) *Suppose that $f_1(t)$ and $f_2(t)$ have Fourier transforms $\hat{f}_1(\omega)$ and $\hat{f}_2(\omega)$, respectively, and let $g(t) = a_1 f_1(t) +$*

$a_2 f_2(t)$, *where* $a_1, a_2 \in \mathbb{C}$. *Then*

$$\hat{g}(\omega) = a_1 \hat{f}_1(\omega) + a_2 \hat{f}_2(\omega)$$

∎

Proof: Using (2.35) we have

$$\hat{g}(\omega) = \frac{1}{\sqrt{2\pi}} \int_{\mathbb{R}} g(t) e^{-i\omega t} \, dt$$

$$= \frac{1}{\sqrt{2}} \int_{\mathbb{R}} \left(a_1 f_1(t) + a_2 f_2(t) \right) e^{-i\omega t} \, dt$$

$$= a_1 \frac{1}{\sqrt{2\pi}} \int_{\mathbb{R}} f_1(t) e^{-i\omega t} \, dt + a_2 \frac{1}{\sqrt{2\pi}} \int_{\mathbb{R}} f_2(t) e^{-i\omega t} \, dt$$

$$= a_1 \hat{f}_1(\omega) + a_2 \hat{f}_2(\omega)$$

∎

Our next proposition says that if you translate a function, then you *modulate* the Fourier transform by a complex exponential.

Proposition 2.6 (Translation Rule for Fourier Transforms) *Assume that* $\hat{f}(\omega)$ *is the Fourier transform of* $f(t)$. *Let* $a \in \mathbb{R}$ *and suppose that* $g(t) = f(t - a)$. *Then*

$$\hat{g}(\omega) = e^{-i\omega a} \hat{f}(\omega)$$

∎

Proof: We start with the defining relation (2.35) and write

$$\hat{g}(\omega) = \frac{1}{\sqrt{2\pi}} \int_{\mathbb{R}} g(t) e^{-i\omega t} \, dt = \frac{1}{\sqrt{2\pi}} \int_{\mathbb{R}} f(t - a) e^{-i\omega t} \, dt$$

Now we make the u-substitution $u = t - a$ so that $du = dt$ and $t = u + a$. Note that the limits of integration are unchanged, so that

$$\hat{g}(\omega) = \frac{1}{\sqrt{2\pi}} \int_{\mathbb{R}} f(t - a) e^{-i\omega t} \, dt$$

$$= \frac{1}{\sqrt{2\pi}} \int_{\mathbb{R}} f(u) e^{-i\omega(u+a)} \, du$$

$$= e^{-i\omega a} \frac{1}{\sqrt{2\pi}} \int_{\mathbb{R}} f(u) e^{-i\omega u} \, du$$

$$= e^{-i\omega a} \hat{f}(\omega)$$

∎

The next proposition tells us how to compute the Fourier transform of $f(at)$.

Proposition 2.7 (Dilation Rule for Fourier Transforms) *Assume that $\hat{f}(\omega)$ is the Fourier transform of $f(t)$. Suppose that $a \in \mathbb{R}, a > 0$ and let $g(t) = f(at)$. Then*

$$\hat{g}(\omega) = \frac{1}{a}\hat{f}\left(\frac{\omega}{a}\right)$$

∎

In Problem 2.32 you will show how to extend this rule to the case where $a \neq 0$.

Proof:
We begin with (2.35) and write

$$\hat{g}(\omega) = \frac{1}{\sqrt{2\pi}}\int_{\mathbb{R}} g(t)e^{-i\omega t}\,dt = \frac{1}{\sqrt{2\pi}}\int_{\mathbb{R}} f(at)e^{-i\omega t}\,dt$$

We now make the u-substitution $u = at$ or $t = \frac{u}{a}$. The differential is $du = a\,dt$ or $dt = \frac{1}{a}\,du$. It is also important to note that since $a > 0$, the limits of integration are unchanged. We have

$$\hat{g}(\omega) = \frac{1}{\sqrt{2\pi}}\int_{\mathbb{R}} f(at)e^{-i\omega t}\,dt$$

$$= \frac{1}{a}\frac{1}{\sqrt{2\pi}}\int_{\mathbb{R}} f(u)e^{-i\omega u/a}\,du$$

$$= \frac{1}{a}\frac{1}{\sqrt{2\pi}}\int_{\mathbb{R}} f(u)e^{-i(\omega/a)u}\,du$$

$$= \frac{1}{a}\hat{f}\left(\frac{\omega}{a}\right)$$

∎

Since the formulas for $\hat{f}(\omega)$ (2.35) and $f(t)$ (2.36) are similar, it is natural to expect that a rule exists for computing $\hat{g}(\omega)$ when $g(t)$ is the Fourier transform of $f(t)$.

Proposition 2.8 (Inversion Rule for Fourier Transforms) *Suppose that $f(t)$ has Fourier transform $\hat{f}(\omega)$ and assume that $g(t) = \hat{f}(t)$. Then $\hat{g}(\omega) = f(-\omega)$.* ∎

Proof: We use (2.35) to write

$$\hat{g}(\omega) = \frac{1}{\sqrt{2\pi}}\int_{\mathbb{R}} g(t)e^{-i\omega t}\,dt = \frac{1}{\sqrt{2\pi}}\int_{\mathbb{R}} \hat{f}(t)e^{i(-\omega)t}\,dt$$

Using (2.36), we see that the right-hand side of the above identity is simply $f(-\omega)$. ∎

The next rule shows us how to find the Fourier transform of a function $f(t)$ scaled by t. The proof is a bit different from the ones we have seen thus far.

Proposition 2.9 (Power Rule for Fourier Transforms) *Suppose that* $g(t) = t \cdot f(t) \in L^2(\mathbb{R})$ *and let* $\hat{f}(\omega)$ *be the Fourier transform of* $f(t)$*. Then* $\hat{g}(\omega) = i \, \hat{f}'(\omega)$. ∎

Proof: We begin by writing the Fourier transform of $g(t)$.

$$\hat{g}(\omega) = \frac{1}{\sqrt{2\pi}} \int_{\mathbb{R}} g(t) e^{-i\omega t}\, dt = \frac{1}{\sqrt{2\pi}} \int_{\mathbb{R}} t \cdot f(t) e^{-i\omega t}\, dt \qquad (2.44)$$

We now compute

$$\hat{f}'(\omega) = \frac{d}{d\omega} \left(\frac{1}{\sqrt{2\pi}} \int_{\mathbb{R}} f(t) e^{-i\omega t}\, dt \right)$$

This is a little tricky since we need to pass the derivative through the improper integral. These steps can be justified and made rigorous, but the analysis necessary to do so is beyond the scope of this text. The interested reader can consult Rudin [48] to learn more about justifying the exchange of the derivative and integral. We formally compute $\hat{f}'(\omega)$ and obtain

$$\begin{aligned}
\hat{f}'(\omega) &= \frac{d}{d\omega} \left(\frac{1}{\sqrt{2\pi}} \int_{\mathbb{R}} f(t) e^{-i\omega t}\, dt \right) \\
&= \frac{1}{\sqrt{2\pi}} \int_{\mathbb{R}} f(t) \frac{d}{d\omega} e^{-i\omega t}\, dt \\
&= \frac{1}{\sqrt{2\pi}} \int_{\mathbb{R}} f(t)(-it) e^{-i\omega t}\, dt \\
&= -i \frac{1}{\sqrt{2\pi}} \int_{\mathbb{R}} t \cdot f(t) e^{-i\omega t}\, dt
\end{aligned}$$

If we compare the right-hand side of this last identity to (2.44), we see that

$$\hat{f}'(\omega) = -i\hat{g}(\omega) \qquad \text{or} \qquad \hat{g}(\omega) = i\hat{f}'(\omega)$$

∎

The last three rules deal with reflections, modulations, and derivatives of functions, respectively. The proofs of these propositions are left as problems.

Proposition 2.10 (Reflection Rule for Fourier Transforms) *Suppose that* $\hat{f}(\omega)$ *is the Fourier transform of* $f(t)$ *and let* $g(t) = f(-t)$*. Then* $\hat{g}(\omega) = \hat{f}(-\omega)$. ∎

Proof: Problem 2.33. ∎

Proposition 2.11 (Modulation Rule for Fourier Transforms) *Assume that* $\hat{f}(\omega)$ *is the Fourier transform of* $f(t)$*. Suppose that* $a \in \mathbb{R}$ *and let* $g(t) = e^{ita} f(t)$*. Then* $\hat{g}(\omega) = \hat{f}(\omega - a)$. ∎

Proof: Problem 2.34. ∎

Proposition 2.12 (Derivative Rule for Fourier Transforms) *Suppose that $f'(t) \in L^2(\mathbb{R})$ satisfies the Dirichlet conditions given in Proposition 2.4 and assume that $\hat{f}(\omega)$ is the Fourier transform of $f(t)$.*

If $g(t) = f'(t)$, then $\hat{g}(\omega) = i\omega \hat{f}(\omega)$. ∎

Proof: Problem 2.35. ∎

Computing Fourier Transforms Using Rules

Now that we have stated all these rules for computing Fourier transforms, let's look at some examples implementing them. We use the two Fourier transform pairs from Example 2.5 and the transform pair from Problem 2.39.

Example 2.6 (Using Rules to Compute Fourier Transforms) *Compute the Fourier transforms of the following functions:*

(a) $f_1(t) = \cos(4\pi t) \sqcap (t)$

(b) $f_2(t) = \begin{cases} t, & 0 \le t < 1 \\ 0, & otherwise \end{cases}$

(c) $f_3(t) = e^{-t^2/2}$

(d) $f_4(t) = \begin{cases} -e^{-t}, & t > 0 \\ e^t, & t < 0 \end{cases}$

(e) $f_5(t) = \dfrac{2it}{4 + t^2}$

Solution
    The idea for finding the Fourier transforms of each of the functions above is to avoid using the defining relation (2.35). Instead, we will try to manipulate each of the functions above and use the appropriate rule(s) (Propositions 2.5–2.12) and a known transform pair.
    We use (2.8) to rewrite $\cos(4\pi t)$ in terms of complex exponentials and thus express $f_1(t)$ as

$$f_1(t) = \frac{e^{4\pi it} + e^{-4\pi it}}{2} \sqcap (t)$$
$$= \frac{1}{2} e^{4\pi it} \sqcap (t) + \frac{1}{2} e^{-4\pi it} \sqcap (t)$$

Since the Fourier transform is linear (see Proposition 2.5), we can concentrate on finding the Fourier transforms of $e^{4\pi it} \sqcap (t)$ and $e^{-4\pi it} \sqcap (t)$, multiplying each by $\frac{1}{2}$, and then adding the results. We can use the modulation rule (Proposition 2.11) twice with $a = \pm 4\pi$ in conjunction with the Fourier transform (2.38) of the box function to write

$$\hat{f}_1(\omega) = \frac{1}{2\sqrt{2\pi}} \left(e^{-i(\omega-4\pi)/2} sinc((\omega - 4\pi)/2) + e^{-i(\omega+4\pi)/2} sinc((\omega + 4\pi)/2) \right)$$

$$= \frac{1}{2\sqrt{2\pi}} e^{-i\omega/2} \left(e^{2\pi i} sinc((\omega - 4\pi)/2) + e^{-2\pi i} sinc((\omega + 4\pi)/2) \right)$$

$$= \frac{1}{2\sqrt{2\pi}} e^{-i\omega/2} \left(sinc((\omega - 4\pi)/2) + sinc((\omega + 4\pi)/2) \right)$$

We can write $f_2(t)$ in terms of the box function. We have $f_2(t) = t \sqcap (t)$, so that we can compute $\hat{f}_2(\omega)$ by using the power rule (Proposition 2.9) in conjunction with the Fourier transform of the box function (2.38). Using the product rule and chain rule from calculus, we have

$$\hat{f}_2(\omega) = i \left(\frac{1}{\sqrt{2\pi}} e^{-i\omega/2} sinc(\omega/2) \right)'$$

$$= \frac{1}{\sqrt{2\pi}} \cdot \frac{i\omega e^{-i\omega} - 1 + e^{-i\omega}}{\omega^2}$$

For the third function we write $f_3(t) = e^{-t^2/2} = e^{-(t/\sqrt{2})^2}$. Now we can use the dilation rule (Proposition 2.7) with $a = \frac{1}{\sqrt{2}}$ in conjunction with the Fourier transform $\hat{f}(\omega) = \frac{\sqrt{2}}{2} e^{-\omega^2/4}$ of $f(t) = e^{-t^2}$ (see Problem 2.39) to write

$$\hat{f}_3(\omega) = \sqrt{2} \cdot \frac{\sqrt{2}}{2} e^{-(\sqrt{2}\omega)^2/4}$$

$$= e^{-\omega^2/2}$$

The function $f_4(t)$ satisfies $f_4(t) = f'(t)$, where $f(t) = e^{-|t|}$. Using the derivative rule (Proposition 2.12) along with the Fourier transform (see (2.43)) $\hat{f}(\omega)$ of $f(t)$, we have

$$\hat{f}_4(\omega) = \sqrt{\frac{2}{\pi}} \frac{i\omega}{1+\omega^2}$$

The function $f_5(t)$ looks closest to $\hat{f}_4(\omega)$. For that reason, we use the inversion rule (Proposition 2.8) with the function $g(t) = \hat{f}_4(t)$ to note that

$$\hat{g}(\omega) = f_4(-\omega) = \begin{cases} -e^{\omega}, & \omega > 0 \\ e^{-\omega}, & \omega < 0 \end{cases} \tag{2.45}$$

Now we manipulate $f_5(t)$ so that it resembles $g(t)$. We factor the 4 from the denominator of $f_5(t)$ and multiply and divide by $\sqrt{\frac{2}{\pi}}$ to obtain

$$f_5(t) = \frac{2it}{4+t^2} = \frac{2it}{4(1+(t/2)^2)} = \frac{i(t/2)}{1+(t/2)^2} = \sqrt{\frac{\pi}{2}} g(t/2)$$

Thus using the dilation rule with $a = \frac{1}{2}$, we see that

$$\hat{f}_5(\omega) = 2\sqrt{\frac{\pi}{2}}\,\hat{g}(2\omega) = \sqrt{2\pi}\begin{cases} -e^{2\omega}, & \omega > 0 \\ e^{-2\omega}, & \omega < 0 \end{cases}$$

■

Parseval's Identity and Plancherel's Identity

We conclude this section with a statement of *Parseval's identity*. This result says that the inner product of two functions $f(t)$ and $g(t)$ is the same as the inner product of their Fourier transforms $\hat{f}(\omega)$ and $\hat{g}(\omega)$. If we replace $g(t)$ by $f(t)$, then we obtain *Plancherel's identity*. Thus the norm of $f(t)$ is the same as the norm of $\hat{f}(\omega)$.

Proposition 2.13 (Parseval's Identity) *Suppose that $f(t), g(t) \in L^2(\mathbb{R})$. Then*

$$\boxed{\langle f(t), g(t)\rangle = \langle \hat{f}(\omega), \hat{g}(\omega)\rangle}$$ (2.46)

or, equivalently,

$$\int_{\mathbb{R}} f(t)\overline{g(t)}\,\mathrm{d}t = \int_{\mathbb{R}} \hat{f}(\omega)\overline{\hat{g}(\omega)}\,\mathrm{d}\omega$$ (2.47)

■

Proof: We start with the right-hand side of (2.46) and write

$$\langle \hat{f}(\omega), \hat{g}(\omega)\rangle = \int_{\mathbb{R}} \hat{f}(\omega)\overline{\hat{g}(\omega)}\,\mathrm{d}\omega$$

$$= \frac{1}{\sqrt{2\pi}}\int_{\omega \in \mathbb{R}} \hat{f}(\omega)\left(\overline{\int_{t \in \mathbb{R}} g(t)e^{-i\omega t}\,\mathrm{d}t}\right)\mathrm{d}\omega$$

$$= \frac{1}{\sqrt{2\pi}}\int_{\omega \in \mathbb{R}} \hat{f}(\omega)\left(\int_{t \in \mathbb{R}} \overline{g(t)e^{-i\omega t}}\,\mathrm{d}t\right)\mathrm{d}\omega$$

$$= \frac{1}{\sqrt{2\pi}}\int_{\omega \in \mathbb{R}} \hat{f}(\omega)\left(\int_{t \in \mathbb{R}} \overline{g(t)}e^{i\omega t}\,\mathrm{d}t\right)\mathrm{d}\omega$$

Note that we have used the defining relation (2.35) to replace $\hat{g}(\omega)$ above. To proceed with the proof, we need to exchange the order of integration. To justify this switch, we need *Fubini's theorem* from analysis (see, e. g., Rudin [48]). It turns out that for the functions we consider, the step is justified and we write

$$\langle \hat{f}(\omega), \hat{g}(\omega) \rangle = \frac{1}{\sqrt{2\pi}} \int_{\omega \in \mathbb{R}} \hat{f}(\omega) \left(\int_{t \in \mathbb{R}} \overline{g(t)} e^{i\omega t} \, dt \right) d\omega$$

$$= \int_{t \in \mathbb{R}} \overline{g(t)} \left(\frac{1}{\sqrt{2\pi}} \int_{\omega \in \mathbb{R}} \hat{f}(\omega) e^{i\omega t} \, d\omega \right) dt$$

$$= \int_{t \in \mathbb{R}} \overline{g(t)} f(t) \, dt$$

$$= \langle f(t), g(t) \rangle$$

We have used the Fourier inversion formula from Proposition 2.4 to obtain the next-to-last identity. ∎

We now state Plancherel's identity.

Corollary 2.1 (Plancherel's Identity) *For $f(t) \in L^2(\mathbb{R})$ we have*

$$\| f(t) \|^2 = \| \hat{f}(\omega) \|^2 \tag{2.48}$$

In integral form, the identity is

$$\int_{\mathbb{R}} | f(t) |^2 \, dt = \int_{\mathbb{R}} | \hat{f}(\omega) |^2 \, d\omega \tag{2.49}$$

∎

Proof: The proof follows immediately by replacing $g(t)$ with $f(t)$ in (2.46). ∎

Although we have obtained Plancherel's identity as a corollary of Parseval's formula, it might surprise you to know that the two are equivalent. That is, we can start with Plancherel's identity and derive Parseval's formula from it. Problem 2.45 leads you through the proof.

PROBLEMS

2.29 In this problem you will show that $\int_{\mathbb{R}} e^{-|t|} \sin \omega t \, dt$ converges. The following steps will help you organize your work.

(a) Show that for any $t, \omega \in \mathbb{R}$, we have

$$-e^{-|t|} \le e^{-|t|} \sin \omega t \le e^{-|t|}$$

(b) Show that $\int_{\mathbb{R}} e^{-|t|} \, dt = 2$.

(c) Use parts (a) and (b) to complete the argument.

2.30 Show that

$$\lim_{t \to \infty} \frac{e^{-t} \left(\omega \sin(\omega t) - \cos(\omega t) \right)}{1 + \omega^2} = 0$$

(a) First show that $|\omega \sin(\omega t) - \cos(\omega t)| \leq |\omega| + 1$.

(b) Use part (a) and the squeeze theorem from calculus to establish the result.

2.31 Suppose that $f(t) \in L^2(\mathbb{R})$ is an odd function. Show that the Fourier transform can be expressed as

$$\hat{f}(\omega) = \sqrt{\frac{2}{\pi}} \int_0^\infty f(t) \sin(\omega t)\, dt$$

2.32 In this problem we generalize the dilation rule (Proposition 2.7).

(a) Let $a < 0$. Show that if $g(t) = f(at)$, then $\hat{g}(\omega) = -\dfrac{1}{a} \hat{f}\left(\dfrac{t}{a}\right)$.

(b) Use Proposition 2.7 and part (a) to write down a dilation rule for $a \neq 0$.

2.33 Prove Proposition 2.10.

2.34 Prove Proposition 2.11.

2.35 Prove Proposition 2.12. (*Hint:* Use integration by parts with $u = e^{-i\omega t}$ and $dv = f'(t)\, dt$. You will also need to use the fact that

$$\lim_{t \to \pm\infty} f(t)e^{-i\omega t} = 0$$

This follows from the fact that $f(t) \in L^2(\mathbb{R})$.)

2.36 Suppose that the nth derivative $f^{(n)}(t) \in L^2(\mathbb{R})$ satisfies the Dirichlet conditions given in Proposition 2.4. Generalize Proposition 2.12 and show that if $g(t) = f^{(n)}(t)$, then $\hat{g}(\omega) = (i\omega)^n \hat{f}(\omega)$. (*Hint:* Use induction on n in conjunction with Proposition 2.12.)

2.37 In this problem you will generalize Proposition 2.9. Suppose that n is a positive integer and assume $g(t) = t^n f(t) \in L^2(\mathbb{R})$. Let $\hat{f}(\omega)$ be the Fourier transform of $f(t)$. Show that $\hat{g}(\omega) = i^n \hat{f}'(\omega)$. (*Hint:* Use induction on n in conjunction with Proposition 2.9.)

2.38 This problem requires some knowledge of multivariable calculus. Show that

$$\int_{\mathbb{R}} e^{-t^2}\, dt = \sqrt{\pi} \tag{2.50}$$

This result is necessary to complete Problem 2.39. The following steps will help organize your work.

(a) First observe that the left-hand side of (2.50) is simply $\sqrt{2\pi}\hat{f}(0)$, where $f(t) = e^{-t^2}$, and show that

$$2\pi\hat{f}(0)^2 = \int_{s\in\mathbb{R}} \int_{t\in\mathbb{R}} e^{-(s^2+t^2)}\, dt\, ds \tag{2.51}$$

(b) Convert (2.51) to polar coordinates and thereby show that

$$2\pi \hat{f}(0)^2 = 2\pi \int_0^\infty e^{-r^2} r \, dr$$

(c) Use the u-substitution $u = -r^2$ in part (b) to obtain the desired result.

2.39   In this problem you will compute the Fourier transform of $f(t) = e^{-t^2}$. The following steps will help you organize your work.

(a) Differentiate $f(t)$ to obtain $f'(t) + 2t \cdot f(t) = 0$.

(b) Take Fourier transforms of both sides of the identity from part (a). You will use the derivative rule (Proposition 2.12) and the power rule (Proposition 2.9) to show that

$$\frac{\hat{f}'(\omega)}{\hat{f}(\omega)} = -\omega/2$$

(c) Integrate both sides of the previous identity from 0 to u and simplify to obtain

$$\hat{f}(u) = \hat{f}(0)e^{-u^2/4}$$

(d) Use part (c) along with Problem 2.38 to show that

$$\hat{f}(\omega) = \frac{1}{\sqrt{2}} e^{-\omega^2/4}$$

★**2.40**   We need to exercise care when utilizing several Fourier transform rules. Suppose that we know the Fourier transform $\hat{f}(\omega)$ of function $f(t)$. Let $a, b \in \mathbb{R}$ with $a > 0$ and define $g(t) = f(at - b)$. It should be clear that we need both the dilation rule and the translation rule to find $\hat{g}(\omega)$.

(a) Show that $g(t) = f(a(t - \frac{b}{a}))$.

(b) Define $h(t) = f(at)$ and find $\hat{h}(\omega)$.

(c) Write $g(t)$ in terms of $h(t)$ and use part (b) to show that $\hat{g}(\omega) = \frac{1}{a} e^{-i\omega b/a} \hat{f}\left(\frac{\omega}{a}\right)$.

(d) Use the result you obtained in part (c) to find $\hat{g}(\omega)$ if $g(t) = e^{-|2x+3|}$.

2.41   Show that if we change $a > 0$ to $a \neq 0$ in the hypothesis of the dilation rule (Proposition 2.7), then we can show that $\hat{g}(\omega) = \frac{1}{|a|} \hat{f}\left(\frac{\omega}{a}\right)$.

2.42   The *Gaussian* or *normal distribution* is an important distribution in statistics and its probability distribution function $f(t)$ is defined by

$$f(t) = \frac{1}{\sqrt{2\pi}\,\sigma} e^{-\frac{(t-\mu)^2}{2\sigma^2}} \tag{2.52}$$

where μ and σ are real numbers with $\sigma > 0$. Find $\hat{f}(\omega)$. (*Hint:* Problems 2.39 and 2.40 will be helpful.)

★**2.43**   We will have need for the Fourier transform of the *triangle function* $\wedge(t)$ defined by (1.5). In this problem we will show that

$$\widehat{\wedge}(\omega) = \frac{1}{\sqrt{2\pi}} e^{-i\omega} \operatorname{sinc}^2(\omega/2) \qquad (2.53)$$

We could compute the Fourier transformation directly via Definition 2.4, but the calculations are tedious. In this problem we use the rules for computing Fourier transformations to find $\widehat{\wedge}(\omega)$. Two possible ways to use Fourier transformation rules are given below.

(a) Show that $\wedge'(t) = \sqcap(t) - \sqcap(t-1)$ (except at the points $= 0, 1, 2$). Now use the derivative rule and the translation rule to find $\widehat{\wedge}(\omega)$.

(b) Show that $\wedge(t) = t\sqcap(t) - (2-t)\sqcap(t-1)$ and use the power rule and the translation rules to find $\widehat{\wedge}(\omega)$. (*Hint:* The computations here are particularly tedious. It might be wise to use a computer algebra system such as Mathematica to check your work.)

In Section 2.4 we will find a much easier way to compute $\widehat{\wedge}(\omega)$.

2.44   Compute the Fourier transforms of the following functions. You will need the transform pairs from Example 2.5, Problem 2.39, and Problem 2.43.

(a) $f_1(t) = \sqcap\left(\frac{t}{2}\right)$

(b) $f_2(t) = \sqcap(t - \frac{1}{2})$

(c) $f_3(t) = \sqcap(3t - 4)$

(d) $f_4(t) = \sin(t)e^{-|t|}$

(e) $f_5(t) = \begin{cases} t, & 0 \le t < 1 \\ 1, & 1 \le t < 3 \\ 4 - t, & 3 \le t < 4 \\ 0, & \text{otherwise} \end{cases}$

(f) $f_6(t) = e^{-(t^2 - 2t)}$

(g) $f_7(t) = \operatorname{sinc}(t)$

(h) $f_8(t) = \sqcap(t) \cdot \sqcap(2t) \cdot \sqcap(3t)$

2.45   In this problem we start with Plancherel's identity and use it to derive Parseval's formula.

(a) Show that

$$f(t) \cdot g(t) = \frac{1}{4} \left(|f(t) + g(t)|^2 + i|f(t) + ig(t)|^2 \right.$$
$$\left. - |f(t) - g(t)|^2 - i|f(t) - ig(t)|^2 \right)$$

We can easily write a similar formula for $\hat{f}(\omega)$ and $\hat{g}(\omega)$ as well.

(b) Use part (a) in conjunction with Plancherel's identity (2.48) to derive Parseval's formula.

2.46   Use Plancherel's identity (2.48) in conjunction with (2.38) to find $\|\text{sinc}(\omega)\|$.

2.47   One feature of the Fourier transform is that we can use it to compute integrals. Suppose that $f(t)$ has the Fourier transform

$$\hat{f}(\omega) = \frac{1}{\sqrt{2\pi}} \int_{\mathbb{R}} f(t) e^{-i\omega t} \, dt$$

Notice that if we evaluate the identity above at $\omega = 0$, we obtain

$$\hat{f}(0) = \frac{1}{\sqrt{2\pi}} \int_{\mathbb{R}} f(t) \, dt$$

Thus if we know the Fourier transform value at 0, we can compute the integral of $f(t)$ over the real line. For example, suppose that we wish to compute[7]

$$\int_{\mathbb{R}} \frac{1}{1 + t^2} \, dt$$

By the inversion rule, we know that the function $f(t) = \frac{1}{1+t^2}$ has Fourier transform $\hat{f}(\omega) = \sqrt{\frac{\pi}{2}} e^{-|\omega|}$, so we can use (2.35) to write

$$\sqrt{\frac{\pi}{2}} e^{-|\omega|} = \frac{1}{\sqrt{2\pi}} \int_{\mathbb{R}} \frac{1}{1 + t^2} e^{-i\omega t} \, dt$$

Substituting $\omega = 0$ into the equation above gives

$$\sqrt{\frac{\pi}{2}} = \frac{1}{\sqrt{2\pi}} \int_{\mathbb{R}} \frac{1}{1 + t^2} \, dt$$

or

$$\int_{\mathbb{R}} \frac{1}{1 + t^2} \, dt = \pi$$

Use the idea outlined above in conjunction with known transform pairs, Plancherel's identity, and the defining relation (2.35) to compute the following integrals.

---

[7]Do you remember computing this integral in calculus?

(a) $\int_{\mathbb{R}} \operatorname{sinc}\left(\frac{t}{2}\right) dt$

(b) $\int_{0}^{\infty} e^{-2t} \cos(t) dt$

(c) $\int_{0}^{\infty} \frac{1}{3+t^2} dt$

(d) $\int_{0}^{\infty} \frac{1}{1+2t^2+t^4} dt$

(e) $\int_{\mathbb{R}} e^{-\pi t^2} dt$

2.48 In this problem you will prove the famous sampling theorem due to C. E. Shannon [54]. Suppose that $f(t) \in L^2(\mathbb{R})$ satisfies the hypotheses of Proposition 2.4 with Fourier transformation $\hat{f}(\omega)$. Further assume that the support of $\hat{f}(\omega)$ is contained in the interval $[-L, L]$ where $L > 0$. Functions whose Fourier transformations are contained in a finite-length interval are called *bandlimited*. Then Shannon's theorem states that

$$f(t) = \sum_{k \in \mathbb{Z}} f\left(\frac{k\pi}{L}\right) \operatorname{sinc}(Lt - k\pi) \qquad (2.54)$$

In other words, $f(t)$ can be completely recovered from its samples $x_k = f\left(\frac{k\pi}{L}\right)$.[8] The following steps will help you organize the proof.

(a) Since the support of $\hat{f}(\omega) \in L^2(\mathbb{R})$ is contained in $[-L, L]$, we can view $\hat{f}(\omega)$ as a function in $L^2([-L, L])$. Thus $\hat{f}(\omega)$ has a Fourier series representation (see Problem 2.28)

$$\hat{f}(\omega) = \sum_{k \in \mathbb{Z}} c_k e^{2\pi i k \omega / 2L}$$

with

$$c_k = \frac{1}{2L} \int_{-L}^{L} \hat{f}(\omega) e^{-2\pi i k \omega / 2L} d\omega$$

where $k \in \mathbb{Z}$. Show that $c_k = \frac{\sqrt{2\pi}}{2L} f\left(-\frac{k\pi}{L}\right)$. (*Hint:* Use the fact that $f(t)$ is bandlimited along with Definition 2.4.)

(b) Use part (a) to show that

$$\hat{f}(\omega) = \frac{\sqrt{2\pi}}{2L} \sum_{k \in \mathbb{Z}} f\left(\frac{k\pi}{L}\right) e^{-\pi i k \omega / L}$$

---

[8] The sampling rate is $\frac{\pi}{L} = \frac{2\pi}{2L}$. The value $2L$ is typically referred to as the *Nyquist sampling rate*. The interested reader is referred to Kammler [38] for more information on Shannon's sampling theorem and the Nyquist rate.

(c) Replace $\hat{f}(\omega)$ in (2.36) with the result in part (b) to show that

$$f(t) = \frac{1}{2L} \sum_{k \in \mathbb{Z}} f\left(\frac{k\pi}{L}\right) \int_{-L}^{L} e^{i\omega(t - k\pi/2)} \, d\omega$$

(d) Show that

$$\int_{-L}^{L} e^{i\omega(t - k\pi/2)} \, d\omega = (2L)\text{sinc}\left(L(t - k\pi/L)\right)$$

(e) Combine parts (c) and (d) to complete the proof.

2.4 CONVOLUTION AND B-SPLINES

In this section, we define convolution for two functions in $L^2(\mathbb{R})$. Our primary goal in this section is to develop the B-splines functions that we will need to construct wavelet functions in Chapter 8. For this reason, we do not provide a detailed introduction to $L^2(\mathbb{R})$ convolution. The interested reader is referred to the detailed presentation of convolution in Kammler [38].

Definition of Convolution

To motivate the definition of the convolution product of two functions in $L^2(\mathbb{R})$, we start with the convolution product of two bi-infinite sequences $\mathbf{v}$ and $\mathbf{h}$. This convolution product was introduced in Problem 2.22. If $\mathbf{v} = (\dots, v_{-2}, v_{-1}, v_0, v_1, v_2, \dots)$ and $\mathbf{h} = (\dots, h_{-2}, h_{-1}, h_0, h_1, h_2, \dots)$, then the convolution product $\mathbf{v} * \mathbf{h}$ is the bi-infinite sequence $\mathbf{y}$ whose components are given by

$$y_n = \sum_{k \in \mathbb{Z}} h_k v_{n-k}$$

The components y_n are obtained by computing an (infinite sum) inner product of $\mathbf{h}$ with a reflected and translated version of $\mathbf{v}$. Let $\mathbf{r}$ be the bi-infinite sequence obtained by reflecting the components of $\mathbf{v}$. That is, $r_k = v_{-k}$ for $k \in \mathbb{Z}$.

To compute y_0 we simply compute the inner product $\mathbf{h} \cdot \mathbf{r}$. To compute y_n we simply shift the elements of $\mathbf{r}$ n units right (left) if n is positive (negative) and then dot the resulting bi-infinite sequence with $\mathbf{h}$.

We will see that the definition of convolution of $L^2(\mathbb{R})$ functions uses the same idea but replaces the infinite sum with an integral. Here is the definition of the convolution of two functions in $L^2(\mathbb{R})$.

Definition 2.5 (Convolution in $\mathbf{L^2(\mathbb{R})}$) *Let $f(t), g(t) \in L^2(\mathbb{R})$. Then we define the convolution of $f(t)$ and $g(t)$ as the function $h(t)$ given by*

$$\boxed{h(t) = (f * g)(t) = \int_{\mathbb{R}} f(u)g(t - u) \, du} \qquad (2.55)$$

Let's analyze this definition. The key here is understanding $g(t - u)$. Remember that the independent variable is u and think of t as some fixed value (even though it is the independent variable of $(f * g)(t)$. Let's define $r(u) = g(-u)$ to be the reflection of $g(u)$ about the u-axis. If we translate $r(u)$ by t units, we have $r(u - t) = g(-(u - t)) = g(t - u)$. Thus we see the process for plotting $g(t - u)$. We first reflect it about the u-axis and then translate the result t units right (left) for t positive (negative). We have plotted this process in Figure 2.10 for $t > 0$. Now that we

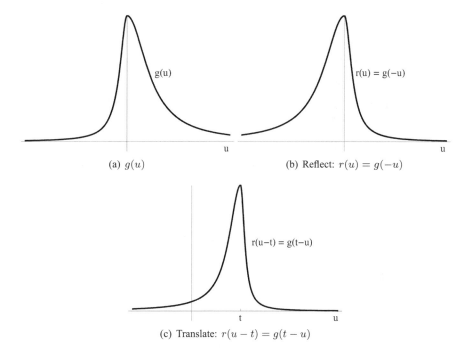

(a) $g(u)$ (b) Reflect: $r(u) = g(-u)$

(c) Translate: $r(u - t) = g(t - u)$

Figure 2.10   Plotting the function $g(t - u)$ for $t > 0$.

know how to plot $g(t - u)$, we can look at the integral in (2.55). For the sake of presentation, suppose that both $f(u)$ and $g(u)$ are nonnegative functions. Then we compute the function $f(u)g(t - u)$ and we can view the convolution as the area under the curve $f(u)g(t - u)$. This area changes as t varies over the real numbers. Figure 2.11 illustrates the computation of the convolution at several values of t where $f(t) = \sqcap(t)$.

Examples of Convolution Products

We are now ready to look at some examples of convolution in $L^2(\mathbb{R})$.

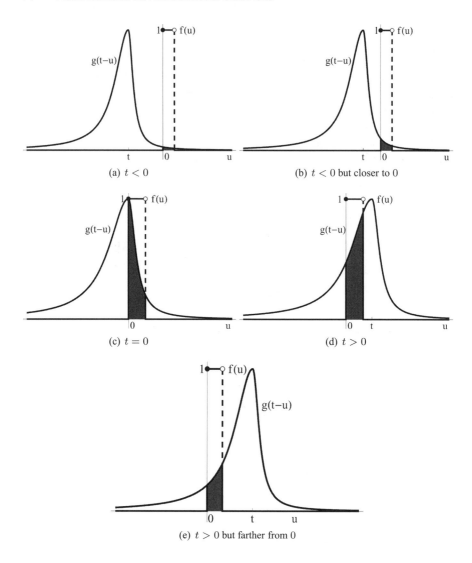

Figure 2.11   Convolution as the area under the curve. We think of letting t range over the reals, plotting the product of $f(u) = \sqcap(u)$ and $g(t - u)$, and then measuring the area under the curve for the convolution value at t.

Example 2.7 (Examples of Convolution in $\mathbf{L^2(\mathbb{R})}$) *Let $f(t) = e^{-|t|}$ (see Problem 1.13(a) in Section 1.2) and $\sqcap(t)$ be the box function defined in (1.4). Compute $(f * \sqcap)(t)$ and $(\sqcap * f)(t)$.*
Solution
    In Problem 2.49 you will show that convolution is a commutative operation. In particular, we have the freedom to compute either $(f * \sqcap)(t)$ or $(\sqcap * f)(t)$. This

*allows us to choose which function we reflect and translate. Since $f(t)$ is even, it
makes more sense to use it as the reflected and translated function.*

*Now both $f(t)$ and $\sqcap(t)$ are piecewise defined and this will influence our strategy
for performing the computation. For $f(u) = e^{-|t-u|}$, this means that we will have
to consider cases where $t - u < 0$ and $t - u \geq 0$, and for $\sqcap(u)$ we need to consider
cases for $u < 0$, $0 \leq u < 1$, and $u \geq 1$. And while we are considering the various
pieces, we must remember that t varies over all reals!*

*Let's first consider $f(t - u)$. Since $f(u) = f(-u)$, we only need to translate by t
in order to plot $f(t - u)$. Let's start with $t < 0$. Both $f(t - u)$ and $\sqcap(u)$ are plotted
in Figure 2.12(a). Notice that we have labeled each piece of $f(u)$. When $u < t$, then
$t - u > 0$ and $-|t - u| = -(t - u) = -t + u$. In this case $f(u) = e^{-t+u}$. In a
similar manner we see that when $u \geq t$, we have $f(u) = e^{t-u}$.*

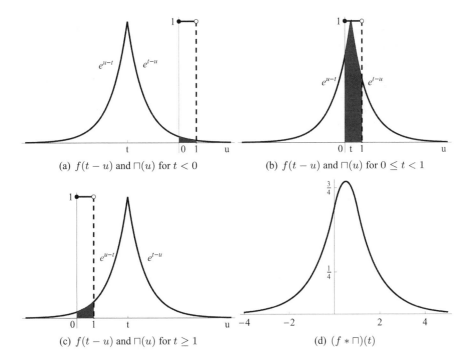

(a) $f(t - u)$ and $\sqcap(u)$ for $t < 0$ (b) $f(t - u)$ and $\sqcap(u)$ for $0 \leq t < 1$

(c) $f(t - u)$ and $\sqcap(u)$ for $t \geq 1$ (d) $(f * \sqcap)(t)$

Figure 2.12   The functions $f(t - u)$ and $\sqcap(u)$ for $t > 0$ and the convolution product
$(f * \sqcap)(t)$.

When $t < 0$ we have

$$(f * \sqcap)(t) = \int_0^1 e^{t-u} \, du = e^t \int_0^1 e^{-u} \, du = e^t(1 - e^{-1}) = e^t - e^{t-1}$$

When $0 \le t < 1$ we have two integrals that contribute to the convolution value:

$$(f * \sqcap)(t) = \int_0^t e^{-t+u}\, du + \int_t^1 e^{t-u}\, du$$
$$= e^{-t}(e^t - 1) + e^t(e^{-t} - e^{-1})$$
$$= 2 - e^{-t} - e^{t-1}$$

Finally, when $t \ge 1$ we have

$$(f * \sqcap)(t) = \int_0^1 e^{-t+u}\, du = e^{-t}(e - 1) = e^{1-t} - e^{-t}$$

We have plotted $(f * \sqcap)(t)$ in Figure 2.12(d).

Notice that although $\sqcap(t)$ is discontinuous at $t = 0,1$, it appears that $(f * \sqcap)(t)$ is a continuous function for all reals. In Problem 2.50 you will show that $(f * \sqcap)(t)$ is indeed a continuous function. Although this is only one example, we will see that the convolution operator is a smoothing operator. That is, the resulting function is at least as smooth as the two convolved functions.

For the convolution product $(\sqcap * \sqcap)(t)$, we must reflect $\sqcap(u)$ about the $u = 0$ axis and then translate it by t units. This process is plotted in Figure 2.13(a) for $t < 0$. As we move t to the right, we note that the product of $\sqcap(t - u)$ and $\sqcap(u)$ is 0 up

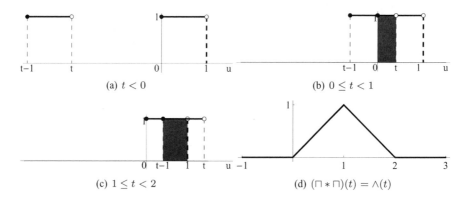

(a) $t < 0$

(b) $0 \le t < 1$

(c) $1 \le t < 2$

(d) $(\sqcap * \sqcap)(t) = \wedge(t)$

Figure 2.13   The convolution product $(\sqcap * \sqcap)(t) = \wedge(t)$.

to $t = 0$. From $t = 0$ to $t = 1$, the convolution product is simply the area of the rectangle with height 1 and width t (see Figure 2.13(b)). At $t = 1$, the boxes align so that the area (and thus the convolution product at $t = 1$) is 1. For $1 < t \le 2$, the boxes still overlap (see Figure 2.13(c)) and the overlap produces a rectangle of height 1 and width $1 - (t - 1) = 2 - t$. Once t moves past 2, the boxes no longer overlap so that the convolution product is 0. Thus we have the piecewise formula for

$(\sqcap * \sqcap)(t)$:

$$(\sqcap * \sqcap)(t) = \begin{cases} t, & 0 \le t < 1 \\ 2 - t, & 1 \le t < 2 \\ 0, & otherwise \end{cases} = \wedge(t) \tag{2.56}$$

Thus we see that $(\sqcap * \sqcap)(t) = \wedge(t)$ *(see Example 1.2). This result is plotted in Figure 2.13(d).*   ■

The Convolution Theorem

One of the most important results involving the convolution product of $L^2(\mathbb{R})$ functions is linked to Fourier transformations. Although computing the convolution product of two functions can be tedious, the convolution theorem tells us that the Fourier transformation of a convolution product can be performed by simple multiplication.

The result below is quite similar to that stated in Problem 2.22. Indeed, the proofs of both results are quite similar.

Theorem 2.1 (The Convolution Theorem) *Suppose that* $f(t), g(t) \in L^2(\mathbb{R})$, *and let* $h(t) = (f * g)(t)$. *Then*

$$\hat{h}(\omega) = \sqrt{2\pi}\hat{f}(\omega)\hat{g}(\omega) \tag{2.57}$$

■

Proof: We insert $h(t) = (f * g)(t)$ into the defining relation (2.35) for $\hat{h}(\omega)$ and then use (2.55) to form the double integral

$$\begin{aligned} \hat{h}(\omega) &= \frac{1}{\sqrt{2\pi}} \int_{\mathbb{R}} h(t)e^{-i\omega t} \, dt \\ &= \frac{1}{\sqrt{2\pi}} \int_{\mathbb{R}} (f * g)(t)e^{-i\omega t} \, dt \\ &= \frac{1}{\sqrt{2\pi}} \int_{t \in \mathbb{R}} \left(\int_{u \in \mathbb{R}} f(u)g(t - u) \, du \right) e^{-i\omega t} \, dt \end{aligned}$$

We now need to switch the order of integration. As was the case with the proof of Parseval's identity (Proposition 2.13), we will perform the switch without rigorously justifying the step. In addition, we will insert a $1 = e^{-i\omega u} \cdot e^{i\omega u}$ into the integrand. We have

$$\hat{h}(\omega) = \int_{u \in \mathbb{R}} f(u) \left(\frac{1}{\sqrt{2\pi}} \int_{t \in \mathbb{R}} g(t - u)e^{-i\omega(t-u)} \, dt \right) e^{-i\omega u} \, du \tag{2.58}$$

We now make the substitution $s = t - u$ in the inner integral of (2.58). Note that $ds = dt$ and the limits of integration are unchanged. The inner integral in (2.58) becomes

$$\frac{1}{\sqrt{2\pi}} \int_{t \in \mathbb{R}} g(t-u)e^{-i\omega(t-u)}\, dt = \frac{1}{\sqrt{2\pi}} \int_{s \in \mathbb{R}} g(s)e^{-i\omega s}\, ds$$
$$= \hat{g}(\omega)$$

We can insert this result into (2.58) to complete the proof:

$$\hat{h}(\omega) = \int_{u \in \mathbb{R}} f(u)\hat{g}(\omega)e^{-i\omega u}\, du$$
$$= \hat{g}(\omega) \int_{u \in \mathbb{R}} f(u)e^{-i\omega u}\, du$$
$$= \hat{g}(\omega)\sqrt{2\pi}\frac{1}{\sqrt{2\pi}} \int_{u \in \mathbb{R}} f(u)e^{-i\omega u}\, du$$
$$= \sqrt{2\pi}\,\hat{g}(\omega)\hat{f}(\omega)$$

∎

Let's look at some examples.

Example 2.8 (Examples Using the Convolution Theorem) *Find the Fourier transforms of the convolution products obtained in Example 2.7.*
Solution
In Example 2.7 we computed the convolution product of $f(t) = e^{-|t|}$ and $\sqcap(t)$:

$$h(t) = \begin{cases} e^t - e^{t-1}, & t < 0 \\ 2 - e^{-t} - e^{t-1}, & 0 \le t < 1 \\ e^{1-t} - e^{-t}, & t \ge 1 \end{cases}$$

Suppose that we wish to compute $\hat{h}(\omega)$. This computation via the defining relation (2.35) is difficult. Instead, we note that from Example 2.5 we know that the Fourier transforms of $f(t)$ and $\sqcap(t)$ are $\hat{f}(\omega) = \sqrt{\dfrac{2}{\pi}}\dfrac{1}{1+\omega^2}$ and $\hat{\sqcap}(\omega) = \dfrac{1}{\sqrt{2\pi}}e^{-i\omega/2}\operatorname{sinc}\left(\frac{\omega}{2}\right)$, respectively. We use the convolution theorem to write

$$\hat{h}(\omega) = \sqrt{\frac{2}{\pi}}\frac{e^{-i\omega/2}\operatorname{sinc}\left(\frac{\omega}{2}\right)}{1+\omega^2}$$

In Problem 2.43 in Section 2.3 we showed that $\hat{\wedge}(\omega) = \dfrac{1}{\sqrt{2\pi}}e^{-i\omega}\operatorname{sinc}^2\left(\frac{\omega}{2}\right)$. The computations used a variety of rules for computing Fourier transforms and required a large amount of algebra to obtain the desired result. Using the convolution theorem and the fact that $\wedge(t) = (\sqcap * \sqcap)(t)$, we can write

$$\hat{\wedge}(\omega) = \sqrt{2\pi}\left(\frac{1}{\sqrt{2\pi}}e^{-i\omega/2}\operatorname{sinc}\left(\frac{\omega}{2}\right)\right)\cdot\left(\frac{1}{\sqrt{2\pi}}e^{-i\omega/2}\operatorname{sinc}\left(\frac{\omega}{2}\right)\right)$$
$$= \frac{1}{\sqrt{2\pi}}e^{-i\omega}\operatorname{sinc}^2\left(\frac{\omega}{2}\right)$$

■

B–Splines

We conclude this section with an introduction to B-splines[9]. These functions are piecewise polynomials that play an important role in the area of *approximation theory*. Applications utilizing B-splines include car body design, computer-aided geometric design, and cartography. We use them in Chapter 8 to build functions that ultimately lead to a class of filters that are widely used in image compression. A nice elementary treatment of B-splines is that of Cheney and Kincaid [12]. More comprehensive treatments of B-splines may be found in de Boor [21] or Schumaker [52].

Definition 2.6 (B–Splines Defined) *We define the B-spline of order 1 to be the box function $B_0(t) = \sqcap(t)$. The B-spline of order $n + 1$ is defined recursively by*

$$B_n(t) = (B_{n-1} * \sqcap)(t) \tag{2.59}$$

■

From Example 2.7 we know that $B_1(t) = (B_0 * \sqcap)(t) = (\sqcap * \sqcap)(t) = \wedge(t)$. In Problem 2.56 you will show that

$$B_2(t) = (\wedge * \sqcap)(t) = \begin{cases} t^2/2, & 0 \le t < 1 \\ (-2t^2 + 6t - 3)/2, & 1 \le t < 2 \\ (t^2 - 6t + 9)/2, & 2 \le t < 3 \\ 0, & \text{otherwise} \end{cases} \tag{2.60}$$

The first four B-splines are plotted in Figure 2.14.

Properties of B–Splines

B-splines satisfy several properties and we state them at this time. The first property notes that $B_n(t)$ is a piecewise polynomial function of degree n.

Proposition 2.14 (A B–Spline Is a Piecewise Polynomial) *The B-spline $B_n(t)$ is a piecewise polynomial of degree n.* ■

Proof: The proof is by induction on n. For $n = 0$ we have $B_0(t) = \sqcap(t)$, and this function is a piecewise constant polynomial.

Now assume that $B_n(t)$ is a piecewise polynomial of degree n. We must show that $B_{n+1}(t)$ is a piecewise polynomial of degree $n + 1$. From Problem 2.55 we know that

$$B_{n+1}(t) = \int_{t-1}^{t} B_n(u)\,\mathrm{d}u \tag{2.61}$$

[9]A *spline* is a flexible tool once used by architects to draw curves.

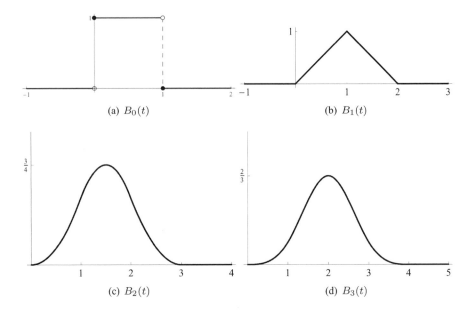

Figure 2.14   The B-spline functions $B_n(t)$ for $n = 0, 1, 2, 3$.

Thus to obtain $B_{n+1}(t)$, we integrate a piecewise polynomial of degree n. The resulting antiderivative is a piecewise polynomial of degree $n + 1$. Since we evaluate the antiderivative at $t - 1$ and t, the result $B_{n+1}(t)$ is a piecewise polynomial of degree $n + 1$. ∎

In Problem 2.61 you will show that the breakpoints for the polynomial pieces of $B_n(t)$ are $0, 1, \ldots, n + 1$.

The next proposition identifies the interval of compact support for $B_n(t)$.

Proposition 2.15 (The Support of a B–Spline) *For all* $n = 0, 1, 2, 3, \ldots$ $B_n(t) \geq 0$ *and* $\overline{\mathrm{supp}(B_n)} = [0, n + 1]$. *Moreover,* $B_n(t) > 0$ *for* $0 < t < n + 1$. ∎

Proof: We prove this proposition by induction on n. For $n = 0$, $B_0(t) = \sqcap(t)$ and this function is nonnegative $\overline{\mathrm{supp}(B_0)} = [0,1]$. Note also that $B_0(t) = 1 > 0$ for $0 < t < 1$.

We now assume that $B_n(t) \geq 0$, $B_n(t) > 0$ for $0 < t < n$, and $\overline{\mathrm{supp}(B_n)} = [0, n + 1]$. We must use this information to show that $B_{n+1}(t) \geq 0$ with $B_{n+1}(t) > 0$ for $0 < t < n + 1$, and $\mathrm{supp}(B_{n+1}) = [0, n + 2]$. Since $B_n(t) \geq 0$, the integral in (2.61) is nonnegative as well. Thus $B_{n+1}(t) \geq 0$. Now $\overline{\mathrm{supp}(B_n)} = [0, n + 1]$ so if $t < 0$, $B_n(t) = 0$. Additionally, if $t - 1 > n$ or $t > n + 1$, $B_n(t) = 0$. Thus $B_n(t) = 0$ for $t \notin [0, n + 1]$. To complete the proof, it suffices to show that $B_{n+1}(t) > 0$ on $(0, n + 2)$. If we could show this, we would have the smallest closed interval containing all the points where $B_{n+1}(t) \neq 0$ is $[0, n + 2]$. Let's consider the definition (2.61) of $B_{n+1}(t)$:

$$B_{n+1}(t) = B_n(t) * \sqcap(t) = \int_0^1 B_n(t - u) \, du \qquad (2.62)$$

Now from the induction step, we know that $B_n(t-u) \geq 0$ whenever $0 \leq t-u \leq n+1$ or when $u \leq t \leq u+n+1$ for all values $0 \leq u \leq 1$. The largest interval, then, where the integrand in (2.62) is nonnegative is $0 \leq t \leq 1+n+1 = n+2$. Outside this interval, we know from the induction step that the integrand is zero so $B_{n+1}(t) = 0$ here as well. Thus $\mathrm{supp}(B_{n+1}) = [0, n+2]$ as desired, and the proof is complete. ∎

The next proposition shows that B-splines are symmetric functions.

Proposition 2.16 (B–Splines are Symmetric Functions) *Let n be a nonnegative integer. Then the B-spline $B_n(t)$ is symmetric about the line $t = (n+1)/2$. That is, $B_n(t) = B_n(n+1-t)$.* ∎

Proof: The proof of this proposition also uses induction on n. For $n = 0$, it is easy to check that $B_0(t) = \sqcap(t)$ satisfies $B_0(t) = B_0(1-t)$ and is thus symmetric about $t = 1/2$.

Now assume that $B_n(t)$ is symmetric about the line $t = (n+1)/2$ so that $B_n(n+1-t) = B_n(t)$. We must show that $B_{n+1}(n+2-t) = B_{n+1}(t)$. We utilize (2.61) and write

$$B_{n+1}(n+2-t) = \int_{n+2-t-1}^{n+2-t} B_n(u) \, du = \int_{n-t+1}^{n-t+2} B_n(u) \, du \qquad (2.63)$$

We now make the substitution $s = n + 1 - u$. This gives $ds = -du$ and $u = n + 1 - s$. We pick this substitution so that we can exploit the symmetry of $B_n(t)$. When $u = n - t + 1$, $s = n + 1 - (n - t + 1) = t$ and when $u = n - t + 2$, $s = n + 1 - (n - t + 2) = t - 1$. Thus (2.63) becomes

$$B_{n+1}(n+2-t) = \int_{n-t+1}^{n-t+2} B_n(u) \, du$$

$$= -\int_t^{t-1} B_n(n+1-s) \, ds$$

$$= \int_{t-1}^t B_n(n+1-s) \, ds$$

We use the induction hypothesis $B_n(n + 1 - t) = B_n(t)$ to write

$$B_{n+1}(n+2-t) = \int_{t-1}^t B_n(s) \, ds$$

Using (2.61), we see that the right side of this identity is simply $B_{n+1}(t)$, and the proof is complete. ∎

Our final proposition makes repeated use of the convolution theorem to find the Fourier transform of a B-spline.

Proposition 2.17 (The Fourier Transform of a B–Spline) *The Fourier transform of the B-spline $B_n(t)$ is*

$$\widehat{B}_n(\omega) = (2\pi)^{-1/2} e^{-i\omega(n+1)/2} \operatorname{sinc}^{n+1}(\omega/2) \qquad (2.64)$$

∎

Proof: The proof of this proposition is left as Problem 2.64.   ∎

PROBLEMS

★**2.49**   Show that convolution is a commutative operation. That is, show that $(f * g)(t) = (g * f)(t)$.

2.50   Show that the function $(f * \sqcap)(t)$ from Example 2.7 is continuous by showing that $\lim_{t \to 0}(f * \sqcap)(t) = (f * \sqcap)(0)$ and $\lim_{t \to 1}(f * \sqcap)(t) = (f * \sqcap)(1)$.

2.51   Compute the following convolution products.

(a) $f(t) = \dfrac{1}{1 + t^2}$ and $\sqcap(t)$.

(b) $\sqcap(t - 1)$ and $\sqcap(t + 1)$.

(c) $\sqcap(t)$ and $\sqcap(t - 1) - \sqcap(t - 3)$.

(d) $(f * f)(t)$, where $f(t) = e^{-t^2}$. (*Hint:* Use the convolution theorem.)

2.52   Find the Fourier transforms of the following functions.

(a) $h(t) = \wedge(t - 1) * e^{-t^2}$.

(b) $h(t) = \int_{-1}^{1} e^{-(t-u)^2}\, du$.

2.53   Let $a > b > 0$. Compute the convolution product $\sqcap(t - a) * \sqcap(t - b)$. (*Hint:* You can either use the defining relation (2.55) or the convolution theorem.)

2.54   Suppose that $f(t)$ and $g(t)$ are even functions. Show that the convolution product $h(t) = (f * g)(t)$ is an even function. (*Hint:* Use the fact that $g(t)$ is even to write $h(-t) = \int_{\mathbb{R}} f(u)g(t + u)\, du$. Then make the substitution $s = t + u$ and use the fact that $f(t)$ is even to show that $h(-t) = h(t)$.)

2.55   Show that $B_n(t) = \int_{t-1}^{t} B_{n-1}(u)\, du$.

2.56   Verify the formula given for $B_2(t)$ in (2.60).

2.57 Find the Fourier transform $\widehat{B_2}(\omega)$ of the function $B_2(t)$ from Problem 2.56.

★**2.58** Show that

$$B_1(t) = (B_1(2t) + 2B_1(2t - 1) + B_1(2t - 2))/2$$

★**2.59** Show that

$$B_2(t) = (B_2(2t) + 3B_2(2t - 1) + 3B_2(2t - 2) + B_2(2t - 3))/4$$

★**2.60** In this problem you will see why B-splines are important in *interpolation* theory. Suppose that you had the ordered pairs

$$(-2,0), (-1,4), (0,-2), (1,5), (2,3), (3,-3), (4,0)$$

Plot these points and connect them with line segments. Call this piecewise linear function $\ell(t)$. If we think of the ordered pairs as points on some unknown curve $f(t)$, then $\ell(t)$ is an approximation to $f(t)$ that interpolates $f(t)$ at $t = -2, \ldots, 4$. In practice, we would like to have a nice representation of $\ell(t)$. We could figure out the formula for each line segment but that would be tedious if the number of points increased. Show that we can easily represent $\ell(t)$ using translates of $B_1(t)$. That is, show that

$$\ell(t) = 4B_1(t + 2) - 2B_1(t + 1) + 5B_1(t) + 3B_1(t - 1) - 3B_1(t - 2)$$

In this way we see that we can easily form $\ell(t)$ as a linear combination of linear B-splines.

2.61 In Proposition 2.14 we showed that $B_n(t)$ is a piecewise polynomial of degree n. Show that the breakpoints for the polynomial pieces of $B_n(t)$ are $0, 1, \ldots, n$. (*Hint:* Use induction on n.)

2.62 Let n and m be nonnegative integers. Show that $(B_n * B_m)(t) = B_{n+m}(t)$.

2.63 Suppose that n is a positive integer. Show that $B_n'(t) = B_{n-1}(t) - B_{n-1}(t - 1)$. (*Hint:* Use Problem 2.55 in conjunction with the fundamental theorem of calculus.

2.64 Prove Proposition 2.17. (*Hint:* First observe that $B_n(t)$ is the n-fold convolution product of box functions $\sqcap(t)$ and then make repeated use of the convolution theorem.)

CHAPTER 3

HAAR SPACES

Now that we have studied $L^2(\mathbb{R})$ and the Fourier transformation, we can start constructing wavelets. Toward this end, we seek to decompose $L^2(\mathbb{R})$ into two nested sequences of subspaces. The sequence $V_j \subset V_{j+1}$ are *approximation spaces*. As j tends to infinity, the space V_j provides a better approximation to an arbitrary function $f(t) \in L^2(\mathbb{R})$. The spaces $W_j \subset W_{j+1}$ are *detail spaces* — W_j is a subset of V_{j+1}, and if we approximate $f_{j+1}(t) \in V_{j+1}$ by $f_j(t) \in V_j$, then $w_j(t) \in W_j$ holds the details we need to combine with $f_j(t)$ to recover $f_{j+1}(t)$. That is, $f_{j+1}(t) = f_j(t) + w_j(t)$.

We can iterate the decomposition process and write $f_M(t) \in V_M$ as a coarse approximation $f_m(t) \in V_m$, $m < M$, and detail functions $w_m(t)$, $w_{m+1}(t)$, $\ldots$, $w_{M-1}(t)$. In applications, we can quantize the detail information to perform signal or image compression. Alternatively, we might analyze the detail information to look for jumps in $f_M(t)$ or one of its derivatives.

While Chapter 5 addresses the general construction of V_j and W_j, we use this chapter to introduce *Haar spaces*. As we will see, a Haar space is comprised of piecewise constant functions. From an applications standpoint, you can think of the heights of these different constant pieces as grayscale intensities or frequencies from an audio signal. The fundamental Haar space V_0 is introduced in Section 3.1. This space is built from piecewise constant functions with possible breaks at the integers. More general Haar spaces, with breakpoints on a coarser or finer grid, are introduced in Section 3.2.

The fundamental idea of our decomposition of $L^2(\mathbb{R})$ into approximation or detail spaces is that a function (or a digital image or audio signal) can be decomposed into an approximation of the original and the details needed to recover the original from the approximate. The Haar spaces V_j will hold our approximations and in Sections 3.3 and 3.4 we define the *Haar wavelet spaces*, which we use to house detail information.

In Section 3.5 we discuss how these spaces relate to each other and how we can do the iterative decomposition and reconstruction of a function using Haar spaces and Haar wavelet spaces. We will see in this section that the continuous model uses the vector pairs $\mathbf{h} = [h_0, h_1]^T = [\sqrt{2}/2, \sqrt{2}/2]^T$ and $\mathbf{g} = [g_0, g_1]^T = [\sqrt{2}/2, -\sqrt{2}/2]^T$ to perform the iterative decomposition and reconstruction of a function. These vector pairs give rise to the *discrete Haar wavelet transformation*, which can be used in applications involving signals and digital images (see Chapter 4).

Since the ideas in this chapter are paramount to understanding the material in Chapter 5, we summarize all the major results in Section 3.6.

3.1 THE HAAR SPACE V_0

A digital grayscale image is comprised of *pixels*. Typically, the gray intensity of these pixels consists of nonnegative integers ranging from 0 (black) to 255 (white). Suppose that we have a small grayscale image and the intensity values of the pixels in the first row are $110, 100, 120, 140, 130, 100, 100$. These intensities are plotted in Figure 3.1.

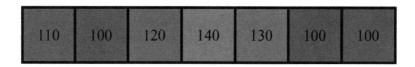

Figure 3.1 Grayscale intensities.

It is natural to ask if there is a function $f(t)$ that can be used to represent the first row of the image. In particular, we are interested in using a single function $\phi(t)$ to build $f(t)$.

One simple way to construct $f(t)$ is to take $\phi(t)$ to be the box function $\sqcap(t)$ from Example 1.2. That is,

$$\phi(t) = \begin{cases} 1, & 0 \le t < 1 \\ 0, & \text{otherwise} \end{cases} \tag{3.1}$$

Now we use ϕ and its integer translates to construct $f(t)$.

$$f(t) = 110\phi(t) + 100\phi(t-1) + 120\phi(t-2) + 140\phi(t-3)$$
$$+ 130\phi(t-4) + 100\phi(t-5) + 100\phi(t-6) \qquad (3.2)$$

Plots of $\phi(t)$ and $f(t)$ appear in Figure 3.2.

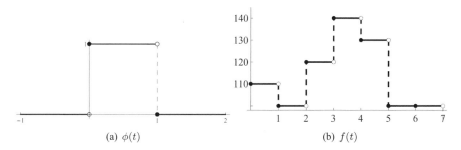

(a) $\phi(t)$ (b) $f(t)$

Figure 3.2   The function $\phi(t)$ defined in (3.1) and $f(t)$ given by (3.2).

Note that $f(t)$ is continuous for all $t \in \mathbb{R}$ except at $t = 1, 2, 3, 4, 5$ and is right continuous for all $t \in \mathbb{R}$.

The Space V_0 Defined

We can use the example above to create a vector space that holds all piecewise constant functions with possible breakpoints at the integers. A tempting way to define this space is as the *linear span* of $\phi(t)$ and its integer translates $\phi(t-k)$, where $k \in \mathbb{Z}$. So a typical element in this space is

$$f(t) = \sum_{k \in \mathbb{Z}} a_k \phi(t-k) \qquad (3.3)$$

where $a_k \in \mathbb{R}$. We have left the summation limits in (3.3) purposely vague. Do we need to put some constraints on the summation limits? Allowing the summation variable k to run from 0 to some $N > 0$ makes sense for modeling rows and columns of digital images, and perhaps it would be convenient to be even more general and allow k to run over any finite set of integers. In such cases the elements in the space are compactly supported.

Insisting that the summation variable k only runs over a finite set of integers does not help us model some useful functions. For example, sound waves are built from linear combinations of sines and cosines and, as such, have infinite support length. Of course, these waves decay over time and our space should model this as well. Thus we need to allow the summation variable to range over all $\mathbb{Z}$, but we need to put conditions on the space so that the signals decay. Toward this end we insist that any element of our space is also an element of $L^2(\mathbb{R})$. We have the following definition:

Definition 3.1 (The Haar Space V_0 and the Haar Function $\phi(t)$) *Let $\phi(t)$ be the box function given in (3.1). We define the space*

$$V_0 = \text{span}\{\ldots, \phi(t+1), \phi(t), \phi(t-1), \ldots\} \cap L^2(\mathbb{R})$$

$$= \text{span}\{\phi(t-k)\}_{k \in \mathbb{Z}} \cap L^2(\mathbb{R}) \qquad (3.4)$$

We call V_0 the Haar space V_0 generated by the Haar function $\phi(t)$.[10] ∎

We can thus describe V_0 as the space of all piecewise constant functions in $L^2(\mathbb{R})$ with possible breakpoints at the integers. In Problem 3.4 you will show that V_0 is a subspace of $L^2(\mathbb{R})$.

By definition, $\phi(t)$ and its integer translates $\phi(t-k)$, span V_0, and in Problem 3.1, you will show that the set $\{\phi(t-k)\}_{k \in \mathbb{Z}}$ is linearly independent. Thus the set is a basis for V_0. We can say even more.

Orthogonality Conditions for $\phi(t)$ and Its Translates

Let's choose integers $j \neq k$ and consider the functions $\phi(t-j)$ and $\phi(t-k)$. The function $\phi(t-j)$ is nonzero on the interval $[j, j+1)$ while $\phi(t-k)$ is nonzero on then interval $[k, k+1)$. Since $j \neq k$, these intervals never intersect (see Figure 3.3).

Figure 3.3   The functions $\phi(t-j)$ and $\phi(t-k)$ for $j \neq k$.

Thus the inner product

$$\langle \phi(t-j), \phi(t-k) \rangle = \int_{\mathbb{R}} \phi(t-j)\phi(t-k)\, dt = 0 \qquad (3.5)$$

so the functions are orthogonal. Note also that $\phi(t-k)^2 = \phi(t-k)$. Since the area under $\phi(t)$ is 1 no matter where we translate the function (see Problem 3.5), we have

$$\int_{\mathbb{R}} \phi(t-k)^2\, dt = \int_{\mathbb{R}} \phi(t-k)\, dt = \int_{k}^{k+1} 1\, dt = 1 \qquad (3.6)$$

[10]V_0 and $\phi(t)$ are so named in honor of the Hungarian mathematician Alfréd Haar (1885–1933), who studied them as part of his 1909 doctoral thesis *Zur Theorie der Orthogonalen Funktionensysteme* (The Theory of Orthogonal Function Systems).

Note: We consider only real-valued functions in this section and the remainder of Chapter 3. Thus we drop the conjugation operator from the inner product given in Definition 1.8.

We have the following proposition:

Proposition 3.1 (An Orthonormal Basis for V_0) *Consider the subspace V_0 given in Definition 3.1. The set $\{\phi(t - k)\}_{k \in \mathbb{Z}}$, where $\phi(t)$ is defined by (3.1), forms an orthonormal basis for V_0.* ∎

Proof: By definition, $\{\phi(t - k)\}_{k \in \mathbb{Z}}$ spans V_0 and in Problem 3.1 you will show that the set $\{\phi(t - k)\}_{k \in \mathbb{Z}}$ is linearly independent. The computations preceding the proposition show that

$$\langle \phi(t - k), \phi(t - j) \rangle = \delta_{j,k}$$

where $\delta_{j,k}$ is given by (2.16). ∎

Let's look at some examples of elements of V_0.

Example 3.1 (Elements in V_0) *Plot each of the following functions and determine if they are elements of V_0.*

(a) $f(t) = \sum\limits_{k \in \mathbb{Z}} a_k \phi(t - k)$, *where*

$$a_k = \begin{cases} 3, & k \in \{-10, \ldots, 9\} \\ 0, & otherwise \end{cases}$$

(b) The stair step function $g(t) = \sum\limits_{k=1}^{\infty} k\,\phi(t - k)$

(c) The function $h(t) = \sum\limits_{k=1}^{\infty} \dfrac{1}{k}\,\phi(t - k)$

(d) The function $\ell(t) = \sum\limits_{k=0}^{3} \phi(4t - k)$

Solution

For part (a) we can reduce $f(t)$ to

$$f(t) = \sum\limits_{k=-10}^{9} 3\phi(t - k) = 3 \sum\limits_{k=-10}^{9} \phi(t - k) = \begin{cases} 3, & -10 \leq t < 10 \\ 0, & otherwise \end{cases}$$

This function is certainly piecewise constant and has two jump discontinuities at $t = \pm 10$. It is easy to see that

$$\int_{\mathbb{R}} f(t)^2 \, dt = \int_{-10}^{10} 3^2 \, dt = 180 < \infty$$

so that $f(t) \in V_0$.

For part (b), $g(t) \geq 1$ for $t \geq 1$ and $k = 1, 2, \ldots$ and in Problem 1.13(b) in Section 1.2 you showed that the function

$$r(t) = \begin{cases} 1, & t \geq 1 \\ 0, & t < 1 \end{cases}$$

is not an element of $L^2(\mathbb{R})$. Since $g(t) \geq r(t)$ on $[1, \infty)$, we have $\int_{\mathbb{R}} g(t)^2 \, \mathrm{d}t = \infty$ as well. Even though $g(t)$ is a piecewise constant function with breakpoints at the integers, $g(t) \notin V_0$.

For part (c) it is easy to see that $h(t)$ is a linear combination of the basis functions $\phi(t - k)$, but it takes some work to see that $h(t) \in L^2(\mathbb{R})$. We need to first simplify

$$h(t)^2 = h(t) \cdot h(t) = \sum_{k=1}^{\infty} \frac{1}{k} \phi(t - k) \cdot \sum_{j=1}^{\infty} \frac{1}{j} \phi(t - j)$$

and then integrate the result over $\mathbb{R}$. In what follows, we will forego the rigor that is typically associated with interchanging infinite series and improper integrals. In an analysis class you will learn how to justify these steps. The interested reader is referred to Rudin [48]. Simplifying $h(t)^2$ gives

$$h(t)^2 = \sum_{k=1}^{\infty} \frac{1}{k} \phi(t - k) \cdot \sum_{j=1}^{\infty} \frac{1}{j} \phi(t - j)$$

$$= \sum_{k=1}^{\infty} \sum_{j=1}^{\infty} \frac{1}{jk} \phi(t - k)\phi(t - j)$$

We now integrate over $\mathbb{R}$

$$\int_{\mathbb{R}} h(t)^2 \, \mathrm{d}t = \int_{\mathbb{R}} \sum_{k=1}^{\infty} \sum_{j=1}^{\infty} \frac{1}{jk} \phi(t - k)\phi(t - j) \, \mathrm{d}t$$

$$= \sum_{k=1}^{\infty} \sum_{j=1}^{\infty} \frac{1}{jk} \int_{\mathbb{R}} \phi(t - k)\phi(t - j) \, \mathrm{d}t$$

From (3.5) we know that $\int_{\mathbb{R}} \phi(t - k)\phi(t - j) \, \mathrm{d}t = 0$ when $j \neq k$, so the values of the inner sum are zero unless $j = k$. Using this fact and (3.6), we have

$$\int_{\mathbb{R}} h(t)^2 \, \mathrm{d}t = \sum_{k=1}^{\infty} \frac{1}{k^2} \int_{\mathbb{R}} \phi(t - k)^2 \, \mathrm{d}t = \sum_{k=1}^{\infty} \frac{1}{k^2} \cdot 1$$

From calculus we know that the series $\sum_{k=1}^{\infty} \frac{1}{k^2}$ converges by the p-series test[11]. We have shown that $h(t) \in L^2(\mathbb{R})$ and thus is in V_0.

The function in part (d) is different from the others since the argument of each term involves a 4t. The first term is $\phi(4t)$, which is nothing more than a contraction of the Haar function. Indeed,

$$\phi(4t) = \begin{cases} 1, & 0 \leq t < 1/4 \\ 0, & otherwise \end{cases}$$

We have plotted $\phi(4t)$ in Figure 3.4. Now $\phi(4t-1) = \phi(4(t-\frac{1}{4}))$, so we can interpret this function as the function $\phi(4t)$ translated 1/4 unit to the right. In a similar way we see that $\phi(4t-2) = \phi(4(t-\frac{1}{2}))$ and $\phi(4t-3) = \phi(4(t-\frac{3}{4}))$ are also translates (by $\frac{1}{2}$ and $\frac{3}{4}$, respectively) of $\phi(4t)$. If we add these four functions (see Figure 3.4), we obtain $\phi(t) \in V_0$.

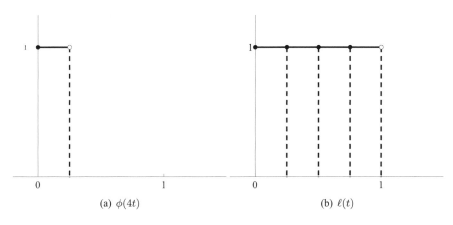

(a) $\phi(4t)$ (b) $\ell(t)$

Figure 3.4   The functions $\phi(4t)$ and $\ell(t)$ from Example 3.1.

■

Projections from $L^2(\mathbb{R})$ into V_0

Since $\{\phi(t-k)\}_{k \in \mathbb{Z}}$ is an orthonormal basis for V_0, we can use Proposition 1.13 to project the functions $g(t) \in L^2(\mathbb{R})$ into V_0. Using (1.31) we have the projection

$$P(g(t)) = \sum_{k \in \mathbb{Z}} \langle g(t), \phi(t-k) \rangle \, \phi(t-k) \tag{3.7}$$

[11] It is a fun exercise to use Fourier series techniques to show that this series sums to $\frac{\pi^2}{6}$. See Kammler [38], Problem 4.4.

Let's look at an example of a projection into V_0 of an $L^2(\mathbb{R})$ function.

Example 3.2 (Projection into V$_0$) *From Problem 1.13(a) in Section 1.2, we know that the function* $g(t) = e^{-|t|} \in L^2(\mathbb{R})$. *Find* $P(g(t))$.
Solution
    For $k \in \mathbb{Z}$, *we must compute* $\langle e^{-|t|}, \phi(t-k) \rangle$. *If* $k \geq 0$, *then* $\phi(t-k) = 1$ *for* $t \in [k, k+1)$ *and zero elsewhere. So we need only integrate over* $[k, k+1)$, *and since* $k \geq 0$ *on this interval, we have* $t \geq 0$, *so that* $e^{-|t|} = e^{-t}$. *The inner product is*

$$\langle e^{-|t|}, \phi(t-k) \rangle = \int_k^{k+1} e^{-t}\, dt = e^{-(1+k)}(e-1)$$

Now for $k < 0$, *we still integrate over the interval* $[k, k+1)$, *but on this interval* $t < 0$, *so that* $e^{-|t|} = e^t$. *The inner product is*

$$\langle e^{-|t|}, \phi(t-k) \rangle = \int_k^{k+1} e^t\, dt = e^k(e-1)$$

Thus we have

$$P(g(t)) = (e-1)\sum_{k=-\infty}^{-1} e^k \phi(t-k) + (e-1)\sum_{k=0}^{\infty} e^{-(k+1)} \phi(t-k)$$

The projection $P(g(t))$ *is plotted in Figure 3.5.*   ∎

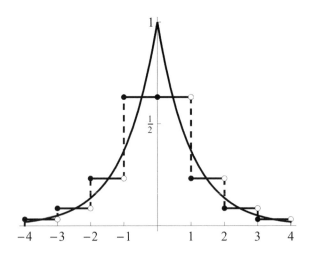

Figure 3.5   The function $g(t) = e^{-|t|}$ and its projection into V_0.

PROBLEMS

Note: *A computer algebra system and the software on the course Web site will be useful for some of these problems. See the Preface for more details.*

3.1   Let $\phi(t)$ be the Haar function from Definition 3.1. Show that the set $\{\phi(t - k)\}_{k \in \mathbb{Z}}$ is linearly independent.

3.2   Determine whether or not the following functions are in V_0.

(a) $f_1(t) = \phi\left(\frac{t}{2}\right) + 3\phi\left(\frac{t}{2} + 1\right)$

(b) $f_2(t) = \phi\left(\frac{t}{1000}\right)$

(c) $f_3(t) = \sum\limits_{k=0}^{5} \phi(8t - k)$

(d) $f_4(t) = \sum\limits_{k=1}^{\infty} \frac{1}{\sqrt{k}}\, \phi(t - k)$

(e) $f_5(t) = \sum\limits_{k \in \mathbb{Z}} e^{-|k|}\phi(t - k)$

3.3   Can $f(t) = \sum\limits_{k \in \mathbb{Z}} a_k \phi(t - k)$ be in V_0 and be continuous for all $t \in \mathbb{R}$?

3.4   Show that V_0 defined in (3.4) satisfies the properties in Definition 1.10 and is thus a subspace of $L^2(\mathbb{R})$.

3.5   Suppose that $\int\limits_{\mathbb{R}} f(t)\, dt = L$ and $\overline{\mathrm{supp}(f)} = [a, b]$. Use a u-substitution to show that for $r \in \mathbb{R}$, we have $\int\limits_{\mathbb{R}} f(t - r)\, dt = L$.

3.6   For each of the following functions, find $P(f(t))$ and plot the projection and function on the same graph.

(a) $f(t) = \dfrac{1}{1 + t^2}$

(b) $f(t) = \begin{cases} \sin(2\pi t), & -4 \le t \le 4 \\ 0, & \text{otherwise} \end{cases}$

(c) $f(t) = 5\phi(2t) + 3\phi(2t - 1) - 4\phi(2t - 2) + 6\phi(2t - 3)$

3.2   THE GENERAL HAAR SPACE V_J

Example 3.2 provides some motivation for what follows next. The projection $P(g(t))$ is a rather crude approximation of the function $g(t) = e^{-|t|}$. We could certainly improve our approximation if we used piecewise constants with possible breaks, say, at either the half-integers $0, \pm\frac{1}{2}, \pm1, \pm\frac{3}{2}, \pm2, \dots$ or even the quarter-integers $0, \pm\frac{1}{4}, \pm\frac{1}{2}, \pm\frac{3}{4}, \pm1, \pm\frac{5}{4}, \pm\frac{3}{2}, \dots$

Piecewise Constant Functions at Different Resolutions

How can we build the vector space of piecewise constant functions with possible breaks at the half-integers? The natural way is to repeat the process we used to construct V_0. Instead of using the box function $\phi(t)$ as a generator, we use the function $\phi(2t)$. Using (3.1), we see that

$$\phi(2t) = \begin{cases} 1, & 0 \le 2t < 1 \\ 0, & \text{otherwise} \end{cases} = \begin{cases} 1, & 0 \le t < 1/2 \\ 0, & \text{otherwise} \end{cases}$$

So $\phi(2t)$ is a contraction of the box function with $\overline{\text{supp}(\phi(2t))} = [0, \frac{1}{2}]$. How about the translates? The function $\phi(2t - 1) = \phi\left(2(t - \frac{1}{2})\right)$ is simply $\phi(2t)$ translated one-half unit right. In general, for $k \in \mathbb{Z}$, $\phi(2t - k) = \phi\left(2(t - \frac{k}{2})\right)$ is simply the function $\phi(2t)$ translated $\frac{k}{2}$ units right (left) when k is positive (negative).

We could build the vector space of piecewise constant functions with possible breaks at the quarter-integers by using the function $\phi(4t) = \phi(2^2t)$. This is a contracted box function with $\overline{\text{supp}(\phi(4t))} = [0, \frac{1}{4}]$. Moreover, for $k \in \mathbb{Z}$, we see that $\phi(4t - k) = \phi\left(4(t - \frac{k}{4})\right)$ is simply the function $\phi(4t)$ translated $\frac{k}{4}$ units right (left) when k is positive (negative).

We have described two spaces that allow us to build piecewise constant functions at a finer resolution than V_0. Suppose we didn't need resolution even as good as V_0. What if we only needed piecewise constant functions with possible breaks at $0, \pm2, \pm4, \pm6, \ldots$ or even $0, \pm4, \pm8, \pm12, \ldots$?

The generator for the vector space of piecewise constant functions with possible breaks at the even integers is $\phi\left(\frac{t}{2}\right) = \phi(2^{-1}t)$. You can easily verify that this is a dilation of the box function with $\overline{\text{supp}\left(\phi\left(\frac{t}{2}\right)\right)} = [0,2]$, and if we take $k \in \mathbb{Z}$, we see that $\phi\left(\frac{t}{2} - k\right) = \phi\left(\frac{1}{2}(t - 2k)\right)$ is simply the function $\phi\left(\frac{t}{2}\right)$ translated $2k$ units right (left) if k is positive (negative). In a similar manner, we can use the generator $\phi\left(\frac{t}{4}\right) = \phi(2^{-2}t)$ and the translates $\phi\left(\frac{t}{4} - k\right)$, where $k \in \mathbb{Z}$, to generate the piecewise constant functions with possible breaks at $0, \pm4, \pm8, \pm12, \ldots$ The functions $\phi(2t)$, $\phi(4t)$, $\phi\left(\frac{t}{2}\right)$, $\phi\left(\frac{t}{4}\right)$ are plotted in Figure 3.6.

The Space V_j Defined

The preceding discussion gives rise to the following definition:

Definition 3.2 (The Space V_j) *Let $\phi(t)$ be the Haar function given in (3.1). We define the vector space*

$$\boxed{\begin{aligned} V_j &= \text{span}\{\ldots, \phi(2^jt + 1), \phi(2^jt), \phi(2^jt - 1), \ldots\} \cap L^2(\mathbb{R}) \\ &= \text{span}\left\{\phi(2^jt - k)\right\}_{k \in \mathbb{Z}} \cap L^2(\mathbb{R}) \end{aligned}}$$

(3.8)

We call V_j the Haar space V_j. ∎

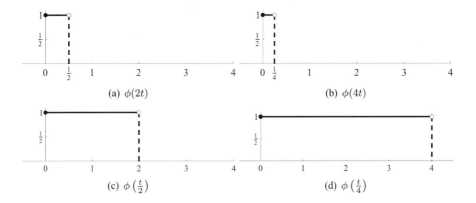

(a) $\phi(2t)$

(b) $\phi(4t)$

(c) $\phi\left(\frac{t}{2}\right)$

(d) $\phi\left(\frac{t}{4}\right)$

Figure 3.6   The functions $\phi(2t)$, $\phi(4t)$, $\phi\left(\frac{t}{2}\right)$, and $\phi\left(\frac{t}{4}\right)$.

In order to describe V_j, $j \in \mathbb{Z}$, we introduce some new notation. Let

$$2\mathbb{Z} = \{\ldots, -4, -2, 0, 2, 4, \ldots\}$$
$$4\mathbb{Z} = \{\ldots, -8, -4, 0, 4, 8, 12, \ldots\}$$
$$\frac{1}{2}\mathbb{Z} = \left\{\ldots, -2, -\frac{3}{2}, -1, -\frac{1}{2}, 0, \frac{1}{2}, 1, \frac{3}{2}, 2, \ldots\right\}$$
$$\frac{1}{4}\mathbb{Z} = \left\{\ldots, -1, -\frac{3}{4}, -\frac{1}{2}, -\frac{1}{4}, 0, \frac{1}{4}, \frac{1}{2}, \frac{3}{4}, 1, \ldots\right\}$$

and for $j \in \mathbb{Z}$,

$$2^j\mathbb{Z} = \left\{\ldots, 2^j(-3), 2^j(-2), 2^j(-1), 0, 2^j(1), 2^j(2), 2^j(3), \ldots\right\}$$

We will refer to $2\mathbb{Z}$ as the even integers, $\frac{1}{2}\mathbb{Z}$ as the half-integers, $\frac{1}{4}\mathbb{Z}$ as the quarter-integers, and so on. Using this notation we see that $V_j \subseteq L^2(\mathbb{R})$ *is a vector space of piecewise constant functions with possible breakpoints at* $2^{-j}\mathbb{Z}$.
   Here are some examples of different elements in V_j.

Example 3.3 (Functions in V_j) *Plot the following functions.*

(a) $f_1(t) = 3\phi\left(\frac{t}{8} + 1\right) + 2\phi\left(\frac{t}{8} - 1\right) + \phi\left(\frac{t}{8} - 2\right) \in V_{-3}$

(b) $f_2(t) = \displaystyle\sum_{k=-1}^{2} k^2\phi\left(\frac{t}{2} - k\right) \in V_{-1}$

(c) $f_3(t) = -2\phi(2t + 3) + \phi(2t) - 3\phi(2t - 1) + 3\phi(2t - 4) \in V_1$

(d) $f_4(t) = \displaystyle\sum_{k=1}^{3} \frac{1}{k}\phi(16t - k) \in V_4$

Solution

For part (a) we know that $\phi\left(\frac{t}{8}\right)$ is a dilated box function with $\overline{\text{supp}\left(\phi\left(\frac{t}{8}\right)\right)} = [0,8]$, so $\phi\left(\frac{t}{8} + 1\right) = \phi\left(\frac{1}{8}(t + 8)\right)$ is simply $\phi\left(\frac{t}{8}\right)$ translated eight units left. Similarly, $\phi\left(\frac{t}{8} - 1\right)$ is $\phi\left(\frac{t}{8}\right)$ translated eight units right, and $\phi\left(\frac{t}{8} - 2\right) = \phi\left(\frac{1}{8}(t - 16)\right)$ is $\phi\left(\frac{t}{8}\right)$ translated 16 units right. We simply multiply these functions by the given heights and add the results together. The function $f_1(t)$ is plotted in Figure 3.7(a).

In part (b) we use the functions $\phi\left(\frac{t}{2} + 1\right)$, $\phi\left(\frac{t}{2}\right)$, $\phi\left(\frac{t}{2} - 1\right)$, and $\phi\left(\frac{t}{2} - 2\right)$. These functions are all translates of $\phi\left(\frac{t}{2}\right)$ and recall that $\phi\left(\frac{t}{2}\right)$ is a dilated box function with $\overline{\text{supp}\left(\phi\left(\frac{t}{2}\right)\right)} = [0,2]$. The first function is translated two units left, the third function is translated two units right, and the fourth function is translated four units right. The function $f_2(t)$ is plotted in Figure 3.7(b).

The function $f_3(t)$ is a piecewise constant function with possible breaks at $\frac{1}{2}\mathbb{Z}$. The first function is $\phi(2t + 3) = \phi\left(2(t + \frac{3}{2})\right)$ and it is $\phi(2t)$ translated $3/2$ units left. Similarly, $\phi(2t - 1)$ is $\phi(2t)$ translated one-half unit right, and $\phi(2t - 4)$ is $\phi(2t)$ translated two units right. The function $f_3(t)$ is plotted in Figure 3.7(c).

The function $f_4(t)$ is a piecewise constant function with possible breaks at $\frac{1}{16}\mathbb{Z}$. We will use translates of $\phi(16t)$ (a contracted box function with $\overline{\text{supp}(f_4)} = \left[0, \frac{1}{16}\right]$). The translates are $\phi(16t - 1)$, $\phi(16t - 2)$, and $\phi(16t - 3)$. These functions are $\phi(16t)$ translated right by $\frac{1}{16}$, $\frac{1}{8}$, and $\frac{3}{16}$ unit, respectively. The function $f_4(t)$ is plotted in Figure 3.7(d). ∎

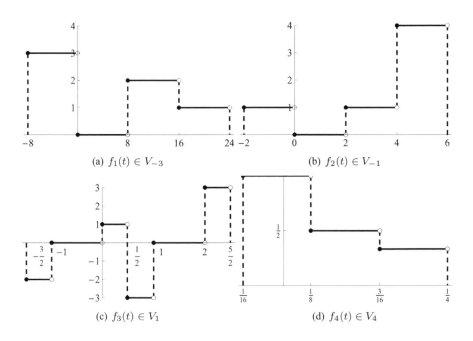

(a) $f_1(t) \in V_{-3}$

(b) $f_2(t) \in V_{-1}$

(c) $f_3(t) \in V_1$

(d) $f_4(t) \in V_4$

Figure 3.7   The functions $f_1(t)$, $f_2(t)$, $f_3(t)$, and $f_4(t)$.

The Functions $\phi_{j,k}(t)$ Defined

In Problem 3.7 you will show that for $j \in \mathbb{Z}$, the set $\{\phi(2^j t - k)\}_{k \in \mathbb{Z}}$ is linearly independent in V_j, and in Problem 3.8 you will prove that V_j is a subspace of $L^2(\mathbb{R})$. Since V_j is defined as a linear combinations of $\{\phi(2^j t - k)\}_{k \in \mathbb{Z}}$, we see that the set $\{\phi(2^j t - k)\}_{k \in \mathbb{Z}}$ is a basis for V_j. In Problem 3.9 you will show that

$$\langle \phi(2^j t - k), \phi(2^j t - \ell) \rangle = 0 \quad \text{for } k \neq \ell$$

$$\|\phi(2^j t - k)\| = 2^{-j/2}$$

We can use these observations to form an orthonormal basis for $V_j, j \in \mathbb{Z}$.

Proposition 3.2 (An Orthonormal Basis for V_j) *Let V_j be given by (3.8) for $j \in \mathbb{Z}$. For each $k \in \mathbb{Z}$, define the function*

$$\boxed{\phi_{j,k}(t) = 2^{j/2}\phi(2^j t - k)} \tag{3.9}$$

Then the set $\{\phi_{j,k}(t)\}_{k \in \mathbb{Z}}$ is an orthonormal basis for V_j. ∎

Proof: The set V_j is defined as linear combinations of $\{\phi(2^j t - k)\}_{k \in \mathbb{Z}}$, so the set spans V_j, and in Problem 3.7 we show that the set $\{\phi(2^j t - k)\}_{k \in \mathbb{Z}}$ is linearly independent in V_j. Multiplying each element of the set by $2^{j/2}$ does not affect the span or linear independence, so the set $\{\phi_{j,k}(t)\}_{k \in \mathbb{Z}}$ is a basis for V_j.
For $k \neq \ell$ we use Problem 3.9(a) to write

$$\langle \phi_{j,k}(t), \phi_{j,\ell}(t) \rangle = \langle 2^{j/2}\phi(2^j t - k), 2^{j/2}\phi(2^j t - \ell) \rangle$$
$$= 2^j \langle \phi(2^j t - k), \phi(2^j t - \ell) \rangle$$
$$= 0$$

If $k = \ell$, we have $\|\phi_{j,k}(t)\|^2 = 2^j \|\phi(2^j t - k)\|^2$. and we know from Problem 3.9(b) that $\|\phi(2^j t - k)\|^2 = 2^{-j}$ so that $\|\phi_{j,k}(t)\|^2 = 1$. Thus $\{\phi_{j,k}(t)\}_{k \in \mathbb{Z}}$ is an orthonormal basis for V_j. ∎

Now that we have introduced the functions $\phi_{j,k}(t)$, let's look at an example about how the indices j and k tell us about the support properties of the function.

Example 3.4 (Graphing $\phi_{j,k}(t)$) *Describe the support and plot of each of the following functions.*

(a) $\phi_{3,7}(t)$

(b) $\phi_{-2,1}(t)$

(c) $\phi_{-4,-6}(t)$

(d) $\phi_{5,18}(t)$

Solution

All the functions are plotted in Figure 3.8. The best way to analyze the support of each function is to write it as a translate of $\phi(2^j t)$. We know that $\phi(2^j t)$ is compactly supported with $\overline{\text{supp}(\phi(2^j t))} = \left[0, \frac{1}{2^j}\right]$.

For part (a), we have

$$\phi_{3,7}(t) = 2^{3/2}\phi(2^3 t - 7)$$
$$= 2\sqrt{2}\,\phi(8t - 7)$$
$$= 2\sqrt{2}\,\phi\left(8\left(t - \frac{7}{8}\right)\right)$$

Thus we are translating the function $\phi(8t)$ to the right $7/8$ of a unit. Since $\overline{\text{supp}(\phi(8t))} = \left[0, \frac{1}{8}\right]$, the function $\phi_{3,7}(t)$ is compactly supported with $\overline{\text{supp}(\phi_{3,7})} = \left[\frac{7}{8}, 1\right]$.

In part (b),

$$\phi_{-2,1}(t) = 2^{-2/2}\phi(2^{-2}t - 1)$$
$$= \frac{1}{2}\,\phi\left(\frac{t}{4} - 1\right)$$
$$= \frac{1}{2}\,\phi\left(\frac{1}{4}(t - 4)\right)$$

Thus $\phi_{-2,1}(t)$ is the function $\phi\left(\frac{t}{4}\right)$ translated four units to the right. Since $\overline{\text{supp}\left(\phi\left(\frac{t}{4}\right)\right)} = [0,4]$, the function $\phi_{-2,1}(t)$ is compactly supported on $[4,8]$.

In part (c), we have

$$\phi_{-4,-6}(t) = 2^{-4/2}\phi(2^{-4}t + 6)$$
$$= \frac{1}{4}\,\phi\left(\frac{t}{16} + 6\right)$$
$$= \frac{1}{4}\,\phi\left(\frac{1}{16}(t + 96)\right)$$

So $\phi_{-4,-6}(t)$ is simply the function $\phi\left(\frac{t}{16}\right)$ translated 96 units to the left. Since $\overline{\text{supp}\left(\phi\left(\frac{t}{16}\right)\right)} = [0,16]$, the function $\phi_{-4,-6}(t)$ satisfies $\overline{\text{supp}(\phi_{-4,-6})} = [-96, -80]$.

For part (d),

$$\phi_{5,18}(t) = 2^{5/2}\phi(2^5 t - 18)$$
$$= 4\sqrt{2}\,\phi(32t - 18)$$
$$= 4\sqrt{2}\,\phi\left(32\left(t - \frac{9}{16}\right)\right)$$

Now $\overline{\text{supp}(\phi(32t))} = \left[0, \frac{1}{32}\right]$, and since $\phi_{5,18}(t)$ is $\phi(32t)$ translated $\frac{9}{16}$ unit to the right, we see that $\overline{\text{supp}(\phi_{5,18})} = \left[\frac{9}{16}, \frac{19}{32}\right]$.   ∎

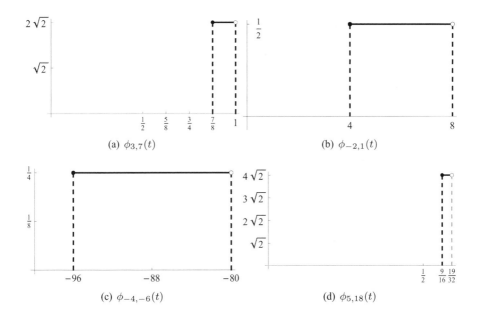

(a) $\phi_{3,7}(t)$

(b) $\phi_{-2,1}(t)$

(c) $\phi_{-4,-6}(t)$

(d) $\phi_{5,18}(t)$

Figure 3.8   The functions from Example 3.4.

Support Properties for $\phi_{j,k}(t)$ and Dyadic Intervals

The functions in Example 3.4 give rise to the following proposition.

Proposition 3.3 (Support Interval for $\phi_{j,k}$) *Let $\phi_{j,k}(t)$ be the function defined by (3.9). Then* $\overline{\mathrm{supp}(\phi_{j,k})} = \left[\frac{k}{2^j}, \frac{k+1}{2^j}\right]$. ∎

Proof: Problem 3.11. ∎

While $\overline{\mathrm{supp}(\phi_{j,k})} = \left[\frac{k}{2^j}, \frac{k+1}{2^j}\right]$, the interval where $\phi_{j,k}(t) = 1$ is $\left[\frac{k}{2^j}, \frac{k+1}{2^j}\right)$. This interval is important and we analyze it at this time. We introduce some new notation for this interval.

Definition 3.3 (Dyadic Intervals $I_{j,k}$) *We define the dyadic interval $I_{j,k}$, $j, k \in \mathbb{Z}$, by*

$$I_{j,k} = \left[\frac{k}{2^j}, \frac{k+1}{2^j}\right) = \left\{ t \in \mathbb{R} \,\bigg|\, \frac{k}{2^j} \le t < \frac{k+1}{2^j} \right\} \qquad (3.10)$$

We will refer to the first index j as the level of the interval. ∎

The adjective *dyadic* describes objects made up of two units. It is natural to use the word here since the the length of each $I_{j,k}$ is 2^j.

Let's look at some examples of different intervals $I_{j,k}$.

Example 3.5 (Examples of Intervals $I_{j,k}$) *Sketch each of the intervals $I_{j,k}$ given below.*

(a) $I_{3,-1}$

(b) $I_{-2,4}$

(c) $I_{0,k},\ k \in \mathbb{Z}$

Solution

For part (a) we have $I_{3,-1} = \left[-\frac{1}{2^3}, \frac{-1+1}{2^3}\right) = \left[-\frac{1}{8}, 0\right)$. We can write the interval in part (b) as $I_{-2,4} = \left[\frac{4}{2^{-2}}, \frac{4+1}{2^{-2}}\right) = [16, 20)$. Finally, we have from part (c) $I_{0,k} = \left[\frac{k}{2^0}, \frac{k+1}{2^0}\right) = [k, k+1)$. Each of the intervals is plotted in Figure 3.9. ∎

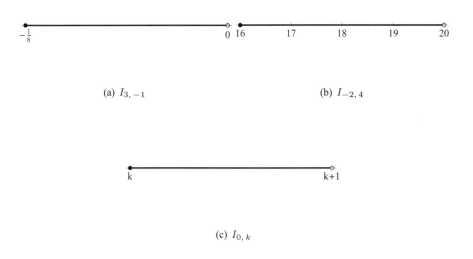

(a) $I_{3,-1}$ (b) $I_{-2,4}$

(c) $I_{0,k}$

Figure 3.9   The intervals $I_{3,-1}$, $I_{-2,4}$, and $I_{0,k}$.

We are particularly interested in the intersection of two dyadic intervals. For instance, in Example 3.5(c) we learned that $I_{0,k} = [k, k+1)$, where $k \in \mathbb{Z}$. These intervals are

$$I_{0,k} = \ldots, [-2,-1), [-1,0), [0,1), [1,2), \ldots \tag{3.11}$$

and note that none of these intervals intersect. For the intervals $I_{1,k},\ k \in \mathbb{Z}$, we have $I_{1,k} = \left[\frac{k}{2}, \frac{k+1}{2}\right) = \left[\frac{k}{2}, \frac{k}{2} + \frac{1}{2}\right)$ and these intervals are

$$I_{1,k} = \ldots, \left[-1, -\frac{1}{2}\right), \left[-\frac{1}{2}, 0\right), \left[0, \frac{1}{2}\right), \left[\frac{1}{2}, 1\right), \ldots \tag{3.12}$$

None of these intervals intersect as well. We can see from (3.11) that the union of all of the dyadic intervals $I_{0,k}$, $k \in \mathbb{Z}$, is the real line $\mathbb{R}$. In a similar manner, we see from (3.12) that the union of the dyadic intervals $I_{1,k}$, $k \in \mathbb{Z}$ is also the real line $\mathbb{R}$. The following proposition says that the union of all dyadic intervals on any level is $\mathbb{R}$.

Proposition 3.4 (Union of Dyadic Intervals) *Let*

$$\mathcal{I} = \cdots \cup I_{j,-2} \cup I_{j,-1} \cup I_{j,0} \cup I_{j,1} \cup I_{j,2} \cup \cdots$$

Then $\mathcal{I} = \mathbb{R}$. ■

Proof: Since $\mathcal{I}$ is a union of subsets of $\mathbb{R}$, then we have $\mathcal{I} \subseteq \mathbb{R}$. Now we much show that $\mathbb{R} \subseteq \mathcal{I}$. To this end, let $t \in \mathbb{R}$ and consider the number $2^j t$. Then there exists an integer[12] k such that $k \leq 2^j t < k + 1$.

Next we divide all three parts of $k \leq 2^j t < k + 1$ by 2^j to obtain $\frac{k}{2^j} \leq t < \frac{k+1}{2^j}$. But this implies that $t \in I_{j,k}$, so that $t \in \mathcal{I}$. Thus $\mathbb{R} \subseteq \mathcal{I}$ and the proof is complete. ■

The next proposition states that two distinct dyadic intervals on the same level never intersect.

Proposition 3.5 (Dyadic Intervals on Level j) *For $k \neq m$, the dyadic intervals $I_{j,k}$ and $I_{j,m}$ satisfy*

$$I_{j,k} \cap I_{j,m} = \varnothing$$

■

Proof: Problem 3.14. ■

Proposition 3.5 tells us that distinct dyadic intervals on the same level never intersect. The next proposition helps us characterize dyadic intervals on different levels.

Proposition 3.6 (Dyadic Intervals on Different Levels) *Suppose that $j, k, \ell,$ and m are integers with $\ell > j$. Then either $I_{j,k} \cap I_{\ell,m} = \varnothing$ or*

$$I_{\ell,m} \subset I_{j,k} \tag{3.13}$$

In the case of (3.13), $I_{\ell,m}$ is totally contained in either the left half of $I_{j,k}$ or the right half of $I_{j,k}$. ■

Proof: We consider two cases. If $I_{j,k} \cap I_{\ell,m} = \varnothing$, then the proof is complete. Thus we assume that $I_{j,k} \cap I_{\ell,m} \neq \varnothing$.

We first note that $\ell > j$ implies that $2^\ell > 2^j$ or $\frac{1}{2^\ell} < \frac{1}{2^j}$. Now $\frac{1}{2^j}, \frac{1}{2^\ell}$ are the lengths of $I_{j,k}, I_{\ell,m}$, respectively, and since 2^ℓ is at least twice as large as 2^j, we see that $\frac{1}{2^\ell}$ is no more than half as large as $\frac{1}{2^j}$. Thus the length of $I_{\ell,m}$ is at most half the length of $I_{j,k}$.

---

[12] Those readers who have taken a course in mathematical logic or elementary analysis know that we can use the Archimedean property to show the existence of the integer k. See Bartle and Sherbert [1].

We now consider two cases. Either (i) the left endpoint $\frac{m}{2^\ell}$ of $I_{\ell,m}$ satisfies $\frac{k}{2^j} \leq \frac{m}{2^\ell} < \frac{k+1}{2^j}$ so that $\frac{m}{2^\ell} \in I_{j,k}$ or (ii) the right endpoint of $I_{\ell,m}$ satisfies $\frac{k}{2^j} < \frac{m+1}{2^\ell} \leq \frac{k+1}{2^j}$ so that $\frac{m+1}{2^\ell} \in I_{j,k}$.

We will prove that (3.13) holds for (i) and leave the proof of (ii) as Problem 3.15. We want to show that $I_{\ell,m}$ is contained entirely in either the left half of $I_{j,k}$ or the right half of $I_{j,k}$. Toward this end, let's first assume (see Figure 3.10) that

$$\frac{k}{2^j} \leq \frac{m}{2^\ell} < \frac{k+1/2}{2^j} \tag{3.14}$$

and note that $\frac{k+1/2}{2^j}$ is the midpoint of $I_{j,k}$. We rewrite (3.14) so that each term has the same denominator:

$$\frac{2^{\ell-j}k}{2^\ell} \leq \frac{m}{2^\ell} < \frac{2^{\ell-j}(k+1/2)}{2^\ell} \tag{3.15}$$

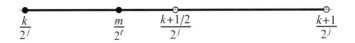

Figure 3.10   The point $m/2^\ell$, located in the left half of $I_{j,k}$.

We can write the numerator of the upper bound of (3.15) as $2^{\ell-j}k + 2^{\ell-j-1}$. Since $\ell > j$, we know that $2^{\ell-j} \geq 2$ and $2^{\ell-j-1} \geq 1$ and both are integers. So $2^{\ell-j}(k+1/2) \in \mathbb{Z}$ and thus $2^{\ell-j}(k+1/2)/2^\ell$ must be an endpoint for some dyadic interval at level ℓ. Thus the right endpoint $\frac{m+1}{2^\ell}$ of $I_{\ell,m}$ must be less than or equal to $(k+1/2)/2^j$. This implies that $I_{\ell,m}$ resides in the left half of $I_{j,k}$.

The case where $\frac{k+1/2}{2^j} \leq \frac{m}{2^\ell} < \frac{k+1}{2^j}$ is similar and is left to the reader.   ∎

Projecting Functions From $L^2(\mathbb{R})$ Into V_j

We can use the basis functions $\phi_{j,k}(t)$, $k \in \mathbb{Z}$ and Proposition 1.13 to create projections of functions $g(t) \in L^2(\mathbb{R})$ onto V_j. The projection is given by

$$P_{g,j}(t) = \sum_{k \in \mathbb{Z}} \langle \phi_{j,k}(t), g(t) \rangle \phi_{j,k}(t) \tag{3.16}$$

We can use (3.9) to rewrite (3.16) as

$$P_{g,j}(t) = \sum_{k \in \mathbb{Z}} \langle 2^{j/2}\phi(2^j t - k), g(t)\rangle 2^{j/2}\phi(2^j t - k)$$

$$= 2^j \sum_{k \in \mathbb{Z}} \langle \phi(2^j t - k), g(t)\rangle \phi(2^j t - k)$$

and use it to find the projections in the following example.

Example 3.6 (Projections into V_j) *Consider the function $g(t) = e^{-|t|}$ from Example 3.2. Compute and plot the projections $P_{g,2}(t)$ and $P_{g,-1}(t)$.*
Solution
For the first projection, we have

$$P_{g,\,2}(t) = 4\sum_{k \in \mathbb{Z}} \langle \phi(4t - k), g(t)\rangle \phi(4t - k)$$

Using Problem 3.11 we know that $\overline{\operatorname{supp}(\phi_{2,\,k})} = \left[\frac{k}{4}, \frac{k+1}{4}\right]$ so that the inner product can be written as

$$\langle \phi(4t - k), g(t)\rangle = \begin{cases} \displaystyle\int_{k/4}^{(k+1)/4} e^{-t}\, dt, & k \geq 0 \\[4mm] \displaystyle\int_{k/4}^{(k+1)/4} e^{t}\, dt, & k < 0 \end{cases}$$

$$= \begin{cases} e^{-k/4}(1 - e^{-1/4}), & k \geq 0 \\[2mm] e^{k/4}(e^{1/4} - 1), & k < 0 \end{cases}$$

Thus

$$P_{g,\,2}(t) = 4(1 - e^{-1/4})\sum_{k=0}^{\infty} e^{-k/4}\phi(4t - k) + 4(e^{1/4} - 1)\sum_{k=-\infty}^{-1} e^{k/4}\phi(4t - k)$$

The projection $P_{g,\,2}(t)$ is plotted in Figure 3.11(a).
For the second projection we have

$$P_{g,-1}(t) = \frac{1}{2}\sum_{k \in \mathbb{Z}} \langle \phi\left(\frac{t}{2} - k\right), g(t)\rangle \phi\left(\frac{t}{2} - k\right)$$

Using Problem 3.11 we know that $\overline{supp(\phi_{-1,\,k})} = [2k,2(k+1)]$ *and thus* $\phi_{-1,k}(t)$
is 0 outside that interval. The inner product is

$$\langle\phi(t/2-k),g(t)\rangle = \begin{cases} \displaystyle\int\limits_{2k}^{2k+2} e^{-t}\,dt, & k\geq 0 \\[4mm] \displaystyle\int\limits_{2k}^{2k+2} e^{t}\,dt, & k<0 \end{cases}$$

$$= \begin{cases} e^{-2k}(1-e^{-2}), & k\geq 0 \\[3mm] e^{2k}(e^{2}-1), & k<0 \end{cases}$$

We can then write the projection as

$$P_{g,-1}(t) = \frac{1}{2}(1-e^{-2})\sum_{k=0}^{\infty}e^{-2k}\phi\left(\frac{t}{2}-k\right) + \frac{1}{2}(e^{2}-1)\sum_{k=-\infty}^{-1}e^{2k}\phi\left(\frac{t}{2}-k\right)$$

The projection $P_{g,-1}(t)$ *is plotted in Figure 3.11(b).*   ∎

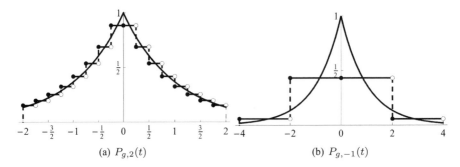

(a) $P_{g,2}(t)$ (b) $P_{g,-1}(t)$

Figure 3.11   The projections $P_{g,2}(t)$ and $P_{g,-1}(t)$ from Example 3.6.

Nestedness of V_j and Other Properties

The V_j spaces satisfy a special *nesting* property. Consider, for example, a function $f(t) \in V_0$. It is a piecewise constant function with possible breaks at $\mathbb{Z}$. Thus it is trivially a piecewise constant function on $\frac{1}{2}\mathbb{Z}$ — we can build any height a_k on $[k, k+1)$ by using a_k times box functions on $[k, k+\frac{1}{2})$ and $[k+\frac{1}{2}, k+1)$. To see this analytically, note that

$$\phi(t) = \phi(2t) + \phi(2t-1) \tag{3.17}$$

We can use this identity to write any function $f(t) \in V_0$ as an element of V_1. Suppose that

$$f(t) = \sum_{k \in \mathbb{Z}} a_k \phi(t - k) \in V_0$$

If we replace t by $t - k$ in (3.17), we see that each term of $f(t)$ is

$$a_k \phi(t - k) = a_k \phi(2(t - k)) + a_k \phi(2(t - k) - 1)$$
$$= a_k \phi(2t - 2k) + a_k \phi(2t - 2k - 1)$$

so that

$$
\begin{aligned}
f(t) &= \sum_{k \in \mathbb{Z}} a_k \phi(t - k) \\
&= \sum_{k \in \mathbb{Z}} a_k \phi(2t - 2k) + a_k \phi(2t - 2k - 1) \\
&= \cdots a_{-1} \phi(2t + 2) + a_{-1} \phi(2t + 1) + a_0 \phi(2t) + a_0 \phi(2t - 1) \\
&\quad + a_1 \phi(2t - 2) + a_1 \phi(2t - 3) + \cdots
\end{aligned}
$$

is a linear combination of $\phi(2t - k)$, $k \in \mathbb{Z}$. Thus $f(t) \in V_0$ implies that $f(t) \in V_1$ so that $V_0 \subseteq V_1$. We can easily generalize this argument to prove the following proposition.

Proposition 3.7 (Nested Property of the V_j Spaces) *The V_j spaces defined by (3.2) satisfy*

$$\boxed{\cdots \subseteq V_{-2} \subseteq V_{-1} \subseteq V_0 \subseteq V_1 \subseteq V_2 \subseteq \cdots} \tag{3.18}$$

∎

Proof: Problem 3.19.   ∎

Not only are the V_j spaces nested, but we can dilate or contract arguments of functions by a power of 2 to move from one V_j space to another. The idea is summarized in the following proposition.

Proposition 3.8 (Moving from One V_j Space to Another) *The function $f(t) \in V_j$ if and only if $f(2t) \in V_{j+1}$.*   ∎

Proof: We assume that $f(t) \in V_j$ and show that $f(2t) \in V_{j+1}$. The proof of the converse of this statement is left as Problem 3.20.

Let $f(t) \in V_j$. Then $f(t)$ can be expressed as a linear combination of the basis functions $\phi_{j,k}(t)$, $k \in \mathbb{Z}$:

$$f(t) = \sum_{k \in \mathbb{Z}} a_k \phi_{j,k}(t) = 2^{j/2} \sum_{k \in \mathbb{Z}} a_k \phi(2^j t - k)$$

Replacing t by $2t$ in the above equation gives

$$f(2t) = 2^{j/2} \sum_{k \in \mathbb{Z}} a_k \phi(2^j(2t) - k)$$

$$= 2^{-1/2} \sum_{k \in \mathbb{Z}} a_k 2^{(j+1)/2} \phi(2^{j+1}t - k)$$

$$= \sum_{k \in \mathbb{Z}} \tilde{a}_k \phi_{j+1, k}(t)$$

where $\tilde{a}_k = 2^{-1/2} a_k$, $k \in \mathbb{Z}$. Since we have written $f(2t)$ as a linear combination of basis functions $\phi_{j+1, k}(t)$ from V_{j+1}, we have $f(t) \in V_{j+1}$.   ∎

We conclude this section with a result that states two important properties obeyed by the spaces $\{V_j\}_{j \in \mathbb{Z}}$.

Theorem 3.1 (Set Properties of the V_j Spaces) *The Haar spaces $\{V_j\}_{j \in \mathbb{Z}}$, defined in (3.8), satisfy the following two properties:*

$$\boxed{\cap_{j \in \mathbb{Z}} V_j = \cdots \cap V_{-1} \cap V_0 \cap V_1 \cap V_2 \cap \cdots = \{0\}} \tag{3.19}$$

and

$$\boxed{\overline{\cup_{j \in \mathbb{Z}} V_j} = \overline{\cdots \cup V_{-1} \cup V_0 \cup V_1 \cup V_2 \cup \cdots} = L^2(\mathbb{R})} \tag{3.20}$$

∎

Proof: The proof of these results requires book. Readers who have some elementary background in analysis are encouraged to see the nice proofs given by Frazier [27] (Lemmas 5.47 and 5.48).   ∎

Some comments are in order regarding the results of Theorem 3.1. The first property (3.19) says that if we intersect all the V_j spaces, then the only common function is the function $f(t) = 0$. This is the *separation property* of the Haar spaces. We can argue in an informal way why this result is true. The nesting property (3.18) says that

$$\cdots \supseteq V_1 \supseteq V_0 \supseteq V_{-1} \supseteq V_{-2} \supseteq \cdots \supseteq V_{-1000} \supseteq \cdots$$

so the intersection is determined by the elements in V_j as j tends to $-\infty$. However, these V_j spaces are constructed using translates of the function $\phi(2^j t)$ compactly supported on $[0, 2^{-j}]$. But as j goes to $-\infty$, the length of the support of $\phi(2^j t)$ tends to ∞. Thus the support length of elements of these spaces become infinitely large, but they are piecewise constant functions that must be square integrable on $\mathbb{R}$. The only constant function to satisfy both of these requirements is $f(t) = 0$.

To understand what the second result is saying, we must explain the bar over the union. The bar represents the *closure* of union. A set V is said to be *closed* if it contains all of its limit points. That is, if $\{\mathbf{v}_n\}_{n \in \mathbb{N}}$ is in V, and $\mathbf{v}_n \to \mathbf{v}$, then $\mathbf{v}$ is in V as well. If V is not closed, we can add the "missing" limit points by forming $\overline{V}$. The second property (3.20) says that *any* function $g(t) \in L^2(\mathbb{R})$ can be approximated to an arbitrarily small tolerance by piecewise constant functions from the V_j spaces. Remember that approximation here is in the L^2 sense. Using the nesting property (3.18), we have

$$\cdots \subseteq V_{-1} \subseteq V_0 \subseteq V_1 \subseteq V_2 \subseteq \cdots \subseteq V_{1000} \subseteq \cdots$$

so the union is determined by the elements in V_j as j tends to ∞. But the elements of these spaces are piecewise constant functions with arbitrarily small support lengths and it seems appropriate that $g(t)$ can be approximated to arbitrary precision using a sequence $v_n(t)$ of these piecewise constant functions with $\lim_{n\to\infty} \|v_n(t) - g(t)\| = 0.$[13]

PROBLEMS

3.7 Let $j \in \mathbb{Z}$ and let V_j be defined by (3.8). Show that the set $\{\phi(2^j t - k)\}_{k \in \mathbb{Z}}$ is a linear independent set in V_j.

3.8 Show that V_j defined in (3.8) satisfies the properties in Definition 1.10 and is thus a subspace of $L^2(\mathbb{R})$.

3.9 Use u-substitutions to show that

(a) $\langle \phi(2^j t - k), \phi(2^j t - \ell) \rangle = 0$ for $k \neq \ell$

(b) $\|\phi(2^j t - k)\| = 2^{-j/2}$

3.10 Using (3.9), plot the following functions.

(a) $\phi_{3,-3}(t)$

(b) $\phi_{6,20}(t)$

(c) $\phi_{-3,1}(t)$

(d) $\phi_{-1,-5}(t)$

3.11 Show that the compact support of $\phi_{j,k}(t)$ given by (3.9) is $\overline{\mathrm{supp}(\phi_{j,k})} = \left[\frac{k}{2^j}, \frac{k+1}{2^j}\right]$.

3.12 Sketch each of the following intervals.

(a) $I_{-2,3}$

(b) $I_{5,-4}$

(c) $I_{3,k}$

[13] With a modest background in analysis, the problem of understanding the formal proofs of (3.19) and (3.20) is indeed tractable. If you have not already done so, we encourage you to take an analysis class and then read the concise proofs given in Frazier [27]!

3.13 What is the length of each of the intervals given in Problem 3.12? What is the length of I_{jk} in general?

3.14 Prove Proposition 3.5. To get you started, since $k \neq m$, you can assume that one is smaller than the other. So suppose that $m < k$. Can you show that $(m+1)/2^j \leq k/2^j$?

3.15 Complete the proof of Proposition 3.6 for the case where $k/2^j < (m+1)/2^\ell \leq (k+1)/2^j$.

3.16 Let $f(t) = \max(4 - t^2, 0)$. Compute and plot the following projections.

(a) $P_{f,0}(t)$

(b) $P_{f,3}(t)$

(c) $P_{f,-1}(t)$

(d) $P_{f,-2}(t)$

3.17 Repeat Problem 3.16 for

$$f(t) = \begin{cases} \sin(2\pi t), & -2 \leq t \leq 2 \\ 0, & \text{otherwise} \end{cases}$$

3.18 Suppose that $g(t) \in L^2(\mathbb{R})$ is an even function.

(a) Show that except for $t \in \mathbb{Z}$, $P_{g,0}(t) = P_{g,0}(-t)$.

(b) Generalize part (a) for projections onto V_j. That is, show that if $j \in \mathbb{Z}$ and $t \notin 2^{-j}\mathbb{Z}$, then $P_{g,j}(t) = P_{g,j}(-t)$.

3.19 Prove Proposition 3.7. That is, for any $j \in \mathbb{Z}$, show that $V_j \subseteq V_{j+1}$.

3.20 Prove the converse of Proposition 3.8. That is, show that if $f(2t) \in V_{j+1}$, then $f(t) \in V_j$.

3.3 THE HAAR WAVELET SPACE W_0

In Section 3.2 we learned many properties regarding the V_j spaces defined in (3.8). In particular, we learned that the spaces are nested and that we can "zoom" from one space to another using dyadic scaling. In this section we learn about another important connection between spaces V_0 and V_1.

Projecting $f_1(t) \in V_1$ Into V_0

Let's begin by considering a function $f_1(t) \in V_1$. Suppose that

$$f_1(t) = 5\phi_{1,0}(t) + 3\phi_{1,1}(t) + 5\phi_{1,2}(t) - \phi_{1,3}(t) + 5\phi_{1,4}(t) + 7\phi_{1,5}(t)$$
$$= 5\sqrt{2}\,\phi(2t) + 3\sqrt{2}\,\phi(2t-1) + 5\sqrt{2}\,\phi(2t-2) - 1\sqrt{2}\phi(2t-3)$$
$$+ 5\sqrt{2}\,\phi(2t-4) + 7\sqrt{2}\,\phi(2t-5) \tag{3.21}$$

The function $f_1(t)$ is plotted in Figure 3.12.

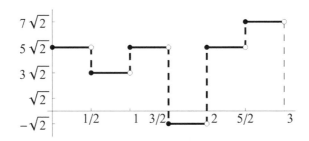

Figure 3.12 The function $f_1(t)$ from (3.21).

From Proposition 3.8, we know that $f_1(t) \in V_j$ for $j \geq 1$. Certainly, $f_1(t) \notin V_0$ since there are jump discontinuities at $\frac{1}{2}, \frac{3}{2}, \frac{5}{2}$. Suppose that we wish to construct a function $f_0(t) \in V_0$ that approximates $f_1(t)$. Since we just learned about projections in Section 3.2, let's project $f_1(t)$ into V_0. We have

$$P\left(f_{1,0}(t)\right) = f_0(t) = \sum_{k \in \mathbb{Z}} \langle f_1(t), \phi(t-k) \rangle \phi(t-k)$$

Since $\overline{\operatorname{supp}(f_1)} = [0,3]$, we can write the projection as a finite sum:

$$f_0(t) = \sum_{k=0}^{2} \langle f_1(t), \phi(t-k) \rangle \phi(t-k)$$
$$= \langle f_1(t), \phi(t) \rangle\, \phi(t) + \langle f_1(t), \phi(t-1) \rangle \phi(t-1) + \langle f_1(t), \phi(t-2) \rangle\, \phi(t-2)$$

The inner products are quite simple to compute. We have

$$\langle f_1(t), \phi(t) \rangle = \int_{\mathbb{R}} f_1(t)\phi(t)\,\mathrm{d}t = \int_0^1 f_1(t)\,\mathrm{d}t$$
$$= 5\sqrt{2} \int_0^{1/2} \phi(2t)\,\mathrm{d}t + 3\sqrt{2} \int_{1/2}^1 \phi(2t-1)\,\mathrm{d}t$$

Now each integral in the previous identity is $1/2$, so the inner product can be written as

$$\langle f_1(t), \phi(t) \rangle = \sqrt{2}\,(5+3)\,/2 = 4\sqrt{2}$$

In a similar manner, we find that

$$\langle f_1(t), \phi(t - 1) \rangle = \sqrt{2}\,(5 + -1)\,/2 = 2\sqrt{2}$$

and

$$\langle f_1(t), \phi(t - 2) \rangle = \sqrt{2}\,(5 + 7)\,/2 = 6\sqrt{2}$$

Thus

$$f_0(t) = 4\sqrt{2}\,\phi(t) + 2\sqrt{2}\,\phi(t - 1) + 6\sqrt{2}\,\phi(t - 2) \qquad (3.22)$$

We have plotted $f_0(t)$ and $f_1(t)$ in Figure 3.13.

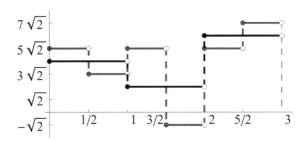

Figure 3.13   The function $f_0(t)$ (black) from (3.22). The function $f_1(t)$, defined in (3.21), is plotted in gray.

From Figure 3.13 we see that the coefficients of $\phi(t - k)$, $k = 0, 1, 2$ of $f_0(t)$ are obtained simply by averaging the coefficients of $f_1(t)$ two at a time and then multiplying the result by $\sqrt{2}$. That is, if we view $f_1(t)$ from (3.21) as

$$f_1(t) = \sum_{k=0}^{5} a_k \phi_{1,k}(t)$$

where $a_0 = 5$, $a_1 = 3$, $a_2 = 5$, $a_3 = -1$, $a_4 = 5$, and $a_5 = 7$, then

$$f_0(t) = \sum_{k=0}^{2} b_k \phi(t - k)$$

where $b_0 = 4\sqrt{2}$, $b_1 = 2\sqrt{2}$, and $b_2 = 5\sqrt{2}$. We see that the b_k obey

$$b_k = \frac{\sqrt{2}}{2}\,(a_{2k} + a_{2k+1}), \qquad k = 0, 1, 2$$

This observation is true for general $f_1(t) \in V_1$, as illustrated by the following proposition.

Proposition 3.9 (Projecting Functions in V_1 into V_0) *Suppose that $f_1(t) \in V_1$ is given by*

$$f_1(t) = \sum_{k \in \mathbb{Z}} a_k \phi_{1,k}(t)$$

Then the projection (see (3.16)) $f_0(t) = P_{f_1,0}(t)$ of $f_1(t)$ into V_0 can be written as

$$f_0(t) = \sum_{k \in \mathbb{Z}} b_k \phi(t - k)$$

where

$$\boxed{b_k = \frac{\sqrt{2}}{2}\left(a_{2k} + a_{2k+1}\right)} \tag{3.23}$$

for $k \in \mathbb{Z}$. ∎

Proof: Let

$$f_0(t) = P_{f_1,0}(t) = \sum_{k \in \mathbb{Z}} b_k \phi(t - k)$$

Using the definition (3.16) for $P_{f_1,0}(t)$, we know that for $k \in \mathbb{Z}$,

$$b_k = \langle \phi(t - k), f_1(t) \rangle = \int_{\mathbb{R}} \phi(t - k) f_1(t)\, dt$$

Now $\overline{\operatorname{supp}(\phi(t - k))} = [k, k + 1]$ and on this interval, the function value is 1 so that the integral above reduces to

$$b_k = \int_k^{k+1} f_1(t)\, dt$$

The function $f_1(t)$ is piecewise constant with possible breakpoints at $\frac{1}{2}\mathbb{Z}$, so there are exactly two basis functions, say $\phi_{1,m}(t) = \sqrt{2}\,\phi(2t - m)$ with $\overline{\operatorname{supp}(\phi_{1,m})} = \left[k, k + \frac{1}{2}\right]$ and $\phi_{1,m+1}(t) = \sqrt{2}\,\phi(2t-(m+1))$ with $\overline{\operatorname{supp}(\phi_{1,m+1})} = \left[k + \frac{1}{2}, k + 1\right]$. We now need to determine the correct value for m.

We can use Problem 3.11 in Section 3.2 to determine that the compact support of $\phi(2t - m)$ is $\left[\frac{m}{2}, \frac{m+1}{2}\right]$. We want the left endpoint $\frac{m}{2}$ to match with the left endpoint of the support of $\phi(t - k)$. That is, we want $\frac{m}{2} = k$ or $m = 2k$. So $\phi(2t - 2k) = \phi_{1,2k}(t)$ is the leftmost function that overlaps $[k, k + 1]$. To find the other function, we simply translate $\phi(2t - 2k)$ right by one-half unit. We have $\phi\left(2\left(t - \frac{1}{2}\right) - 2k\right) = \phi(2t - 2k - 1) = \phi(2t - (2k + 1)) = \phi_{1,2k+1}(t)$.

Thus on the interval $[k, k + 1]$,

$$f_1(t) = a_{2k}\phi(2t - 2k) + a_{2k+1}\phi(2t - (2k + 1))$$

with $\overline{\operatorname{supp}(\phi(2t - 2k))} = \left[k, k + \frac{1}{2}\right]$ and $\operatorname{supp}(\phi(2t - (2k + 1))) = \left[k + \frac{1}{2}, k + 1\right]$. Our inner product becomes

$$b_k = \int_k^{k+1} f_1(t)\, dt$$
$$= \int_k^{k+1} \left(a_{2k}\sqrt{2}\,\phi(2t - 2k) + a_{2k+1}\sqrt{2}\,\phi(2t - (2k + 1))\right) dt$$
$$= \sqrt{2}\int_k^{k+\frac{1}{2}} a_{2k}\phi(2t - 2k)\, dt + \sqrt{2}\int_{k+\frac{1}{2}}^{k+1} a_{2k+1}\phi(2t - (2k + 1))\, dt$$

Now $\phi(2t - 2k) = 1$ on $[k + 1/2, k + 1)$ and $\phi(2t - (2k + 1)) = 1$ on $\left[k + \frac{1}{2}, k + 1\right)$, so that inner product reduces to

$$b_k = \sqrt{2} \int_k^{k+\frac{1}{2}} a_{2k} \phi(2t - 2k) \, dt + \sqrt{2} \int_{k+\frac{1}{2}}^{k+1} a_{2k+1} \phi(2t - (2k + 1)) \, dt$$

$$= \sqrt{2} \, a_{2k} \int_k^{k+\frac{1}{2}} 1 \, dt + \sqrt{2} \, a_{2k+1} \int_{k+\frac{1}{2}}^{k+1} 1 \, dt$$

$$= \frac{\sqrt{2}}{2} (a_{2k} + a_{2k+1})$$

and the proof is complete.   ∎

The Residual Function $g_0(t) = f_1(t) - f_0(t)$

We constructed $f_0(t)$ as an approximation in V_0 of $f_1(t) \in V_1$. Suppose instead that we were given $f_0(t)$ and $f_1(t)$ and asked to construct a residual function $g_0(t)$ so that $f_0(t) + g_0(t) = f_1(t)$. What does $g_0(t)$ look like?

We can easily compute $g_0(t) = f_1(t) - f_0(t)$. Using Figure 3.13, we see that $g_0(t)$ is going to be a piecewise constant function with breaks at $\frac{1}{2}, 1, \frac{3}{2}, 2, \frac{5}{2}$. Thus $g_0(t)$ is a function in V_1 that can be written as a linear combination of $\phi_{1,k}(t)$. Since $\text{supp}(f_0) = \text{supp}(f_1) = [0,3]$, we only need $\phi_{1,k}(t)$, for $k = 0, \ldots, 5$. Consider $f_1(t)$ and $f_0(t)$ on $[0,1)$. On this interval, $f_0(t)$ is simply the average of the two constant heights of $f_1(t)$. Thus $g_0(t)$ should represent the directed distance from $f_0(t)$ to $f_1(t)$ on the interval. The directed distance is $\sqrt{2}$ on $[0, \frac{1}{2})$ and $-\sqrt{2}$ on $[\frac{1}{2}, 1)$. We can repeat this process for the intervals $[1,2)$ and $[2,3)$ and obtain

$$g_0(t) = f_1(t) - f_0(t)$$
$$= \sqrt{2} \, \phi(2t) - \sqrt{2} \, \phi(2t - 1) + 3\sqrt{2} \, \phi(2t - 2) - 3\sqrt{2} \, \phi(2t - 3)$$
$$\quad - \sqrt{2} \, \phi(2t - 4) + \sqrt{2} \, \phi(2t - 5)$$
$$= \sqrt{2} \, (\phi(2t) - \phi(2t - 1)) + 3\sqrt{2} \, (\phi(2t - 2) - \phi(2t - 3))$$
$$\quad - \sqrt{2} \, (\phi(2t - 4) - \phi(2t - 5)) \tag{3.24}$$

The residual function $g_0(t)$ is plotted in Figure 3.14.

The Wavelet Function $\psi(t)$ Defined

Let's take a closer look at $g_0(t)$. In particular, consider $g_0(t)$ on $[0,1)$. It is $\sqrt{2}$ times the function $\psi(t)$, where $\psi(t)$ is given by

$$\psi(t) = \phi(2t) - \phi(2t - 1) = \begin{cases} 1, & 0 \le t < \frac{1}{2} \\ -1, & \frac{1}{2} \le t < 1 \\ 0, & \text{otherwise} \end{cases} \tag{3.25}$$

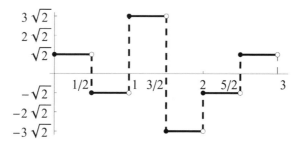

Figure 3.14   The function $g_0(t)$ given by (3.24).

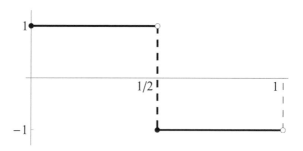

Figure 3.15   The function $\psi(t)$ given by (3.25).

The function $\psi(t)$ is plotted in Figure 3.15. Now look at $g_0(t)$ on $[1,2)$. We can view $g_0(t)$ here as a translate (by one unit right) of $\psi(t)$ multiplied by $3\sqrt{2}$. Thus on $[1,2)$, $g_0(t) = 3\sqrt{2}\,\psi(t-1)$. In an analogous way, we see that on $[2,3)$, $g_0(t) = -\psi(t-2)$. Thus

$$g_0(t) = \sqrt{2}\,\psi(t) + 3\sqrt{2}\,\psi(t-1) - \sqrt{2}\,\psi(t-2) \qquad (3.26)$$

We also note that there is a relationship between the coefficients of $\psi(t-k)$, $k = 0,1,2$ in (3.24) and the coefficients of $f_1(t)$ given in (3.21). Recall that the coefficients of $f_1(t)$ are $a_0 = 5$, $a_1 = 3$, $a_2 = 5$, $a_3 = -1$, $a_4 = 5$, and $a_5 = 7$. If we let

$$g_0(t) = \sum_{k=0}^{2} c_k \psi(t-k)$$

where $c_0 = \sqrt{2}$, $c_1 = 3\sqrt{2}$, and $c_2 = -\sqrt{2}$, we see that the c_k obey

$$c_k = \frac{\sqrt{2}}{2}(a_{2k} - a_{2k+1})$$

Writing $g_0(t)$ in Terms of Translates of $\psi(t)$

This observation is true for general $f_1(t) \in V_1$, as illustrated by the following proposition.

Proposition 3.10 (A Formula for the Residual Function $g_0(t)$) *Suppose that $f_1(t)$ is an element of V_1 is given by*

$$f_1(t) = \sum_{k \in \mathbb{Z}} a_k \phi_{1,k}(t)$$

and assume that $f_0(t)$ is the projection (see (3.16)) $P_{f_1,0}(t)$ of $f_1(t)$ into V_0. If $g_0(t) = f_1(t) - f_0(t)$ is the residual function in V_1, then $g_0(t)$ is given by

$$g_0(t) = \sum_{k \in \mathbb{Z}} c_k \psi(t - k)$$

where $\psi(t)$ is given by (3.25) and

$$c_k = \frac{\sqrt{2}}{2} (a_{2k} - a_{2k+1}) \tag{3.27}$$

for $k \in \mathbb{Z}$.   ∎

Proof: The proof of this proposition is left as Problem 3.22.   ∎

The Space $\mathbf{W_0}$ Defined

Recall that we defined V_0 as the linear span (in $L^2(\mathbb{R})$) of $\phi(t)$ and its integer translates. In Proposition 3.10 we see that we are interested in linear combinations of $\psi(t)$ and its integer translates. We thus define the following space:

Definition 3.4 (The Residual Space $\mathbf{W_0}$) *Let $\psi(t)$ be the function given in (3.25). We define the space*

$$W_0 = \text{span} \{\dots, \psi(t+1), \psi(t), \psi(t-1), \dots\} \cap L^2(\mathbb{R})$$

$$= \text{span} \{\psi(t-k)\}_{k \in \mathbb{Z}} \cap L^2(\mathbb{R}) \tag{3.28}$$

We call W_0 the Haar wavelet space *generated by the* Haar wavelet function $\psi(t)$.   ∎

We can see where the word *wavelet* originates. Given the nature of the approximation $f_0(t)$ of $f_1(t)$ (averaging consecutive values a_{2k} and a_{2k+1}), it makes sense to model this error using "small" waves like $\psi(t)$ and its integer translates.

By definition, the set $\{\psi(t-k)\}_{k \in \mathbb{Z}}$ spans W_0, and in Problem 3.23 you will show that $\{\psi(t-k)\}_{k \in \mathbb{Z}}$ is a set of linearly independent functions. It is also a simple task (see Problem 3.25) to show that

$$\langle \psi(t-k), \psi(t-m) \rangle = \begin{cases} 0, & k \neq m \\ 1, & k = m \end{cases}$$

We have the following proposition.

Proposition 3.11 (An Orthonormal Basis for $\mathbf{W_0}$) *Let W_0 be the space defined by (3.28) and $\psi(t)$ be defined by (3.25). Then the set $\{\psi(t-k)\}_{k \in \mathbb{Z}}$ forms an orthonormal basis for W_0.* ∎

Proof: The proof is left as Problems 3.23 and 3.25. ∎

Scaling Function and Dilation Equations

From (3.23) we can see that the vector

$$\mathbf{h} = [h_0, h_1]^T = \left[\frac{\sqrt{2}}{2}, \frac{\sqrt{2}}{2} \right]^T \tag{3.29}$$

plays an important role in projecting a function $f_1(t) \in V_1$ into V_0. Indeed, we can rewrite the projection coefficients in (3.23) using (3.29). We have, for $k \in \mathbb{Z}$,

$$b_k = \mathbf{h} \cdot \mathbf{a}^k$$

where $\mathbf{a}^k = [a_{2k}, a_{2k+1}]^T$. In particular, we can use $\mathbf{h}$ to write $\phi(t)$ in terms of basis functions of V_1:

$$\phi(t) = \phi(2t) + \phi(2t - 1)$$
$$= \left(\frac{\sqrt{2}}{2} \right) \sqrt{2}\,\phi(2t) + \left(\frac{\sqrt{2}}{2} \right) \sqrt{2}\,\phi(2t - 1)$$
$$= h_0\,\phi_{1,0}(t) + h_1\phi_{1,1}(t) \tag{3.30}$$

In a similar manner, we can use the vector

$$\mathbf{g} = [g_0, g_1]^T = \left[\frac{\sqrt{2}}{2}, -\frac{\sqrt{2}}{2} \right]^T \tag{3.31}$$

to project $f_1(t) \in V_1$ into W_0. As we can see by (3.25),

$$\psi(t) = \phi(2t) - \phi(2t - 1)$$
$$= \left(\frac{\sqrt{2}}{2} \right) \sqrt{2}\,\phi(2t) + \left(-\frac{\sqrt{2}}{2} \right) \sqrt{2}\,\phi(2t - 1)$$
$$= g_0\,\phi_{1,0}(t) + g_1\phi_{1,1}(t) \tag{3.32}$$

Thus when we write $\psi(t)$ in terms of the basis functions of V_1, the filter coefficients g_0 and g_1 appear as the coefficients of the linear combination.

The equations (3.30) and (3.32) are paramount to the development of general wavelet theory. The function $\phi(t)$ given by (3.1) basically is used to generate a sequence of spaces V_j that allow us to approximate functions from $L^2(\mathbb{R})$. Equation (3.30) gives us a way to connect V_0 and V_1. In a similar way, the function $\psi(t)$ given by (3.25) generates the space W_0 and is used to measure the error when approximating functions in V_1 by functions in V_0. Equation (3.32) shows how $\psi(t)$ can be expressed as a function in V_1.

We have already designated $\psi(t)$ as the *Haar wavelet function*. We now give names to $\phi(t)$ and the equations (3.30) and (3.32).

Definition 3.5 (Haar Scaling Function and Dilation Equation) *We call the equations*

$$\phi(t) = \frac{\sqrt{2}}{2}\phi_{1,0}(t) + \frac{\sqrt{2}}{2}\phi_{1,1}(t) = \phi(2t) + \phi(2t-1) \qquad (3.33)$$

and

$$\psi(t) = \frac{\sqrt{2}}{2}\phi_{1,0}(t) - \frac{\sqrt{2}}{2}\phi_{1,1}(t) = \phi(2t) - \phi(2t-1) \qquad (3.34)$$

dilation equations. *The Haar function $\phi(t)$ used to generate these dilation equations is typically called a* scaling function[14].   ∎

Note: The vectors h and g defined in (3.29) and (3.31), respectively, play an important role in many applications involving *discrete data*. We will see in Chapter 4 that the vectors h and g can be used to form a matrix that allows us to decompose a signal (or digital image) into an approximation of the original signal and the details needed to combine with the approximation to recover the original signal.

An Alternative Proof of Proposition 3.9

Writing $\phi(t)$ in terms of the basis functions $\phi_{1,0}(t)$ and $\phi_{1,1}(t)$ of V_1 allows us to give an alternative proof of Proposition 3.9.

Proof: Let $f_1(t) = \sum_{k \in \mathbb{Z}} a_k \phi_{1,k}(t) \in V_1$ and suppose that we project $f_1(t)$ into V_0 using (3.16). We have

$$f_0(t) = \sum_{m \in \mathbb{Z}} b_m \phi(t-m) = \sum_{m \in \mathbb{Z}} \langle f_1(t), \phi(t-m)\rangle \phi(t-m)$$

---

[14]The functions $\phi(t)$ and $\psi(t)$ are sometimes called the *father wavelet* and the *mother wavelet*, respectively.

But $\phi(t) = \frac{\sqrt{2}}{2}\phi_{1,0}(t) + \frac{\sqrt{2}}{2}\phi_{1,1}(t)$, so that

$$
\begin{aligned}
\phi(t-m) &= \frac{\sqrt{2}}{2}\phi_{1,0}(t-m) + \frac{\sqrt{2}}{2}\phi_{1,1}(t-m) \\
&= \phi(2(t-m)) + \phi(2(t-m)-1) \\
&= \phi(2t-2m) + \phi(2t-(2m+1)) \\
&= \frac{\sqrt{2}}{2}\phi_{1,2m}(t) + \frac{\sqrt{2}}{2}\phi_{1,2m+1}(t)
\end{aligned}
$$

Thus

$$
\begin{aligned}
\langle f_1(t), \phi(t-m)\rangle &= \left\langle f_1(t), \frac{\sqrt{2}}{2}\phi_{1,\,2m}(t) + \frac{\sqrt{2}}{2}\phi_{1,\,2m+1}(t)\right\rangle \\
&= \frac{\sqrt{2}}{2}\left\langle f_1(t), \phi_{1,\,2m}(t)\right\rangle + \frac{\sqrt{2}}{2}\left\langle f_1(t), \phi_{1,\,2m+1}(t)\right\rangle
\end{aligned}
$$

Since $f_1(t) = \sum\limits_{k\in\mathbb{Z}} a_k\phi_{1,\,k}(t)$, we compute the first inner product to be

$$
\begin{aligned}
\frac{\sqrt{2}}{2}\langle f_1(t), \phi_{1,\,2m}(t)\rangle &= \frac{\sqrt{2}}{2}\left\langle \sum_{k\in\mathbb{Z}} a_k\phi_{1,\,k}(t), \phi_{1,\,2m}(t)\right\rangle \\
&= \frac{\sqrt{2}}{2}\sum_{k\in\mathbb{Z}} a_k\langle \phi_{1,\,k}(t), \phi_{1,\,2m}(t)\rangle
\end{aligned}
$$

But $\phi_{1,\,k}(t)$ and $\phi_{1,\,2m}(t)$ are orthogonal, so that the only nonzero inner product results when $k = 2m$. Thus we see that

$$
\frac{\sqrt{2}}{2}\langle f_1(t), \phi_{1,\,2m}(t)\rangle = \frac{\sqrt{2}}{2}a_{2k}
$$

In a similar manner, we can show that

$$
\frac{\sqrt{2}}{2}\langle f_1(t), \phi_{1,\,2m+1}(t)\rangle = \frac{\sqrt{2}}{2}a_{2k+1}
$$

and thus obtain (3.23).   ∎

In Problem 3.27 you will use (3.34) to give an alternative proof of Proposition 3.10.

Direct Sums and Direct Orthogonal Sums

In Problem 3.25(c) you will show that $\langle \phi(t-k), \psi(t-m)\rangle = 0$ for all $k, m \in \mathbb{Z}$. Since the basis functions for V_0 and W_0 are orthogonal to one another, the proof of Proposition 3.12 follows immediately.

Proposition 3.12 (Orthogonality and the Spaces V_0, W_0) *Suppose that* $f(t) \in$ V_0 *and* $g(t) \in W_0$. *Then* $\langle f(t), g(t) \rangle = 0$. ∎

Proof: Since $f(t) \in V_0$, we can write it as a linear combination of $\phi(t)$ and its integer translates. We have

$$f(t) = \sum_{k \in \mathbb{Z}} b_k \phi(t - k)$$

In a similar fashion, we write

$$g(t) = \sum_{m \in \mathbb{Z}} c_m \psi(t - m)$$

Thus the inner product becomes

$$\langle f(t), g(t) \rangle = \left\langle \sum_{k \in \mathbb{Z}} b_k \phi(t - k), \sum_{m \in \mathbb{Z}} c_m \psi(t - m) \right\rangle$$
$$= \sum_{k \in \mathbb{Z}} b_k \sum_{m \in \mathbb{Z}} c_m \langle \phi(t - k), \psi(t - m) \rangle$$

In Problem 3.25(c) you will show $\langle \phi(t - k), \psi(t - m) \rangle = 0$ to complete the proof. ∎

We will be interested in spaces that are orthogonal to each other in the remainder of the chapter, so we give a formal definition at this time.

Definition 3.6 (Perpendicular Spaces) *Suppose that V and W are two subspaces of $L^2(\mathbb{R})$. We say that V and W are* perpendicular *to each other and write $V \perp W$ if $\langle f(t), g(t) \rangle = 0$ for all functions $f(t) \in V$ and $g(t) \in W$.* ∎

In terms of Definition 3.6, Proposition 3.12 shows that $V_0 \perp W_0$.

We saw in Proposition 3.10 that we could write $f_1(t) \in V_1$ as $f_1(t) = f_0(t) + g_0(t)$, where $f_0(t) \in V_0$ and $g_0(t) \in W_0$ and $f_0(t)$ is the projection of $f_1(t)$ into V_0 and $g_0(t)$ is the projection of $f_1(t)$ into V_1 (see Problem 3.26). Thus we have constructed a function from V_1 by adding functions from V_0 and W_0. This gives rise to the following definition.

Definition 3.7 (Direct Sum and Direct Orthogonal Sum) *Suppose that V and W are subspaces of $L^2(\mathbb{R})$. We define the* direct sum *of V and W as the subspace*

$$X = V + W = \{f(t) + g(t) \mid f(t) \in V, g(t) \in W\}$$

If $V \perp W$, then we call X the direct orthogonal sum *and write*

$$X = V \oplus W = \{f(t) + g(t) \mid f(t) \in V, g(t) \in W, \langle f(t), g(t) \rangle = 0\}$$

∎

Note that if $X = V + W$, then $V \subseteq X$ since $0 \in W$ and $V = \{f(t) + 0 \mid f(t) \in V\}$. In a similar manner, we see that $W \subseteq X$ as well. Thus both V and W are subspaces of X.

Connecting V_0 and W_0 to V_1

Since $V_0 \perp W_0$, we can look at the direct orthogonal sum $V_0 \oplus W_0$. The following result characterizes this direct orthogonal sum.

Proposition 3.13 (The Direct Orthogonal Sum $V_0 \oplus W_0$) *Suppose that V_0 and V_1 are defined by (3.8) and W_0 is defined by (3.28). Then*

$$V_1 = V_0 \oplus W_0$$

∎

Proof: In Proposition 3.10, we saw that an arbitrary $f_1(t) \in V_1$ could be expressed as a sum of $f_0(t) \in V_0$ and $g_0(t) \in W_0$. Since $f_0(t)$ and $g_0(t)$ are both elements of V_1, we see that $V_1 = V_0 + W_0$. To complete the proof we need to show that if $f_1(t) \in V_1$ and perpendicular to all functions $h(t) \in V_0$, then $f_1(t) \in W_0$. Assume $f_1(t) = f_0(t) + g_0(t)$ and let $h(t)$ be any function in V_0. Then

$$\begin{aligned}
0 &= \langle f_1(t), h(t) \rangle \\
&= \langle f_0(t), h(t) \rangle + \langle g_0(t), h(t) \rangle \\
&= \langle f_0(t), h(t) \rangle + 0 \\
&= \langle f_0(t), h(t) \rangle
\end{aligned}$$

Since $h(t)$ is arbitrary, it must be that $f_0(t) = 0$ so that $f_1(t) = g_0(t) \in W_0$ and the proof is complete. ∎

PROBLEMS

3.21   Find $f_0(t)$ and $g_0(t)$ for each of the following functions from V_1.

(a) $f_1(t) = 10\phi_{1,0}(t) - 4\phi_{1,1}(t) + 4\phi_{1,2}(t) - 6\phi_{1,4}(t) + 2\phi_{1,5}(t)$

(b) $f_1(t) = 20\phi_{1,0}(t) - 20\phi_{1,1}(t) + 10\phi_{1,2}(t) - 10\phi_{1,3}(t) - 5\phi_{1,4}(t) + 5\phi_{1,5}(t)$

(c) $f_1(t) = 5\phi_{1,0}(t) + 5\phi_{1,1}(t) + 8\phi_{1,2}(t) + 8\phi_{1,3}(t) - 4\phi_{1,4}(t) - 4\phi_{1,5}(t)$

(d) $f_1(t) = \sum\limits_{k=1}^{\infty} \dfrac{1}{k}\phi_{1,k}(t)$

3.22   Prove Proposition 3.10. The following steps will help you organize your work.

(a) We know that $g_0(t) = f_1(t) - f_0(t)$. Since we need to show that $g_0(t)$ is constructed as a linear combinations of the functions $\psi(t{-}k)$ with $\mathrm{supp}(\psi(t-k)) = [k, k+1]$, first find formulas for $f_1(t)$ and $f_0(t)$ on $[k, k+1)$. The proof of Proposition 3.9 will be helpful.

(b) Compute $f_1(t) - f_0(t)$ on $\left[k, k + \tfrac{1}{2}\right)$ and $\left[k + \tfrac{1}{2}, k+1\right)$.

(c) Verify that the results you obtained in part (b) can be written as $c_k \psi(t - k)$, where the c_k are defined in (3.27).

3.23   Show that the set $\{\psi(t - k)\}_{k \in \mathbb{Z}}$ is a linearly independent set in W_0.

3.24   Show that W_0 is a subspace of V_1.

3.25   Show that

(a) $\int_{\mathbb{R}} \psi(t) \, dt = 0$

(b) $\langle \psi(t - k), \psi(t - m) \rangle = \begin{cases} 0, & k \neq m \\ 1, & k = m \end{cases}$

(c) $\langle \phi(t - k), \psi(t - m) \rangle = \int_{\mathbb{R}} \phi(t - k)\psi(t - m) \, dt = 0$

3.26   Show that $g_0(t)$ given in Proposition 3.10 can be obtained by projecting $f_1(t)$ into W_0. That is, show that

$$g_0(t) = \sum_{k \in \mathbb{Z}} \langle f_1(t), \psi(t - k) \rangle \psi(t - k)$$

3.27   Use the ideas from the alternative proof of Proposition 3.9 with the vector **g** defined in (3.31) to give an alternative proof of Proposition 3.10.

3.4   THE GENERAL HAAR WAVELET SPACE W_J

We have analyzed the connection between V_1 and V_0, W_0. Suppose that we were given a function $f_j(t) \in V_j$ and we wished to approximate it with a function $f_{j-1}(t) \in V_{j-1}$ or even $f_0(t) \in V_0$. We could certainly project $f_j(t)$ into the desired space, but how do we measure the residual? Since V_j is constructed using the functions $\phi(2^j t - k)$, $k \in \mathbb{Z}$, is natural to expect that our residuals will be built from translates and dilations of the function $\psi(t)$. We have the following definition.

Definition 3.8 (The Function $\psi_{j,k}(t)$)   *Let $\psi(t)$ be the Haar wavelet function given by (3.25). Then for $j, k \in \mathbb{Z}$, we define the function*

$$\boxed{\psi_{j,k}(t) = 2^{j/2}\psi(2^j t - k)} \qquad (3.35)$$

∎

The next proposition notes that the multiplier $2^{j/2}$ in the definition of $\psi_{j,k}(t)$ serves as a normalizing factor.

Proposition 3.14 ($\|\psi_{j,k}(t)\| = 1$)   *For $\psi_{j,k}(t)$ defined by (3.35), $\|\psi_{j,k}(t)\| = 1$.* ∎

Proof: In order to compute $\|\psi_{j,k}(t)\|^2$, we make the substitution $u = 2^j t - k$. Then the limits of integration remain unchanged and $du = 2^j\,dt$ or $2^{-j}\,du = dt$. We have

$$\|\psi_{j,k}(t)\|^2 = \int_{\mathbb{R}} 2^j \psi^2(2^j t - k)\,dt = 2^{-j} \int_{\mathbb{R}} 2^j \psi^2(u)\,du = \int_{\mathbb{R}} \psi^2(u)\,du$$

But $\psi^2(t) = \phi(t)$ and we know from (3.6) that

$$\int_{\mathbb{R}} \psi^2(u)\,du = \int_{\mathbb{R}} \phi(t)\,dt = 1$$

so that $\|\psi_{j,k}(t)\|^2 = 1$.   ∎

In Problem 3.28 you will show that $\int_{\mathbb{R}} \psi_{j,k}(t)\,dt = 0$. Let's look at some examples of $\psi_{j,k}(t)$.

Example 3.7 (Plotting $\psi_{j,k}(t)$) *Plot each of the following functions. In addition, describe the support of each function.*

(a) $\psi_{1,-2}(t)$

(b) $\psi_{-2,4}(t)$

(c) $\psi_{4,-5}(t)$

Solution
    For the function in part (a) we have

$$\psi_{1,-2}(t) = 2^{1/2}\psi(2t + 2) = \sqrt{2}\,\psi(2(t+1))$$

Thus we are contracting the function $\psi(t)$ by a factor of 2 and then shifting the result one unit to the left. Since $\mathrm{supp}(\psi) = [0,1]$, $\psi(2t)$ satisfies $\overline{\mathrm{supp}(\psi(2t))} = [0,\frac{1}{2}]$ and the shift of one unit left results in $\overline{\mathrm{supp}(\psi(2t + 2))} = [-1,-\frac{1}{2}]$. The function $\psi_{1,-2}(t)$ is plotted in Figure 3.16(a).
    In part (b)

$$\psi_{-2,4}(t) = 2^{-2/2}\psi\left(\frac{1}{4}t - 4\right) = \frac{1}{2}\psi\left(\frac{1}{4}(t - 16)\right)$$

In this case we are expanding $\psi(t)$ by a factor of 4 and then shifting the result right 16 units. The compact support of $\psi\left(\frac{1}{4}t\right)$ is $\overline{\mathrm{supp}\left(\psi\left(\frac{1}{4}t\right)\right)} = [0,4]$, so that $\overline{\mathrm{supp}(\psi_{-2,4})} = [16,20]$. We have plotted $\psi_{-2,4}(t)$ in Figure 3.16(b).
    For the function in part (c), we have

$$\psi_{4,-5}(t) = 2^{4/2}\psi(16t + 5) = 4\psi\left(16\left(t + \frac{5}{16}\right)\right)$$

Here we have contracted the function $\psi(t)$ by a factor of 16 and then shifted the result $\frac{5}{16}$ unit to the left. Since $\text{supp}(\psi(16t)) = \left[0, \frac{1}{16}\right]$, we have $\text{supp}(\psi_{4,-5}) = \left[-\frac{5}{16}, -\frac{1}{4}\right]$. Note that $\psi_{4,-5}(t)$ assumes the value 1 on the interval $\left[-\frac{5}{16}, -\frac{9}{32}\right)$, value -1 on the interval $\left[-\frac{9}{32}, -\frac{1}{4}\right]$, and zero elsewhere. The function is plotted in Figure 3.16(c). ∎

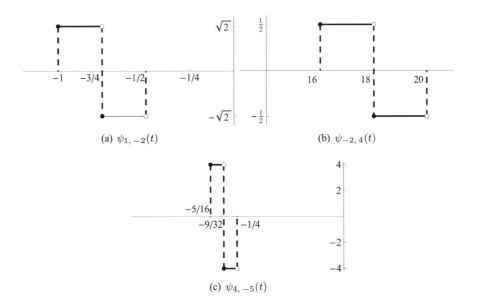

(a) $\psi_{1,-2}(t)$ (b) $\psi_{-2,4}(t)$

(c) $\psi_{4,-5}(t)$

Figure 3.16   The functions $\psi_{1,-2}(t)$, $\psi_{-2,4}(t)$, and $\psi_{4,-5}(t)$.

Let's analyze the function $\psi_{j,k}(t) = 2^{j/2}\psi(2^j t - k)$ further. Using (3.25), we see that

$$2^{j/2}\psi(2^j t - k) = \begin{cases} 2^{j/2}, & 0 \le 2^j t - k < \frac{1}{2} \\ -2^{j/2}, & \frac{1}{2} \le 2^j t - k < 1 \\ 0, & \text{otherwise} \end{cases} \qquad (3.36)$$

Now $2^{j/2}$ scales the height of $\psi(t)$ and we also know that $\psi(2^j t - k) = \psi\left(2^j\left(t - \frac{k}{2^j}\right)\right)$ so that 2^j contracts (expands) $\psi(t)$ if $j > 0$ ($j < 0$) and we then translate by $\frac{k}{2^j}$. To get a better understanding of exactly where the function is equal to $2^{j/2}$, we solve the inequality that defines the first part of $\psi_{j,k}(t)$ in (3.36) for t. We start with $0 \le 2^j t - k < \frac{1}{2}$, add k to all three terms, and then multiply each term by 2^{-j} to obtain

$$2^{-j}k \le t < 2^{-j}k + 2^{-(j+1)}$$

In a similar manner, we solve the inequality that defines the second part of $\psi_{j,k}(t)$ in (3.36) for t to obtain

$$2^{-j}k + 2^{-(j+1)} \le t < 2^{-j}k + 2^{-j}$$

Combining, we can rewrite (3.36) as

$$\psi_{j,k}(t) = \begin{cases} 2^{j/2}, & 2^{-j}k \le t < 2^{-j}k + 2^{-(j+1)} \\ -2^{j/2}, & 2^{-j}k + 2^{-(j+1)} \le t < 2^{-j}k + 2^{-j} \\ 0, & \text{otherwise} \end{cases} \tag{3.37}$$

The preceding discussion gives rise to the following proposition.

Proposition 3.15 (Support for $\psi_{j,k}(t)$) *The compact support for the function $\psi_{j,k}(t)$ given by (3.35) is $\overline{supp(\psi_{j,k})} = [2^{-j}k, 2^{-j}(k+1)]$.* ∎

Proof: From the discussion preceding Proposition 3.15 we see that $\psi_{j,k}(t)$ takes on the value 2^j on the interval $[2^{-j}k, 2^{-j}k + 2^{-(j+1)})$ and the value 2^{-j} on the interval $[2^{-j}k + 2^{-(j+1)}, 2^{-j}k + 2-j)$. Thus $\overline{supp(\psi_{j,k})} = [2^{-j}k, 2^{-j}(k+1)]$. ∎

In Definition 3.4 we introduced the space W_0 as the linear span (in $L^2(\mathbb{R})$) of the functions $\psi(t-k)$, $k \in \mathbb{Z}$. Just as we constructed the space V_j from the functions $\phi_{j,k}(t)$, $k \in \mathbb{Z}$, we can build spaces generated by $\psi_{j,k}(t)$, $k \in \mathbb{Z}$.

Definition 3.9 (The Space W_j) *Let $\psi(t)$ be the Haar function given in (3.25). We define the vector space*

$$\boxed{\begin{aligned} W_j &= span\left\{\ldots, \psi(2^j t + 1), \psi(2^j t), \psi(2^j t - 1), \ldots\right\} \cap L^2(\mathbb{R}) \\ &= span\left\{\psi(2^j t - k)\right\}_{k \in \mathbb{Z}} \cap L^2(\mathbb{R}) \end{aligned}} \tag{3.38}$$

We call W_j the Haar wavelet space W_j. ∎

Let's look at examples of typical elements of W_j.

Example 3.8 (Elements of W_j) *Plot the following functions.*

(a) $f_1(t) = 2\psi_{-2,-1}(t) - 6\psi_{-2,1}(t) + 4\psi_{-2,2}(t) \in W_{-2}$

(b) $f_2(t) = \sum\limits_{k=-4}^{-1} (-1)^k k^2 \psi(8t - k) \in W_3$

Solution
For the function in part (a) we have

$$\begin{aligned} f_1(t) &= 2\psi_{-2,-1}(t) - 6\psi_{-2,1}(t) + 4\psi_{-2,2}(t) \\ &= 2 \cdot 2^{-2/2}\psi\left(\frac{t}{4} + 1\right) - 6 \cdot 2^{-2/2}\psi\left(\frac{t}{4} - 1\right) + 4 \cdot 2^{-2/2}\psi\left(\frac{t}{4} - 2\right) \\ &= \psi\left(\frac{1}{4}(t + 4)\right) - 3\psi\left(\frac{1}{4}(t - 4)\right) + 2\psi\left(\frac{1}{4}(t - 8)\right) \end{aligned}$$

Each of the three basis functions used to build $f_1(t)$ are translated versions of $\psi\left(\frac{t}{4}\right)$ with $\overline{\text{supp}\left(\psi\left(\frac{t}{4}\right)\right)} = [0,4]$. *The first function is $\psi\left(\frac{t}{4}\right)$ shifted four units left, the second is $\psi\left(\frac{t}{4}\right)$ shifted four units right and multiplied by -3, and the final function is $\psi\left(\frac{t}{4}\right)$ shifted eight units right and multiplied by 2.*

We can write $f_2(t)$ as

$$f_2(t) = 16\psi\left(8(t+\frac{1}{2})\right) - 9\psi\left(8(t+\frac{3}{8})\right) + 4\psi\left(8(t+\frac{1}{4})\right) - \psi\left(8(t+\frac{1}{8})\right)$$

and in this way see that the fundamental building block for $f_2(t)$ is the function $\psi(8t)$ with $\overline{\text{supp}(\psi(8t))} = \left[0,\frac{1}{8}\right]$. *We use left shifts of $\frac{1}{2}$, $\frac{3}{8}$, $\frac{1}{4}$, and $\frac{1}{8}$, respectively, of $\psi(8t)$ to build $f_2(t)$. The functions are plotted in Figure 3.17.* ∎

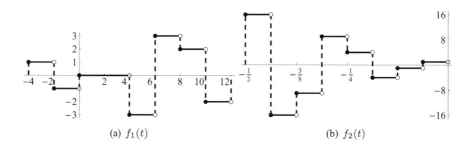

(a) $f_1(t)$ (b) $f_2(t)$

Figure 3.17   The functions $f_1(t)$ and $f_2(t)$.

In Problem 3.32 you will show that for $j \in \mathbb{Z}$, the set $\{\psi(2^j t - k)\}_{k \in \mathbb{Z}}$ is linearly independent in W_j, and in Problem 3.33 you will prove that W_j is a subspace of $L^2(\mathbb{R})$. Since W_j is defined as linear combinations of $\{\psi(2^j t - k)\}_{k \in \mathbb{Z}}$, we see that the set $\{\psi(2^j t - k)\}_{k \in \mathbb{Z}}$ is a basis for W_j. In Problem 3.34 you will show that

$$\left\langle \psi\left(2^j t - k\right), \psi\left(2^j t - \ell\right) \right\rangle = 0 \text{ for } k \neq \ell$$

$$\|\psi(2^j t - k)\| = 2^{-j/2}$$

We can use these observations to form an orthonormal basis for W_j, $j \in \mathbb{Z}$.

Proposition 3.16 (An Orthonormal Basis for W_j) *Let W_j be given by (3.38) for $j \in \mathbb{Z}$. Then the set $\{\psi_{j,k}(t)\}_{k \in \mathbb{Z}}$ where $\psi_{j,k}(t)$ is defined in Definition 3.8 is an orthonormal basis for W_j.* ∎

Proof: The proof is essentially the same as that given for Proposition 3.2. ∎

We have just seen that $\psi_{j,k}(t)$ is orthogonal to $\psi_{j,m}(t)$ for $k \neq m$. It turns out there are more orthogonality properties obeyed by $\psi_{j,k}(t)$.

Our first property says that $\psi_{\ell,m}(t)$ is orthogonal to $\phi_{j,k}(t)$ provided that $\psi_{\ell,m}$ resides on a finer level than $\phi_{j,k}$.

Proposition 3.17 (Orthogonality of $\phi_{j,k}$ and $\psi_{\ell,m}$) *Suppose that j, k, ℓ, m are integers with $\ell \geq j$. Then $\langle \phi_{j,k}(t), \psi_{\ell,m}(t) \rangle = 0$.* ∎

Proof: Let's first consider the case where $\ell = j$. We then have $\phi_{j,k}(t) = 2^{j/2}$ on $I_{j,k}$ and $\psi_{j,m}(t) = \pm 2^{j/2}$ on $I_{j,m}$. By Proposition 3.5, if $k \neq m$, then $I_{j,k} \cap I_{j,m} = \varnothing$. Thus the intervals where both functions are nonzero do not intersect and the inner product is zero. In the case where $k = m$, we have

$$\langle \phi_{j,k}(t), \psi_{j,k}(t) \rangle = \int_{I_{jk}} \phi_{j,k}(t) \psi_{j,k}(t) \, dt$$

$$= 2^{j/2} \int_{I_{jk}} \psi_{j,k}(t) \, dt$$

and by Problem 3.28, we know that this last integral is zero.

Now we consider the case where $\ell > j$. Note that $\phi_{j,k}(t)$ and $\psi_{\ell,m}(t)$ are nonzero on $I_{j,k}, I_{\ell,m}$, respectively. We consider two cases. When $I_{j,k} \cap I_{\ell,m} = \varnothing$, then we are finished since both $\phi_{j,k}(t)$ and $\psi_{\ell,m}(t)$ are zero in this case. Now consider $I_{j,k} \cap I_{\ell,m} \neq \varnothing$. Then by Proposition 3.6 we see that $I_{\ell,m}$ is contained entirely in either the left half or the right half of $I_{j,k}$. Since $\phi_{j,k}(t) = 2^{j/2}$ on $I_{j,k}$, it is equal to $2^{j/2}$ on $I_{\ell,m}$ as well. Thus

$$\langle \phi_{j,k}(t), \psi_{\ell,m}(t) \rangle = 2^{j/2} \int_{I_{\ell,m}} \psi_{\ell,m}(t) \, dt$$

In Problem 3.28 you will that show this last integral is 0. ∎

Proposition 3.17 immediately gives rise to the following characterization of V_j and W_ℓ.

Proposition 3.18 (Orthogonality of V_j and W_ℓ) *Suppose that j, ℓ are integers with $\ell \geq j$. Then $V_j \perp W_\ell$.* ∎

Proof: Let $f(t) \in V_j$ and $g(t) \in W_\ell$. We use the fact that $\{\phi_{j,k}(t)\}_{k \in \mathbb{Z}}$ and $\{\psi_{\ell,m}(t)\}_{m \in \mathbb{Z}}$ are bases for V_j and W_ℓ, respectively, to write

$$f(t) = \sum_{k \in \mathbb{Z}} c_k \phi_{j,k}(t) \qquad \text{and} \qquad g(t) = \sum_{m \in \mathbb{Z}} d_m \psi_{j,m}(t)$$

We next compute the inner product of $f(t)$ and $g(t)$. We have

$$\langle f(t), g(t) \rangle = \int_{\mathbb{R}} f(t) g(t) \, dt$$

$$= \int_{\mathbb{R}} \sum_{k \in \mathbb{Z}} c_k \phi_{j,k}(t) \cdot \sum_{m \in \mathbb{Z}} d_m \psi_{j,m}(t) \, dt$$

$$= \sum_{k \in \mathbb{Z}} c_k \sum_{m \in \mathbb{Z}} d_m \int_{\mathbb{R}} \phi_{j,k}(t) \psi_{\ell,m}(t) \, dt$$

But by Proposition 3.17, each of the integrals in this last equation is 0, so that $f(t)$ and $g(t)$ are orthogonal. Since $f(t)$ and $g(t)$ are arbitrary functions in V_j, W_ℓ, respectively, we conclude that $V_j \perp W_\ell$. ∎

We have shown in Proposition 3.16 that $\psi_{j,k}(t)$ and $\psi_{j,m}(t)$ are orthogonal for $k \neq m$. Thus we can say that the $\psi_{j,k}(t)$ are orthogonal on the same level. It turns out that we can say quite a bit more, as the following proposition illustrates.

Proposition 3.19 (The Orthogonality of $\psi_{j,k}(t)$ Across Levels) *Let j, k, ℓ, m be integers. Then*

$$\langle \psi_{j,k}(t), \psi_{\ell,m}(t) \rangle = \begin{cases} 1, & \text{if } j = \ell \text{ and } k = m \\ 0, & \text{otherwise} \end{cases}$$

∎

Proof: Let's first consider the case when $j = \ell$ and $k = m$. Then we have

$$\langle \psi_{j,k}(t), \psi_{\ell,m}(t) \rangle = \int_{\mathbb{R}} \psi_{j,k}^2(t)\, dt = \|2^{j/2}\psi(2^j t - k)\|^2 = 2^j \|\psi(2^j t - k)\|^2$$

Using Problem 3.34(b) we see that $\|\psi(2^j t - k)\|^2 = 2^{-j}$, so that $2^j \|\psi(2^j t - k)\|^2 = 1$.

Now if $k \neq m$, we have two cases. The first case, $j = \ell$, is true by Problem 3.34(a). We thus consider the case where $j \neq \ell$. Since the values are different, we will say that $\ell > j$. Then by Proposition 3.6, $I_{\ell,m}$ is contained entirely in either the left half or the right half of $I_{j,k}$. If $I_{\ell,m}$ is contained in the left half of $I_{j,k}$, then $\psi_{j,k}(t) = 2^{j/2}$, else $\psi_{j,k}(t) = -2^{j/2}$. We can combine these facts to write

$$\langle \psi_{j,k}(t), \psi_{\ell,m}(t) \rangle = \pm 2^{j/2} \int_{\mathbb{R}} \psi_{\ell,m}(t)\, dt$$

and note that by Problem 3.28, this integral has value 0. ∎

Since the $\psi_{j,k}(t)$ are orthogonal across levels, we can infer that for $j \neq \ell$, any function in W_j must be orthogonal to any function in W_ℓ. We state this result as Proposition 3.20 and leave the proof as Problem 3.35.

Proposition 3.20 (The Spaces W_j and W_ℓ Are Orthogonal) *Let j and ℓ be integers with $j \neq \ell$. Then $W_j \perp W_\ell$.* ∎

Proof: Problem 3.35. ∎

Let's return to the dilation equations (3.33) and (3.34) for $\phi(t)$ and $\psi(t)$, respectively. We know (see (3.30)) that we can write (3.33) as

$$\phi(t) = \phi(2t) + \phi(2t - 1)$$

We now replace t by $2^j t$ in this equation to obtain

$$\phi(2^j t) = \phi(2^{j+1} t) + \phi(2^{j+1} t - 1) \tag{3.39}$$

We will now try to build $\phi_{j,0}(t)$ on the left-hand side of (3.39). Toward this end, we multiply (3.39) by $2^{j/2}$ to obtain

$$2^{j/2}\phi(2^j t) = \phi_{j,0}(t) = 2^{j/2}\phi(2^{j+1}t) + 2^{j/2}\phi(2^{j+1}t - 1) \qquad (3.40)$$

Apart from the normalizing factor $2^{(j+1)/2}$, the terms on the right-hand side of (3.40) look like our basis elements for V_{j+1}. Indeed, if we multiply and divide both terms on the right hand-side of (3.40) by $2^{1/2}$, we have

$$\begin{aligned}
\phi_{j,0}(t) &= 2^{-1/2}2^{1/2}2^{j/2}\phi(2^{j+1}t) + 2^{-1/2}2^{1/2}2^{j/2}\phi(2^{j+1}t - 1) \\
&= 2^{-1/2}2^{(j+1)/2}\phi(2^{j+1}t) + 2^{-1/2}2^{(j+1)/2}\phi(2^{j+1}t - 1) \\
&= \frac{\sqrt{2}}{2}\phi_{j+1,0}(t) + \frac{\sqrt{2}}{2}\phi_{j+1,1}(t)
\end{aligned}$$

The discussion above serves as a proof of the following proposition.

Proposition 3.21 (Dilation Equation for $\phi_{j,0}(t)$) *The function $\phi_{j,0}(t)$ defined by (3.9) satisfies the dilation equation*

$$\boxed{\phi_{j,0}(t) = \frac{\sqrt{2}}{2}\phi_{j+1,0}(t) + \frac{\sqrt{2}}{2}\phi_{j+1,1}(t)} \qquad (3.41)$$

∎

The dilation equation given in (3.41) indicates how $\phi_{j,0}$ can be written in V_{j+1}. What happens if we use an arbitrary integer k rather than 0? The next proposition gives the resulting dilation equation.

Proposition 3.22 (Dilation Equation for $\phi_{j,k}(t)$) *For $j, k \in \mathbb{Z}$, we have*

$$\boxed{\phi_{j,k}(t) = \frac{\sqrt{2}}{2}\phi_{j+1,2k}(t) + \frac{\sqrt{2}}{2}\phi_{j+1,2k+1}(t)} \qquad (3.42)$$

∎

Proof: Recall from (3.9) that

$$\begin{aligned}
\phi_{j,k}(t) &= 2^{j/2}\phi(2^j t - k) \\
&= 2^{j/2}\phi\left(2^j\left(t - \frac{k}{2^j}\right)\right) \\
&= \phi_{j,0}\left(t - \frac{k}{2^j}\right)
\end{aligned}$$

Thus we begin our proof by replacing t with $t - \frac{k}{2^j}$ in (3.41).

$$\phi_{j,k}(t) = \phi_{j,0}\left(t - \frac{k}{2^j}\right) = \frac{\sqrt{2}}{2}\phi_{j+1,0}\left(t - \frac{k}{2^j}\right) + \frac{\sqrt{2}}{2}\phi_{j+1,1}\left(t - \frac{k}{2^j}\right) \qquad (3.43)$$

Next we expand the term $\phi_{j+1,0}\left(t - \frac{k}{2^j}\right)$. We have

$$
\begin{aligned}
\phi_{j+1,0}\left(t - \frac{k}{2^j}\right) &= 2^{(j+1)/2}\phi\left(2^{j+1}\left(t - \frac{k}{2^j}\right)\right) \\
&= 2^{(j+1)/2}\phi\left(2^{j+1}t - \frac{2^{j+1}k}{2^j}\right) \\
&= 2^{(j+1)/2}\phi(2^{j+1}t - 2k) \\
&= \phi_{j+1,2k}(t)
\end{aligned}
\tag{3.44}
$$

Expanding the term $\phi_{j+1,1}\left(t - \frac{k}{2^j}\right)$ in a similar manner gives

$$
\begin{aligned}
\phi_{j+1,1}(t - k/2^j) &= 2^{(j+1)/2}\phi\left(2^{j+1}\left(t - \frac{k}{2^j}\right) - 1\right) \\
&= 2^{(j+1)/2}\phi\left(2^{j+1}t - \frac{2^{j+1}k}{2^j} - 1\right) \\
&= 2^{(j+1)/2}\phi(2^{j+1}t - 2k - 1) \\
&= 2^{(j+1)/2}\phi(2^{j+1}t - (2k+1)) \\
&= \phi_{j+1,2k+1}(t)
\end{aligned}
\tag{3.45}
$$

Inserting the expressions for $\phi_{j+1,0}(t)$ and $\phi_{j+1,1}(t)$ given in (3.44) and (3.45), respectively, into (3.43) completes the proof. ∎

We can obtain results analogous to those obtained in Propositions 3.21 and 3.22 for $\psi_{j,0}(t)$ and $\psi_{j,k}(t)$. We summarize the dilation equations for these functions in the following proposition and leave the proofs for Problem 3.36.

Proposition 3.23 (Dilation Equations for $\psi_{j,0}(t)$ and $\psi_{j,k}(t)$) *For $j, k \in \mathbb{Z}$, we have*

$$
\boxed{\psi_{j,k}(t) = \frac{\sqrt{2}}{2}\phi_{j+1,2k}(t) - \frac{\sqrt{2}}{2}\phi_{j+1,2k+1}(t)}
\tag{3.46}
$$

In particular, when $k = 0$, we have

$$
\boxed{\psi_{j,0}(t) = \frac{\sqrt{2}}{2}\phi_{j+1,0}(t) - \frac{\sqrt{2}}{2}\phi_{j+1,1}(t)}
\tag{3.47}
$$

∎

Proof: This proof is left as Problem 3.36. ∎

Why are we interested in the dilation equations on the arbitrary level? Remember that we used (3.33) to provide an alternative proof of Proposition 3.9. This proposition shows that the vector **h** (3.29) relates the coefficients a_k of $f_1(t) = \sum_{k \in \mathbb{Z}} a_k \phi_{1,0}(t) \in V_1$ to the projection $f_0(t) \in V_0$. The more general dilation equation (3.42) allows us to state and prove a result analogous to Proposition 3.9 for projecting functions $f_{j+1}(t) \in V_{j+1}$ into V_j.

Proposition 3.24 (Projecting Functions in V_{j+1} into V_j) *Let $f_{j+1}(t) \in V_{j+1}$ be defined as*

$$f_{j+1}(t) = \sum_{m \in \mathbb{Z}} a_m \phi_{j,m}(t)$$

*and suppose that **h** is the vector given by (3.29). Then the projection $f_j(t) = P_{f_{j+1},j}(t)$ (see (3.16)) of $f_{j+1}(t)$ into V_j can be written as*

$$f_j(t) = \sum_{k \in \mathbb{Z}} b_k \phi_{j,k}(t) = \sum_{k \in \mathbb{Z}} \langle f_{j+1}(t), \phi_{j,k}(t) \rangle \phi_{j,k}(t)$$

where

$$\boxed{b_k = \frac{\sqrt{2}}{2}(a_{2k} + a_{2k+1}) = \mathbf{h} \cdot \mathbf{a}^k} \tag{3.48}$$

for $k \in \mathbb{Z}$. Here $\mathbf{a}^k = [a_{2k}, a_{2k+1}]^T$. ∎

Proof: We begin the proof by writing

$$b_k = \langle f_{j+1}(t), \phi_{j,k}(t) \rangle$$
$$= \left\langle \sum_{m \in \mathbb{Z}} a_m \phi_{j+1,m}(t), \phi_{j,k}(t) \right\rangle$$
$$= \sum_{m \in \mathbb{Z}} a_m \langle \phi_{j+1,m}(t), \phi_{j,k}(t) \rangle$$

We now use the dilation equation (3.42) for $\phi_{j,k}(t)$ to write

$$b_k = \sum_{m \in \mathbb{Z}} a_m \langle \phi_{j+1,m}(t), \phi_{j,k}(t) \rangle$$
$$= \sum_{m \in \mathbb{Z}} a_m \left\langle \phi_{j+1,m}(t), \frac{\sqrt{2}}{2}\phi_{j+1,2k}(t) + \frac{\sqrt{2}}{2}\phi_{j+1,2k+1}(t) \right\rangle$$
$$= \frac{\sqrt{2}}{2} \sum_{m \in \mathbb{Z}} a_m \langle \phi_{j+1,m}(t), \phi_{j+1,2k}(t) \rangle + \frac{\sqrt{2}}{2} \sum_{m \in \mathbb{Z}} a_m \langle \phi_{j+1,m}, \phi_{j+1,2k+1}(t) \rangle$$

Now the functions $\phi_{j+1,m}(t)$ and $\phi_{j+1,2k}(t)$ in the inner products in the first infinite sum above are members of an orthonormal basis for V_{j+1} (see Proposition 3.2), so the only nonzero inner product occurs when $m = 2k$. Thus the first term above reduces to $\frac{\sqrt{2}}{2}a_{2k}$. In a similar manner we see the second term in the equation above reduces to $\frac{\sqrt{2}}{2}a_{2k+1}$. This completes the proof. ∎

As you might expect, the general dilation equation (3.46) for $\psi_{j,k}(t)$ can be used to produce a result analogous to Proposition 3.10. That is, we can use the vector **g** (3.31) to project a function $f_{j+1}(t) \in V_{j+1}$ into W_j. We have the following proposition.

Proposition 3.25 (The Residual Function $g_j(t)$ in W_j) *Suppose that $f_{j+1}(t) \in V_{j+1}$ is given by*

$$f_{j+1}(t) = \sum_{m \in \mathbb{Z}} a_m \phi_{j+1,\,m}(t)$$

and assume that $f_j(t)$ is the projection $P_{f_{j+1},\,j}(t)$ (see (3.16)) of $f_{j+1}(t)$ into V_j. Let g be given by (3.31). If $g_j(t) = f_{j+1}(t) - f_j(t)$ is the residual function in V_{j+1}, then $g_j(t) \in W_j$ and is thus given by

$$g_j(t) = \sum_{k \in \mathbb{Z}} c_k \psi_{j,\,k}(t)$$

Moreover,

$$\boxed{c_k = \frac{\sqrt{2}}{2} (a_{2k} - a_{2k+1}) = \mathbf{g} \cdot \mathbf{a}^k} \tag{3.49}$$

for $k \in \mathbb{Z}$. Here $\mathbf{a}^k = [a_{2k}, a_{2k+1}]^T$. ∎

Proof: The proof of this proposition is somewhat technical and is based on the compact support properties of $\phi_{j,\,k}(t)$ and $\phi_{j+1,\,k}(t)$. We will analyze $f_{j+1}(t) - f_j(t)$ on an interval-by-interval basis. The basis elements $\phi_{j,\,k}(t)$ used to build $f_j(t)$ are nonzero on $I_{jk} = \left[\frac{k}{2^j}, \frac{k+1}{2^j} \right)$ (see (3.3)) and these intervals are of length 2^{-j}. The basis elements $\phi_{j+1,\,k}(t)$ used to build $f_{j+1}(t)$ are nonzero on an interval exactly half as long as 2^{-j}. Thus if we analyze $f_{j+1}(t) - f_j(t)$ on an interval-by-interval basis, we need to consider the larger intervals $\left[\frac{k}{2^j}, \frac{k+1}{2^j} \right)$, where $\phi_{j,\,k}(t)$ is nonzero and the two intervals $\left[\frac{k}{2^j}, \frac{k+1/2}{2^j} \right), \left[\frac{k+1/2}{2^j}, \frac{k+1}{2^j} \right)$ where the two basis functions on level $j + 1$ are nonzero. We need to first identify these basis functions.

The easiest way to identify the basis function for the interval $\left[\frac{k}{2^j}, \frac{k+1/2}{2^j} \right)$ is to start with the fact that the basis function assumes the value $2^{(j+1)/2}$ for t satisfying

$$\frac{k}{2^j} \le t < \frac{k + 1/2}{2^j} \tag{3.50}$$

and try to rewrite this inequality so that the lower bound is 0 and the upper bound is 1. Then the translation and dilation parameters will be apparent. If we multiply all three parts of (3.50) by 2^j, we obtain

$$k \le 2^j t < k + \frac{1}{2}$$

Now, multiply all three parts of this inequality by 2 and then subtract $2k$ from all parts of the result:

$$2k \le 2^{j+1} t < 2k + 1 \quad \Rightarrow \quad 0 \le 2^{j+1} t - 2k < 1$$

Thus the basis function we seek is $\phi_{j+1,\,2k}(t)$ (see (3.9)). In a similar manner, we can see that the basis function that is $2^{(j+1)/2}$ on $\left[\frac{k+1/2}{2^j}, \frac{k+1}{2^j} \right)$ is $\phi_{j+1,\,2k+1}(t)$. By

Proposition 3.24 we know that $\phi_{j,k}(t)$ assumes the value

$$\frac{2^{j/2}\sqrt{2}}{2}(a_{2k}+a_{2k+1}) = 2^{(j-1)/2}(a_{2k}+a_{2k+1})$$

on the interval $\left[\frac{k}{2^j}, \frac{k+1}{2^j}\right)$. Thus on the left half–interval $\left[\frac{k}{2^j}, \frac{k+1/2}{2^j}\right)$, the function $f_{j+1}(t) - f_j(t)$ has the value (see Figure 3.18)

$$2^{(j+1)/2}a_{2k} - 2^{(j-1)/2}(a_{2k}+a_{2k+1}) = 2^{j/2}a_{2k}(2^{1/2} - 2^{-1/2}) - 2^{(j-1)/2}a_{2k+1}$$
$$= 2^{(j-1)/2}(a_{2k} - a_{2k+1}) \tag{3.51}$$

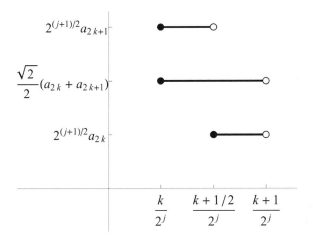

Figure 3.18   The interval $I_{jk} = \left[\frac{k}{2^j}, \frac{k+1}{2^j}\right)$ and the functions values for $f_{j+1}(t)$ and $f_j(t)$ on $I_{j,k}$.

On the right half–interval $\left[\frac{k+1/2}{2^j}, \frac{k+1}{2^j}\right)$, the function $f_{j+1}(t) - f_j(t)$ assumes the value

$$2^{(j+1)/2}a_{2k+1} - 2^{(j-1)/2}(a_{2k}+a_{2k+1}) = 2^{j/2}a_{2k+1}(2^{1/2} - 2^{-1/2}) - 2^{(j-1)/2}a_{2k}$$
$$= -2^{(j-1)/2}(a_{2k} - a_{2k+1}) \tag{3.52}$$

We recall that $\overline{\text{supp}(\psi_{j,k})} = \left[\frac{k}{2^j}, \frac{k+1}{2^j}\right]$ (see Proposition 3.15) and write

$$\psi_{j,k}(t) = 2^{j/2}\psi(2^j t - k) = \begin{cases} 2^{j/2}, & \frac{k}{2^j} \leq t < \frac{k+1/2}{2^j} \\ -2^{j/2}, & \frac{(k+1/2)}{2^j} \leq t < \frac{k+1}{2^j} \end{cases}$$

Finally, we can complete the proof by using (3.51) and (3.52) to summarize our findings on $\left[\frac{k}{2^j}, \frac{k+1}{2^j}\right)$. On this interval, we have

$$
f_{j+1}(t) - f_j(t) = \begin{cases} 2^{(j-1)/2}(a_{2k} - a_{2k+1}), & \frac{k}{2^j} \le t < \frac{k+1/2}{2^j} \\ -2^{(j-1)/2}(a_{2k} - a_{2k+1}), & \frac{k+1/2}{2^j} \le t < \frac{k+1}{/2^j} \end{cases}
$$

$$
= \frac{\sqrt{2}}{2}(a_{2k} - a_{2k+1}) \begin{cases} 2^{j/2}, & \frac{k}{2^j} \le t < \frac{k+1/2}{2^j} \\ -2^{j/2}, & \frac{k+1/2}{2^j} \le t < \frac{k+1}{2^j} \end{cases}
$$

$$
= \frac{\sqrt{2}}{2}(a_{2k} - a_{2k+1})\psi_{j,\,k}(t)
$$

∎

In Problem 3.37 you will show that $g_j(t)$ is the projection of $f_{j+1}(t)$ into W_j.

In Proposition 3.18 we showed that V_j is orthogonal to W_ℓ, where $\ell \ge j$. In particular we can write $V_j \oplus W_j$. We also learned in Proposition 3.25 that an arbitrary function $f_{j+1}(t) \in V_{j+1}$ can be written as the sum of a function $f_j(t) \in V_j$ and $g_j(t) \in W_j$. Moreover (see Problem 3.37), the functions $f_j(t)$ and $g_j(t)$ are projections of $f_{j+1}(t)$ into V_j, W_j, respectively. In our next proposition, which is a generalization of Proposition 3.13, we show that V_{j+1} can be constructed from V_j and W_j.

Proposition 3.26 (The Direct Orthogonal Sum $\mathbf{V_j \oplus W_j}$) *Suppose that V_j, V_{j+1} are defined by (3.8) and W_j is defined by (3.38). Then*

$$
\boxed{V_{j+1} = V_j \oplus W_j} \tag{3.53}
$$

∎

Proof: The proof of Proposition 3.26 is very similar to the proof given for Proposition 3.13 and thus left as Problem 3.38.   ∎

Summarizing Propositions 3.24–3.26

Proposition 3.26 tells us that every function $f_{j+1}(t) \in V_{j+1}$ can be written as the sum of an *approximation function* $f_j(t) \in V_j$ and a *residual function* (or *detail function*) $g_j(t) \in W_j$. Moreover, the functions $f_j(t)$ and $g_j(t)$ are orthogonal to each other and if

$$
f_{j+1}(t) = \sum_{k \in \mathbb{Z}} b_k \phi_{j+1,\,k}(t)
$$

$$
f_j(t) = \sum_{k \in \mathbb{Z}} a_k \phi_{j+1,\,k}(t), \quad g_j(t) = \sum_{k \in \mathbb{Z}} c_k \psi_{j+1,\,k}(t)
$$

and for $k \in \mathbb{Z}$,

$$\mathbf{h} = \left[\frac{\sqrt{2}}{2}, \frac{\sqrt{2}}{2} \right]^T, \quad \mathbf{g} = \left[\frac{\sqrt{2}}{2}, -\frac{\sqrt{2}}{2} \right]^T, \quad \mathbf{a}^k = [a_{2k}, a_{2k+1}]^T$$

we have

$$b_k = \mathbf{h} \cdot \mathbf{a}^k, \qquad c_k = \mathbf{g} \cdot \mathbf{a}^k \tag{3.54}$$

The formulas in (3.54) will serve as the basis for creating a *discrete Haar wavelet transformation* that can be used to process digital signals and images.

PROBLEMS

3.28   Show that $\int_{\mathbb{R}} \psi_{\ell, m}(t) \, dt = \int_{I_{\ell, m}} \psi_{\ell, m}(t) \, dt = 0$.

3.29   Plot each of the following functions. In addition, describe the compact support of each function.

(a) $\psi_{3, -6}(t)$

(b) $\psi_{-4, 2}(t)$

(c) $\psi_{6, 12}(t)$

3.30   Plot each of the following functions.

(a) $f_1(t) = 3\psi(t/8 - 2) - 5\psi(t/8 - 1) + \psi(t/8) + 2\psi(t/8 + 1)$

(b) $f_2(t) = \sum_{k=1}^{4} k\psi(16t - k)$

(c) $f_3(t) = \sum_{k=-3}^{3} \cos(\pi k)\psi(t/2 - k)$

(d) $f_4(t) = \sum_{k=0}^{\infty} \langle e^{-t}, \psi(2t - k) \rangle \, \psi(2t - k)$

3.31   We learned in Proposition 3.7 that the Haar spaces V_j are nested. That is, $V_j \subseteq V_{j+1}$ for $j \in \mathbb{Z}$. Is the same true for the W_j?

3.32   Let $j \in \mathbb{Z}$ and let W_j be defined by (3.38). Show that the set $\{\psi(2^j t - k)\}_{k \in \mathbb{Z}}$ is a linear independent set in W_j.

3.33   Show that W_j defined in (3.38) satisfies the properties in Definition 1.10 and is thus a subspace of $L^2(\mathbb{R})$.

3.34   Use u-substitutions to show that

(a) $\langle \psi(2^j t - k), \psi(2^j t - \ell) \rangle = 0$ for $k \neq \ell$

(b) $\|\psi(2^j t - k)\| = 2^{-j/2}$

3.35   Prove Proposition 3.20. (*Hint:* The proof will utilize Proposition 3.19 and be similar to the proof of Proposition 3.18.)

3.36   Prove Proposition 3.23. (*Hint:* Your proof will be similar to that given for Proposition 3.22.)

3.37   Show that the residual function defined in Proposition 3.25 is the projection of $f_{j+1}(t)$ into W_j.

3.38   Prove Proposition 3.26.

3.5   DECOMPOSITION AND RECONSTRUCTION

We learned in Section 3.4 that we can decompose $f_{j+1}(t) \in V_{j+1}$ into an approximation function $f_j(t) \in V_j$ and a detail function $g_j(t) \in W_j$. In this section we learn that this decomposition step can be performed multiple times. We also develop a method for recovering the $f_{j+1} \in V_{j+1}$ from the functions used to create the general decomposition.

Decomposition

In this section we learn how to iterate the decomposition process. The key is repeated use of (3.53). For purposes of illustration, consider the space V_5. Using Proposition 3.26, we can write

$$V_5 = V_4 \oplus W_4$$

Now apply (3.53) to V_4:

$$\begin{aligned} V_5 &= V_4 \oplus W_4 \\ &= V_3 \oplus W_3 \oplus W_4 \end{aligned}$$

Thus we can write a function $f_5(t) \in V_5$ in terms of a coarse approximation $f_3(t) \in V_3$ and two residual functions in W_3, W_4. We can also continue this iterative process as often as desired. For example, if we wanted an approximation of $f_5(t)$ from V_0, we would write

$$\begin{aligned} V_5 &= V_4 \oplus W_4 \\ &= V_3 \oplus W_3 \oplus W_4 \\ &= V_2 \oplus W_2 \oplus W_3 \oplus W_4 \\ &= V_1 \oplus W_1 \oplus W_2 \oplus W_3 \oplus W_4 \\ &= V_0 \oplus W_0 \oplus W_1 \oplus W_2 \oplus W_3 \oplus W_4 \end{aligned}$$

Let's look at an example.

Example 3.9 (Decomposing a Function $f_3(t) \in V_3$) *Suppose that $f_3(t) \in V_3$ is given by*

$$
\begin{aligned}
f_3(t) &= \sum_{k=0}^{7} a_k \phi_{3,\,k}(t) \\
&= 3\phi_{3,\,0}(t) + \phi_{3,\,1}(t) - 2\phi_{3,\,2}(t) \\
&\quad + 4\phi_{3,\,3}(t) + 5\phi_{3,\,4}(t) + \phi_{3,\,5} - 2\phi_{3,\,6}(t) - 4\phi_{3,\,7}(t)
\end{aligned}
\tag{3.55}
$$

The function $f_3(t)$ is plotted in Figure 3.19.

Remember that $\phi_{3,\,k}(t) = 2^{\frac{3}{2}} \phi(8t - k)$, so the coefficients are each multiplied by $2^{3/2}$. In this example, we decompose $f_3(t)$ into an approximation $f_0(t)$ from V_0 and detail functions $g_0(t), g_1(t), g_2(t)$ from W_0, W_1, W_2, respectively.

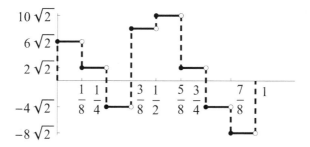

Figure 3.19 The function $f_3(t) \in V_3$.

We know that $V_3 = V_2 \oplus W_2$. So $f_3(t) = f_2(t) + g_2(t)$, where $f_2(t) \in V_2$ and $g_2(t) \in W_2$. Moreover, $f_2(t)$ and $g_2(t)$ are projections of $f_3(t)$ into V_2 and W_2, respectively. Thus, to obtain formulas for these projections, we use Proposition 3.24 to construct $f_2(t)$ and Proposition 3.25 to build $g_2(t)$. Since only $a_0, \ldots, a_7$ are nonzero in (3.55), formula (3.48) tells us that the only nonzero coefficients in

$$
f_2(t) = \sum_{k \in \mathbb{Z}} b_k \phi_{2,\,k}(t)
$$

are

$$
\begin{aligned}
b_0 &= \frac{\sqrt{2}}{2}(a_0 + a_1) = \frac{\sqrt{2}}{2}(3 + 1) &= 2\sqrt{2} \\
b_1 &= \frac{\sqrt{2}}{2}(a_2 + a_3) = \frac{\sqrt{2}}{2}(-2 + 4) &= \sqrt{2} \\
b_2 &= \frac{\sqrt{2}}{2}(a_4 + a_5) = \frac{\sqrt{2}}{2}(5 + 1) &= 3\sqrt{2} \\
b_3 &= \frac{\sqrt{2}}{2}(a_6 + a_7) = \frac{\sqrt{2}}{2}(-2 - 4) &= -3\sqrt{2}
\end{aligned}
$$

In a similar manner, we can compute the nonzero coefficients c_k for

$$g_2(t) = \sum_{k \in \mathbb{Z}} c_k \psi_{2,k}(t)$$

as

$$c_0 = \frac{\sqrt{2}}{2}(a_0 - a_1) = \frac{\sqrt{2}}{2}(3 - 1) = \sqrt{2}$$

$$c_1 = \frac{\sqrt{2}}{2}(a_2 - a_3) = \frac{\sqrt{2}}{2}(-2 - 4) = -3\sqrt{2}$$

$$c_2 = \frac{\sqrt{2}}{2}(a_4 - a_5) = \frac{\sqrt{2}}{2}(5 - 1) = 2\sqrt{2}$$

$$c_3 = \frac{\sqrt{2}}{2}(a_6 - a_7) = \frac{\sqrt{2}}{2}(-2 + 4) = \sqrt{2}$$

The functions $f_2(t)$ and $g_2(t)$ are plotted in Figure 3.20. Note that on this level, the coefficients obtained for $f_2(t)$ and $g_2(t)$ are multiplied by $2^{2/2} = 2$.

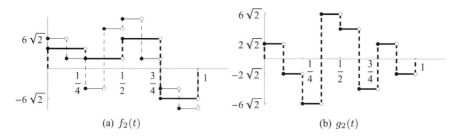

(a) $f_2(t)$ (b) $g_2(t)$

Figure 3.20 The functions $f_2(t) \in V_2$ and $g_2(t) \in W_2$. In Figure 3.20(a) the function $f_3(t)$ is plotted (with thinner lines) as well.

We now use Propositions 3.24 and 3.25 to create $f_1(t) \in V_1$ and $g_1(t) \in W_1$ from $f_2(t) \in V_2$. The nonzero coefficients for $f_1(t)$ are

$$b_0 = \frac{\sqrt{2}}{2}(2\sqrt{2} + \sqrt{2}) = 3, \qquad b_1 = \frac{\sqrt{2}}{2}(3\sqrt{2} - 3\sqrt{2}) = 0$$

and the nonzero coefficients for $g_1(t)$ are

$$c_0 = \frac{\sqrt{2}}{2}(2\sqrt{2} - \sqrt{2}) = 1 \qquad c_1 = \frac{\sqrt{2}}{2}(3\sqrt{2} - (-3)\sqrt{2}) = 6$$

The functions $f_1(t)$ and $g_1(t)$ are plotted in Figure 3.21. Note that on this level, the coefficients obtained for $f_1(t)$ and $g_1(t)$ are multiplied by $2^{1/2} = \sqrt{2}$.
Finally, we project $f_1(t) \in V_1$ into V_0 and W_0. We have

$$b_0 = \frac{\sqrt{2}}{2}(3 + 0) = \frac{3}{2}\sqrt{2} \qquad c_0 = \frac{\sqrt{2}}{2}(3 - 0) = \frac{3}{2}\sqrt{2}$$

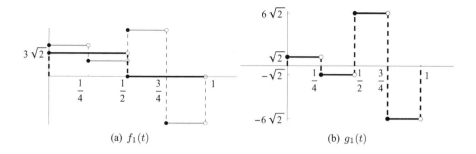

(a) $f_1(t)$

(b) $g_1(t)$

Figure 3.21   The functions $f_1(t) \in V_1$ and $g_1(t) \in W_1$. In Figure 3.21(a) the function $f_2(t)$ is plotted (with thinner lines) as well.

so that $f_0(t) = \frac{3\sqrt{2}}{2} \phi(t)$ *and* $g_0(t) = \frac{3\sqrt{2}}{2} \psi(t)$. *These two projections are plotted in Figure 3.22.*

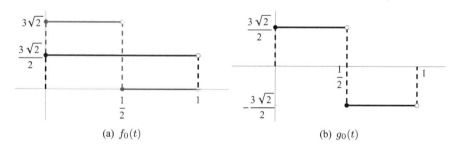

(a) $f_0(t)$

(b) $g_0(t)$

Figure 3.22   The functions $f_0(t) \in V_0$ and $g_0(t) \in W_0$. In Figure 3.22(a), the function $f_1(t)$ is plotted as well.

Thus we can build $f_3(t)$ *using the sum* $f_0(t)+g_0(t)+g_1(t)+g_2(t)$. *In Problem 3.39 you are asked to compute this sum and verify that the result is* $f_3(t)$.   ∎

Reconstruction

In applications we often start with some function $f_{j+L}(t) \in V_{j+L}$ and then perform L decomposition steps so that we can write

$$f_{j+L}(t) = f_j(t) + g_j(t) + g_{j+1}(t) + \cdots + g_{j+L-1}(t)$$

Particular applications dictate what is done with the decomposition or how it is analyzed, but in many cases, we might be given the decomposition and asked to recover $f_{j+L}(t)$. We close this section by developing formulas for performing such a reconstruction.

Suppose that we have $f_j(t) \in V_j$ and $g_j(t) \in W_j$ and that these functions have been obtained by projecting $f_{j+1}(t) \in V_{j+1}$ into V_j and W_j, respectively. Propositions 3.24 and 3.25 tell us that we can write $f_{j+1}(t) = \sum_{k \in \mathbb{Z}} a_k \phi_{j+1, k}(t)$ as

$$f_{j+1}(t) = \sum_{k \in \mathbb{Z}} b_k \phi_{j, k}(t) + c_k \psi_{j, k}(t) \tag{3.56}$$

To reconstruct $f_{j+1}(t)$ from $f_j(t)$ and $g_j(t)$, we need to write the a_k in terms of the b_k and c_k. Propositions 3.24 and 3.25 also give us formulas for computing b_k and c_k. We know that

$$b_k = \frac{\sqrt{2}}{2}(a_{2k} + a_{2k+1}) \quad \text{and} \quad c_k = \frac{\sqrt{2}}{2}(a_{2k} - a_{2k+1})$$

Let's consider a single term in (3.56). We know that both $\phi_{j, k}(t)$ and $\psi_{j, k}(t)$ are nonzero on $\left[\frac{k}{2^j}, \frac{k+1}{2^j}\right)$. In fact, on the half-interval $\left[\frac{k}{2^j}, \frac{k+1/2}{2^j}\right)$, both $\phi_{j, k}(t)$ and $\psi_{j, k}(t)$ assume the value $2^{j/2}$, so simplification of $b_k \phi_{j, k}(t) + c_k \psi_{j, k}(t)$ on $\left[\frac{k}{2^j}, \frac{k+1/2}{2^j}\right)$ gives

$$\begin{aligned} b_k \phi_{j, k}(t) + c_k \psi_{j, k}(t) &= b_k \cdot 2^{j/2} + c_k \cdot 2^{j/2} \\ &= 2^{j/2}(b_k + c_k) \end{aligned} \tag{3.57}$$

As we saw in the proof of Proposition 3.25, the only function at level $j + 1$ with support on the interval $\left[\frac{k}{2^j}, \frac{k+1/2}{2^j}\right)$ is $\phi_{j+1, 2k}(t)$. On this interval, $\phi_{j+1, 2k}(t)$ has value $2^{(j+1)/2}$, so that

$$f_{j+1}(t) = 2^{(j+1)/2} a_{2k} \quad \text{for} \quad t \in \left[\frac{k}{2^j}, \frac{k+1/2}{2^j}\right) \tag{3.58}$$

Comparing (3.57) and (3.58), we see that

$$2^{j/2}(b_k + c_k) = 2^{(j+1)/2} a_{2k}$$

so that $a_{2k} = \frac{\sqrt{2}}{2}(b_k + c_k)$.

In Problem 3.41 you will show that a similar analysis on the right half-interval $\left[\frac{k+1/2}{2^j}, \frac{k+1}{2^j}\right)$ produces the identity $a_{2k+1} = \frac{\sqrt{2}}{2}(b_k - c_k)$. Thus we can reconstruct a_{2k} by adding b_k and c_k and scaling the resulting sum by $\sqrt{2}/2$, and we can obtain a_{2k+1} by subtracting c_k from b_k and again scaling by $\sqrt{2}/2$. These computations are very similar to those used to obtain b_k and c_k in the decomposition process! We summarize the foregoing discussion with the following proposition.

Proposition 3.27 (Recovering $\mathbf{f_{j+1}(t)}$ From $\mathbf{f_j(t)}$ and $\mathbf{g_j(t)}$) *Suppose that*

$$f_{j+1}(t) = \sum_{k \in \mathbb{Z}} a_k \phi_{j+1, k}(t)$$

and further assume that

$$f_j(t) = \sum_{k \in \mathbb{Z}} b_k \phi_{j,k}(t) \quad and \quad g_j(t) = \sum_{k \in \mathbb{Z}} c_k \psi_{j,k}(t)$$

are the projections (given by (3.16)) of $f_{j+1}(t)$ into V_j and W_j, respectively. Then

$$\boxed{a_{2k} = \frac{\sqrt{2}}{2}(b_k + c_k) = \mathbf{h} \cdot \mathbf{d}^k \quad and \quad a_{2k+1} = \frac{\sqrt{2}}{2}(b_k - c_k) = \mathbf{h} \cdot \mathbf{d}^k}$$

where $\mathbf{h}, \mathbf{g}$ are given by (3.29) and (3.31), respectively, and $\mathbf{d}^k = [b_k, c_k]^T$.   ∎

To see the proposition in action, let's return to the decomposition that we obtained in Example 3.9.

Example 3.10 (Reconstructing $\mathbf{f_3(t)} \in \mathbf{V_3}$) *In Example 3.9, we decomposed a function $f_3(t) \in V_3$ into $f_0(t) \in V_0$ and $g_2(t), g_1(t),$ and $g_0(t)$ in $W_2, W_1,$ and W_0, respectively. In particular, we found*

$$f_0(t) = \frac{3\sqrt{2}}{2}\phi(t) \quad and \quad g_0(t) = \frac{3\sqrt{2}}{2}\psi(t)$$

If we add and subtract $b_0 = c_0 = \frac{3\sqrt{2}}{2}$ and scale the results by $\sqrt{2}/2$, we obtain 3 and 0. These are exactly the coefficients of the function $f_1(t)$. Now if we take $b_0 = 3$ and $b_1 = 0$ and combine them with the coefficients $c_0 = 1$ and $c_1 = 6$ of $g_1(t)$, we have

$$\frac{\sqrt{2}}{2}(b_0 + c_0) = \frac{\sqrt{2}}{2}(3 + 1) = 2\sqrt{2}$$

$$\frac{\sqrt{2}}{2}(b_0 - c_0) = \frac{\sqrt{2}}{2}(3 - 1) = \sqrt{2}$$

$$\frac{\sqrt{2}}{2}(b_1 + c_1) = \frac{\sqrt{2}}{2}(0 + 6) = 3\sqrt{2}$$

$$\frac{\sqrt{2}}{2}(b_1 - c_1) = \frac{\sqrt{2}}{2}(0 - 6) = -3\sqrt{2}$$

Note that these numbers are the coefficients of $f_2(t) \in V_2$. If we combine these numbers with the coefficients of $g_2(t) \in W_2$ we obtain

$$\frac{\sqrt{2}}{2}(b_0 + c_0) = \frac{\sqrt{2}}{2}(2\sqrt{2} + \sqrt{2}) \quad = \quad 3$$

$$\frac{\sqrt{2}}{2}(b_0 - c_0) = \frac{\sqrt{2}}{2}(2\sqrt{2} - \sqrt{2}) \quad = \quad 1$$

$$\frac{\sqrt{2}}{2}(b_1 + c_1) = \frac{\sqrt{2}}{2}\left(\sqrt{2} + (-3\sqrt{2})\right) = -2$$

$$\frac{\sqrt{2}}{2}(b_1 - c_1) = \frac{\sqrt{2}}{2}\left(\sqrt{2} - (-3\sqrt{2})\right) = \quad 4$$

$$\frac{\sqrt{2}}{2}(b_2 + c_2) = \frac{\sqrt{2}}{2}(3\sqrt{2} + 2\sqrt{2}) \quad = \quad 5$$

$$\frac{\sqrt{2}}{2}(b_2 - c_2) = \frac{\sqrt{2}}{2}(3\sqrt{2} - 2\sqrt{2}) \quad = \quad 1$$

$$\frac{\sqrt{2}}{2}(b_3 + c_3) = \frac{\sqrt{2}}{2}(-3\sqrt{2} + \sqrt{2}) \quad = -2$$

$$\frac{\sqrt{2}}{2}(b_3 - c_3) = \frac{\sqrt{2}}{2}(-3\sqrt{2} - \sqrt{2}) \quad = -4$$

and these are the coefficients of our original $f_3(t) \in V_3$.   ∎

PROBLEMS

3.39 Compute the sum $f_0(t) + g_0(t) + g_1(t) + g_2(t)$ in Example 3.9 and verify the result is $f_3(t)$. The best way to do this is consider the sum on each of the intervals $\left[k, \frac{k+1}{8}\right)$ for $k = 0, 1, \ldots, 7$. For each of these intervals simply add each term's contribution and verify that the total is a_k.

3.40 For each of the following functions in V_2, decompose them (see Example 3.9) into functions $f_0(t)$, $g_0(t)$, and $g_1(t)$ from V_0, W_0, and W_1, respectively.

(a) $f_2(t) = 3\phi_{2,0}(t) - 4\phi_{2,1}(t) + 2\phi_{2,4}(t) - 2\phi_{2,5}(t)$

(b) $f_2(t) = \sum_{k=0}^{7} \phi_{2,k}(t)$

(c) $f_2(t) = \sum_{k=0}^{7} (k+1)\phi_{2,k}(t)$

3.41 Complete the proof (given in the discussion preceding the proposition) of Proposition 3.27 by showing that $a_{2k+1} = \frac{\sqrt{2}}{2}(b_k - c_k)$. (*Hint:* Consider the contributions of both $b_k\phi_{j,k}(t) + c_k\psi_{j,k}(t)$ and $f_{j+1}(t)$ on the half–interval $\left[\frac{k+1/2}{2^j}, \frac{k+1}{2^j}\right)$. Equate these contributions and simplify to obtain the desired result.)

3.42   For each of the functions in Problem 3.40, use Proposition 3.27 and the ideas of Example 3.10 to reproduce $f_2(t)$ from the functions $f_0(t)$, $g_0(t)$, and $g_1(t)$.

3.6 SUMMARY

It is worthwhile before proceeding to the next chapter to summarize all the ideas that we have covered in this chapter. They are very important and we use them to build a general framework for constructing scaling functions, wavelet functions, and even discrete wavelet transforms.

The fundamental tool in all we constructed in this chapter was the box function $\phi(t)$ given by (3.1). This function and its integer translates constitute an orthonormal basis for the space V_0. We learned that $V_0 \in L^2(\mathbb{R})$ is the space of all piecewise constant functions with possible breakpoints at the integers.

The Approximation Spaces V_j

We took the idea of spaces of piecewise constant functions a bit further by defining the *approximation space* V_j, $j \in \mathbb{Z}$. Recall that the space $V_j \in L^2(\mathbb{R})$ is the space of piecewise constant functions with possible breakpoints at $2^{-j}\mathbb{Z}$.

An orthonormal basis for V_j is the set $\{\phi_{j,k}(t)\}_{k \in \mathbb{Z}}$ where

$$\phi_{j,k}(t) = 2^{j/2}\phi(2^j t - k)$$

We learned that the V_j spaces are *nested*. That is,

$$\cdots \subset V_{-2} \subset V_{-1} \subset V_0 \subset V_1 \subset V_2 \subset \cdots$$

and the closure of the union of the V_j spaces is the space $L^2(\mathbb{R})$. Furthermore, the only function in the intersection of each set is the function $f(t) = 0$.

We are able to move between V_j spaces as follows:

$$f(t) \in V_j \iff f(2t) \in V_{j+1}$$

The Scaling Function $\phi(t)$ and Dilation Equation

The reason we are able to move between spaces in this manner is due the fact that the function $\phi(t)$ satisfies the *dilation equation*

$$\phi(t) = \frac{\sqrt{2}}{2}\phi_{1,0}(t) + \frac{\sqrt{2}}{2}\phi_{1,1}(t)$$

For this reason, we call $\phi(t)$ a *scaling function*.

Note also that coefficients $h_0 = h_1 = \frac{\sqrt{2}}{2}$ are exactly the same numbers as those used in the averages portion of the discrete Haar wavelet transform.

Projections

We learned that if we wish to project

$$f_{j+1}(t) = \sum_{k \in \mathbb{Z}} a_k \phi_{j+1,\,k}(t)$$

into V_j, the result was

$$f_j(t) = \sum_{k \in \mathbb{Z}} b_k \phi_{j+1,\,k}(t)$$

where

$$b_k = \frac{\sqrt{2}}{2} a_{2k} + \frac{\sqrt{2}}{2} a_{2k+1}$$

The Wavelet Spaces W_j

We also looked at the residual function $g_j(t) = f_{j+1}(t) - f_j(t)$. We learned that this function is a member of the *wavelet space* W_j. We showed that the spaces W_j and W_ℓ are orthogonal to each other (when $j \neq \ell$) and that an orthonormal basis for W_j is the set $\{\psi_{j,\,k}(t)\}_{k \in \mathbb{Z}}$ where

$$\psi_{j,\,k}(t) = 2^{j/2} \psi(2^j t - k)$$

The Wavelet Function $\psi(t)$ and the Dilation Equation

We call the function $\psi(t)$ (see 3.25) a *wavelet function*. The function $\psi(t)$ also satisfies a dilation equation:

$$\psi(t) = \frac{\sqrt{2}}{2} \phi_{j,\,0}(t) - \frac{\sqrt{2}}{2} \phi_{j,\,1}(t)$$

As was the case with the scaling function for $\phi(t)$, the coefficients $g_0 = \frac{\sqrt{2}}{2}$ and $g_1 = -\frac{\sqrt{2}}{2}$ are exactly the same numbers as those used in the differences portion of the discrete Haar wavelet transformation.

Projections

We learned that if we wish to project

$$f_{j+1}(t) = \sum_{k \in \mathbb{Z}} a_k \phi_{j+1,\,k}(t)$$

into W_j, the result was

$$g_j(t) = \sum_{k \in \mathbb{Z}} c_k \phi_{j+1,\,k}(t)$$

where

$$c_k = \frac{\sqrt{2}}{2} a_{2k} - \frac{\sqrt{2}}{2} a_{2k+1}$$

Direct Orthogonal Sums

There is a further connection with regard to the spaces V_{j+1}, V_j, and W_j. We learned that V_j and W_j are orthogonal and the direct orthogonal sum of the two spaces is

$$V_{j+1} = V_j \oplus W_j$$

In Section 3.5 we learned that we could "iterate" the decomposition process. That is, if $f_{j+1}(t) \in V_j$, then we could use (3.53) repeatedly and decompose $f_{j+1}(t)$ as

$$f_{j+1}(t) = f_0(t) + g_0(t) + g_1(t) + \cdots + g_j(t)$$

where $f_0 \in V_0$, and $g_\ell(t) \in W_\ell, \ell = 0, \ldots, j$.

We learned to reconstruct $f_{j+1}(t)$ from these pieces. That is, combining $f_0(t)$ and $g_0(t)$ gives $f_1(t)$. We can in turn use $f_1(t)$ with $g_1(t)$ to reconstruct $f_2(t)$ and continue this process until $f_{j+1}(t)$ is recovered.

Conclusion

We will learn in Chapter 4 that for functions $f_{j+1}(t)$ composed of finitely many $a_k \neq 0$, a matrix transformation can be used to perform the projection of $f_{j+1}(t)$ into V_j and W_j. This transformation is the discrete Haar wavelet transformation and can be used in applications such as signal or image processing.

It always helps when moving through abstract mathematical theory to have an example that helps you understand new results. In the next chapter we study scaling functions, wavelet functions, and the spaces V_j and W_j in a general setting. All the properties we derive for the general case apply to the Haar scaling function $\phi(t)$ and wavelet $\psi(t)$ as well as the Haar spaces V_j and the Haar wavelet spaces W_j. Keep this example in mind as you move through the material in the remainder of the book.

CHAPTER 4

THE DISCRETE HAAR WAVELET TRANSFORM AND APPLICATIONS

In previous sections we developed the ability to take a piecewise constant function $f_j(t), j > 0$, with possible breakpoints at points in $2^j \mathbb{Z}$ and learned how to decompose it into a piecewise constant approximation function $f_{j-1}(t)$ with possible breakpoints on the coarser grid $2^{j-1}\mathbb{Z}$ and a piecewise constant detail function $g_{j-1}(t)$ with possible breakpoints at points in $2^j \mathbb{Z}$. We can iterate the process and ultimately write $f_j(t)$ as the sum of an approximation function $f_0(t)$ with possible breakpoints at the integers and detail functions $g_0(t), g_1(t), \ldots, g_{j-1}(t)$ with possible breakpoints at $\frac{1}{2}\mathbb{Z}, \frac{1}{4}\mathbb{Z}, \ldots, 2^j \mathbb{Z}$, respectively. We also learned that the vectors

$$\mathbf{h} = [h_0, h_1]^T = \left[\frac{\sqrt{2}}{2}, \frac{\sqrt{2}}{2} \right]^T$$

and

$$\mathbf{g} = [g_0, g_1]^T = \left[\frac{\sqrt{2}}{2}, -\frac{\sqrt{2}}{2} \right]^T$$

play important roles in this decomposition. In fact, to move from level j to the next coarser level $j-1$, we learned from Propositions 3.24 and 3.25 that we simply need to form inner products of two-vectors built from the expansion of $f_j(t)$ in $V_j(t)$ with the vectors **h** and **g**. While the process allows us to decompose functions in subspaces of $L^2(\mathbb{R})$, the computations for doing so are *discrete*. Thus it is natural to wonder if it possible to model the decomposition in terms of linear transformations (matrices). Moreover, since digital signals and images are composed of discrete data, we will need a discrete analog of the decomposition algorithm so that we can process signal and image data. In this section we learn how to process discrete data via a discrete version of the decomposition process outlined in Propositions 3.24 and 3.25. The resulting transformation is called the *discrete Haar wavelet transformation*.

4.1   THE ONE-DIMENSIONAL TRANSFORM

In this section we introduce the one-dimensional discrete Haar wavelet transformation. We will also construct the inverse transformation and develop an iterative implementation of both the discrete Haar wavelet transformation and its inverse. We also introduce the *cumulative energy vector* and *entropy*. Both of these measures provide us with some understanding of the effectiveness of the transformation. These measures are used again in Section 4.3, where we look at applications of the discrete Haar wavelet transformation.

We motivate the construction of the discrete Haar wavelet transformation by returning to the decomposition of $f_3(t) \in V_3$ in Example 3.9.

Example 4.1 (Motivating the Discrete Haar Wavelet Transformation) *The function $f_3(t) \in V_3$ from Example 3.9 is defined by (3.55) so that $a_k = 0$ for $k < 0, k \geq 8$ and*

$$a_0 = 3, \quad a_1 = 1, \quad a_2 = -2, \quad a_3 = 4$$
$$a_4 = 5, \quad a_5 = 1, \quad a_6 = 4, \quad a_7 = -4$$

To project $f_3(t)$ into V_2, we use Proposition 3.24 to write

$$b_0 = \mathbf{h} \cdot \mathbf{a}^0 = [h_0, h_1] \cdot \begin{bmatrix} a_0 \\ a_1 \end{bmatrix} = \left[\frac{\sqrt{2}}{2}, \frac{\sqrt{2}}{2}\right] \cdot \begin{bmatrix} 3 \\ 1 \end{bmatrix} = 2\sqrt{2}$$

$$b_1 = \mathbf{h} \cdot \mathbf{a}^1 = [h_0, h_1] \cdot \begin{bmatrix} a_2 \\ a_3 \end{bmatrix} = \left[\frac{\sqrt{2}}{2}, \frac{\sqrt{2}}{2}\right] \cdot \begin{bmatrix} -2 \\ 4 \end{bmatrix} = \sqrt{2}$$

$$b_2 = \mathbf{h} \cdot \mathbf{a}^2 = [h_0, h_1] \cdot \begin{bmatrix} a_4 \\ a_5 \end{bmatrix} = \left[\frac{\sqrt{2}}{2}, \frac{\sqrt{2}}{2}\right] \cdot \begin{bmatrix} 5 \\ 1 \end{bmatrix} = 3\sqrt{2}$$

$$b_3 = \mathbf{h} \cdot \mathbf{a}^3 = [h_0, h_1] \cdot \begin{bmatrix} a_6 \\ a_7 \end{bmatrix} = \left[\frac{\sqrt{2}}{2}, \frac{\sqrt{2}}{2}\right] \cdot \begin{bmatrix} 4 \\ -4 \end{bmatrix} = 0$$

We could easily formulate the computations above as a matrix product. Indeed, we can write $H_4\mathbf{a} = \mathbf{b}$ *or*

$$
\begin{bmatrix}
h_0 & h_1 & 0 & 0 & 0 & 0 & 0 & 0 \\
0 & 0 & h_0 & h_1 & 0 & 0 & 0 & 0 \\
0 & 0 & 0 & 0 & h_0 & h_1 & 0 & 0 \\
0 & 0 & 0 & 0 & 0 & 0 & h_0 & h_1
\end{bmatrix}
\cdot
\begin{bmatrix}
a_0 \\ a_1 \\ a_2 \\ a_3 \\ a_4 \\ a_5 \\ a_6 \\ a_7
\end{bmatrix}
=
\begin{bmatrix}
b_0 \\ b_1 \\ b_2 \\ b_4
\end{bmatrix}
$$

$$
\begin{bmatrix}
\frac{\sqrt{2}}{2} & \frac{\sqrt{2}}{2} & 0 & 0 & 0 & 0 & 0 & 0 \\
0 & 0 & \frac{\sqrt{2}}{2} & \frac{\sqrt{2}}{2} & 0 & 0 & 0 & 0 \\
0 & 0 & 0 & 0 & \frac{\sqrt{2}}{2} & \frac{\sqrt{2}}{2} & 0 & 0 \\
0 & 0 & 0 & 0 & 0 & 0 & \frac{\sqrt{2}}{2} & \frac{\sqrt{2}}{2}
\end{bmatrix}
\cdot
\begin{bmatrix}
3 \\ 1 \\ -2 \\ 4 \\ 5 \\ 1 \\ 4 \\ -4
\end{bmatrix}
=
\begin{bmatrix}
2\sqrt{2} \\ \sqrt{2} \\ 3\sqrt{2} \\ 0
\end{bmatrix}
\qquad (4.1)
$$

In a similar manner, we can compute the detail coefficients c_0, c_1, c_2, c_3 *using the matrix equation* $G_4\mathbf{a} = \mathbf{c}$ *or*

$$
\begin{bmatrix}
g_0 & g_1 & 0 & 0 & 0 & 0 & 0 & 0 \\
0 & 0 & g_0 & g_1 & 0 & 0 & 0 & 0 \\
0 & 0 & 0 & 0 & g_0 & g_1 & 0 & 0 \\
0 & 0 & 0 & 0 & 0 & 0 & g_0 & g_1
\end{bmatrix}
\cdot
\begin{bmatrix}
a_0 \\ a_1 \\ a_2 \\ a_3 \\ a_4 \\ a_5 \\ a_6 \\ a_7
\end{bmatrix}
=
\begin{bmatrix}
c_0 \\ c_1 \\ c_2 \\ c_4
\end{bmatrix}
$$

$$
\begin{bmatrix}
\frac{\sqrt{2}}{2} & -\frac{\sqrt{2}}{2} & 0 & 0 & 0 & 0 & 0 & 0 \\
0 & 0 & \frac{\sqrt{2}}{2} & -\frac{\sqrt{2}}{2} & 0 & 0 & 0 & 0 \\
0 & 0 & 0 & 0 & \frac{\sqrt{2}}{2} & -\frac{\sqrt{2}}{2} & 0 & 0 \\
0 & 0 & 0 & 0 & 0 & 0 & \frac{\sqrt{2}}{2} & -\frac{\sqrt{2}}{2}
\end{bmatrix}
\cdot
\begin{bmatrix}
3 \\ 1 \\ -2 \\ 4 \\ 5 \\ 1 \\ 4 \\ -4
\end{bmatrix}
=
\begin{bmatrix}
\sqrt{2} \\ -3\sqrt{2} \\ 2\sqrt{2} \\ 4\sqrt{2}
\end{bmatrix}
$$

$$(4.2)$$

We can concatenate the matrix equations in (4.1) and (4.2) to write $W_8 \mathbf{a} = \begin{bmatrix} \mathbf{b} \\ \mathbf{c} \end{bmatrix}$

where

$$
W_8 = \begin{bmatrix} H_4 \\ \hline G_4 \end{bmatrix}
$$

$$
= \left[\begin{array}{cccccccc}
h_0 & h_1 & 0 & 0 & 0 & 0 & 0 & 0 \\
0 & 0 & h_0 & h_1 & 0 & 0 & 0 & 0 \\
0 & 0 & 0 & 0 & h_0 & h_1 & 0 & 0 \\
0 & 0 & 0 & 0 & 0 & 0 & h_0 & h_1 \\
\hline
g_0 & g_1 & 0 & 0 & 0 & 0 & 0 & 0 \\
0 & 0 & g_0 & g_1 & 0 & 0 & 0 & 0 \\
0 & 0 & 0 & 0 & g_0 & g_1 & 0 & 0 \\
0 & 0 & 0 & 0 & 0 & 0 & g_0 & g_1
\end{array} \right]
$$

$$
= \left[\begin{array}{cccccccc}
\frac{\sqrt{2}}{2} & \frac{\sqrt{2}}{2} & 0 & 0 & 0 & 0 & 0 & 0 \\
0 & 0 & \frac{\sqrt{2}}{2} & \frac{\sqrt{2}}{2} & 0 & 0 & 0 & 0 \\
0 & 0 & 0 & 0 & \frac{\sqrt{2}}{2} & \frac{\sqrt{2}}{2} & 0 & 0 \\
0 & 0 & 0 & 0 & 0 & 0 & \frac{\sqrt{2}}{2} & \frac{\sqrt{2}}{2} \\
\hline
\frac{\sqrt{2}}{2} & -\frac{\sqrt{2}}{2} & 0 & 0 & 0 & 0 & 0 & 0 \\
0 & 0 & \frac{\sqrt{2}}{2} & -\frac{\sqrt{2}}{2} & 0 & 0 & 0 & 0 \\
0 & 0 & 0 & 0 & \frac{\sqrt{2}}{2} & -\frac{\sqrt{2}}{2} & 0 & 0 \\
0 & 0 & 0 & 0 & 0 & 0 & \frac{\sqrt{2}}{2} & -\frac{\sqrt{2}}{2}
\end{array} \right] \qquad (4.3)
$$

∎

Let's have a closer look at the matrix W_8 in (4.3). First, note that the inner product of the first row with itself is

$$
\left(\frac{\sqrt{2}}{2} \right)^2 + \left(\frac{\sqrt{2}}{2} \right)^2 + 0^2 + \cdots + 0^2 = 1
$$

The inner product of the first row with any other row is zero — it is trivially zero for rows two to four and rows six to eight, and it easy to check that the inner product of rows one and five is zero as well. As a matter of fact, it is easy to see that *any* row $\mathbf{w}^k$, $k = 1, \ldots, 8$ in W_8 satisfies $\|\mathbf{w}^k\| = 1$ and $\mathbf{w}^k \cdot \mathbf{w}^j = 0, j \neq k$. Thus W_8 is an *orthogonal matrix* so that $W_8^{-1} = W_8^T$.

If we form the vector $\mathbf{y} = \begin{bmatrix} \mathbf{b} \\ \mathbf{c} \end{bmatrix}$, where $\mathbf{b}$ and $\mathbf{c}$ are defined in Example 4.1, we see that

$$W_8^T \mathbf{y} = W_8^T \begin{bmatrix} \mathbf{b} \\ \mathbf{c} \end{bmatrix}$$

$$= \begin{bmatrix} \frac{\sqrt{2}}{2} & 0 & 0 & 0 & \frac{\sqrt{2}}{2} & 0 & 0 & 0 \\ \frac{\sqrt{2}}{2} & 0 & 0 & 0 & -\frac{\sqrt{2}}{2} & 0 & 0 & 0 \\ 0 & \frac{\sqrt{2}}{2} & 0 & 0 & 0 & \frac{\sqrt{2}}{2} & 0 & 0 \\ 0 & \frac{\sqrt{2}}{2} & 0 & 0 & 0 & -\frac{\sqrt{2}}{2} & 0 & 0 \\ 0 & 0 & \frac{\sqrt{2}}{2} & 0 & 0 & 0 & \frac{\sqrt{2}}{2} & 0 \\ 0 & 0 & \frac{\sqrt{2}}{2} & 0 & 0 & 0 & -\frac{\sqrt{2}}{2} & 0 \\ 0 & 0 & 0 & \frac{\sqrt{2}}{2} & 0 & 0 & 0 & \frac{\sqrt{2}}{2} \\ 0 & 0 & 0 & \frac{\sqrt{2}}{2} & 0 & 0 & 0 & -\frac{\sqrt{2}}{2} \end{bmatrix} \cdot \begin{bmatrix} b_0 \\ b_1 \\ b_2 \\ b_3 \\ c_0 \\ c_1 \\ c_2 \\ c_3 \end{bmatrix}$$

$$= \begin{bmatrix} \frac{\sqrt{2}}{2} b_0 + \frac{\sqrt{2}}{2} c_0 \\ \frac{\sqrt{2}}{2} b_0 - \frac{\sqrt{2}}{2} c_0 \\ \frac{\sqrt{2}}{2} b_1 + \frac{\sqrt{2}}{2} c_1 \\ \frac{\sqrt{2}}{2} b_1 - \frac{\sqrt{2}}{2} c_1 \\ \frac{\sqrt{2}}{2} b_2 + \frac{\sqrt{2}}{2} c_2 \\ \frac{\sqrt{2}}{2} b_2 - \frac{\sqrt{2}}{2} c_2 \\ \frac{\sqrt{2}}{2} b_3 + \frac{\sqrt{2}}{2} c_3 \\ \frac{\sqrt{2}}{2} b_3 - \frac{\sqrt{2}}{2} c_3 \end{bmatrix} = \begin{bmatrix} \frac{\sqrt{2}}{2} \cdot (2\sqrt{2}) + \frac{\sqrt{2}}{2} \cdot (\sqrt{2}) \\ \frac{\sqrt{2}}{2} \cdot (2\sqrt{2}) - \frac{\sqrt{2}}{2} \cdot (\sqrt{2}) \\ \frac{\sqrt{2}}{2} \cdot (\sqrt{2}) + \frac{\sqrt{2}}{2} \cdot (-3\sqrt{2}) \\ \frac{\sqrt{2}}{2} \cdot (\sqrt{2}) - \frac{\sqrt{2}}{2} \cdot (-3\sqrt{2}) \\ \frac{\sqrt{2}}{2} \cdot (3\sqrt{2}) + \frac{\sqrt{2}}{2} \cdot (2\sqrt{2}) \\ \frac{\sqrt{2}}{2} \cdot (3\sqrt{2}) - \frac{\sqrt{2}}{2} \cdot (2\sqrt{2}) \\ \frac{\sqrt{2}}{2} \cdot (0) + \frac{\sqrt{2}}{2} \cdot (4\sqrt{2}) \\ \frac{\sqrt{2}}{2} \cdot (0) - \frac{\sqrt{2}}{2} \cdot (4\sqrt{2}) \end{bmatrix} = \begin{bmatrix} 3 \\ 1 \\ -2 \\ 4 \\ 5 \\ 1 \\ 4 \\ -4 \end{bmatrix} = \begin{bmatrix} a_0 \\ a_1 \\ a_2 \\ a_3 \\ a_4 \\ a_5 \\ a_6 \\ a_7 \end{bmatrix}$$

So the matrix W_8 gives us a way to take a finite-length vector and decompose it using formulas (3.48) and (3.49). Since W_8 is orthogonal, we can use W_8^T to recover the original vector. We are now ready to define the discrete Haar wavelet transformation matrix.

Definition 4.1 (The Discrete Haar Wavelet Transformation) *Suppose that N is an even positive integer. We define the* discrete Haar wavelet transformation *as*

$$W_N = \begin{bmatrix} H_{N/2} \\ \hline G_{N/2} \end{bmatrix} = \begin{bmatrix} \frac{\sqrt{2}}{2} & \frac{\sqrt{2}}{2} & 0 & 0 & & 0 & 0 \\ 0 & 0 & \frac{\sqrt{2}}{2} & \frac{\sqrt{2}}{2} & & 0 & 0 \\ \vdots & & & & \ddots & & \vdots \\ 0 & 0 & 0 & 0 & \cdots & \frac{\sqrt{2}}{2} & \frac{\sqrt{2}}{2} \\ \hline \frac{\sqrt{2}}{2} & -\frac{\sqrt{2}}{2} & 0 & 0 & & 0 & 0 \\ 0 & 0 & \frac{\sqrt{2}}{2} & -\frac{\sqrt{2}}{2} & & 0 & 0 \\ \vdots & & & & \ddots & & \vdots \\ 0 & 0 & 0 & 0 & \cdots & \frac{\sqrt{2}}{2} & -\frac{\sqrt{2}}{2} \end{bmatrix} \qquad (4.4)$$

The $\frac{N}{2} \times N$ block $H_{N/2}$ is called the averages *block and the $\frac{N}{2} \times N$ block $G_{N/2}$ is called the* details *block.* ∎

If we apply $H_{N/2}$ to vector $\mathbf{a}$, we can see why it is called the averages block. We have

$$
H\mathbf{a} = \begin{bmatrix} \frac{\sqrt{2}}{2} & \frac{\sqrt{2}}{2} & 0 & 0 & & 0 & 0 \\ 0 & 0 & \frac{\sqrt{2}}{2} & \frac{\sqrt{2}}{2} & & 0 & 0 \\ \vdots & & & & \ddots & & \vdots \\ 0 & 0 & 0 & 0 & \cdots & \frac{\sqrt{2}}{2} & \frac{\sqrt{2}}{2} \end{bmatrix} \cdot \begin{bmatrix} a_0 \\ a_1 \\ \vdots \\ a_{N-2} \\ a_{N-1} \end{bmatrix}
$$

$$
= \sqrt{2} \begin{bmatrix} \frac{1}{2} & \frac{1}{2} & 0 & 0 & & 0 & 0 \\ 0 & 0 & \frac{1}{2} & \frac{1}{2} & & 0 & 0 \\ \vdots & & & & \ddots & & \vdots \\ 0 & 0 & 0 & 0 & \cdots & \frac{1}{2} & \frac{1}{2} \end{bmatrix} \cdot \begin{bmatrix} a_0 \\ a_1 \\ \vdots \\ a_{N-2} \\ a_{N-1} \end{bmatrix}
$$

$$
= \sqrt{2} \begin{bmatrix} \frac{a_0+a_1}{2} \\ \frac{a_2+a_3}{2} \\ \vdots \\ \frac{a_{N-2}+a_{N-1}}{2} \end{bmatrix}
$$

and thus see that $H_{N/2}\mathbf{a}$ computes pairwise averages of consecutive values of $\mathbf{a}$ and weights the result by $\sqrt{2}$.

The ability to decompose a signal into averages and details makes discrete wavelet transformations desirable tools in applications such as signal denoising, image compression, and image edge detection. While the discrete Haar wavelet transformation is not typically the wavelet transformation matrix used to process signals and images in applications — we develop "better" scaling and wavelet functions for use in applications in Chapters 6, 8, and 9 — it is, however, a useful tool for understanding how wavelets are used in applications. We consider an application of the discrete Haar wavelet transformation to estimate the noise level of a noisy signal in the next example.

Example 4.2 (The Haar Transform and Noise-Level Estimation) *Use of the discrete wavelet transformation to estimate the noise level of a noisy signal is one application of wavelets to signal processing. Suppose that*

$$\mathbf{y} = \mathbf{v} + \mathbf{e}$$

where $\mathbf{v}$ *is the true signal and* $\mathbf{e}$ *is a noise vector. In practice, we do not know* $\mathbf{v}$ *or* $\mathbf{e}$, *but we are given* $\mathbf{y}$ *and our task is to estimate* $\mathbf{v}$. *One of the steps in this process is to estimate the noise level* σ *that occurs in* $\mathbf{e}$. *For our application,* $\mathbf{e}$ *is composed of independent samples from a normal distribution with mean zero and variance* σ^2. *Such noise is called* Gaussian white noise. *Readers unfamiliar with the normal distribution are encouraged to consult an elementary statistics book such as DeGroot and Schervish [22].*

Suppose that $\mathbf{y} \in \mathbb{R}^N$. *We apply the discrete Haar wavelet transform to obtain*

$$\begin{bmatrix} \mathbf{s} \\ \mathbf{d} \end{bmatrix} = W_N \mathbf{y} = W_N(\mathbf{v} + \mathbf{e}) = W_N \mathbf{v} + W_N \mathbf{e}$$

Here $\mathbf{s} = H_{N/2}\mathbf{y}$ *is the approximation portion of the transform and* $\mathbf{d} = G_{N/2}\mathbf{y}$ *is the detail portion of the transform. Using the fact that* W_N *is an orthogonal matrix, it can be shown (see Problem 9.1 in [60]) that the elements of the vector* $W_N \mathbf{e}$ *are normally distributed with mean zero and variance* σ^2. *As Donoho and Johnstone point out in [24], the bulk of this transformed noise* $W_N \mathbf{e}$ *ends up in the detail portion* $\mathbf{d}$. *Thus the vector* $\mathbf{d}$ *is an excellent candidate for use in estimating* σ.

In 1974, Hampel [32] showed that the median absolute deviation *(MAD)* [15] *of a sample can be used to estimate* σ. *Let* a_{med} *denote the median of samples* $\mathbf{a} = [a_1, \ldots, a_N]^T$. *Then we define the median absolute deviation as*

$$\text{MAD}(\mathbf{a}) = \text{median}(|a_1 - a_{\text{med}}|, |a_2 - a_{\text{med}}|, \ldots, |a_N - a_{\text{med}}|) \tag{4.5}$$

Hampel showed that

$$\text{MAD}(\mathbf{a}) \to 0.6745\sigma$$

as the sample size tends to infinity. Combining this result with the fact that most of the noise in $W_N\mathbf{y}$ *resides in* $\mathbf{d}$ *leads to the following estimate* $\hat{\sigma}$ *to* σ:

$$\hat{\sigma} = \text{MAD}(\mathbf{d})/0.6745 \tag{4.6}$$

We illustrate this wavelet-based method for estimating the noise level of a signal by considering the signal formed by evaluating the heavisine *function and adding some white noise to it. The heavisine function was introduced by Donoho and Johnstone [24] for use in analyzing discrete wavelet transformations in signal-denoising applications. The function is defined as*

$$h(t) = 4\sin(4\pi t) - \text{sgn}(t - 0.3) - \text{sgn}(0.72 - t) \tag{4.7}$$

for $t \in [0,1]$. *Here the* sign *function is defined by*

$$\text{sgn}(t) = \begin{cases} 1, & t > 0 \\ 0, & t = 0 \\ -1, & t < 0 \end{cases}$$

We form $\mathbf{v} \in \mathbb{R}^{2048}$ *using the formula* $v_k = h(k/2048)$, $k = 0, \ldots, 2047$. *The elements of* $\mathbf{v}$ *are plotted in Figure 4.1(a). We next create a* noise *vector* $\mathbf{e}$, *where* e_k, *for* $k = 0, \ldots, 2047$, *are independent samples from a normal distribution with mean zero and variance* σ^2. *For our application, we take* $\sigma = 0.5$. *The* noisy *vector* $\mathbf{y} = \mathbf{v} + \mathbf{e}$ *is plotted in Figure 4.1(b).*

We first compute the discrete Haar wavelet transformation of $\mathbf{y}$. *We use (4.4) for this task. The* 1024-vector $\mathbf{s} = H_{1024}\mathbf{y}$ *(averages) is plotted in Figure 4.2(a) and the detail vector* $\mathbf{d} = G_{1024}\mathbf{y}$ *is plotted in Figure 4.2(b).*

---

[15]In statistics, MAD is a simple measure used to measure variation. See De Groot and Schervish [22] for more details.

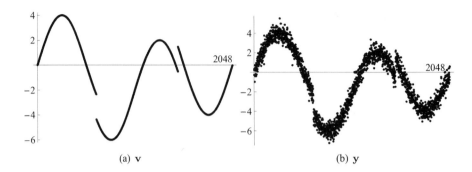

(a) v   (b) y

Figure 4.1   The vector **v** constructed from the heavisine function and the noisy vector **y**.

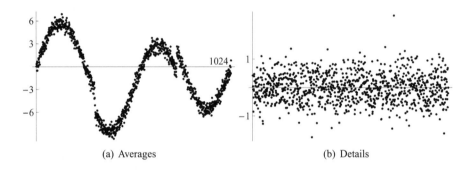

(a) Averages   (b) Details

Figure 4.2   The discrete Haar wavelet transformation — the averages $\mathbf{s} = H\mathbf{y}$ are plotted on the left and the details $\mathbf{d} = G\mathbf{y}$ are plotted on the right.

Using (4.5) we compute the median absolute deviation of $\mathbf{d}$ *and find that* $\mathrm{MAD}(\mathbf{d}) = 0.356043$. *Using Hampel's result (4.6), the noise level* σ *can be estimated by dividing this number by* 0.6745. *We have*

$$\sigma \approx \hat{\sigma} = \frac{0.356043}{0.6745} = 0.527862$$

The absolute error in this approximation is $|\sigma - \hat{\sigma}| = 0.027862$, *and the percentage error is about* 05.57%.

While the error in estimating the noise level is low, we can improve our approximation error by using a "better" discrete wavelet transformation. The filters we develop in Chapters 6 and 8 can be used to actually denoise *the vector* **y**. *For more information on the* wavelet shrinkage method *for signal/image denoising, please Donoho [23] or Chapter 9 of Van Fleet [60].* ∎

Cumulative Energy and Entropy

To determine the effectiveness of discrete wavelet transformations in applications, we introduce two measures. The first, *cumulative energy*, is a vector-valued function that returns information about how energy is stored in the input vector. The second, *entropy*, is used to assess the performance of image compression algorithms.

Definition 4.2 (Cumulative Energy) *Let* $\mathbf{v} \in \mathbb{R}^N$ *with* $\mathbf{v} \neq \mathbf{0}$*. Suppose that* $\mathbf{y}$ *is the vector formed by taking the absolute value of each component of* $\mathbf{v}$ *and sorting the results from largest to smallest. Then we define the* cumulative energy vector $\mathbf{C}(\mathbf{v})$ *as a vector in* $\mathbb{R}^N$ *whose components are given by*

$$\mathbf{C}(\mathbf{v})_k = \sum_{i=1}^{k} \frac{y_i^2}{\|\mathbf{y}\|^2}, \qquad k = 1, 2, \ldots, N \tag{4.8}$$

■

Note that $\|\mathbf{v}\| = \|\mathbf{y}\|$ and if we use the fact that $\|\mathbf{y}\|^2 = y_1^2 + y_2^2 + \cdots + y_N^2$, then we see that the components of $\mathbf{C}(\mathbf{v})$ are

$$\mathbf{C}(\mathbf{v})_1 = \frac{y_1^2}{y_1^2 + y_2^2 + \cdots + y_N^2}$$

$$\mathbf{C}(\mathbf{v})_2 = \frac{y_1^2 + y_2^2}{y_1^2 + y_2^2 + \cdots + y_N^2}$$

$$\vdots$$

$$\mathbf{C}(\mathbf{v})_{N-1} = \frac{y_1^2 + y_2^2 + y_{N-1}^2}{y_1^2 + y_2^2 + \cdots + y_N^2}$$

$$\mathbf{C}(\mathbf{v})_N = \frac{y_1^2 + y_2^2 + \cdots + y_N^2}{y_1^2 + y_2^2 + \cdots + y_N^2} = 1$$

so that $\mathbf{C}(\mathbf{v})_k$ is simply the percentage of the kth largest components (in absolute value) of $\mathbf{v}$ in $\|\mathbf{v}\|$. Note also that $0 \leq \mathbf{C}(\mathbf{v})_k \leq 1$ for all $k = 1, 2, \ldots, N$.

Example 4.3 (Cumulative Energy) *Find the cumulative energy of the following vectors.*

(a) $\mathbf{u} = [1, 2, 3, 4, 5, 6, 7, 8]^T$

(b) $\mathbf{v} = [1, 1, 1, 1, 1, 1, 1, 1]^T$

(c) $\mathbf{w} = \left[\sqrt{2}, \sqrt{2}, \sqrt{2}, \sqrt{2}, 0, 0, 0, 0\right]^T$

Solution

For part (a) we have $\|\mathbf{u}\|^2 = 1^2 + \cdots + 8^2 = 204$. *The components of* $\mathbf{C}(\mathbf{u})$ *are*

$$\mathbf{C}(\mathbf{u})_k = \sum_{i=0}^{k-1} \frac{(8-i)^2}{204}, \quad k = 1, \dots, 8$$

so that

$$\mathbf{C}(\mathbf{u}) = \left[\frac{64}{204}, \frac{113}{204}, \frac{149}{204}, \frac{174}{204}, \frac{190}{204}, \frac{199}{204}, \frac{203}{204}, 1 \right]^T$$

For part (b) note that $\|\mathbf{v}\|^2 = 8$ *and*

$$\mathbf{C}(\mathbf{v})_k = \sum_{i=1}^{k} \frac{1}{8} = \frac{k}{8}, \quad k = 1, \dots, 8$$

so that

$$\mathbf{C}(\mathbf{v}) = \left[\frac{1}{8}, \frac{2}{8}, \frac{3}{8}, \frac{4}{8}, \frac{5}{8}, \frac{6}{8}, \frac{7}{8}, 1 \right]^T$$

For part (c) note that $\mathbf{w}$ *is obtained by applying the discrete Haar wavelet transformation to* $\mathbf{v}$. *Since* W_8 *is orthogonal, we know that* $\|\mathbf{w}\|^2 = \|\mathbf{v}\|^2 = 8$. *The cumulative energy vector is*

$$\mathbf{C}(\mathbf{w}) = \left[\frac{2}{8}, \frac{4}{8}, \frac{6}{8}, 1, 1, 1, 1, 1 \right]^T$$

∎

To better understand how we will use cumulative energy, look at the plot in Figure 4.3. We can easily retrieve $\mathbf{w}$ from $\mathbf{v}$ since $\mathbf{v} = W_8^T \mathbf{w}$, but as we can see from the plot, all of the energy of $\mathbf{w}$ is stored in the four largest (in absolute value) components whereas the energy is uniformly distributed among the elements in $\mathbf{v}$. Thus the cumulative energy vector can be viewed as an indicator of which components might be important contributors to a signal. In applications such as image compression, those components not deemed important can be converted to zero.

Whereas cumulative energy returns a vector that indicates how the energy of a vector is distributed, *entropy* tells us the amount of information on average held by each unit of measure. In the case of digital images comprised of pixels, the unit of measure is typically *bits*. Thus the entropy of a digital image (stored as a vector) tells us on average how many bits are need to represent or encode each pixel.

Definition 4.3 (Entropy) *Let* $\mathbf{v} = [v_1, v_2, \dots, v_n]^T \in \mathbb{R}^n$ *and suppose that there are k distinct values in* $\mathbf{v}$. *Denote these distinct values by* a_i, $i = 1, \dots, k$, *and set* $p(a_i)$ *to be the relative frequency of* a_i *in* $\mathbf{v}$. *That is,* $p(a_i)$ *is the number of times* a_i *occurs in* $\mathbf{v}$ *divided by n. We define the* entropy *of* $\mathbf{v}$ *by*

$$\text{Ent}(\mathbf{v}) = \sum_{i=1}^{k} p(a_i) \log_2(1/p(a_i)) \tag{4.9}$$

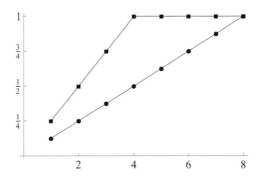

Figure 4.3   The cumulative energy vectors for **v** (circles) and **w** (squares) in Example 4.3.

∎

For now, let's think of **v** as the first row of a digital image. Then the entries are integers ranging from 0 (black) to 255 (white). The term $\log_2(1/p(a_i))$ is an exponent — it is measuring the power of 2 we need to represent $1/p(a_i)$. This exponent can also be viewed as a bit length. We then multiply this exponent by the relative frequency of a_i in **v** and sum over all distinct values a_i. Thus entropy gives us an average of the number of bits that we need encode the elements of **v**.

Claude Shannon was a leading researcher in the area of information theory, and in his seminal 1948 paper [53] he showed that the best compression rate (say, in bits per pixel) that we could hope for when using method S to perform lossless compression [16] on **v**, as the length n of **v** goes to infinity, is Ent(**v**).

Example 4.4 (Entropy Examples) *Compute the entropy of*

(a) $\mathbf{v} = [1, 2, 3, 4, 5, 6, 7, 8]^T$

(b) $\mathbf{y} = \left[\frac{3\sqrt{2}}{2}, \frac{7\sqrt{2}}{2}, \frac{11\sqrt{2}}{2}, \frac{15\sqrt{2}}{2}, -\frac{\sqrt{2}}{2}, -\frac{\sqrt{2}}{2}, -\frac{\sqrt{2}}{2}, -\frac{\sqrt{2}}{2} \right]^T$

Solution
For part (a) we note there are eight distinct elements in the 8-vector **v**, *so each element has relative frequency* $\frac{1}{8}$. *We have*

$$\text{Ent}(\mathbf{v}) = \sum_{k=1}^{8} \frac{1}{8} \log_2 8 = 3$$

Note that **y** *is the discrete Haar wavelet transformation of* **v**. *There are five distinct values in* **y** *— four have relative frequency* $\frac{1}{8}$ *and one has relative frequency* $\frac{1}{2}$. *So*

---

[16]Lossless compression is a form of compression where the original input can be recovered completely from the compressed data.

the entropy is

$$\text{Ent}(\mathbf{y}) = 4 \cdot \frac{1}{8} \log_2 8 + \frac{1}{2} \log_2 2 = \frac{3}{2} + \frac{1}{2} = 2$$

∎

Example 4.4 is helpful for understanding entropy. From a compression standpoint, the vector $\mathbf{v}$ is the worst possible scenario. It is composed of distinct elements and since there are $2^3 = 8$ elements, we expect that 3 bits are needed to encode each element. We can see the advantage of computing the discrete Haar wavelet transformation of $\mathbf{v}$. The result $\mathbf{y}$ has a constant value in the detail portion which allows the entropy of $\mathbf{y}$ to be one less than that of $\mathbf{v}$. Transformations that can convert large blocks of elements to zero (or near zero so that they can be quantized to zero in a lossy compression scheme) are the best for encoding methods.

Iterating the Process

In practice, it is common to *iterate* the discrete wavelet transformation. Consider the vector $\mathbf{v}$ from and its discrete Haar wavelet transformation $\mathbf{y}$ from Example 4.4. The entropies for these vectors are 3 and 2, respectively. The first four elements of $\mathbf{y}$ form the averages portion of the transformation. We iterate the process by applying W_4 to the averages portion of $\mathbf{y}$. If $\mathbf{y}^a = \left[\frac{3\sqrt{2}}{2}, \frac{7\sqrt{2}}{2}, \frac{11\sqrt{2}}{2}, \frac{15\sqrt{2}}{2} \right]^T$, then

$$W_4 \mathbf{y}^a = \begin{bmatrix} \frac{\sqrt{2}}{2} & \frac{\sqrt{2}}{2} & 0 & 0 \\ 0 & 0 & \frac{\sqrt{2}}{2} & \frac{\sqrt{2}}{2} \\ \frac{\sqrt{2}}{2} & -\frac{\sqrt{2}}{2} & 0 & 0 \\ 0 & 0 & \frac{\sqrt{2}}{2} & -\frac{\sqrt{2}}{2} \end{bmatrix} \cdot \begin{bmatrix} \frac{3\sqrt{2}}{2} \\ \frac{7\sqrt{2}}{2} \\ \frac{11\sqrt{2}}{2} \\ \frac{15\sqrt{2}}{2} \end{bmatrix} = \begin{bmatrix} 5 \\ 13 \\ -2 \\ -2 \end{bmatrix}$$

We overwrite $\mathbf{y}^a$ in $\mathbf{y}$ with the values above to obtain the iterated discrete Haar wavelet transformation

$$\mathbf{y}^2 = \left[5, 13, -2, -2 \,\middle|\, -\frac{\sqrt{2}}{2}, -\frac{\sqrt{2}}{2}, -\frac{\sqrt{2}}{2}, -\frac{\sqrt{2}}{2} \right]^T$$

The entropy of $\mathbf{y}^2$

$$\text{Ent}(\mathbf{y}^2) = 2\frac{1}{8} \log_2 8 + \frac{1}{4} \log_2 4 + \frac{1}{2} \log_2 2 = \frac{7}{4}$$

is smaller than that computed in Example 4.4 for $\mathbf{y}$. Since the averages portion of $\mathbf{y}^2$ has length 2, we could iterate one more time to obtain

$$\mathbf{y}^3 = \left[9\sqrt{2}, -4\sqrt{2} \,\middle|\, -2, -2 \,\middle|\, -\frac{\sqrt{2}}{2}, -\frac{\sqrt{2}}{2}, -\frac{\sqrt{2}}{2}, -\frac{\sqrt{2}}{2} \right]^T$$

but there is no improvement in entropy.

The General Form of the Iterated Haar Wavelet Transformation

Suppose that $\mathbf{v} \in \mathbf{R}^N$, where N is a positive integer divisible by 2^p. We compute the discrete Haar wavelet transformation of $\mathbf{v}$, and we obtain

$$\mathbf{y}^1 = W_N \mathbf{v} = \begin{bmatrix} \mathbf{y}^{1,a} \\ \mathbf{y}^{1,d} \end{bmatrix} \tag{4.10}$$

where $\mathbf{y}^{1,a}$ represents the $N/2$-length averages portion of the transformation and $\mathbf{y}^{1,d}$ denotes the $N/2$-length details portion of the transformation. The 1 that appears in the superscripts indicates that output is the result of one application of the discrete Haar wavelet transformation. The second iteration of the discrete Haar wavelet transformation is

$$\mathbf{y}^2 = \begin{bmatrix} W_{N/2} \mathbf{y}^{1,a} \\ \hline \mathbf{y}^{1,d} \end{bmatrix} = \begin{bmatrix} \mathbf{y}^{2,a} \\ \mathbf{y}^{2,d} \\ \hline \mathbf{y}^{1,d} \end{bmatrix} \tag{4.11}$$

Here the vectors $\mathbf{y}^{2,a}$, $\mathbf{y}^{2,d}$ are length-$N/4$ vectors that represent the averages and details portions, respectively, of the discrete Haar wavelet transformation applied to $\mathbf{y}^{1,a}$.

Since N is divisible by 2^p, we can iterate a total of p times if we so desire, and the jth iteration of the discrete Haar wavelet transformation, where $j = 1, \ldots, p$, is

$$\mathbf{y}^j = \begin{bmatrix} \mathbf{y}^{j,a} \\ \hline \mathbf{y}^{j,d} \\ \hline \mathbf{y}^{j-1,d} \\ \hline \mathbf{y}^{j-2,d} \\ \hline \vdots \\ \hline \mathbf{y}^{2,d} \\ \hline \mathbf{y}^{1,d} \end{bmatrix} \tag{4.12}$$

with $\mathbf{y}^{j,a} \in \mathbf{R}^{N/2^j}$ and $\mathbf{y}^{k,d} \in \mathbf{R}^{N/2^k}$, $k = 1, \ldots, j$.

We can also write the iterated discrete Haar wavelet transformation in terms of products of block matrices. If we let I_n and 0_n denote the $n \times n$ identity and zero matrices, respectively, we can rewrite (4.11) in matrix block product form as

$$
\mathbf{y}^2 = \left[\frac{W_{N/2}\mathbf{y}^{1,a}}{\mathbf{y}^{1,d}} \right] = \left[\begin{array}{c|c} W_{N/2} & 0_{N/2} \\ \hline 0_{N/2} & I_{N/2} \end{array} \right] \cdot \left[\begin{array}{c} \mathbf{y}^{1,a} \\ \mathbf{y}^{1,d} \end{array} \right]
$$

$$
= \left[\begin{array}{c|c} W_{N/2} & 0_{N/2} \\ \hline 0_{N/2} & I_{N/2} \end{array} \right] \cdot \mathbf{y}^1 = \left[\begin{array}{c|c} W_{N/2} & 0_{N/2} \\ \hline 0_{N/2} & I_{N/2} \end{array} \right] \cdot W_N \mathbf{v} \quad (4.13)
$$

and the third iteration of the transformation can be written as

$$
\mathbf{y}^3 = \left[\frac{W_{N/4}\mathbf{y}^{2,a}}{\mathbf{y}^{2,d}} \right] = \left[\begin{array}{cc|c} \begin{array}{c|c} W_{N/4} & 0_{N/4} \\ \hline 0_{N/4} & I_{N/4} \end{array} & 0_{N/2} \\ \hline 0_{N/2} & I_{N/2} \end{array} \right] \cdot \left[\begin{array}{c} \mathbf{y}^{2,a} \\ \mathbf{y}^{2,d} \\ \mathbf{y}^{1,d} \end{array} \right]
$$

$$
= \left[\begin{array}{cc|c} \begin{array}{c|c} W_{N/4} & 0_{N/4} \\ \hline 0_{N/4} & I_{N/4} \end{array} & 0_{N/2} \\ \hline 0_{N/2} & I_{N/2} \end{array} \right] \cdot \mathbf{y}^2
$$

$$
= \left[\begin{array}{cc|c} \begin{array}{c|c} W_{N/4} & 0_{N/4} \\ \hline 0_{N/4} & I_{N/4} \end{array} & 0_{N/2} \\ \hline 0_{N/2} & I_{N/2} \end{array} \right] \cdot \left[\begin{array}{c|c} W_{N/2} & 0_{N/2} \\ \hline 0_{N/2} & I_{N/2} \end{array} \right] \cdot W_N \mathbf{v} \quad (4.14)
$$

In general, if we let

$$
W_{N,0} = W_N, \quad W_{N,1} = \left[\begin{array}{c|c} W_{N/2} & 0_{N/2} \\ \hline 0_{N/2} & I_{N/2} \end{array} \right], \quad W_{N,2} = \left[\begin{array}{cc|c} \begin{array}{c|c} W_{N/4} & 0_{N/4} \\ \hline 0_{N/4} & I_{N/4} \end{array} & 0_{N/2} \\ \hline 0_{N/2} & I_{N/2} \end{array} \right]
$$

We can rewrite (4.13) and (4.14) as

$$
\mathbf{y}^2 = W_{N,1}W_{N,0}\mathbf{v} \quad \text{and} \quad \mathbf{y}^3 = W_{N,2}W_{N,1}W_{N,0}\mathbf{v}
$$

In general, if N is divisible by 2^p, then we define

$$
W_{N,j} = \text{diag}\left[W_{N/2^j}, I_{N/2^j}, I_{N/2^{j-1}}, \dots, I_{N/4}, I_{N/2} \right] \quad (4.15)
$$

for $j = 1, \dots, p$, where $\text{diag}[C_1, C_2, \dots, C_n]$ is a matrix with blocks $C_1, \dots, C_n$ on the main diagonal and zeros elsewhere. The representation (4.15) allows us to express the jth iteration of the discrete Haar wavelet transformation as

$$
\mathbf{y}^j = W_{N,j-1} \cdots W_{N,1}W_{N,0}\mathbf{v} \quad (4.16)
$$

Iterating the Inverse Discrete Haar Wavelet Transformation

In Problem 4.12 you will show that the matrix given in (4.15) is orthogonal so that $W_{N,j}^{-1} = W_{N,j}^T$. This fact allows us to solve (4.16) for $\mathbf{v}$ and thus invert the iterated discrete Haar wavelet transformation. We multiply both sides of (4.16) by

$W_{N,0}^T W_{N,1}^T \cdots W_{N,j-2}^T W_{N,j-1}^T$ to obtain

$$
\begin{aligned}
W_{N,0}^T W_{N,1}^T &\cdots W_{N,j-1}^T \mathbf{y}^j \\
&= W_{N,0}^T W_{N,1}^T \cdots \left(W_{N,j-1}^T W_{N,j-1} \right) W_{N,j-2} \cdots W_{N,1} W_{N,0} \mathbf{v} \\
&= W_{N,0}^T W_{N,1}^T \cdots W_{N,j-2}^T I_N W_{N,j-2} \cdots W_{N,1} W_{N,0} \mathbf{v} \\
&\vdots \\
&= W_{N,0}^T \left(W_{N,1}^T W_{N,1} \right) W_{N,0} \mathbf{v} \\
&= W_{N,0}^T W_{N,0} \mathbf{v} \\
&= \mathbf{v}
\end{aligned}
$$

PROBLEMS

Note: *A computer algebra system and the software on the course Web site will be useful for some of these problems. See the Preface for more details.*

★**4.1**   In later chapters we make extensive use of the Fourier series constructed from the vectors **h** and **g** that give rise to the discrete wavelet transformation. For the discrete Haar wavelet transformation, these vectors are given by (3.29) and (3.31). Show that

(a) The Fourier series $\mathcal{H}(\omega)$ constructed from **h** given by (3.29) is given by

$$
\mathcal{H}(\omega) = \frac{\sqrt{2}}{2} + \frac{\sqrt{2}}{2} e^{i\omega} = \sqrt{2} \, e^{i\omega/2} \cos\left(\frac{\omega}{2}\right)
$$

(b) Use part (a) to show that $|\mathcal{H}(\omega)| = \sqrt{2} \cos\left(\frac{\omega}{2}\right)$ for $-\pi \leq \omega \leq \pi$.

(c) Show that the Fourier series $\mathcal{G}(\omega)$ constructed from the vector **g** given by (3.31) is given by

$$
\mathcal{G}(\omega) = \frac{\sqrt{2}}{2} - \frac{\sqrt{2}}{2} e^{i\omega} = -\sqrt{2} \, i \, e^{i\omega/2} \sin\left(\frac{\omega}{2}\right)
$$

(d) Show that $|\mathcal{G}(\omega)| = \sqrt{2} \, |\sin\left(\frac{\omega}{2}\right)|$ where $-\pi \leq \omega \leq \pi$.

(e) Use parts (b) and (d) to show that

$$
|\mathcal{H}(\omega)|^2 + |\mathcal{H}(\omega + \pi)|^2 = |\mathcal{G}(\omega)|^2 + |\mathcal{G}(\omega + \pi)|^2 = 2
$$

and

$$
\mathcal{H}(\omega)\overline{\mathcal{G}(\omega)} + \mathcal{H}(\omega + \pi)\overline{\mathcal{G}(\omega + \pi)} = 0
$$

4.2   Compute the discrete Haar wavelet transformation for each vector.

(a) $\mathbf{u} = [1, 1, 1, 1, 1, 1, 1, 1]^T$

(b) $\mathbf{v} = [1, 2, 3, 4, 5, 6, 7, 8]^T$

(c) $\mathbf{w} = [1, 4, 9, 16, 25, 36, 49, 64]^T$

4.3 Compute three iterations of the discrete Haar wavelet transformation for each of the vectors in Problem 4.2.

4.4 Suppose that we compute $\mathbf{y} = W_N \mathbf{v}$ where N is an even positive integer. We next quantize $\mathbf{y} = \begin{bmatrix} \mathbf{a} \\ \mathbf{d} \end{bmatrix}$ by replacing $\mathbf{d}$ by the $\frac{N}{2}$-length zero matrix $\mathbf{0}$ to obtain $\tilde{\mathbf{y}} = \begin{bmatrix} \mathbf{a} \\ \mathbf{0} \end{bmatrix}$. We next apply W_N^T to $\tilde{\mathbf{y}}$ to obtain $\tilde{\mathbf{v}} = W_N^T \tilde{\mathbf{y}}$. Write the elements of $\tilde{\mathbf{v}}$ in terms of the elements of $\mathbf{v}$.

4.5 Suppose that $\mathbf{v} \in \mathbb{R}^{16}$. In this exercise you will investigate the effects of quantizing the averages portion of the iterated Haar wavelet transformation.

(a) Compute one iteration of the Haar wavelet transformation to obtain $\mathbf{y}^1 = \begin{bmatrix} \mathbf{y}^{1,a} \\ \mathbf{y}^{1,d} \end{bmatrix}$. Replace the averages portion $\mathbf{y}^{1,a}$ with the zero vector $\mathbf{0} \in \mathbb{R}^8$ to obtain $\tilde{\mathbf{y}}^1 = \begin{bmatrix} \mathbf{0} \\ \mathbf{y}^{1,d} \end{bmatrix}$. Now compute the inverse transform of $\tilde{\mathbf{y}}^1$ to obtain $\tilde{\mathbf{v}}$. How many elements of $\tilde{\mathbf{v}}$ are the same as their corresponding elements of $\mathbf{v}$? Can you describe the elements that are different?

(b) Repeat part (a), but this time perform two iterations of the Haar wavelet transformation and replace $\mathbf{y}^{2,a}$ with $\mathbf{0} \in \mathbb{R}^4$.

(c) Repeat part (a), but this time perform three iterations of the Haar wavelet transformation and replace $\mathbf{y}^{3,a}$ with $\mathbf{0} \in \mathbb{R}^2$.

(d) Repeat part (a), but this time perform four iterations of the Haar wavelet transformation and replace $\mathbf{y}^{4,a}$ with 0.

(e) Use a CAS and rework this problem for different vectors in $\mathbf{v} \in \mathbf{R}^{16}$. Plot both $\mathbf{v}$ and $\tilde{\mathbf{v}}$.

4.6 For this problem we define $\mathcal{W}_N = W_{N,2} W_{N,1} W_{N,0}$, where $N = 2^p$ and $p \geq 3$ is a positive integer. Suppose that $\mathbf{v} \in \mathbb{R}^8$ and we wish to apply three iterations of the discrete Haar wavelet transformation to $\mathbf{v}$. Then we use (4.16) to write

$$\mathbf{y}^3 = \mathcal{W}_8 \mathbf{v} = W_{8,2} W_{8,1} W_{8,0} \mathbf{v}$$

(a) Show that

$$\mathbf{y}^3 = \mathcal{W}_8 \cdot \mathbf{v} = \begin{bmatrix} \frac{\sqrt{2}}{4} & \frac{\sqrt{2}}{4} & \frac{\sqrt{2}}{4} & \frac{\sqrt{2}}{4} & \frac{\sqrt{2}}{4} & \frac{\sqrt{2}}{4} & \frac{\sqrt{2}}{4} & \frac{\sqrt{2}}{4} \\ -\frac{\sqrt{2}}{4} & -\frac{\sqrt{2}}{4} & -\frac{\sqrt{2}}{4} & -\frac{\sqrt{2}}{4} & \frac{\sqrt{2}}{4} & \frac{\sqrt{2}}{4} & \frac{\sqrt{2}}{4} & \frac{\sqrt{2}}{4} \\ -\frac{1}{2} & -\frac{1}{2} & \frac{1}{2} & \frac{1}{2} & 0 & 0 & 0 & 0 \\ 0 & 0 & 0 & 0 & -\frac{1}{2} & -\frac{1}{2} & \frac{1}{2} & \frac{1}{2} \\ -\frac{\sqrt{2}}{2} & \frac{\sqrt{2}}{2} & 0 & 0 & 0 & 0 & 0 & 0 \\ 0 & 0 & -\frac{\sqrt{2}}{2} & \frac{\sqrt{2}}{2} & 0 & 0 & 0 & 0 \\ 0 & 0 & 0 & 0 & -\frac{\sqrt{2}}{2} & \frac{\sqrt{2}}{2} & 0 & 0 \\ 0 & 0 & 0 & 0 & 0 & 0 & -\frac{\sqrt{2}}{2} & \frac{\sqrt{2}}{2} \end{bmatrix} \cdot \mathbf{v}$$

(b) Repeat part (a) where now $\mathbf{v} \in \mathbb{R}^{16}$ and we compute four iterations of the discrete Haar wavelet transformation. What is $\mathcal{W}_{16}$ in this case?

(c) Describe the general pattern of $\mathcal{W}_N$ where $N = 2^p$.

4.7   Suppose that $c \in \mathbb{R}$, $c \neq 0$, is fixed and let N be an even positive integer. Construct the vector $\mathbf{v} \in \mathbb{R}^N$ element–wise by the rule $v_k = c$, $k = 1, \ldots, N$.

(a) Compute one iteration $\mathbf{y}$ of the discrete Haar wavelet transformation of $\mathbf{v}$.

(b) Find $\mathrm{Ent}(\mathbf{v})$ and $\mathrm{Ent}(\mathbf{y})$.

(c) Compute the cumulative energy vectors $\mathbf{C}(\mathbf{v})$ and $\mathbf{C}(\mathbf{y})$ and plot the components of each vector on the same axes.

4.8   Repeat Problem 4.7 for the vector $\mathbf{v} \in \mathbb{R}^N$, where $v_k = mk + b$ now. Here $m, b \in \mathbb{R}$ are fixed values with $m \neq 0$.

4.9   Let $\mathbf{v} \in \mathbb{R}^N$ and $c \in \mathbb{R}$ with $c \neq 0$. Define $\mathbf{w} = c\mathbf{v}$. Show that $\mathbf{C}(\mathbf{v}) = \mathbf{C}(\mathbf{w})$ and $\mathrm{Ent}(\mathbf{v}) = \mathrm{Ent}(\mathbf{w})$.

4.10   Show that $\mathrm{Ent}(\mathbf{v}) \geq 0$ with equality if and only if $\mathbf{v}$ is a constant vector with $v_k = c$, for $k = 1, \ldots, N$ and $c \in \mathbb{R}$.

4.11   Suppose that $N \geq 4$ is an even integer and let $\mathbf{v} \in \mathbb{R}^N$ be a vector with N distinct elements.

(a) Show that $\mathrm{Ent}(\mathbf{v}) = \log_2(N)$.[17]

(b) The discrete Haar transformation has the form $\mathbf{y} = W_N\mathbf{v} = \begin{bmatrix} \mathbf{y}^{1,a} \\ \mathbf{y}^{1,d} \end{bmatrix}$. Create

the vector $\tilde{\mathbf{y}} = \begin{bmatrix} \mathbf{y}^{1,a} \\ \mathbf{0} \end{bmatrix}$ where $\mathbf{0} \in \mathbb{R}^{N/2}$ is the zero vector. Show that if the

---

[17]It can be shown that all vectors $\mathbf{v} \in \mathbb{R}^N$ satisfy $\mathrm{Ent}(\mathbf{v}) \leq \log_2(N)$. See Problems 3.29–3.32 in Van Fleet [60] for details.

elements of **y** are distinct, then

$$\text{Ent}(\tilde{\mathbf{y}}) = \frac{1}{2}\text{Ent}(\mathbf{y}) + \frac{1}{2}$$

(c) Why do we not consider the case where $N = 2$?

4.12   In this problem you will show that the block diagonal matrix given in (4.15) is orthogonal. Suppose that N is divisible by 2^p, where p is a positive integer.

(a) Use the fact that $W_{N/2}$ is orthogonal to show via direct multiplication that $W_{N,2}^T W_{N,2} = I_N$.

(b) Use the fact that $W_{N/4}$ is orthogonal to show via direct multiplication that $W_{N,3}^T W_{N,3} = I_N$.

(c) Generalize the results in parts (a) and (b) to show that $W_{N,j}^T W_{N,j} = I_N$ for $j = 1, \ldots, p$.

4.13   In this problem you will learn how the discrete Haar wavelet transformation relates to calculus. You will need a CAS and functions from the `DiscreteWavelets` package to solve this problem.
    Choose N to be an even positive integer (start with $N = 100$) and define the points $x_k = k/N$, for $k = 1, \ldots, N$. Using the function $f_0(x) = \cos(2\pi x)$, create the vector $\mathbf{v} \in \mathbb{R}^N$ using the rule $v_k = f_0(x_k)$, $k = 1, \ldots, N$.

(a) Compute one iteration $\mathbf{y}^1 = \begin{bmatrix} \mathbf{y}^{1,a} \\ \mathbf{y}^{1,d} \end{bmatrix} = W_N \mathbf{v}$ of the discrete Haar wavelet transformation.

(b) Plot $\mathbf{y}^{1,a}$ and $\mathbf{y}^{1,d}$ on separate axes. How does $\mathbf{y}^{1,d}$ relate to $\mathbf{y}^{1,a}$?

(c) Repeat parts (a) and (b) but build $\mathbf{v}$ using the functions $f_1(t) = \sin(2\pi t)$, $f_2(t) = e^t$, and $f_3(t) = t^2$.

(d) Can you explain why the relationship between $\mathbf{y}^{1,d}$ and $\mathbf{y}^{1,a}$ makes sense in each case?

4.14   Suppose that $\mathbf{v} \in \mathbb{R}^N$, where N is an even positive integer. Show that $\mathbf{y} = W_N \mathbf{v} = \begin{bmatrix} X\mathbf{h} \\ X\mathbf{g} \end{bmatrix}$, where $\mathbf{h}, \mathbf{g}$ are given by (3.29), (3.31), respectively, and

$$X = \begin{bmatrix} v_1 & v_2 \\ v_3 & v_4 \\ & \vdots \\ v_{N-1} & v_N \end{bmatrix}$$

This reformulation of the one-dimensional transformation leads to an efficient algorithm (in Mathematica below) for computing $W_N \mathbf{v}$:

```
X = Partition[ v, 2 ];
y = Join[ X.h, X.g ];
```

4.15   Suppose that $\mathbf{y} = W_N \mathbf{v}$ where N is an even positive integer. Then we can recover $\mathbf{v}$ using the inverse transformation $\mathbf{v} = W_N^T \mathbf{y}$. Can you reformulate this matrix product in terms of the vectors $\mathbf{h}, \mathbf{g}$ given by (3.29), (3.31), respectively, and

$$
Y = \begin{bmatrix} y_1 & y_{N/2+1} \\ y_2 & y_{N/2+2} \\ & \vdots \\ y_{N/2} & y_N \end{bmatrix} ?
$$

Write code in a computer algebra system such as Mathematica to implement your result.

4.2   THE TWO-DIMENSIONAL TRANSFORM

Suppose that A is an $M \times N$ matrix where M, N are even positive integers. If we denote the columns of A by $\mathbf{a}^1, \ldots, \mathbf{a}^N$, then the matrix product $W_M A$ can be written as

$$
W_M A = W_M \left[\mathbf{a}^1, \mathbf{a}^2, \ldots, \mathbf{a}^N \right] = \left[W_M \mathbf{a}^1, W_M \mathbf{a}^2, \ldots, W_M \mathbf{a}^N \right]
$$

and we see that the product $W_M A$ simply applies the discrete Haar wavelet transformation to each column of A. Let's look at an example.

Example 4.5 (Applying the HWT to the Columns of a Matrix) *The matrix product $W_4 A$ where*

$$
A = \begin{bmatrix} 10 & 16 & 0 & 4 \\ 2 & 24 & 0 & 4 \\ 9 & 10 & 3 & 3 \\ 5 & 10 & 3 & 3 \end{bmatrix}
$$

is

$$
W_4 A = \sqrt{2} \begin{bmatrix} \frac{1}{2} & \frac{1}{2} & 0 & 0 \\ 0 & 0 & \frac{1}{2} & \frac{1}{2} \\ \frac{1}{2} & -\frac{1}{2} & 0 & 0 \\ 0 & 0 & \frac{1}{2} & -\frac{1}{2} \end{bmatrix} \cdot \begin{bmatrix} 10 & 16 & 0 & 4 \\ 2 & 24 & 0 & 4 \\ 9 & 10 & 3 & 3 \\ 5 & 10 & 3 & 3 \end{bmatrix} = \sqrt{2} \begin{bmatrix} 6 & 20 & 0 & 4 \\ 7 & 10 & 3 & 3 \\ 4 & -4 & 0 & 0 \\ 2 & 0 & 0 & 0 \end{bmatrix}
$$

The $\sqrt{2}$ has been factored from the product so that we can better see that each column of the product is indeed the pairwise averages and details of the corresponding columns of A. ■

In applications, the discrete wavelet transformation is used to process digital images. The fundamental element of such an image is a *pixel*.[18] A pixel is a small rectangle that ranges in grayscale from black to white, and digital grayscale images are represented as a table of $M \times N$ pixels. Thus we can view a digital grayscale image mathematically as an $M \times N$ matrix whose elements are integers from the set $\{0, \ldots, 255\}$. Figure 4.4 illustrates how a digital image is stored as a matrix.

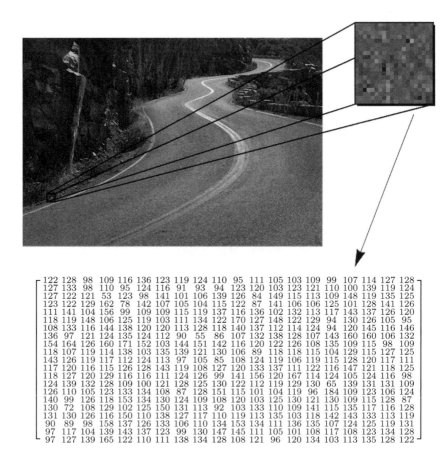

$$
\begin{bmatrix}
122 & 128 & 98 & 109 & 116 & 136 & 123 & 119 & 124 & 110 & 95 & 111 & 105 & 103 & 109 & 99 & 107 & 114 & 127 & 128 \\
127 & 133 & 98 & 110 & 95 & 124 & 116 & 91 & 93 & 94 & 123 & 120 & 103 & 123 & 121 & 110 & 100 & 139 & 119 & 124 \\
127 & 122 & 121 & 53 & 123 & 98 & 141 & 101 & 106 & 139 & 126 & 84 & 149 & 115 & 113 & 109 & 148 & 119 & 135 & 125 \\
123 & 122 & 129 & 162 & 78 & 142 & 107 & 105 & 104 & 115 & 122 & 87 & 141 & 106 & 106 & 125 & 101 & 128 & 141 & 126 \\
111 & 141 & 104 & 156 & 99 & 109 & 109 & 115 & 119 & 137 & 116 & 136 & 102 & 132 & 113 & 117 & 143 & 137 & 126 & 120 \\
118 & 119 & 148 & 106 & 125 & 119 & 103 & 111 & 134 & 122 & 170 & 127 & 148 & 122 & 129 & 94 & 130 & 126 & 105 & 95 \\
108 & 133 & 116 & 144 & 138 & 120 & 120 & 113 & 128 & 118 & 140 & 137 & 112 & 114 & 124 & 94 & 120 & 145 & 116 & 146 \\
136 & 97 & 121 & 124 & 135 & 124 & 112 & 90 & 55 & 86 & 107 & 132 & 138 & 128 & 107 & 143 & 160 & 160 & 106 & 132 \\
154 & 164 & 126 & 160 & 171 & 152 & 103 & 144 & 151 & 142 & 116 & 120 & 122 & 126 & 108 & 135 & 109 & 115 & 98 & 109 \\
118 & 107 & 119 & 114 & 138 & 103 & 135 & 139 & 121 & 130 & 106 & 89 & 118 & 118 & 115 & 104 & 129 & 115 & 127 & 125 \\
143 & 126 & 119 & 117 & 112 & 124 & 113 & 97 & 105 & 85 & 108 & 124 & 119 & 106 & 119 & 115 & 128 & 120 & 117 & 111 \\
117 & 120 & 116 & 115 & 126 & 128 & 143 & 119 & 108 & 127 & 120 & 133 & 137 & 111 & 122 & 116 & 147 & 121 & 118 & 125 \\
118 & 127 & 120 & 129 & 116 & 116 & 111 & 124 & 126 & 99 & 141 & 156 & 120 & 167 & 114 & 124 & 105 & 124 & 116 & 98 \\
124 & 139 & 132 & 128 & 109 & 100 & 121 & 128 & 125 & 130 & 122 & 112 & 119 & 129 & 130 & 65 & 139 & 131 & 131 & 109 \\
126 & 110 & 105 & 123 & 133 & 134 & 108 & 87 & 128 & 151 & 115 & 101 & 104 & 119 & 96 & 184 & 109 & 123 & 106 & 124 \\
140 & 99 & 126 & 118 & 153 & 134 & 130 & 124 & 109 & 108 & 120 & 103 & 125 & 130 & 121 & 130 & 109 & 115 & 128 & 87 \\
130 & 72 & 108 & 129 & 102 & 125 & 150 & 131 & 113 & 92 & 103 & 133 & 110 & 109 & 141 & 115 & 135 & 117 & 116 & 128 \\
131 & 130 & 126 & 116 & 150 & 110 & 138 & 127 & 117 & 110 & 119 & 113 & 135 & 103 & 118 & 142 & 143 & 133 & 113 & 119 \\
90 & 89 & 98 & 158 & 137 & 126 & 133 & 106 & 110 & 134 & 153 & 134 & 111 & 136 & 135 & 107 & 124 & 125 & 119 & 131 \\
97 & 117 & 104 & 139 & 143 & 137 & 123 & 99 & 130 & 147 & 145 & 111 & 105 & 101 & 108 & 117 & 108 & 123 & 134 & 128 \\
97 & 127 & 139 & 165 & 122 & 110 & 111 & 138 & 134 & 128 & 108 & 121 & 96 & 120 & 134 & 103 & 113 & 135 & 128 & 122
\end{bmatrix}
$$

Figure 4.4   Storing a digital image as a matrix. A 20×20 portion of the 768×512 image is enlarged and the pixel intensity values are displayed.

[18]Pixel is short for *picture* (pix) *element* (el).

Example 4.6 illustrates the product of $W_M A$ where A represents an $M \times N$ digital grayscale image and M is an even positive integer.

Example 4.6 (Transforming the Columns of an Image) *In this example we compute the product $B = W_{768}A$ where A is the 768×512 digital grayscale image plotted in Figure 4.4. Each column of B is the result of applying the discrete Haar wavelet transformation to the corresponding column of A. Thus the top half of B is pairwise averages along columns of A and the bottom half of B are the details needed to recover the original values of A from the averages. Note that if there is little change in column elements, then the corresponding elements in B are either zero or near zero. On the other hand, large horizontal changes in the columns of the original image can be identified by (near) white pixels in the details portion of B. The matrix B is plotted in Figure 4.5.*

Figure 4.5 The discrete Haar wavelet transformation matrix W_{768} applied to the columns of a digital grayscale image.

■

Instead of processing the columns of a digital grayscale image, we could just as well process the rows.

Example 4.7 (Transforming the Rows of an Image) *To process the rows of A, we compute $B = AW_N^T$, where A is an $M \times N$ matrix with N an even positive integer. Figure 4.6 illustrates the product $B = AW_{512}^T$, where A is the 768×512 digital grayscale image plotted in Figure 4.4.*

Figure 4.6   The discrete Haar wavelet transformation matrix W_{512}^T applied to the rows of a digital grayscale image.

Each row of B is the transformation of the corresponding row in A. Note that when there is little change in the row elements of A, the corresponding elements in B are (near) zero. On the other hand, large vertical changes in rows of the original image results in (near) white pixels in B.   ■

The Two–Dimensional Transformation as a Matrix Product

In practice, both columns and rows of a digital grayscale image are processed. That is, the *two-dimensional discrete Haar wavelet transformation* of $M \times N$ matrix A where both M, N are even positive integers is defined as

$$\boxed{B = W_M A W_N^T} \tag{4.17}$$

where W_M, W_N are defined by (4.4). Figure 4.7 shows the two-dimensional discrete Haar wavelet transformation of the 768×512 digital grayscale image plotted in Figure 4.4.

Notice that the transformed image $B = W_{768} A W_{512}^T$ consists of four blocks. In order to interpret the role of each block, let's look at the transformation in block form. Suppose that A is an $M \times N$ matrix with M, N positive even integers. We write W_M, W_N^T in block form using (4.4), we see that (4.17) can be written in block form

Figure 4.7   The two-dimensional discrete Haar wavelet transformation matrix $B = W_{768} A W_{512}^T$.

as

$$B = W_M A W_N^T$$

$$= \begin{bmatrix} H_{M/2} \\ \hline G_{M/2} \end{bmatrix} \cdot A \cdot \begin{bmatrix} H_{N/2} \\ \hline G_{N/2} \end{bmatrix}^T$$

$$= \begin{bmatrix} H_{M/2} \\ \hline G_{M/2} \end{bmatrix} \cdot A \cdot \left[H_{N/2}^T \middle| G_{N/2}^T \right]$$

$$= \begin{bmatrix} H_{M/2} A \\ \hline G_{M/2} A \end{bmatrix} \cdot \left[H_{N/2}^T \middle| G_{N/2}^T \right]$$

$$= \begin{bmatrix} H_{M/2} A H_{N/2} & H_{M/2} A G_{N/2}^T \\ \hline G_{M/2} A H_{N/2}^T & G_{M/2} A G_{N/2}^T \end{bmatrix}$$

$$= \begin{bmatrix} \mathcal{A} & \mathcal{V} \\ \hline \mathcal{H} & \mathcal{D} \end{bmatrix} \tag{4.18}$$

So we see that the construction of the matrices W_M, W_N contributes directly to the block structure of the two-dimensional transformation.

We can also analyze each block and learn more about the information they hold. The upper left block is $\mathcal{A} = H_{M/2}AH_{N/2}^T$. Recall that $H_{M/2}A$ produces (weighted) column averages. We right-multiply this product by $H_{N/2}^T$ and recall that this operation produces (weighted) averages along rows, so $\mathcal{A}$ is an *approximation* of the original input matrix A. Indeed, in Problem 4.18 you will show that if α_{ij}, $i = 1, \ldots, M/2$, $j = 1, \ldots, N/2$, is an element of $\mathcal{A}$, we have

$$\alpha_{ij} = 2 \cdot \frac{a_{2i-1,2j-1} + a_{2i-1,2j} + a_{2i,2j-1} + a_{2i,2j}}{4} \tag{4.19}$$

Thus if we partition A into 2×2 blocks A_{ij}, $i = 1, \ldots, M/2$, $j = 1, \ldots, N/2$,

$$A = \begin{bmatrix} A_{11} & A_{12} & \cdots & A_{1,N/2} \\ A_{21} & A_{22} & & A_{2,N/2} \\ & & \ddots & \\ A_{M/2,1} & A_{M/2,2} & & A_{M/2,N/2} \end{bmatrix}$$

we see that α_{ij} is twice the average of the four elements of A_{ij}.

In a similar manner, you will show in Problem 4.18 that if β_{ij}, γ_{ij} and δ_{ij} are the elements of $\mathcal{V} = H_{M/2}AG_{N/2}^T$, $\mathcal{H} = G_{M/2}AH_{N/2}^T$, and $\mathcal{D} = G_{M/2}AG_{N/2}^T$, respectively, then for $i = 1, \ldots, M/2$, $j = 1, \ldots, N/2$,

$$\beta_{ij} = \frac{(a_{2i-1,2j-1} + a_{2i,2j-1}) - (a_{2i-1,2j} + a_{2i,2j})}{2} \tag{4.20}$$

$$\gamma_{ij} = \frac{(a_{2i-1,2j-1} + a_{2i-1,2j}) - (a_{2i,2j-1} + a_{2i,2j})}{2} \tag{4.21}$$

$$\delta_{ij} = \frac{(a_{2i-1,2j-1} + a_{2i,2j}) - (a_{2i-1,2j} + a_{2i,2j-1})}{2} \tag{4.22}$$

Thus β_{ij}, γ_{ij} and δ_{ij} measure weighted differences in the column sums, row sums, and diagonal sums of A_{ij}, respectively. We can rewrite (4.18) as

$$B = W_M A W_N^T = \left[\begin{array}{c|c} \mathcal{A} & \mathcal{V} \\ \hline \mathcal{H} & \mathcal{D} \end{array} \right] = \left[\begin{array}{c|c} \text{averages} & \begin{array}{c}\text{vertical} \\ \text{differences}\end{array} \\ \hline \begin{array}{c}\text{horizontal} \\ \text{differences}\end{array} & \begin{array}{c}\text{diagonal} \\ \text{differences}\end{array} \end{array} \right] \tag{4.23}$$

Inverting the Two-Dimensional Transform and Iteration

It is straightforward to invert the two-dimensional discrete Haar wavelet transformation. Since W_M and W_N are orthogonal, we can left-multiply both sides of (4.17) by W_M^T to obtain

$$W_M^T B = \left(W_M^T W_M \right) A W_N^T = A W_N^T$$

and then right-multiply the identity above by W_N to obtain

$$W_M^T B W_N = A \left(W_N^T W_N \right) = A$$

Iterating the two-dimensional transform is a simple extension of the one–dimensional case. If A is an $M \times N$ matrix with both M, N divisible by 4, we first compute one iteration of the discrete Haar wavelet transformation via (4.17). We then replace the averages portion $\mathcal{A}$ in (4.18) with $W_{M/2}\mathcal{A}W_{N/2}^T$. We can continue this process a total of p times if both M and N are divisible by 2^p. The following example shows why the iterative transformation is useful in applications.

Example 4.8 (Iterating the Haar Wavelet Transform) *Compute three iterations of the discrete Haar wavelet transformation of the 768×512 image A plotted in Figure 4.4 and plot the cumulative energy vectors for A, one iteration of the transformation, and three iterations of the transformation.*

Solution

One iteration of the transformation is plotted in Figure 4.7. We extract the 384×256 upper left-hand corner $\mathcal{A}$ of this transformation and compute $W_{384}\mathcal{A}W_{256}^T$. We replace $\mathcal{A}$ with the product. The result is plotted in Figure 4.8.

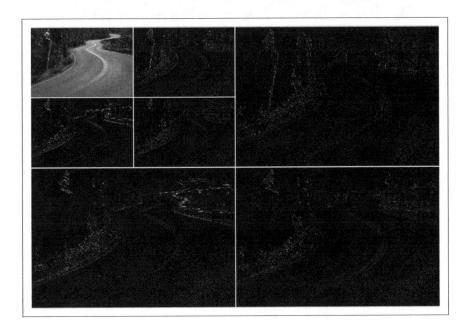

Figure 4.8 Two iterations of the discrete Haar wavelet transformation applied to the image from Figure 4.4.

We next extract the 192×128 upper left-hand corner $\mathcal{A}$ from Figure 4.8 and replace it with the product $W_{192}\mathcal{A}W_{128}^T$. The resulting (3-iteration) transformation is plotted in Figure 4.9.

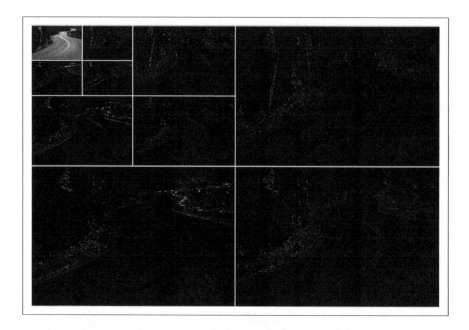

Figure 4.9   Three iterations of the discrete Haar wavelet transformation applied to the image from Figure 4.4.

The cumulative energy vectors[19] for the original image, one iteration of the transformation, and three iterations of the transformation are plotted in Figure 4.10. Note the improved conservation of energy in the third iteration versus the first iteration. ■

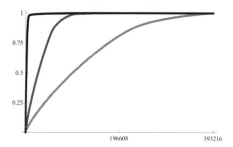

Figure 4.10   The cumulative energy of the original image (light gray), one iteration (dark gray), and three iterations (black) of the discrete Haar wavelet transformation. The horizontal axis denotes the number of elements $768 \times 512 = 393{,}216$ in each vector.

[19]To compute the cumulative energy of a matrix of size $M \times N$, we simply create a vector of length MN by concatenating the columns of the matrix.

PROBLEMS

Note: *A computer algebra system and the software on the course Web site will be useful for some of these problems. See the Preface for more details.*

4.16   Suppose that A is a 6×6 matrix whose entries are $a_{ij} = c$, $i, j = 1, \ldots, 6$, where $c \in \mathbb{R}$. Compute the two-dimensional discrete Haar wavelet transformation $B = W_6 A W_6^T$.

4.17   Suppose that A is an $N \times N$ matrix with N an even positive integer. Further suppose that for each $i = 1, \ldots, N$, we have $a_{ij} = c_i$, where c_i is a real number. So the rows of A are constant. Describe the elements in the blocks $\mathcal{A}$, $\mathcal{V}$, $\mathcal{H}$, and $\mathcal{D}$ of the Haar wavelet transform. Repeat the problem for A^T. (*Hint:* Use a CAS to create A for a small size and then look at the output of the wavelet transform.)

4.18   In this problem you will verify the formulas (4.19)–(4.22) for $M \times N$ matrix A, where M, N are even positive integers.

(a) Let $A = \begin{bmatrix} a & b \\ c & d \end{bmatrix}$. Verify (4.19)–(4.22) for A. Here

$$H_1 = \begin{bmatrix} \frac{\sqrt{2}}{2} & \frac{\sqrt{2}}{2} \end{bmatrix} \quad \text{and} \quad G_1 = \begin{bmatrix} \frac{\sqrt{2}}{2} & -\frac{\sqrt{2}}{2} \end{bmatrix}$$

(b) Now suppose that A is an $M \times N$ matrix with M, N even positive integers. Partition A into 2×2 blocks A_{ij}, $i = 1, \ldots, M/2$, $j = 1, \ldots, N/2$, where

$$A_{ij} = \begin{bmatrix} a_{2i-1,2j-1} & a_{2i-1,2j} \\ a_{2i,2j-1} & a_{2i,2j} \end{bmatrix}$$

Show that

$$\mathcal{A} = H_{M/2} \cdot \begin{bmatrix} A_{11} & A_{12} & \cdots & A_{1,N/2} \\ A_{21} & A_{22} & & A_{2,N/2} \\ & & \ddots & \\ A_{M/2,1} & A_{M/2,2} & & A_{M/2,N/2} \end{bmatrix} \cdot H_{N/2}^T$$

$$= \begin{bmatrix} H_2 A_{11} H_2^T & H_2 A_{12} H_2^T & \cdots & H_2 A_{1,N/2} H_2^T \\ H_2 A_{21} H_2^T & H_2 A_{22} H_2^T & & H_2 A_{2,N/2} H_2^T \\ & & \ddots & \\ H_2 A_{M/2,1} H_2^T & H_2 A_{M/2,2} H_2^T & & H_2 A_{M/2,N/2} H_2^T \end{bmatrix}$$

and then use part (a) to verify (4.19).

(c) Use the idea of part (b) to verify (4.21)–(4.22).

4.19   Suppose that $B = W_M A W_N^T$ where M and N are even positive integers. Since both W_M and W_N^T are orthogonal, we know that $A = W_M^T B W_N$. Use (4.19)–(4.22) to find closed formulas for $a_{2i-1,2j-1}, a_{2i-1,2j}, a_{2i,2j-1},$ and $a_{2i,2j}, i = 1, \ldots, M/2,$ $j = 1, \ldots, N/2$ in terms of $\alpha_{ij}, \beta_{ij}, \gamma_{ij},$ and δ_{ij}.

4.20   This is a two-dimensional analog of Problem 4.4. Suppose that we compute $B = W_M A W_N^T$, where M, N are even positive integers. We next quantize $B = \begin{bmatrix} A & V \\ \hline H & D \end{bmatrix}$ by replacing each of $V, H,$ and D by the $\frac{M}{2} \times \frac{N}{2}$ zero matrix $\mathcal{O}$ to obtain $\tilde{B} = \begin{bmatrix} A & \mathcal{O} \\ \hline \mathcal{O} & \mathcal{O} \end{bmatrix}$. We next "invert" the matrix $\tilde{B}$ via the inverse Haar wavelet transform to obtain $\tilde{A} = W_M^T \tilde{B} W_N$. Write the elements of $\tilde{A}$ in terms of the elements of A.

4.3   EDGE DETECTION AND NAIVE IMAGE COMPRESSION

One straightforward application of the two-dimensional discrete Haar wavelet transformation is the detection of edges in digital images. Edges in digital images appear when there is an abrupt change in pixel intensities. The natural mathematical tool for measuring rate of change is the derivative. In Problem 4.13 we see that if we view vector components as function values of equally spaced points, then the details portion of the one-dimensional Haar wavelet transformation is a scaled approximation to the derivative of the function. Since the details portion of a two-dimensional Haar wavelet transformation are $V, H,$ and D, it is natural to assume that image edge information is probably contained in these parts of the transformation. Indeed, the naive algorithm for detecting edges in digital images is as follows:

(a) Given $M \times N$ image matrix A where M, N are both divisible by 2^p, compute i iterations of the two-dimensional discrete Haar wavelet transformation where $i = 1, \ldots, p$.

(b) Replace the final averages portion $\mathcal{A}$ of the transformation matrix by a matrix $\mathcal{O}$ whose elements are zero.

(c) Perform i iterations of the inverse discrete Haar wavelet transformation.

Let's look at an example.

Example 4.9 (Image Edge Detection)   *Find edges in the 512×512 digital grayscale image plotted in Figure 4.11(a).*
Solution
   We next compute two iterations of the discrete Haar wavelet transformation on the image. The averages portion is the block $\mathcal{A}$ of size 128×128 in the upper left-hand corner of the transformation matrix. We convert this block to zero. Both the transformation and modified transformation are plotted in Figure 4.12.
   The last step in the process is to compute two iterations of the inverse discrete Haar wavelet transformation to obtain the edge matrix E. This matrix is plotted in

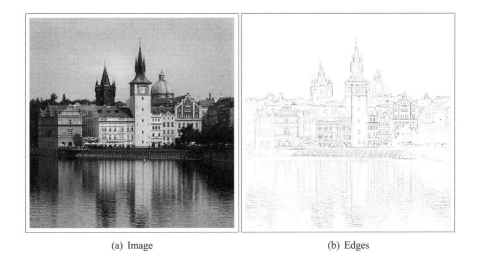

(a) Image   (b) Edges

Figure 4.11   A 512×512 digital grayscale image and the edge matrix $\tilde{E}$.

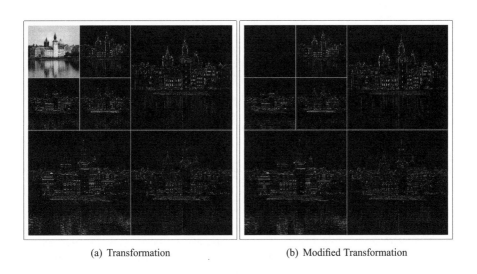

(a) Transformation   (b) Modified Transformation

Figure 4.12   Two iterations of the transformation and the modified transformation.

Figure 4.11(b). In order to make the edges more visible, we have plotted the negative image $\tilde{E}$ where $\tilde{e}_{ij} = 255 - e_{ij}$, $i, j = 1, \ldots, 512$.   ■

Naive Image Compression – An Application of the Haar Wavelet Transformation

We conclude this section with a look at how the discrete wavelet transformation can be used in digital image compression. This is the application in which wavelets have proved most useful — the JPEG2000 image compression format uses a discrete wavelet transformation to perform compression. This compression format is a vast improvement over the immensely popular and effective JPEG compression format. See Gonzalez and Woods [31] or Van Fleet [60] for more information on the JPEG2000 image compression format.

We will describe a very naive approach to image compression. Actually, the algorithm we describe is basic to most compression methods, but at this point in the book, we have not developed the sophisticated wavelet functions used in compression schemes such as JPEG2000.

Algorithm 4.1 (Naive Image Compression) *Suppose that A is an $M \times N$ grayscale image with M, N both divisible by 2^p. The basic image compression algorithm is as follows:*

1. *Subtract 128 from each element of A. This centers the intensities of the image around 0.*

2. *Compute i iterations, $1 \leq i \leq p$, of the modified discrete Haar wavelet transformation to A. We denote this transformation as B. It is desirable that the output of our transformation fall in some finite-size range since irrational numbers present difficulties for computers. The coding methods that actually rearrange the quantized data in the last step of the algorithm are much more effective if the data are output in this way. In our case we modify the Haar wavelet transform matrix W_N from (4.4) by scaling it by $1/\sqrt{2}$. That is, the modified HWT matrix is $\tilde{W}_N = \frac{1}{\sqrt{2}} W_N$. The filter used to construct the transformation is $\left[\frac{1}{2}, \frac{1}{2}\right]^T$ instead of the Haar scaling filter. This filter creates true pair-wise averages and differences and produces output contained in $\frac{1}{2}\mathbb{Z}$.*

3. *Quantize the elements in B. As we see in Figures 4.7, 4.8, and 4.9, the details portions of the wavelet transformation are comprised largely of elements that are either zero or near zero. The quantization process converts elements that are near zero to zero. Denote the quantized transformation by $\tilde{B}$. The quantization step means that our compression method is lossy. We can never recover the original image from the compressed image. For lossless compression, we omit this step.*

4. *Apply a coding method to restructure how the intensities in $\tilde{B}$ are stored in terms of bits. An uncompressed image has intensities ranging from 0 to 255. Since there are $256 = 2^8$ possible intensity values, each intensity is assigned a length 8 bitstream. For example, the intensity $139 = 10001011_2$. When we code the elements in B, we typically assign new (and variable-length) bitstreams*

to each element. The coding methods are most effective when there are a
relatively small number of distinct values in $\tilde{B}$. Since the quantized wavelet
transformation contains a large number of zeros, we expect the coded version
of $\tilde{B}$ to require much fewer bits than the original image A.

■

To view the compressed image, we uncode the data and apply i iterations of the
inverse modified Haar wavelet transform. Since we used $\tilde{W}_N = \frac{1}{\sqrt{2}}W_N$ for our
transformation, we use $\tilde{W}_N^{-1} = \sqrt{2}\,W_N^T$ as our inverse transformation. The last
step is to add 128 to each element in the compressed image matrix. Let's look at an
example.

Example 4.10 (Image Compression and the Haar Wavelet Transform) *We apply*
Algorithm 4.1 to the digital grayscale image (stored in 512×512 matrix A) plotted in
Figure 4.11(a). After subtracting 128 from each element in A, we compute four iter-
ations of the modified discrete Haar wavelet transformation and denote the resulting
matrix by B. This transformation is plotted in Figure 4.13(a).

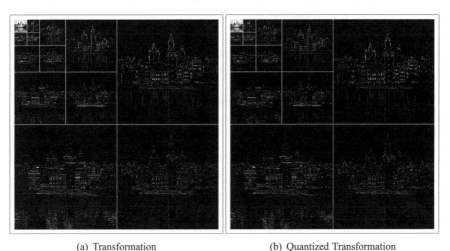

(a) Transformation   (b) Quantized Transformation

Figure 4.13   Four iterations of the modified Haar wavelet transformation and the quantized
transformation.

We use cumulative energy to quantize the transformation. We compute the cumu-
lative energy vector for B and plot this vector in Figure 4.14. Note the horizontal line
at 0.995 (chosen somewhat arbitrarily) in Figure 4.14. The horizontal coordinate
for the intersection point of this line and the cumulative energy vector components is
$72{,}996$. This means that 99.5% of the energy in B resides in the largest (in absolute
value) $72{,}996$ elements of B. For our quantization method, we convert the remaining
$512 \times 512 - 72{,}996 = 189{,}148$ in B to zero to obtain $\tilde{B}$. The modified transformation

is plotted in Figure 4.13(b). There is not much difference visually between B and $\tilde{B}$. Indeed, the mean squared error

$$\text{MSE}(B, \tilde{B}) = \frac{1}{512^2} \sum_{i=1}^{512} \sum_{j=1}^{512} (b_{ij} - \tilde{b}_{ij})^2 \qquad (4.24)$$

between the two matrices is 1.01411.

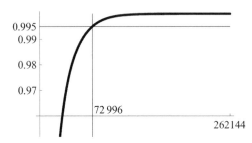

Figure 4.14   The cumulative energy vector for the transformation matrix B.

We can use the entropy measure to determine how well a coding method will work to compress the size of $\tilde{B}$. We find that $\text{Ent}(\tilde{B}) = 2.53$. Since each pixel intensity in the original image required 8 bits of storage space, we can hope for at best a savings of about 68% of our original storage size.

A basic coding tool is Huffman coding. For more details regarding Huffman coding, see Appendix A, [31], or [60]. Although not the most effective coding method available, forms of Huffman coding are used in the JPEG image compression format and the FBI Fingerprint Image Compression Standard (see Chapter 9). Huffman's idea was to create bitstreams of small length for intensities that appear often in data and bitstreams of large length for intensities that rarely appear in the data. If we apply Huffman coding to the elements in $\tilde{B}$, we find that a total of $705{,}536$ bits are needed to represent the elements. The number of bits needed to represent the elements in A is $512 \times 512 \times 8 = 2{,}097{,}152$. We can compute the average bits per pixel needed to store $\tilde{B}$ as $865{,}933/512^2 = 2.69$ bpp. Thus the Huffman coding method produces results not much larger than that predicted by the entropy.

In Figure 4.15(b), we plot the compressed version of the image from Figure 4.11(a). It is difficult to see the differences in the images — there are some noticeable discrepancies in the sky above the buildings.

The compressed image is a fair approximate to the original image. One way to measure the effectiveness of compression methods is to compute the peak signal-to-noise ratio PSNR between the original image matrix A and the compressed image matrix $\tilde{A}$. The peak signal-to-noise ratio (measured in dB) between two $M \times N$ matrices A and B is defined as

$$\text{PSNR}(A, B) = 10 \log_{10} \left(\frac{255^2}{\text{MSE}(A, B)} \right) \qquad (4.25)$$

(a) Original Image   (b) Compressed Image

Figure 4.15   Four iterations of the modified Haar wavelet transformation and the quantized transformation.

where the mean squared error $\text{MSE}(A, B)$ *is given by (4.24). The peak signal-to-noise ratio of A and* $\tilde{A}$ *is* $\text{PSNR}(A, \tilde{A}) = 36.71$. *Typical values of peak signal-to-noise ratio range from* 35 *to* $50\,dB$ *with higher values representing better approximations. See Gonzalez and Woods [31] for more details.*

If we omit the quantization step and perform Huffman coding on the transformed image plotted in Figure 4.13(a), we find that the transform can be coded using ,$1069,617$ *bits or* 4.08 bpp. *Thus lossless coding results in a* 50% *savings!*

■

In the chapters that follow we develop scaling and wavelet functions that give rise to discrete wavelet transformations that vastly outperform the Haar transformation in image compression.

Computer Software. *Although there are no problems for this section, the ideas of edge detection and image compression may be investigated via the software package* DiscreteWavelets *at the text Web site. Labs and projects involving edge detection and image compression may be found on the Web site as well. See the Preface for more details about the Web site.*

CHAPTER 5

MULTIRESOLUTION ANALYSIS

In this chapter we generalize the ideas introduced in Chapter 3. That is, we seek to find nested spaces $V_j \subseteq L^2(\mathbb{R})$, $j \in \mathbb{Z}$, that we can use to approximate functions in $L^2(\mathbb{R})$. Recall that the Haar space V_0 was generated by the Haar scaling function $\phi(t) = \sqcap(t)$. Although this function has many desirable properties (orthogonal to its integer translates, short support, symmetry about the line $t = 1/2$, easy to integrate), it is not continuous and its derivative is zero almost everywhere. In many applications, it is important that the scaling function that generates V_0 possess an appropriate number of continuous derivatives. The mathematical structure we introduce in this chapter allows us to construct such scaling functions. Moreover, the theoretical tools we develop ultimately lead us in Chapter 8 to the construction of filters that are used in the FBI fingerprint image standard and the JPG compression algorithm.

In the first section we introduce the concept of a *multiresolution analysis*. A multiresolution analysis is basically a set of nested subspaces that obey the properties listed for the Haar spaces in Chapter 3.

The development of filters for use in applications depends on writing down the *dilation equation* that the scaling function $\phi(t)$ for V_0 satisfies. To find the dilation equation, it is convenient to transform some of the properties listed in Section 5.1 to the Fourier domain. We discuss these relationships in Section 5.2.

Wavelet Theory: An Elementary Approach with Applications. By D. K. Ruch and P. J. Van Fleet
Copyright © 2009 John Wiley & Sons, Inc.

We describe two examples of multiresolution analyses in Section 5.3. In addition to providing a better understanding of a multiresolution analysis of $L^2(\mathbb{R})$, these examples also point to several desired properties that lead to the constructions in Chapters 6 and 8.

In Section 5.4 we summarize the main results from the chapter and describe the method we use for constructing a multiresolution analysis of $L^2(\mathbb{R})$ in subsequent chapters.

5.1   MULTIRESOLUTION ANALYSIS

The concept of *multiresolution analysis* is due to Stéphane Mallat [41] and Yves Meyer [44].

Definition 5.1 (Multiresolution Analysis) *Let V_j, $j \in \mathbb{Z}$, be a sequence of subspaces of $L^2(\mathbf{R})$. We say that $\{V_j\}_{j \in \mathbb{Z}}$ is a* multiresolution analysis (MRA) *of $L^2(\mathbb{R})$ if*

$$
\begin{array}{ll}
V_j \subset V_{j+1} & \text{(nested)} \\[2mm]
\overline{\cup_{j \in \mathbb{Z}} V_j} = L^2(\mathbb{R}) & \text{(density)} \\[2mm]
\cap_{j \in \mathbb{Z}} V_j = \{0\} & \text{(separation)} \\[2mm]
f(t) \in V_0 \iff f(2^j t) \in V_j & \text{(scaling)}
\end{array}
\tag{5.1}
$$

and there exists a function $\phi(t) \in V_0$, with $\int_{\mathbb{R}} \phi(t)\,dt \neq 0$, called a scaling function, *such that the set $\{\phi(t - k)\}_{k \in \mathbb{Z}}$ is an orthonormal basis for V_0.* ∎

Note: While all the results that follow in this chapter are valid for complex-valued functions, we assume hereafter that all scaling functions $\phi(t)$ are real-valued functions.

The requirement in Definition 5.1 that $\int_{\mathbb{R}} \phi(t)\,dt \neq 0$ is a necessary condition for obtaining many results in this chapter. If we remove this condition from the definition and insist instead that $\int_{x \in \mathbb{R}} |\phi(t)|\,dt < \infty$, then it can be shown (see, e. g., Walnut) [62] that

$$
\left| \int_{\mathbb{R}} \phi(t)\,dt \right| = 1
$$

so that $\int_{\mathbb{R}} \phi(t)\,dt \neq 0$. Since the integral is not zero, then we can normalize it however we choose. Thus we adopt the following convention for the remainder of the book.

Convention 5.1 (Requirements for the Scaling Function $\phi(t)$) *We assume that all scaling functions $\phi(t)$ from Definition 5.1 are real-valued functions with*

$$\boxed{\int_{\mathbb{R}} \phi(t)\,\mathrm{d}t = \sqrt{2\pi}\,\hat{\phi}(0) = 1}$$ (5.2)

■

A multiresolution analysis gives us a nice way to decompose $L^2(\mathbb{R})$. We have nested approximation spaces V_j and the ability to zoom from one to another using the scaling property from (5.1). We have an orthonormal basis for V_0 generated by a single scaling function $\phi(t) \in V_0$ and we will soon see that we can use this basis to form an orthonormal basis for V_j. While the V_j spaces are nested, they are not *too* redundant since the only element common to all the spaces is the zero function. Finally, the density property tells us that the multiresolution analysis covers all of $L^2(\mathbb{R})$.

Example 5.1 (The Haar Multiresolution Analysis) *The Haar spaces V_j (3.8) introduced in Chapter 3 form a multiresolution analysis of $L^2(\mathbb{R})$. These spaces are nested (see Proposition 3.7) and by Theorem 3.1, the V_j spaces satisfied the density and separation properties given in (5.1). The scaling property in (5.1) for the Haar spaces was given by Propositions 3.8 and 3.1, and we showed that the set $\{\phi(t-k)\}_{k\in\mathbb{Z}}$ is an orthonormal basis for V_0 with scaling function $\phi(t) = \sqcap(t)$.* ■

The Functions $\phi_{j,k}(t)$ and Related Results

In the remainder of this section we develop various properties of multiresolution analyses. It is useful to keep the Haar multiresolution analysis in mind as we learn about these properties. We begin by defining the following functions.

Definition 5.2 (The Functions $\phi_{j,k}(t)$) *Suppose that $\phi(t)$ is a scaling function from the multiresolution analysis $\{V_j\}_{j\in\mathbb{Z}}$ of $L^2(\mathbb{R})$. For $k \in \mathbb{Z}$, we define*

$$\phi_{j,k}(t) = 2^{j/2}\phi(2^j t - k)$$ (5.3)

■

A couple of facts follow immediately from Definition 5.2.

Proposition 5.1 (Properties Obeyed by $\phi_{j,k}(t)$) *Let $\phi_{j,k}(t)$ be defined by (5.3). Then for all $k \in \mathbb{Z}$, $\phi_{j,k}(t) \in V_j$ and $\|\phi_{j,k}(t)\| = 1$.* ■

Proof: Let $k \in \mathbb{Z}$. Since $f(t) = \phi(t-k)$ is a basis element for V_0, it is in V_0, so by the scaling property of Definition 5.1, we must have $f(2^j t) \in V_j$. But $f(2^j t) = \phi(2^j t - k)$, so that $\phi_{j,k}(t) \in V_j$. You are asked to show that $\|\phi_{j,k}(t)\| = 1$ in Problem 5.1.

■

The scaling property and the fact that $\{\phi(t-k)\}_{k\in\mathbb{Z}}$ can be used in conjunction with Definition 5.2 to show that the set $\{\phi_{j,k}(t)\}_{k\in\mathbb{Z}}$ forms an orthonormal basis for V_j.

Proposition 5.2 (An Orthonormal Basis for V_j) *Suppose that $\{V_\ell\}_{\ell\in\mathbb{Z}}$ is a multiresolution analysis for $L^2(\mathbb{R})$ with scaling function $\phi(t)$. Suppose $j \in \mathbb{Z}$ with $j \neq 0$ and consider the set $S = \{\phi_{j,k}(t)\}_{k\in\mathbb{Z}}$, where $\phi_{j,k}(t)$ is defined by (5.3). Then S is an orthornormal basis for V_j.* ∎

Proof: From Proposition 5.1 we know that $\phi_{j,k}(t) \in V_j$ and $\|\phi_{j,k}(t)\| = 1$. Now suppose that $k, \ell \in \mathbb{Z}$ with $k \neq \ell$ and consider the inner product

$$\langle \phi_{j,k}(t), \phi_{j,\ell}(t) \rangle = \int_{\mathbb{R}} 2^{j/2}\phi(2^j t - k)\overline{2^{j/2}\phi(2^j t - \ell)}\, dt$$

$$= 2^j \int_{\mathbb{R}} \phi(2^j t - k)\phi(2^j t - \ell)\, dt \qquad (5.4)$$

We drop the conjugation since $\phi_{j,\ell}(t)$ is a real-valued function. Using (1.21) from Proposition 1.7, we can write (5.4) as

$$\langle \phi_{j,k}(t), \phi_{j,\ell}(t) \rangle = \int_{\mathbb{R}} \phi(u)\phi(u - (\ell - k))\, du$$

$$= \langle \phi(u), \phi(u - (\ell - k)) \rangle$$

But we know that $\ell - k \neq 0$ and since $\phi(u)$ and $\phi(u-(\ell-k))$ are two distinct elements of an orthonormal basis for V_0, their inner product is 0. Thus $\langle \phi_{j,k}(t), \phi_{j,\ell}(t) \rangle = 0$ and we see that the set $\{\phi_{j,k}(t)\}_{k\in\mathbb{Z}}$ is an orthonormal set of functions in V_j.

To show that S forms a basis for V_j, we must show that the elements of S are linearly independent and that we can write any function $f(t) \in V_j$ as a linear combination of the elements from S. We leave to Problem 5.2 the proof that the elements of S are linearly independent.

To see that S spans V_j, let $f(t)$ be any element in V_j. Then by the scaling property of Definition 5.1, we know that $f(2^{-j}t) \in V_0$. Since the set $\{\phi(t-k)\}_{k\in\mathbb{Z}}$ is an orthonormal basis for V_0, we know that for $k \in \mathbb{Z}$ there exists $a_k \in \mathbb{R}$ so that

$$f(2^{-j}t) = \sum_{k\in\mathbb{Z}} a_k\, \phi(t - k)$$

Replacing t by $2^j t$ and multiplying and dividing each term on the right-hand side by $2^{j/2}$ gives

$$f(t) = \sum_{k\in\mathbb{Z}} \frac{a_k}{2^{j/2}} \cdot 2^{j/2}\phi(2^j t - k) = \sum_{k\in\mathbb{Z}} \frac{a_k}{2^{j/2}} \phi_{j,k}(t)$$

Thus we have written $f(t) \in V_j$ as a linear combination of the elements of S, and the proof is complete. ∎

The following corollary gives us an explicit representation of $f(t) \in V_j$ in terms of the basis functions $\phi_{j,k}(t)$.

Corollary 5.1 (Explicit Form for f(t) ∈ V$_j$) *Suppose that $f(t) \in V_j$ and*

$$f(t) = \sum_{k \in \mathbb{Z}} c_k \phi_{j,k}(t) \tag{5.5}$$

Then

$$c_k = \langle f(t), \phi_{j,k}(t) \rangle$$

■

Proof: The proof of this corollary is left as Problem 5.3.   ■

The Dilation Equation and Related Results

Our next result shows that the scaling function $\phi(t)$ from a multiresolution analysis satisfies a *dilation equation*.

Proposition 5.3 (The Dilation Equation for ϕ(t)) *Suppose that $\phi(t)$ is the scaling function for a multiresolution analysis $\{V_j\}_{j \in \mathbb{Z}}$ of $L^2(\mathbb{R})$. Then $\phi(t)$ satisfies the dilation equation*

$$\phi(t) = \sqrt{2} \sum_{k \in \mathbb{Z}} h_k \phi(2t - k) \tag{5.6}$$

where

$$h_k = \langle \phi(t), \phi_{1,k}(t) \rangle \tag{5.7}$$

In addition, for $\phi_{j,k}(t)$ defined in (5.3), we have

$$\phi_{j,\ell}(t) = \sum_{k \in \mathbb{Z}} h_{k-2\ell}\, \phi_{j+1,k}(t) \tag{5.8}$$

The coefficients h_k, $k \in \mathbb{Z}$ that appear in (5.6) form what is called the scaling filter. *We typically denote this filter by $\mathbf{h} = (\dots, h_{-1}, h_0, h_1, \dots)$.*   ■

Proof: Since the V_j spaces are nested, we have in particular, $V_0 \subset V_1$. Thus $\phi(t) \in V_0$ implies that $\phi(t) \in V_1$. From Proposition 5.2 we know that $\phi(t)$ can be expressed as a linear combination of the basis elements $\{\phi_{1,k}(t)\}_{k \in \mathbb{Z}}$. Thus there exists real numbers h_k, $k \in \mathbb{Z}$, such that

$$\phi(t) = \sum_{k \in \mathbb{Z}} h_k \phi_{1,k}(t) = \sqrt{2} \sum_{k \in \mathbb{Z}} h_k \phi(2t - k)$$

From Corollary 5.1 we know that

$$h_k = \langle \phi(t), \phi_{1,k}(t) \rangle = \sqrt{2} \int_{\mathbb{R}} \phi(t)\phi(2t - k)\, dt$$

Thus we have established the validity of (5.6). If we replace t by $2^j t - \ell$ in (5.6), we obtain

$$\phi(2^j t - \ell) = \sqrt{2} \sum_{k \in \mathbb{Z}} h_k \, \phi\left(2(2^j t - \ell) - k\right)$$

$$= \sqrt{2} \sum_{k \in \mathbb{Z}} h_k \, \phi\left(2^{j+1} t - (k + 2\ell)\right) \tag{5.9}$$

We now make the change of variable $m = k + 2\ell$. The limits on the sum remain the same and we also note that $k = m - 2\ell$. Inserting these relations and multiplying both sides of (5.9) by $2^{j/2}$ gives

$$2^{j/2}\phi(2^j t - \ell) = \phi_{j,\ell}(t) = 2^{(j+1)/2} \sum_{m \in \mathbb{Z}} h_{m-2\ell} \, \phi\left(2^{j+1} t - m\right)$$

$$= \sum_{m \in \mathbb{Z}} h_{m-2\ell} \, \phi_{j+1,m}(t)$$

and completes the proof. ∎

Conversely, we can show that if scaling function $\phi(t)$ satisfies a dilation equation (5.6) then the scaling and nested properties of a multiresolution analysis (5.1) are satisfied.

Proposition 5.4 (Scaling and Nested Spaces) *Suppose that $V_j = \text{span}\{\phi(2^j t - k)\}_{k \in \mathbb{Z}} \subset L^2(\mathbb{R})$ and that $\phi(t) \in V_0$ satisfies a dilation equation (5.6). Then $V_j \subset V_{j+1}$ and $f(t) \in V_0$ if and only if $f(2^j t) \in V_j$.* ∎

Proof: We prove that $V_j \subset V_{j+1}$ and leave the proof that the scaling property is satisfied to Problem 5.8. Suppose that $f(t) \in V_j$. Since V_j is spanned by $\phi\left(2^j t - k\right)$, $k \in \mathbb{Z}$, we can write

$$f(t) = \sum_{k \in \mathbb{Z}} a_k \phi\left(2^j t - k\right) \tag{5.10}$$

But $\phi(t)$ satisfies a dilation equation (5.6), so we can write

$$\phi\left(2^j t - k\right) = \sqrt{2} \sum_{\ell \in \mathbb{Z}} h_\ell \phi(2\left(2^j t - k\right) - \ell)$$

$$= \sqrt{2} \sum_{\ell \in \mathbb{Z}} h_\ell \phi\left(2^{j+1} t - 2k - \ell\right) \tag{5.11}$$

Inserting (5.11) into (5.10) gives

$$f(t) = \sum_{k \in \mathbb{Z}} a_k \left(\sqrt{2} \sum_{\ell \in \mathbb{Z}} h_\ell \phi\left(2^{j+1} t - 2k - \ell\right)\right)$$

$$= \sqrt{2} \sum_{k \in \mathbb{Z}} a_k \left(\sum_{\ell \in \mathbb{Z}} h_\ell \phi\left(2^{j+1} t - (\ell + 2k)\right)\right)$$

Now we make the substitution $m = \ell + 2k$ on the inner sum. Then $\ell = m - 2k$ and we have

$$f(t) = \sqrt{2} \sum_{k \in \mathbb{Z}} a_k \left(\sum_{m \in \mathbb{Z}} h_{m-2k} \phi \left(2^{j+1} t - m \right) \right)$$

$$= \sum_{m \in \mathbb{Z}} \left(\sqrt{2} \sum_{k \in \mathbb{Z}} a_k h_{m-2k} \right) \phi \left(2^{j+1} t - m \right)$$

Thus we have written $f(t)$ as a linear combination of functions $\phi \left(2^{j+1} t - m \right), m \in \mathbb{Z}$, where the coefficients are $p_m = \sqrt{2} \sum_{k \in \mathbb{Z}} a_k h_{m-2k}$, so that $f(t) \in V_{j+1}$. ∎

We can use the dilation equation (5.6) to obtain properties satisfied by elements of the scaling filter **h**.

Proposition 5.5 (Properties of the Scaling Filter) *Suppose that $\{V_j\}_{j \in \mathbb{Z}}$ is a multiresolution analysis of $L^2(\mathbb{R})$ with scaling function $\phi(t)$. If **h** is the scaling filter with real-valued elements from the dilation equation (5.6), then:*

(i) $\displaystyle\sum_{k \in \mathbb{Z}} h_k = \sqrt{2}$

(ii) $\displaystyle\sum_{k \in \mathbb{Z}} h_k h_{k-2\ell} = \delta_{0,\ell} \text{ for all } \ell \in \mathbb{Z}$

(iii) $\displaystyle\sum_{k \in \mathbb{Z}} h_k^2 = 1$

∎

Proof: The proofs of (i) and (ii) are left as Problems 5.5 and 5.6, respectively. Part (iii) follows easily from (ii) by taking $\ell = 0$ in (ii). ∎

Interpreting the Scaling Filter

The scaling filter is an important tool for applications involving wavelets. Indeed we learned in Chapter 4 how the scaling filter **h** (3.29) was used to develop the discrete Haar wavelet transformation. Engineers typically refer to the scaling filter as a *lowpass filter*. To understand this terminology, let's return to the Haar scaling filter $\mathbf{h} = [h_0, h_1]^T = \left[\frac{\sqrt{2}}{2}, \frac{\sqrt{2}}{2} \right]^T$. In Problem 4.1(a) you showed that the Fourier series constructed using this vector as the Fourier coefficients is

$$\mathcal{H}(\omega) = \frac{\sqrt{2}}{2} + \frac{\sqrt{2}}{2} e^{i\omega} = \sqrt{2} e^{i\omega/2} \cos\left(\frac{\omega}{2} \right)$$

with (using part (b) of Problem 4.1)

$$|\mathcal{H}(\omega)| = \sqrt{2} \left| \cos\left(\frac{\omega}{2} \right) \right|$$

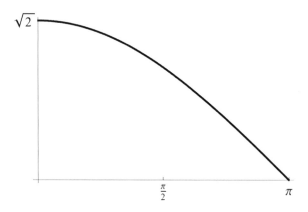

Figure 5.1   A plot of $|\mathcal{H}(\omega)|$ on $[0, \pi]$.

Using Problem 2.15, we know that $p(\omega) = |\mathcal{H}(\omega)|$ is an even, 2π-periodic function. Thus we consider a plot of $|\mathcal{H}(\omega)|$ on the interval $[0, \pi]$ in Figure 5.1.

Given input $\mathbf{x} = (\ldots, x_{-2}, x_{-1}, x_0, x_1, x_2, \ldots)$ in the form of a bi-infinite sequence, it is often processed via convolution with $\mathbf{h}$. That is, $\mathbf{y} = \mathbf{h} * \mathbf{x}$ with

$$y_n = \sum_{k \in \mathbb{Z}} h_k x_{n-k} = \sum_{k=0}^{1} \frac{\sqrt{2}}{2} x_{n-k} = \sqrt{2}\left(\frac{x_n + x_{n-1}}{2}\right) \qquad (5.12)$$

The plot of $|\mathcal{H}(\omega)|$ indicates how $\mathbf{h}$ will process $\mathbf{x}$ with respect to the oscillatory nature of the elements of $\mathbf{x}$. At frequency $\omega = 0$ (no oscillation in consecutive elements of $\mathbf{x}$), $|\mathcal{H}(\omega)|$ attains a maximum, so we expect the values y_n to be a scaled version of x_n. Indeed, if $x_n = x_{n-1}$, then using (5.12), we see that $y_n = \sqrt{2} x_n$. On the other hand, $\omega = \pi$ represents the largest frequency (most oscillation in consecutive values of $\mathbf{x}$) and $|\mathcal{H}(\pi)| = 0$ here. Indeed, if $x_n = -x_{n-1}$, then $y_n = 0$. Such filters that preserve homogeneous trends in $\mathbf{x}$ and attenuate highly oscillatory trends in $\mathbf{x}$, are called *lowpass filters*. Using the modulus of the Fourier series associated with a filter, we can quickly determine if a filter is lowpass — it achieves a maximum at $\omega = 0$ and has value zero at $\omega = \pi$.

As you might have already guessed, the scaling filter will be exactly the elements we use to form the averages portion of a discrete wavelet transform. At first glance, implementation might seem to be numerically intractable since the sum in (5.6) is infinite. For practical applications we construct scaling functions that are finitely supported. In this case the infinite sum (5.6) will reduce to a finite sum. See Problem 5.4 for more details.

We link convolution to the scaling filter $\mathbf{h}$ later in the section (see (5.16)).

Note: In this book, we consider only scaling filters $\mathbf{h}$ with real-valued elements.

Projections into the Space V_j

As was the case with the Haar spaces in Chapter 3, we want to think of the V_j spaces for a general multiresolution analysis of $L^2(\mathbb{R})$ as approximation spaces. Thus we have need for projecting an arbitrary function $f(t) \in L^2(\mathbb{R})$ into V_j. Recall that the projection of $f(t)$ into V_j can be written as

$$P_{f,j}(t) = \sum_{k \in \mathbb{Z}} \langle f(t), \phi_{j,k}(t) \rangle \phi_{j,k}(t) \qquad (5.13)$$

Since the V_j spaces are nested, we can move from a "finer" approximation space, V_{j+1}, to a "coarser" one, V_j. We do this via a projection and the following result.

Proposition 5.6 (Projecting Functions from V_{j+1} into V_j) *Assume that $\{V_j\}_{j \in \mathbb{Z}}$ is a multiresolution analysis of $L^2(\mathbb{R})$ with associated scaling function $\phi(t)$. Suppose that $f_{j+1}(t) \in V_{j+1}$ is of the form*

$$f_{j+1}(t) = \sum_{k \in \mathbb{Z}} a_k \phi_{j+1,k}(t) \qquad (5.14)$$

where $\phi_{j+1,k}(t)$ is given by (5.3). If $f_j(t)$ is the projection of $f_{j+1}(t)$ into V_j, then $f_j(t)$ has the form

$$f_j(t) = \sum_{\ell \in \mathbb{Z}} b_\ell \phi_{j,\ell}(t) = \sum_{\ell \in \mathbb{Z}} \left(\sum_{k \in \mathbb{Z}} a_k h_{k-2\ell} \right) \phi_{j,\ell}(t) \qquad (5.15)$$

where h_k, $k \in \mathbb{Z}$ are the scaling filter coefficients given in Proposition 5.3. ∎

Proof: We begin by writing down the projection of $f_{j+1}(t)$ into V_j. We have

$$f_j(t) = \sum_{\ell \in \mathbb{Z}} \langle f_{j+1}(t), \phi_{j,\ell}(t) \rangle \phi_{j,\ell}(t)$$

We now insert (5.14) into the identity above and simplify to obtain

$$f_j(t) = \sum_{\ell \in \mathbb{Z}} \langle f_{j+1}(t), \phi_{j,\ell}(t) \rangle \phi_{j,\ell}(t)$$

$$= \sum_{\ell \in \mathbb{Z}} \left\langle \sum_{k \in \mathbb{Z}} a_k \phi_{j+1,k}(t), \phi_{j,\ell}(t) \right\rangle \phi_{j,\ell}(t)$$

$$= \sum_{\ell \in \mathbb{Z}} \sum_{k \in \mathbb{Z}} a_k \left\langle \phi_{j+1,k}(t), \phi_{j,\ell}(t) \right\rangle \phi_{j,\ell}(t)$$

If we can show that $\langle \phi_{j+1,k}(t), \phi_{j,\ell}(t) \rangle = h_{k-2\ell}$, the proof is complete. We use (5.8) and write

$$\langle \phi_{j+1,k}(t), \phi_{j,\ell}(t) \rangle = \left\langle \phi_{j+1,k}(t), \sum_{m \in \mathbb{Z}} h_{m-2\ell} \phi_{j+1,m}(t) \right\rangle$$

$$= \sum_{m \in \mathbb{Z}} h_{m-2\ell} \left\langle \phi_{j+1,k}(t), \phi_{j+1,m}(t) \right\rangle$$

Now $\phi_{j+1, k}(t)$ and $\phi_{j+1, m}(t)$ are part of an orthonormal basis for V_{j+1} so that their inner product is 0 when $k \neq m$ and 1 when $k = m$. Thus there is only one nonzero term in the sum above and that is when $k = m$. The value of this term is $h_{k-2\ell}$ and the proof is complete. ∎

Projection as a Matrix Product

It is interesting to look further at the projection coefficients $b_\ell = \sum_{k \in \mathbb{Z}} a_k h_{k-2\ell}$. For each $\ell \in \mathbb{Z}$, we can think of b_ℓ as an inner product of the bi-infinite sequence $\mathbf{a} = (\ldots, a_{-1}, a_0, a_1, \ldots)$ and a shifted version of the bi–infinite sequence $\mathbf{h} = (\ldots, h_{-1}, h_0, h_1, \ldots)$. For example, when $\ell = 0$, we compute $b_0 = \sum_{k \in \mathbb{Z}} h_k a_k$. When $\ell = -1, 1$, the computations are $b_{-1} = \sum_{k \in \mathbb{Z}} a_k h_{k+2}$ and $b_1 = \sum_{k \in \mathbb{Z}} a_k h_{k-2}$, respectively. The computation $b_\ell = \sum_{k \in \mathbb{Z}} a_k h_{k-2\ell}$ is very similar to the discrete convolution element (see Problem 2.22) $y_{2n} = \sum_{k \in \mathbb{Z}} a_k h_{2n-k}$. Indeed, if $h_m = h_{-m}, m \in \mathbb{Z}$, then the computations are the same.

We can also formulate the computation of the projection coefficients as a matrix product. We form the infinite-dimensional matrix $\mathcal{H}$ by placing $\mathbf{h}$ in row zero with h_0 in column zero. Row $\ell, \ell > 0$ ($\ell < 0$), is formed by shifting row zero 2ℓ units to the right (left). We have

$$
\mathbf{b} = \mathcal{H}\mathbf{a} =
\begin{bmatrix}
\ddots & \ddots & & & & & & & \\
\cdots & h_1 & \mathbf{h_2} & h_3 & h_4 & h_5 & h_6 & h_7 & h_8 & \cdots \\
\cdots & h_{-1} & h_0 & \mathbf{h_1} & h_2 & h_3 & h_4 & h_5 & h_6 & \cdots \\
\cdots & h_{-3} & h_{-2} & h_{-1} & \mathbf{h_0} & h_1 & h_2 & h_3 & h_4 & \cdots \\
\cdots & h_{-5} & h_{-4} & h_{-3} & h_{-2} & \mathbf{h_{-1}} & h_0 & h_1 & h_2 & \cdots \\
\cdots & h_{-7} & h_{-6} & h_{-5} & h_{-4} & h_{-3} & \mathbf{h_{-2}} & h_{-1} & h_0 & \cdots \\
& & \ddots & & & & \ddots & & & \ddots
\end{bmatrix}
\begin{bmatrix}
\vdots \\
a_{-2} \\
a_{-1} \\
a_0 \\
a_1 \\
a_2 \\
\vdots
\end{bmatrix}
$$

$$(5.16)$$

The boldface elements in $\mathcal{H}$ lie along the main diagonal with the row zero containing the boldface h_0.

In some cases (see Problem 5.9) we can use $\rho(t) = \phi(-t)$ as our scaling function for V_0, and in this case, the scaling filter is $\tilde{\mathbf{h}}$, where $\tilde{h}_k = h_{-k}, k \in \mathbb{Z}$. Then the projection coefficients are

$$
b_\ell = \sum_{k \in \mathbb{Z}} a_k \tilde{h}_{k-2\ell} = \sum_{k \in \mathbb{Z}} a_k h_{2\ell-k}
$$

the same as the even elements $y_{2\ell}$ in the discrete convolution product $\mathbf{y} = \mathbf{a} * \mathbf{h}$ (see Problem 2.22). In this case we *downsample* the convolution product $\mathbf{y}$ (i. e., remove the odd elements) to obtain the projection coefficients.

Except for some examples that appear in Section 5.3, all of our scaling functions $\phi(t)$ are supported on a finite interval. In such cases, the scaling filter **h** will contain only finitely many nonzero terms and the matrix $\mathcal{H}$ will be a *band or a bandlimited matrix* (see, e. g., Meyer [43] for example).

The matrix product (5.16) gives us an idea of how to construct the top half $H_{N/2}$, N even, of the discrete wavelet transformation matrix W_N. Let's look at an example in the case where $\phi(t) = \sqcap(t)$ for the Haar multiresolution analysis.

Example 5.2 (The Haar Scaling Filter and Projection Matrix) *Let $\phi(t) = \sqcap(t)$ be the Haar scaling function. From (3.33) we know that the only nonzero elements in the scaling filter* **h** *are $h_0 = h_1 = \frac{\sqrt{2}}{2}$. Then the projection coefficients are $b_\ell = \sum_{k\in\mathbb{Z}} h_{k-2\ell}a_k$, and we know by Proposition 5.6 that the sum reduces to two terms. The only nonzero terms result when $k - 2\ell = 0$ (so that $k = 2\ell$) and $k - 2\ell = 1$ (so that $k = 2\ell + 1$). Thus the project coefficients are*

$$\frac{\sqrt{2}}{2}a_{2\ell} + \frac{\sqrt{2}}{2}a_{2\ell+1}$$

This is exactly the result that we obtained in Proposition 3.24.

We can compute the projection coefficients b_ℓ using the matrix product $\mathbf{b} = \mathcal{H}\mathbf{a}$ where $\mathcal{H}$ is formed using (5.16). In this case we have

$$
\mathbf{b} = \mathcal{H}\mathbf{a} =
\begin{bmatrix}
\ddots & & \ddots & & & & & & & \\
\cdots & \frac{\sqrt{2}}{2} & 0 & 0 & 0 & 0 & 0 & 0 & \cdots \\
\cdots & 0 & \frac{\sqrt{2}}{2} & \frac{\sqrt{2}}{2} & 0 & 0 & 0 & 0 & \cdots \\
\cdots & 0 & 0 & 0 & \frac{\sqrt{2}}{2} & \frac{\sqrt{2}}{2} & 0 & 0 & \cdots \\
\cdots & 0 & 0 & 0 & 0 & 0 & \frac{\sqrt{2}}{2} & \frac{\sqrt{2}}{2} & 0 & \cdots \\
\cdots & 0 & 0 & 0 & 0 & 0 & 0 & 0 & \frac{\sqrt{2}}{2} & \cdots \\
& & & & \ddots & & & \ddots & & \ddots
\end{bmatrix}
\begin{bmatrix}
\vdots \\ a_{-2} \\ a_{-1} \\ a_0 \\ a_1 \\ a_2 \\ \vdots
\end{bmatrix}
$$

Here row zero is boldface with the first $h_0 = \frac{\sqrt{2}}{2}$, the first nonzero element in the row, on the main diagonal. If $a_0, \ldots, a_{N-1}$ are the only nonzero elements in $\mathbf{a}$, then we can truncate $\mathcal{H}$ to form the $N/2 \times N$ matrix $H_{N/2}$ (see Definition 4.1). ∎

The Wavelet Function

We can generalize the notion of the Haar wavelet spaces W_j developed in Section 3.4 for a general multiresolution analysis $\{V_j\}_{j\in\mathbb{Z}}$. Since $V_j \subset V_{j+1}$ for $j \in \mathbb{Z}$, we can build better approximations to some function $f(t) \in L^2(\mathbb{R})$ by increasing j. In many applications, it is necessary that we record the details (or error) as we move from one approximate space to the next. This is what the the Haar wavelet space W_j does for us — if $f_j(t), f_{j+1}(t) \in L^2(\mathbb{R})$ are projections of some $f(t) \in L^2(\mathbb{R})$ into Haar spaces V_j, V_{j+1}, respectively, then the residual function $g_j(t) = f_{j+1}(t) - f_j(t)$ is not only an element of the Haar wavelet space W_j, but it is also the orthogonal projection of $f(t)$ into W_j.

For the Haar spaces, we are actually able to calculate the residual function $g_j(t)$ and verify that it was indeed the projection of $f(t)$ into W_j (see Proposition 3.25 and Problem 3.37 in Section 3.4). The fundamental tool for this computation is the Haar wavelet function $\psi(t) = \phi(2t) - \phi(2t - 1) = \sqcap(2t) - \sqcap(2t - 1) \in W_0$ and more generally, the family of functions $\psi_{j,k}(t) = 2^{j/2}\psi\left(2^j t - k\right) \in W_j$ (see (3.35)). In order to construct general wavelet spaces, we first define the *wavelet function*.

Definition 5.3 (The Wavelet Function and Wavelet Filter) *Suppose that* $\{V_j\}_{j \in \mathbb{Z}}$ *forms a multiresolution analysis for* $L^2(\mathbb{R})$ *with scaling function* $\phi(t)$ *that satisfies the dilation equation (5.6). We define the* wavelet function $\psi(t) \in V_1$ *by*

$$\psi(t) = \sqrt{2} \sum_{k \in \mathbb{Z}} g_k \phi(2t - k) \qquad (5.17)$$

and more generally, the functions $\psi_{j,k}(t) \in V_{j+1}$ *by*

$$\psi_{j,k}(t) = 2^{j/2}\psi(2^j t - k) \qquad (5.18)$$

for integers $j, k \in \mathbb{Z}$. *The* wavelet filter *is given by* $\mathbf{g} = (\ldots, g_{-1}, g_0, g_1, \ldots)$, *where*

$$g_k = (-1)^k h_{1-k}, \ k \in \mathbb{Z} \qquad (5.19)$$

∎

Motivation for the definition of the wavelet filter will be given in the proof of Theorem 5.1 and Problems 5.12 and 5.13. For now, let's look at an example.

Example 5.3 (The Haar Wavelet Function and Wavelet Filter) *Consider the Haar multiresolution analysis with scaling function* $\phi(t)$ *given by (3.1). Then the scaling filter* $\mathbf{h}$ *is given by the dilation equation (3.33) so that* $h_0 = h_1 = \sqrt{2}/2$ *and all other* $h_k = 0$. *We can use (5.17) to find the Haar wavelet* $\psi(t)$. *We have*

$$\psi(t) = \sqrt{2} \sum_{k \in \mathbb{Z}} g_k \phi(2t - k) = \sqrt{2} \sum_{k \in \mathbb{Z}} (-1)^k h_{1-k} \phi(2t - k)$$

The only terms that are nonzero in the identity above occur when $k = 0, 1$. *We have*

$$\psi(t) = \sqrt{2} \left((-1)^0 h_{1-0}\phi(2t - 0) + (-1)^1 h_{1-1}\phi(2t - 1)\right)$$
$$= \sqrt{2} \left(h_1\phi(2t) - h_0\phi(2t - 1)\right)$$
$$= \sqrt{2} \left(\frac{\sqrt{2}}{2}\phi(2t) - \frac{\sqrt{2}}{2}\phi(2t - 1)\right)$$
$$= \phi(2t) - \phi(2t - 1)$$

Note that $\phi(2t) - \phi(2t-1)$ is exactly the way that we defined $\psi(t)$ in (3.25). Moreover, the dilation equation with the boldface coefficients matches the dilation equation (3.34) derived in Section 3.3. ∎

Like the scaling functions $\phi_{j,k}(t)$, the wavelet functions $\psi_{j,k}(t)$ also satisfy a dilation equation.

Proposition 5.7 (Dilation Equation for $\psi_{j,k}(t)$) *Let $\psi_{j,k}(t)$ be given by (5.18), where $j, k \in \mathbb{Z}$. For any integer $\ell \in \mathbb{Z}$ we have*

$$\psi_{j,\ell}(t) = \sum_{k \in \mathbb{Z}} g_{k-2\ell}\phi_{j+1,k}(t) = \sum_{k \in \mathbb{Z}} (-1)^k h_{1+2\ell-k}\phi_{j+1,k}(t) \tag{5.20}$$

∎

Proof: The proof of Proposition 5.7 is similar to the argument given to establish (5.8) in Proposition 5.3 and is left as Problem 5.10. ∎

Interpreting the Wavelet Filter

We can analyze the wavelet filter in a way quite similar to that done for the scaling filter (see Figure 5.1). Engineers often refer to the wavelet filter as a *highpass filter*. To understand this term, let's first create the Fourier series $\mathcal{G}(\omega)$ whose only nonzero coefficients are wavelet filter coefficients g_0 and g_1 and then plot $|\mathcal{G}(\omega)|$ on $[0, \pi]$. We have

$$\begin{aligned}
\mathcal{G}(\omega) &= \frac{\sqrt{2}}{2} - \frac{\sqrt{2}}{2}e^{i\omega} \\
&= \frac{\sqrt{2}}{2}e^{i\omega/2}\left(e^{-i\omega/2} - e^{i\omega/2}\right) \\
&= -\sqrt{2}\,ie^{i\omega/2}\left(\frac{e^{i\omega/2} - e^{-i\omega/2}}{2i}\right) \\
&= -\sqrt{2}\,ie^{i\omega/2}\sin\left(\frac{\omega}{2}\right)
\end{aligned}$$

We can then compute $|\mathcal{G}(\omega)| = \sqrt{2}\left|\sin\left(\frac{\omega}{2}\right)\right|$. We know that $\mathcal{G}(\omega)$ is an even, 2π-periodic function, so a graph of $|\mathcal{G}(\omega)|$ on $[0, \pi]$ is sufficient to analyze the modulus of the series.

Note that this graph is a reflection about $\omega = \frac{\pi}{2}$ of $|\mathcal{H}(\omega)|$ on $[0, \pi]$. Our interpretation is as follows: Where **h**, when convolved with an input vector, tends to preserve consecutive values that are homogeneous, the convolution of these values with **g** will produce zero or near-zero values. On the other hand, **h** will attenuates oscillatory data where **g** will preserve this feature of the input. So when we convolve data with a highpass filter, the filter will preserve oscillatory trends in the data and dampen or annihilate homogeneous trends in the data. We link convolution to the wavelet filter **g** later in the section (see (5.27)). Here we will see that the wavelet filter **g** is what we use to construct the details portion of the discrete wavelet transformation.

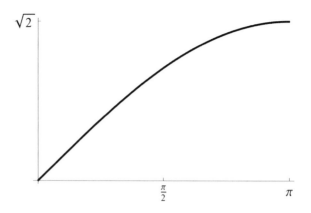

Figure 5.2   A plot of $|\mathcal{G}(\omega)|$ on $[0, \pi]$.

The Wavelet Spaces W_j

The following result, due to Stephane Mallat [41] and Yves Meyer [44], identifies the orthogonal complement W_j to V_j in V_{j+1} and establishes an orthonormal basis for W_j.

Theorem 5.1 (An Orthonormal Basis for W_j) *Suppose that $\{V_j\}_{j \in \mathbb{Z}}$ is a multiresolution analysis for $L^2(\mathbb{R})$ with scaling function $\phi(t)$ that satisfies the dilation equation*

$$\phi(t) = \sqrt{2} \sum_{k \in \mathbb{Z}} h_k \phi(2t - k)$$

Here, we assume that the scaling filter $\mathbf{h} = (\ldots, h_{-1}, h_0, h_1, \ldots)$ is comprised of real numbers.

Let the wavelet function $\psi(t) \in V_1$ be given by Definition 5.3. If

$$W_j = \mathrm{span}\{\psi_{j,k}(t)\}_{k \in \mathbb{Z}} \tag{5.21}$$

then W_j is the orthogonal complement of V_j in V_{j+1}. In other words,

$$\boxed{V_{j+1} = V_j \oplus W_j} \tag{5.22}$$

Moreover, the set $\{\psi_{j,k}(t)\}_{k \in \mathbb{Z}}$ forms an orthonormal basis for W_j.   ∎

Proof: The proof of this theorem is quite technical, and the arguments that appear in [41] and [44] utilize the Fourier transform. A very nice proof of this theorem that does not utilize the Fourier transform appears in the book by Boggess and Narcowich [4]. Here the authors establish the result for $j = 0$ and then use the scaling condition from Definition 5.1 to complete the proof for an arbitrary j. As noted in [4], to prove the result for $j = 0$, it must first be shown that

$$\langle \psi_{0,m}(t), \psi_{0,n}(t) \rangle = \langle \psi(t - m), \psi(t - n) \rangle = \delta_{m,n} \tag{5.23}$$

You will establish this result in Problem 5.12.

Next, we show that $V_0 \perp W_0$ by showing that the respective basis functions $\phi_{0,m}(t) = \phi(t-m)$ and $\psi_{0,n}(t) = \psi(t-n)$ satisfy

$$\langle \phi(t-m), \psi(t-n) \rangle = 0 \qquad (5.24)$$

You will establish the proof of this result in Problem 5.13.

Recall $W_0 \subset V_1$ since by virtue of (5.18), the basis functions $\psi_{0,\ell}(t) = \psi(t-\ell)$, $\ell \in \mathbb{Z}$ are all elements of V_1. Once we have verified (5.24) we know that W_0 is contained in the orthogonal complement of V_0 in V_1. The final piece of the proof is to establish the fact that the orthogonal complement of V_0 in V_1 is contained in W_0. Then we have set equality and the result holds. We refer the reader to Appendix A.2.1 in Boggess and Narcowich [4] for a nice proof of this result. ∎

Suppose that $m, M \in \mathbb{Z}$ with $m < M$ and $\{V_j\}_{j, \in \mathbb{Z}}$ is a multiresolution analysis of $L^2(\mathbb{R})$. Using Theorem 5.1 we can write, $V_M = V_{M-1} \oplus W_{M-1}$. We can apply Theorem 5.1 recursively to the spaces $V_{M-1}, \ldots, V_m$:

$$\begin{aligned} V_M &= V_{M-1} \oplus W_{M-1} \\ &= V_{M-2} \oplus W_{M-2} \oplus W_{M-1} \\ &= V_{M-3} \oplus W_{M-3} \oplus W_{M-2} \oplus W_{M-1} \\ &\vdots \\ &= V_m \oplus W_m \oplus W_{m+1} \oplus \cdots \oplus W_{M-1} \qquad (5.25) \end{aligned}$$

Now the V_j spaces are nested so that V_M holds all V_j where $j < M$. Thus as $M \to \infty$, the density property from Definition 5.1 says that W_M and thus the sum of spaces in (5.25) becomes a better approximation of $L^2(\mathbb{R})$. On the other hand, the separation property of Definition 5.1 and the nested nature of the V_j spaces indicates that as $m \to -\infty$, the spaces V_m resemble the space consisting of only the function $f(t) = 0$. This heuristic argument leads to the following important theorem in wavelet theory.

Theorem 5.2 (Infinite Orthogonal Sum of the Wavelet Spaces) *Let $\{V_j\}_{j \in \mathbb{Z}}$ be a multiresolution analysis for $L^2(\mathbb{R})$ and that the wavelet spaces $\{W_j\}_{j \in \mathbb{Z}}$ are as defined in (5.22). Then*

$$L^2(\mathbb{R}) = \cdots \oplus W_{-2} \oplus W_{-1} \oplus W_1 \oplus W_2 \oplus \cdots$$

and the set $\{\psi_{j,k}(t)\}_{j,k \in \mathbb{Z}}$, where $\psi_{j,k}(t)$ is given by (5.18), is an orthonormal basis for $L^2(\mathbb{R})$. ∎

Proof: The proof is quite technical and beyond the scope of this book. The interested reader is referred to Walnut [62]. ∎

Theorem 5.2 tells us that any $f(t) \in L^2(\mathbb{R})$ can be written as

$$f(t) = \sum_{j \in \mathbb{Z}} w_j(t)$$

where $w_j(t) \in W_j$ and $w_j(t)$ is orthogonal to $w_\ell(t)$ whenever $j \neq \ell$. Moreover, we can also write $L^2(\mathbb{R})$ as

$$L^2(\mathbb{R}) = V_J \oplus W_J \oplus W_{J-1} \oplus W_{J-2} \oplus \cdots$$

Projecting from V_{j+1} into W_j

To decompose a function $f_{j+1}(t) \in V_{j+1}$, we use the fact that $V_{j+1} = V_j \oplus W_j$ and write $f_{j+1} = f_j(t) + w_j(t)$ where $f_j(t)$, $w_j(t)$ are orthogonal, $f_j(t) \in V_j$ is the orthogonal projection of $f_{j+1}(t)$ into V_j and $w_j(t) \in W_j$ is the orthogonal projection of f_{j+1} into W_j. We already have a formula (5.15) for $f_j(t)$. Now we seek a formula for $w_j(t)$. We have the following proposition:

Proposition 5.8 (Projecting Functions from V_{j+1} into W_j) *Assume that $\{V_j\}_{j \in \mathbb{Z}}$ is a multiresolution analysis of $L^2(\mathbb{R})$ with associated scaling function $\phi(t)$. Let the spaces W_j be given by (5.21) with associated wavelet function $\psi(t)$. Suppose that $f_{j+1}(t) \in V_{j+1}$ is of the form*

$$f_{j+1}(t) = \sum_{k \in \mathbb{Z}} a_k \phi_{j+1,\,k}(t)$$

where $\phi_{j+1,\,k}(t)$ is given by (5.3) and $a_k = \langle f_{j+1}(t), \phi_{j+1,k}(t) \rangle$. If $w_j(t)$ is the projection of $f_{j+1}(t)$ into W_j, then $w_j(t)$ has the form

$$w_j(t) = \sum_{\ell \in \mathbb{Z}} c_\ell \psi_{j,\,\ell}(t) = \sum_{\ell \in \mathbb{Z}} \left(\sum_{k \in \mathbb{Z}} a_k g_{k-2\ell} \right) \psi_{j,\,\ell}(t)$$

$$= \sum_{\ell \in \mathbb{Z}} \left(\sum_{k \in \mathbb{Z}} a_k (-1)^k h_{1+2\ell-k} \right) \psi_{j,\,\ell}(t) \qquad (5.26)$$

where g_k, $k \in \mathbb{Z}$ are the wavelet filter coefficients given by (5.19) and $\psi_{j,k}(t)$ is given by (5.18). ∎

Proof: The proof of Proposition 5.8 and is left as Problem 5.15. ∎

Projection as a Matrix Product

In (5.16) we saw how to write the projection coefficients for sending $f_{j+1}(t) \in V_{j+1}$ into V_j as a matrix product. We can do the same for the coefficients c_ℓ, $\ell \in \mathbb{Z}$ for the coefficients needed to write the projection of $f_{j+1}(t)$ into W_j. If $\mathbf{c} = [\ldots, c_{-2}, c_{-1}, c_0, c_1, c_2, \ldots]^T$ and $\mathbf{a} = [\ldots, a_{-2}, a_{-1}, a_0, a_1, a_2, \ldots]^T$, then we can write **c** in terms of **a** using the infinite-dimensional matrix $\mathcal{G}$ and write

$$
\mathbf{c} = \mathcal{G}\mathbf{a} =
\begin{bmatrix}
\ddots & \ddots & & & & & & & \\
\cdots & g_1 & \mathbf{g_2} & g_3 & g_4 & g_5 & g_6 & g_7 & g_8 & \cdots \\
\cdots & g_{-1} & g_0 & \mathbf{g_1} & g_2 & g_3 & g_4 & g_5 & g_6 & \cdots \\
\cdots & g_{-3} & g_{-2} & g_{-1} & \mathbf{g_0} & g_1 & g_2 & g_3 & g_4 & \cdots \\
\cdots & g_{-5} & g_{-4} & g_{-3} & g_{-2} & \mathbf{g_{-1}} & g_0 & g_1 & g_2 & \cdots \\
\cdots & g_{-7} & g_{-6} & g_{-5} & g_{-4} & g_{-3} & \mathbf{g_{-2}} & g_{-1} & g_0 & \cdots \\
& & & \ddots & & & \ddots & & & \ddots
\end{bmatrix}
\begin{bmatrix}
\vdots \\ a_{-2} \\ a_{-1} \\ a_0 \\ a_1 \\ a_2 \\ \vdots
\end{bmatrix}
$$

$$(5.27)$$

The boldface elements in $\mathcal{G}$ lie along the main diagonal with the row zero containing the boldface $\mathbf{g_0}$. It is also possible to use $\tilde{\mathbf{g}}$ where $\tilde{g}_k = g_{-k}$, $k \in \mathbb{Z}$ to form the matrix $\mathcal{G}$, so that $\mathcal{G}$ represents a downsampled convolution product. The details are considered in Problem 5.16.

Reconstruction

It is a straightforward process to recover $f_{j+1}(t) \in V_{j+1}$ from the projections of $f_{j+1}(t)$ into V_j and W_j. Suppose that f_j and w_j are the projections of $f_{j+1}(t)$ into V_j, W_j, respectively. Then

$$f_{j+1}(t) = \sum_{k \in \mathbb{Z}} a_k \phi_{j+1,k}(t) \tag{5.28}$$

$$f_j(t) = \sum_{\ell \in \mathbb{Z}} b_\ell \phi_{j,\ell}(t) = \sum_{\ell \in \mathbb{Z}} \left(\sum_{k \in \mathbb{Z}} a_k h_{k-2\ell} \right) \phi_{j,\ell}(t) \tag{5.29}$$

$$w_j(t) = \sum_{\ell \in \mathbb{Z}} c_\ell \psi_{j,\ell}(t) = \sum_{\ell \in \mathbb{Z}} \left(\sum_{k \in \mathbb{Z}} a_k (-1)^k h_{1+2\ell-k} \right) \psi_{j,\ell}(t) \tag{5.30}$$

with $f_{j+1}(t) = f_j(t) + w_j(t)$. So we obtain

$$b_\ell = \langle f_j(t), \phi_{j,\ell}(t) \rangle = \sum_{k \in \mathbb{Z}} a_k h_{k-2\ell}$$

$$c_\ell = \langle w_j(t), \psi_{j,\ell}(t) \rangle = \sum_{k \in \mathbb{Z}} a_k (-1)^k h_{1+2\ell-k}$$

from

$$a_k = \langle f_{j+1}(t), \phi_{j+1,k}(t) \rangle \tag{5.31}$$

For reconstruction, we assume that b_ℓ, c_ℓ are known and we use these values to recover a_k, $k, \ell \in \mathbb{Z}$. Since $f_{j+1}(t) = f_j(t) + w_j(t)$, we can use (5.28–5.30) to write

$$f_{j+1}(t) = \sum_{\ell \in \mathbb{Z}} b_\ell \phi_{j,\ell}(t) + c_\ell \psi_{j,\ell}(t)$$

Since a_k is given by (5.31), we can dot the identity above with $\phi_{j+1,k}(t)$ to obtain

$$a_k = \langle f_{j+1}(t), \phi_{j+1,k}(t)\rangle$$

$$= \Big\langle \sum_{\ell \in \mathbb{Z}} b_\ell \phi_{j,\ell}(t) + c_\ell \psi_{j,\ell}(t), \phi_{j+1,k}(t)\Big\rangle$$

$$= \sum_{\ell \in \mathbb{Z}} b_\ell \langle \phi_{j,\ell}(t), \phi_{j+1,k}(t)\rangle + c_\ell \langle \psi_{j,\ell}(t), \phi_{j+1,k}(t)\rangle$$

From the proof of Proposition 5.6 we know that $\langle \phi_{j,\ell}(t), \phi_{j+1,k}(t)\rangle = h_{k-2\ell}$, and from Problem 5.14 we have $\langle \psi_{j,\ell}(t), \phi_{j+1,k}(t)\rangle = g_{k-2\ell} = (-1)^k h_{1+2\ell-k}$, so that

$$\boxed{a_k = \sum_{\ell \in \mathbb{Z}} b_\ell h_{k-2\ell} + c_\ell g_{k-2\ell} = \sum_{\ell \in \mathbb{Z}} b_\ell h_{k-2\ell} + c_\ell (-1)^k h_{1+2\ell-k}} \qquad (5.32)$$

PROBLEMS

★5.1 In this exercise you will complete the proof of Proposition 5.1 by showing that $\|\phi_{j,k}(t)\| = 1$. (*Hint:* Since $\{\phi(t-k)\}_{k\in\mathbb{Z}}$ is an orthonormal basis for V_0, we know that $\|\phi(t)\| = 1$. Use this fact and an appropriate u-substitution to complete the proof.)

5.2 Show that the elements in $S = \{\phi_{j,k}(t)\}_{k\in\mathbb{Z}}$ are linearly independent.

5.3 Prove Corollary 5.1. (*Hint:* Multiply both sides of (5.5) by $\phi_{j,m}(t)$ and integrate over $\mathbb{R}$.)

5.4 Let $\{V_j\}_{j\in\mathbb{Z}}$ be a multiresolution analysis of $L^2(\mathbb{R})$ with associated scaling function $\phi(t)$. Further suppose that for integer $L > 0$, the support of $\phi(t)$ is $[0, L]$. Use (5.7) to show that (5.6) reduces to a finite sum, and determine the number of terms in the finite sum.

★5.5 Let **h** be the scaling filter associated with scaling function $\phi(t)$ where $\phi(t)$ satisfies Convention 5.1. Show that

$$\sum_{k \in \mathbb{Z}} h_k = \sqrt{2}$$

(*Hint:* Integrate both sides of the dilation equation (5.6) over $\mathbb{R}$ and then make an appropriate u-substitution. You do not need to justify passing the integral through the infinite sum in (5.6).) Note that the Haar scaling filter **h** with nonzero $h_0 = h_1 = \frac{\sqrt{2}}{2}$ satisfies this result.

★5.6 Let **h** be the scaling filter associated with scaling function $\phi(t)$. Show that for each $\ell \in \mathbb{Z}$,

$$\sum_{k \in \mathbb{Z}} h_k h_{k-2\ell} = \delta_{0,\ell}$$

(*Hint:* Recall from the proof of Proposition 5.6 that for integers j, k, ℓ,

$$\langle \phi_{j,\,\ell}(t), \phi_{j+1,\,k}(t) \rangle = h_{k-2\ell}$$

Multiply each side of the dilation equation (5.6) by $\phi_{0,\,\ell}(t)$ and integrate over $\mathbb{R}$. Then consider the cases where $\ell = 0$ and $\ell \neq 0$.)

5.7   Verify the result of Problem 5.6 for the Haar scaling filter.

5.8   Complete the proof of Proposition 5.4. That is, show that $f(t) \in V_0$ if and only if $f(2^j t) \in V_j$.

★**5.9**   Suppose that $\{V_j\}_{j \in \mathbb{Z}}$ is a multiresolution analysis with scaling function $\phi(t)$ satisfying the dilation equation (5.6). Finally, assume that the reflection $\rho(t) = \phi(-t) \in V_0$.

(a) Show that $\langle \rho(t-k), \rho(t-\ell) \rangle = \delta_{k\ell}$. That is, show that the set of functions $\{\rho(t-k)\}_{k \in \mathbb{Z}}$ is an orthonormal set.

(b) Show that $\{\rho(t-k)\}_{k \in \mathbb{Z}}$ spans V_0 and is thus an orthonormal basis for V_0.

(c) Show that $\rho(t)$ satisfies a dilation equation with scaling filter $\tilde{\mathbf{h}}$ where $\tilde{h}_k = h_{-k}, k \in \mathbb{Z}$.

(d) Write down the matrix product (5.16) needed to compute the projection coefficients of $f_{j+1}(t) \in V_{j+1}$ into W_j. Verify that the computations are the downsampled discrete convolution products y_{2n} (see (2.24)).

5.10   Prove Proposition 5.7.

★**5.11**   Suppose that $\mathbf{h}$ is a scaling filter and $\mathbf{g}$ is a wavelet filter with elements $g_k = (-1)^k h_{1-k}$. Assume that $\ell \in \mathbb{Z}$ and use Proposition 5.5(ii) to show that

$$\sum_{k \in \mathbb{Z}} g_k g_{k-2\ell} = \delta_{0,\ell} \qquad (5.33)$$

(*Hint:* Start with the left side of (5.33) and make the substitutions $g_k = (-1)^k h_{1-k}$. Simplify and then make the index substitution $m = 1 - k - 2\ell$. Then use Proposition 5.5(ii).)

5.12   In this exercise you will verify (5.23). The following steps will help you organize the proof.

(a) Let $n, m \in \mathbb{Z}$. Use the scaling function (5.17) to write

$$\psi(t-m) = \sqrt{2} \sum_{k \in \mathbb{Z}} (-1)^k h_{1-k} \phi(2t - (k + 2m))$$

and

$$\psi(t-n) = \sqrt{2} \sum_{\ell \in \mathbb{Z}} (-1)^\ell h_{1-\ell} \phi(2t - (\ell + 2n))$$

(b) Use the relations in part (a) to show that

$$\langle \psi(t-m), \psi(t-n) \rangle$$

$$= 2 \sum_{k \in \mathbb{Z}} \sum_{\ell \in \mathbb{Z}} (-1)^{k+\ell} h_{1-k} h_{1-\ell} \langle \phi(2t - (k+2m)), \phi(2t - (\ell + 2n)) \rangle$$

(c) Now consider the case where $n = m$. Use the fact that $\sqrt{2}\phi(2t - (k+2m))$ and $\sqrt{2}\phi(2t - (\ell + 2n))$ are members of an orthonormal basis to reduce the inner product in part (b) to

$$\langle \psi(t-m), \psi(t-m) \rangle = \sum_{k \in \mathbb{Z}} h_{1-k}^2$$

Now make a change of variable on the summation index k and use Problem 5.6 to show that the inner product is 1.

(d) Now consider the case where $n \neq m$ so that $n - m \neq 0$. Using the substitution $j = k+2m$ and the fact that $\sqrt{2}\phi(2t-j)$ and $\sqrt{2}\phi(2t-(\ell+2n))$ are members of an orthonormal basis to show that the inner product in part (b) reduces to

$$\langle \psi(t-m), \psi(t-n) \rangle = \sum_{\ell \in \mathbb{Z}} h_{1-\ell} h_{1-\ell-2(m-n)}$$

Now make the index substitution $p = 1 - \ell$ and use Problem 5.6 to show that the inner product is 0.

5.13   In this problem you will verify (5.24). In order to show $V_0 \perp W_0$, we must show that $\langle \phi(t-m), \psi(t-n) \rangle$ for integers m, n. The following steps will help you complete the proof.

(a) Use dilation equations (5.6) and (5.17) to write

$$\phi(t-m) = \sqrt{2} \sum_{k \in \mathbb{Z}} h_k \phi(2t - (k+2m))$$

and

$$\psi(t-n) = \sqrt{2} \sum_{\ell \in \mathbb{Z}} (-1)^{\ell} h_{1-\ell} \phi(2t - (\ell + 2n))$$

(b) Use part (a) to show that

$$\langle \phi(t-m), \psi(t-n) \rangle$$

$$= 2 \sum_{k \in \mathbb{Z}} \sum_{\ell \in \mathbb{Z}} (-1)^{\ell} h_k h_{1-\ell} \langle \phi(2t - (k+2m)), \phi(2t - (\ell + 2n)) \rangle$$

(c) Use the fact that $\sqrt{2}\phi(2t - (\ell + 2n))$ and $\sqrt{2}\phi(2t - (k+2m))$ are orthogonal to show that the inner product in part (b) is 0 when $2n - \ell \neq 2m - k$. In the case where $2n - \ell = 2m - k$, show that the inner product in part (b) reduces to

$$\langle \phi(t - m), \psi(t - n) \rangle = 2 \sum_{k \in \mathbb{Z}} (-1)^k h_k h_{1-k-2m+2n}$$

(d) Make the substitution $p = k + 2m$ in part (c) to show that

$$\langle \phi(t - m), \psi(t - n) \rangle = 2 \sum_{p \in \mathbb{Z}} (-1)^p h_{p-2m} h_{1-p+2n}$$

This substitution makes it easier to see why the inner product is 0 for any integers m and n.

(e) The trick to see that the inner product in part (d) is always zero is to realize that the indices $p - 2m$ and $1 - p + 2n$ are symmetric about some half-integer. For example, when $n = 1$ and $m = 2$, we have

$$\langle \phi(t - 2), \psi(t - 1) \rangle = 2 \sum_{p \in \mathbb{Z}} (-1)^p h_{p-4} h_{3-p}$$

the indices $p - 4$ and $3 - p$ are symmetric about $\frac{7}{2}$. So we pair integers p that are equidistant to $\frac{7}{2}$. For example, when $p = 3$, the term in the inner product is $(-1)^3 h_{-1} h_0 = -h_{-1} h_0$, and when $p = 4$, the term in the inner product is $(-1)^4 h_0 h_{-1} = h_0 h_{-1}$. These two terms cancel, as will the terms for $p = 2,5$, $p = 1,6$, and so on.

Find the line of symmetry for arbitrary m and n and show that paired terms cancel. In this way, we see that the inner product is zero for all integers m and n.

5.14 Suppose that $\psi_{j,k}(t)$ is a wavelet function given by (5.18). Use (5.20) to show that $\langle \psi_{j,\ell}(t), \phi_{j+1,k}(t) \rangle = g_{k-2\ell} = (-1)^k h_{1+2\ell-k}$.

5.15 Prove Proposition 5.8. (*Hint:* The proof is nearly identical to that of Proposition 5.6. Use (5.20) in this proof instead of (5.8) used in Proposition 5.6. Problem 5.14 will be useful as well.)

5.16 This problem is related to Problem 5.9. Suppose that $\{V_j\}_{j \in \mathbb{Z}}$ is a multiresolution analysis with scaling function $\phi(t)$ and further suppose $\rho(t) = \phi(-t) \in V_0$. Let W_j be the wavelet space (5.21), $\psi(t)$ be the wavelet function given by Definition 5.3, and $\mu(t) = \psi(-t)$.

(a) Show that $\mu(t) \in W_1$.

(b) Show that $\langle \mu(t - \ell), \mu(t - k) \rangle = \delta_{k,\ell}$, where $k, \ell \in \mathbb{Z}$.

(c) Show that $\langle \mu(t - \ell), \rho(t - k) \rangle = \delta_{k,\ell}$, where $k, \ell \in \mathbb{Z}$.

(d) Show that $\mu(t)$ satisfies the dilation equation

$$\mu(t) = \sqrt{2} \sum_{k \in \mathbb{Z}} \tilde{g}_k \rho(2t - k)$$

where $\tilde{g}_k = g_{-k} = (-1)^k h_{k-1}$, $k \in \mathbb{Z}$.

(d) Write the matrix product (5.27) needed to compute the projection coefficients of $f_{j+1}(t) \in V_{j+1}$ into V_j. Verify that the computations are the downsampled discrete convolution products y_{2n} (see (2.24)).

5.17   Write $\mathcal{G}(\omega)$ from (5.27) in the case where **g** is the Haar wavelet filter. Verify that in the case where $a_0, \ldots, a_{N-1}$ are the only nonzero elements of **a**, we can truncate $\mathcal{G}(\omega)$ and form matrix $G_{N/2}$ (see Definition 4.1) to perform the computation.

5.18   Show that (5.32) gives the same results as those obtained in Proposition 3.27 in the case where **h** and **g** are the Haar scaling and wavelet filters, respectively.

5.2   THE VIEW FROM THE TRANSFORM DOMAIN

Certain problems are easier to solve in transform domain. Often, we can convert difficult problems into easier problems of a more algebraic nature. Working in the transform domain gives insight into frequency analysis for some applications. We first transform the dilation equation (5.6) and define a related *symbol* $H(\omega)$ for scaling function $\phi(t)$, which will help in our studies. After deriving some properties of $\hat{\phi}(\omega)$ and $H(\omega)$, we develop a pair of results for the multiresolution analysis scaling and orthonormality properties in the transform domain. We also consider the ramifications for the Fourier transformation $\hat{\psi}(\omega)$ of the wavelet function. These abstract results are examined for the Haar multiresolution analysis throughout the section.

The Symbol for $\phi(t)$

Our first theorem characterizes the dilation equation (5.6) in the transform domain.

Theorem 5.3 (The Dilation Equation in the Transform Domain) *A function $\phi(t)$ in $L^2(\mathbb{R})$ satisfies the dilation equation*

$$\phi(t) = \sqrt{2} \sum_{k \in \mathbb{Z}} h_k \phi(2t - k)$$

if and only if its Fourier transform $\hat{\phi}(\omega)$ satisfies the equation

$$\hat{\phi}(\omega) = \frac{1}{\sqrt{2}} \sum_{k \in \mathbb{Z}} h_k e^{-ik\omega/2} \hat{\phi}\left(\frac{\omega}{2}\right) \tag{5.34}$$

∎

Proof: We set $f(t) = \phi(2t - k)$ and use Propositions 2.2 and 2.3 to see that

$$\hat{f}(\omega) = \frac{1}{2}e^{-ik\omega/2}\hat{\phi}\left(\frac{\omega}{2}\right)$$

Then using the invertibility and linearity of the Fourier transformation, $\phi(t) = \sqrt{2}\sum_{k\in\mathbb{Z}} h_k\phi(2t - k)$ if and only if

$$\hat{\phi}(\omega) = \sqrt{2}\sum_{k\in\mathbb{Z}} h_k\frac{1}{2}e^{-ik\omega/2}\hat{\phi}\left(\frac{\omega}{2}\right)$$

$$= \frac{1}{\sqrt{2}}\sum_{k\in\mathbb{Z}} h_k e^{-ik\omega/2}\hat{\phi}\left(\frac{\omega}{2}\right)$$

which is the right-hand side of (5.34), so we have proved that $\phi(t)$ satisfies the time domain dilation equation if and only if $\hat{\phi}(\omega)$ satisfies (5.34).   ■

Note that the transform domain version of the dilation equation (5.34) implies that $\hat{\phi}(\omega)$ is equal to itself (dilated) times a trigonometric polynomial:

$$\hat{\phi}(\omega) = H\left(\frac{\omega}{2}\right)\hat{\phi}\left(\frac{\omega}{2}\right) \tag{5.35}$$

where

$$\boxed{H(\omega) = \frac{1}{\sqrt{2}}\sum_{k\in\mathbb{Z}} h_k e^{-ik\omega}} \tag{5.36}$$

The function appearing in (5.36) plays an important role in wavelet theory. We formally define it at this time:

Definition 5.4 (The Symbol of $\phi(\mathbf{t})$) *For scaling function $\phi(t)$, the function $H(\omega)$ given by (5.36) is called the* symbol *of $\phi(t)$.*   ■

It is often convenient to think of H as a polynomial, which we see by setting $z = e^{-i\omega}$ and writing

$$H(z) = \frac{1}{\sqrt{2}}\sum_{k\in\mathbb{Z}} h_k z^k \tag{5.37}$$

Notational Warning: We frequently use the notations $H(z)$ and $H(\omega)$ interchangeably, since each is a useful way of looking at symbol H, depending on the context. The context should make it clear whether we are using the ω or the $z = e^{-i\omega}$ notation.

It turns out that many interesting properties of ϕ are encoded in $H(\omega)$. For example, the degree of polynomial $H(z)$ is finite whenever the support of $\phi(t)$ is finite.

Theorem 5.4 (The Degree of the Symbol and the Support of $\phi(\mathbf{t})$) *Suppose that $\phi(t)$ has compact support and generates a multiresolution analysis. Then the compact support of $\phi(t)$ is $\overline{\mathrm{supp}(\phi)} = [0, N]$ for some integer $N \geq 1$ if and only if* $H(z) = \frac{1}{\sqrt{2}}\sum_{k=0}^{N} h_k z^k.$ ■

Proof: First recall that the form of $H(z)$ says that the dilation equation for $\phi(t)$ is
$$\phi(t) = \sqrt{2} \sum_{k=0}^{N} h_k \phi(2t - k).$$

($\Leftarrow$) We first assume that $H(z) = \frac{1}{\sqrt{2}} \sum_{k=0}^{N} h_k z^k$, and let $[A, B]$ denote the compact support of $\phi(t)$. Suppose for the sake of contradiction that $B > N$. Consider $\phi(y)$ for $\frac{N+B}{2} < y < B$. For $k \leq N$ we have $2y - k > 2(N+B)/2 - k > (N+B) - N = B$, so $\phi(2y - k) = 0$ since $2y - k \notin [A, B]$. Thus

$$\phi(y) = \sqrt{2} \sum_{k=0}^{N} h_k \phi(2y - k) = 0$$

for all y where $\frac{N+B}{2} < y < B$. This implies that the right endpoint of the compact support of ϕ is at most $\frac{N+B}{2}$. This contradicts our assumption that $\overline{\operatorname{supp}(\phi)} = [A, B]$ with $B > N$. Thus, by contradiction, we have shown that $B \leq N$. A similar argument in Problem 5.21 shows that $B \geq N$, so we conclude $B = N$. A similar argument shows $A = 0$, and thus completes the proof in this direction.

($\Rightarrow$) Now assume the compact support of $\phi(t)$ is $[0, N]$ for some integer $N \geq 1$. From Problem 5.4 we know that the dilation equation has a finite sum, so

$$\phi(t) = \sqrt{2} \sum_{k=L}^{M} h_k \phi(2t - k) \tag{5.38}$$

for some L and M with $L < M$. Since $\overline{\operatorname{supp}(\phi)} = [0, N]$ we know that the compact support of each $\phi(2t - k)$ is $\left[\frac{k}{2}, \frac{k+N}{2}\right]$. Thus the compact support of the right-hand side of (5.38) is $\left[\frac{L}{2}, \frac{M+N}{2}\right]$. Equating the support endpoints $0 = L/2$ and $N = \frac{M+N}{2}$, we deduce that $L = 0$ and $M = N$. Thus the symbol of $\phi(t)$ must be
$$H(z) = \frac{1}{\sqrt{2}} \sum_{k=0}^{N} h_k z^k. \qquad \blacksquare$$

For the proof that $H(z) = \frac{1}{\sqrt{2}} \sum_{k=0}^{N} h_k z^k$ implies that $\overline{\operatorname{supp}(\phi)} = [0, N]$, note that we did not use the orthogonality of $\{\phi(x - k)\}_{k \in \mathbb{Z}}$. This will be helpful later when we look at B-splines. We also note that stronger versions of this theorem can be proved. We revisit this with the cascade algorithm in Chapter 6. The interested reader should consult Walnut [62] for more on support properties.

Example 5.4 (Haar Support) *We know that the dilation equation for the Haar multiresolution analysis is (3.33), so the symbol for the Haar scaling function must be*

$$H(\omega) = \frac{1}{2}(1 + e^{-i\omega}) \quad or \quad H(z) = \frac{1+z}{2} \tag{5.39}$$

By Theorem 5.4 we can deduce the support of the Haar scaling function to be $[0, 1]$, which we learned in Chapter 3. $\blacksquare$

Another remarkable fact about the symbol $H(\omega)$ of scaling function ϕ is that we can determine whether $\{\phi_{jk}\}_{k\in\mathbb{Z}}$ generates an *orthonormal* basis of V_j and satisfies a dilation equation based purely on an equation involving only $H(\omega)$. Finding and proving this equation is a challenge and will be our focus for most of this section. As you might guess, properties of the Fourier transform will be very important tools for our journey. We need the following useful facts about $\hat{\phi}$ and symbol $H(\omega)$.

Proposition 5.9 (Properties of Scaling Functions) *Suppose that $\phi(t)$ is a scaling function generating a multiresolution analysis $\{V_j\}$ of $L^2(\mathbb{R})$. Then the following are true:*

(a) $\|\hat{\phi}\| = 1$

(b) $\hat{\phi}(0) = \dfrac{1}{\sqrt{2\pi}}$

(c) $H(0) = 1$

(d) $H(\omega)$ *is a 2π-periodic function. That is,* $H(\omega + 2\pi n) = H(\omega)$ *for all $\omega \in \mathbb{R}$, $n \in \mathbb{Z}$.*

∎

Proof: To prove (a), recall that Plancherel's identity (Corollary 2.1) tells us that $\|\hat{\phi}\| = \|\phi\|$. By Proposition 5.1, $\|\phi\| = 1$ so $\|\hat{\phi}\| = 1$.

For part (b), from the definition of the Fourier transform,

$$\hat{\phi}(0) = \frac{1}{\sqrt{2\pi}} \int_{\mathbb{R}} \phi(t) e^{-i\omega \cdot 0} \, d\omega = \frac{1}{\sqrt{2\pi}} \int_{\mathbb{R}} \phi(t) \, d\omega$$

which is equal to $\frac{1}{\sqrt{2\pi}}$ since $\int_{\mathbb{R}} \phi(t) \, dt = 1$ by Convention 5.1.

To prove part (c), substitute $\omega = 0$ into (5.35) and use (b) to obtain

$$\frac{1}{\sqrt{2\pi}} = H(0) \cdot \frac{1}{\sqrt{2\pi}}$$

from which we deduce the result.

Finally, for part (d), we first look at the definition of $H(\omega)$ and note that

$$e^{-i(\omega+2\pi n)k} = e^{-ik\omega}(e^{2\pi ni})^{-k} = e^{-ik\omega}(1)^{-k} = e^{-ik\omega}$$

since n is an integer. Thus

$$H(\omega + 2\pi n) = \frac{1}{\sqrt{2}} \sum_{k\in\mathbb{Z}} h_k e^{-i(\omega+2\pi n)k} = \frac{1}{\sqrt{2}} \sum_{k\in\mathbb{Z}} h_k e^{-i\omega k} = H(\omega)$$

∎

The Stability Function

We now derive a remarkable condition that allows us to restate the orthogonality of $\{\phi(t - k)\}_{k \in \mathbb{Z}}$ in terms of $\hat{\phi}(\omega)$. To do this, we consider the *stability function*[20]

$$A(\omega) = \sum_{\ell \in \mathbb{Z}} \left| \hat{\phi}(\omega + 2\pi\ell) \right|^2 \tag{5.40}$$

which we can show is a 2π-periodic function (see Problem 5.23). This function appears frequently in the development of wavelet theory. Let's first examine this function for the Haar scaling function $\phi(t) = \sqcap(t)$.

Example 5.5 ($A(\omega)$ for the Haar Scaling Function) *We consider $A(\omega)$ for the Haar scaling function $\phi(t) = \sqcap(t)$. From Chapter 1 and (2.38) we have*

$$A(\omega) = \sum_{\ell \in \mathbb{Z}} \left| \hat{\phi}(\omega + 2\pi\ell) \right|^2 = \sum_{\ell \in \mathbb{Z}} \left| \hat{\sqcap}(\omega + 2\pi\ell) \right|^2 = \frac{1}{2\pi} \sum_{\ell \in \mathbb{Z}} \left| \frac{\sin\left((\omega + 2\pi\ell)/2\right)}{(\omega + 2\pi\ell)/2} \right|^2$$

where the last step holds since $\left| e^{-i(\omega+2\pi\ell)/2} \right| = 1$ for any $\omega, \ell \in \mathbb{R}$. It is not clear how to simplify much further, so let's graph some partial sums

$$\sum_{\ell=-M}^{M} \left| \hat{\phi}(\omega + 2\pi\ell) \right|^2$$

for $M = 2, 3, 11$ in Figure 5.3. This plot suggests that the partial sums are converging to a constant; that is, $A(\omega) \approx 0.16$ for all $\omega \in \mathbb{R}$. Amazingly, we will see in the statement of the next theorem that this is true for any scaling function, and find the exact constant value $c \approx 0.16$. ∎

Theorem 5.5 ($A(\omega)$ is a Constant Function) *Suppose that $\phi(t)$ is in $L^2(\mathbb{R})$. Then the set $\{\phi(t - k)\}_{k \in \mathbb{Z}}$ is orthonormal if and only if the stability function $A(\omega)$ is a constant function:*

$$A(\omega) = \sum_{\ell \in \mathbb{Z}} \left| \hat{\phi}(\omega + 2\pi\ell) \right|^2 = \frac{1}{2\pi} \tag{5.41}$$

for all $\omega \in \mathbb{R}$. ∎

Proof: We will assume that $\{\phi(t - k)\}_{k \in \mathbb{Z}}$ is an orthonormal set and prove that $A(\omega) = \frac{1}{2\pi}$. The proof of the converse is left to Problem 5.25.

We begin by recalling that $\int_{\mathbb{R}} \phi(t)\overline{\phi(t - k)}\, dt = \delta_{0,k}$ since $\{\phi(t - k)\}_{k \in \mathbb{Z}}$ is an orthonormal set. We want to apply Parseval's identity (2.46), to this integral, so

---

[20]We call $A(\omega)$ the stability function since it is often considered in the inequality $0 < A \le A(\omega) \le B < \infty$. This inequality is referred to by some authors as the *stability condition* (see, e. g., example, Hong, et al. [36]). We investigate the stability condition in more detail in Chapter 8.

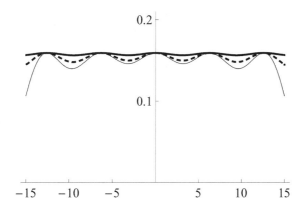

Figure 5.3 The partial sums $\sum\limits_{\ell=-M}^{M} |\hat{\phi}(\omega + 2\pi\ell)|^2$ shown for $M = 2, 3, 11$ are converging to a constant as $M \to \infty$. The heavy curve is $M = 11$, the dashed curve is $M = 3$, and the thin curve is $M = 2$.

we need the Fourier transform of $\overline{\phi(t - k)}$. Using a variation of the translation rule (Proposition 2.6), we see that $\widehat{\overline{\phi(t - k)}} = e^{iwk}\overline{\hat{\phi}(\omega)}$ (details are left to Problem 5.26). Thus by Parseval's identity (2.46) we have

$$\delta_{0,k} = \int_{\mathbb{R}} \phi(t)\overline{\phi(t - k)}\,\mathrm{d}t = \int_{\mathbb{R}} \hat{\phi}(\omega)e^{iwk}\overline{\hat{\phi}(\omega)}\,\mathrm{d}\omega = \int_{\mathbb{R}} \left|\hat{\phi}(\omega)\right|^2 e^{iwk}\,\mathrm{d}\omega \quad (5.42)$$

Now we split $\mathbb{R}$ into a union of intervals of length 2π:

$$\mathbb{R} = \cdots \cup [-4\pi, -2\pi] \cup [-2\pi, 0] \cup [0, 2\pi] \cup [2\pi, 4\pi] \cup \cdots$$

and rewrite (5.42) as

$$\delta_{0,k} = \int_{\mathbb{R}} \left|\hat{\phi}(\omega)\right|^2 e^{iwk}\,\mathrm{d}\omega = \sum_{\ell=-\infty}^{\infty} \int_{2\pi\ell}^{2\pi(\ell+1)} \left|\hat{\phi}(\omega)\right|^2 e^{iwk}\,\mathrm{d}\omega \quad (5.43)$$

Make the substitution $u = \omega - 2\pi\ell$ in each integral to obtain

$$\int_{2\pi\ell}^{2\pi(\ell+1)} \left|\hat{\phi}(\omega)\right|^2 e^{iwk}\,\mathrm{d}\omega = \int_{0}^{2\pi} \left|\hat{\phi}(u + 2\pi\ell)\right|^2 e^{iuk}(e^{2\pi\ell})^u\,\mathrm{d}u$$

$$= \int_{0}^{2\pi} \left|\hat{\phi}(u + 2\pi\ell)\right|^2 e^{iuk}\,\mathrm{d}u$$

and substitute back into (5.43) to obtain

$$\int_{\mathbb{R}} \left|\hat{\phi}(\omega)\right|^2 e^{iwk}\,\mathrm{d}\omega = \sum_{\ell=-\infty}^{\infty} \int_{0}^{2\pi} \left|\hat{\phi}(t + 2\pi\ell)\right|^2 e^{itk}\,\mathrm{d}t$$

We reverse the order of the infinite sum and integration here (see Rudin [48] for a justification of this exchange) and use (5.42) to obtain

$$\int_0^{2\pi} \sum_{\ell \in \mathbb{Z}} |\hat{\phi}(t + 2\pi\ell)|^2 e^{itk} \, dt = \delta_{0,k} \tag{5.44}$$

Notice that (5.44) is beginning to resemble what we are trying to prove. Indeed, if we substitute $k = 0$ and simplify, we obtain

$$\int_0^{2\pi} \sum_{\ell \in \mathbb{Z}} |\hat{\phi}(t + 2\pi\ell)|^2 \, dt = 1 \tag{5.45}$$

This is consistent with what we wish to prove, but our argument does not yet show that $\mathcal{A}(\omega)$ is a *constant* function. To complete the argument, we note that (5.44) implies that $\mathcal{A}(t)$ can be viewed as a 2π-periodic function that is orthogonal to all complex exponential functions e^{itk} for all integers k except $k = 0$. We can thus use Problem 2.17 to see that $\mathcal{A}(t) = c$ for all $t \in \mathbb{R}$!

To find c, substitute $\mathcal{A}(t) = \sum_{\ell \in \mathbb{Z}} |\hat{\phi}(t + 2\pi\ell)|^2 = c$ into (5.45) and carry out the integration:

$$\int_0^{2\pi} c \, dt = 1 \quad \Rightarrow \quad c = \frac{1}{2\pi}$$

This completes the proof.   ∎

From Proposition 5.9(b) we know that $\hat{\phi}(0) = \frac{1}{\sqrt{2\pi}}$. A nice corollary to Theorem 5.5 is that we can also determine the values of $\hat{\phi}(2\pi\ell)$ when $\ell \neq 0$ is an integer.

Corollary 5.2 ($\hat{\phi}(\omega)$ at Integer Multiples of 2π) *Suppose that $\phi(t)$ is a scaling function generating a multiresolution analysis $\{V_j\}_{j \in \mathbb{Z}}$ of $L^2(\mathbb{R})$. Then $\hat{\phi}(2\pi\ell) = 0$ for all $\ell \neq 0$.*   ∎

Proof: The proof of this corollary is left to Problem 5.29.   ∎

Another consequence of Theorem 5.5 is that we can write any constant as a sum of $\phi(t - k)$ terms, for *any* scaling function $\phi(t)$. This is a useful approximation property of scaling functions.

Proposition 5.10 (Integer Translates Sum to 1) *Suppose that $\phi(t)$ is a scaling function generating a multiresolution analysis $\{V_j\}_{j \in \mathbb{Z}}$ of $L^2(\mathbb{R})$. Then*

$$\sum_{k \in \mathbb{Z}} \phi(t - k) = 1$$

∎

Proof: The proof of this proposition is left to Problem 5.30.   ∎

Orthogonality and the Symbol $\phi(t)$

Reflecting for a moment about Theorem 5.5, the key property of $\phi(t)$ for this result was the *orthogonality* of its translates. We did not need the fact that $\phi(t)$ satisfied the dilation equation (5.6). If we want a condition involving the symbol $H(\omega)$ that characterizes the orthogonality of $\phi(t)$ and its integer translates and the fact that $\phi(t)$ satisfies a dilation equation, then we surely need to use the frequency-domain version of the dilation equation (5.35) along with the theorem we just proved. Here is the desired condition, which will be very helpful in Chapter 6 when we construct other orthogonal scaling functions and wavelets.

Theorem 5.6 (Orthogonal Symbol Condition) *Suppose that $\phi(t)$ is a scaling function generating a multiresolution analysis $\{V_j\}_{j\in\mathbb{Z}}$ of $L^2(\mathbb{R})$. Then its symbol $H(\omega)$ satisfies*

$$\left|H(\omega)\right|^2 + \left|H(\omega + \pi)\right|^2 = 1 \tag{5.46}$$

for all $\omega \in \mathbb{R}$. ∎

Proof: In definition (5.40) of $\mathcal{A}(\omega)$, use (5.35) to replace each $\hat{\phi}(\omega + 2\pi\ell)$ by $H\left(\frac{\omega+2\pi\ell}{2}\right) \hat{\phi}\left(\frac{\omega+2\pi\ell}{2}\right)$ to obtain

$$\sum_{\ell\in\mathbb{Z}} \left|H\left(\frac{\omega}{2} + \pi\ell\right)\right|^2 \left|\hat{\phi}\left(\frac{\omega}{2} + \pi\ell\right)\right|^2 = \frac{1}{2\pi}$$

for all $\omega \in \mathbb{R}$. Now split this sum over all integers into the sum over all even integers plus the sum over all odd integers:

$$\sum_{k\in\mathbb{Z}} \left|H\left(\frac{\omega}{2} + \pi(2k)\right)\right|^2 \left|\hat{\phi}\left(\frac{\omega}{2} + \pi(2k)\right)\right|^2$$

$$+ \sum_{j\in\mathbb{Z}} \left|H\left(\frac{\omega}{2} + \pi(2j+1)\right)\right|^2 \left|\hat{\phi}\left(\frac{\omega}{2} + \pi(2j+1)\right)\right|^2 = \frac{1}{2\pi} \tag{5.47}$$

As proved in Proposition 5.9, $H(\omega)$ is a 2π-periodic function, so $H\left(\frac{\omega}{2} + \pi(2k)\right) = H\left(\frac{\omega}{2}\right)$ and $H\left(\frac{\omega}{2} + \pi(2j+1)\right) = H\left(\frac{\omega}{2} + \pi\right)$. We can factor these terms out of the sums in (5.47), resulting in

$$\left|H\left(\frac{\omega}{2}\right)\right|^2 \sum_{k\in\mathbb{Z}} \left|\hat{\phi}\left(\frac{\omega}{2} + 2\pi k\right)\right|^2 + \left|H\left(\frac{\omega}{2} + \pi\right)\right|^2 \sum_{j\in\mathbb{Z}} \left|\hat{\phi}\left(\frac{\omega + 2\pi}{2} + 2\pi j\right)\right|^2 = \frac{1}{2\pi}$$

and apply Theorem 5.5 to obtain

$$\left|H\left(\frac{\omega}{2}\right)\right|^2 \cdot \frac{1}{2\pi} + \left|H\left(\frac{\omega}{2} + \pi\right)\right|^2 \cdot \frac{1}{2\pi} = \frac{1}{2\pi}$$

Replacing $\frac{\omega}{2}$ by ω and multiplying through by 2π yields the desired result. ∎

In Problem 5.31 you will show that the symbol for the Haar scaling function $\phi(t) = \sqcap(t)$ satisfies (5.46). We saw that the Haar scaling function had symbol $H(z) = \frac{1+z}{2}$. It turns out that *every* scaling function will have $1 + z$ as a factor of its symbol $H(z)$, as we see in the corollary below.

Corollary 5.3 (Theorem 5.6 in Terms of z) *Suppose that $\phi(t)$ is a scaling function generating a multiresolution analysis $\{V_j\}_{j \in \mathbb{Z}}$ of $L^2(\mathbb{R})$. For $z = e^{-i\omega}$, the symbol $H(z)$ of $\phi(t)$ satisfies*

$$\left|H(z)\right|^2 + \left|H(-z)\right|^2 = 1 \qquad (5.48)$$

and $H(z)$ must have factor $1 + z$. ■

Proof: Problem 5.27. ■

Theorem 5.6 gives a necessary condition for scaling functions. You might wonder if it is also a sufficient condition. Sufficiency is desirable, for it would mean that finding new and interesting scaling functions can be done by building symbols $H(\omega)$ satisfying (5.46), as we shall see in Chapter 6. It turns out that proving the converse of the theorem is quite difficult and requires another condition on $H(\omega)$. We therefore postpone the proof but state the result here because it is crucial for building important multiresolution analyses in Chapter 6.

Theorem 5.7 (Converse of Theorem 5.6) *Suppose that a finite degree symbol $H(\omega)$ satisfies (5.46), $H(0) = 1$, and $H(z)$ has form $H(z) = \left(\frac{1+z}{2}\right)^N S(z)$, where $S(z)$ satisfies*

$$\max_{|z|=1} |S(z)| \leq 2^{N-1} \qquad (5.49)$$

Then there is a scaling function $\phi(t)$ generating a multiresolution analysis $\{V_j\}_{j \in \mathbb{Z}}$ of $L^2(\mathbb{R})$ that has $H(\omega)$ as its symbol. ■

In the next chapter, we will create symbols that satisfy this theorem, and thus be guaranteed that the scaling functions associated with the symbols will generate multiresolution analyses.

The Symbol $G(\omega)$ for the Wavelet Function

We have seen that the symbol $H(\omega)$ can be used to determine if $\phi(t)$ is orthogonal to its integer translates and solves a dilation equation. It turns out that the symbol $G(\omega)$ for the wavelet function $\psi(t)$ can be used in a similar fashion. Following Definition 5.4, we define the symbol for the wavelet function $\psi(t)$.

Definition 5.5 (The Symbol for the Wavelet Function) *Suppose that wavelet function $\psi(t)$ satisfies the dilation equation (5.17). We define the symbol $G(\omega)$ associated with $\psi(t)$ as*

$$G(\omega) = \frac{1}{\sqrt{2}} \sum_{k \in \mathbb{Z}} g_k e^{-ik\omega} = \frac{1}{\sqrt{2}} \sum_{k \in \mathbb{Z}} (-1)^k h_{1-k} e^{-ik\omega} \qquad (5.50)$$

Where $z = e^{-i\omega}$, the symbol becomes

$$G(z) = \frac{1}{\sqrt{2}} \sum_{k \in \mathbb{Z}} (-1)^k h_{1-k} z^k \tag{5.51}$$

∎

We can immediately derive several properties obeyed by $G(\omega)$.

Proposition 5.11 (Properties of the Symbol $G(\omega)$) *Suppose that $\phi(t)$ is a scaling function generating a multiresolution analysis $\{V_j\}_{j \in \mathbb{Z}}$ of $L^2(\mathbb{R})$ and the wavelet function $\psi(t)$ is defined by (5.17). Then the following are true:*

(a) $\hat{\psi}(\omega) = G\left(\dfrac{\omega}{2}\right) \hat{\phi}\left(\dfrac{\omega}{2}\right)$ *for all $\omega \in \mathbb{R}$.*

(b) $G(\omega) = -e^{-i\omega} \overline{H(\omega + \pi)}$

(c) $G(0) = 0$

(d) For $z = e^{-i\omega}$, the polynomial $G(z)$ must have the factor $1 - z$.

∎

Proof: The proof of parts (a) and (b) are left as Problems 5.33 and 5.34, respectively. To prove (c), we plug $\omega = 0$ into the identity in part(b) to obtain $G(0) = \overline{H(\pi)}$. From Problem 5.32 we know $H(\pi) = 0$ and the proof is complete. For part (d), note that $G(0) = 0$ from part (c). But $\omega = 0$ is the same as $z = e^{-i \cdot 0} = 1$. So in terms of z, we have $G(1) = 0$, which means $(1 - z)$ must be a factor of $G(z)$. ∎

In Proposition 5.11(c) we learned that $G(0) = 0$. This identity has consequences not only on the wavelet filter **g**, but also the scaling filter **h**.

Proposition 5.12 (More Properties of the Scaling Filter) *Suppose that $\{V_j\}_{j \in \mathbb{Z}}$ is a multiresolution analysis of $L^2(\mathbb{R})$ with scaling function $\phi(t)$ and wavelet function $\psi(t)$. For scaling filter **h** and wavelet filter **g**, we have the following identities:*

(i) $\sum_{k \in \mathbb{Z}} g_k = \sum_{k \in \mathbb{Z}} (-1)^k h_{1-k} = 0$

(ii) $\sum_{k \in \mathbb{Z}} h_{2k} = \sum_{k \in \mathbb{Z}} h_{2k+1} = \dfrac{\sqrt{2}}{2}$

∎

Proof: To prove (i), insert $\omega = 0$ into the symbol $G(\omega)$ and use Proposition 5.11(c) to obtain

$$0 = G(0) = \frac{1}{2} \sum_{k \in \mathbb{Z}} g_k e^{-i \cdot 0 \cdot k} = \sum_{k \in \mathbb{Z}} g_k = \sum_{k \in \mathbb{Z}} (-1)^k h_{1-k}$$

For (ii) we use (i) and the identity (5.5(i):

$$\sqrt{2} = \sum_{k \in \mathbb{Z}} h_k \qquad \text{and} \qquad 0 = \sum_{k \in \mathbb{Z}} (-1)^k h_{1-k} \qquad (5.52)$$

If we add the identities in (5.52), we obtain

$$\sqrt{2} = \sum_{k \in \mathbb{Z}} h_k + \sum_{k \in \mathbb{Z}} (-1)^k h_{1-k}$$

Make the substitution $\ell = 1 - k$ in the second sum above to obtain

$$\sqrt{2} = \sum_{k \in \mathbb{Z}} h_k + \sum_{\ell \in \mathbb{Z}} (-1)^{1-\ell} h_\ell = \sum_{k \in \mathbb{Z}} h_k - \sum_{\ell \in \mathbb{Z}} (-1)^\ell h_\ell$$

Returning to index k in the second sum and then combining sums gives

$$\sqrt{2} = \sum_{k \in \mathbb{Z}} h_k - \sum_{k \in \mathbb{Z}} (-1)^k h_k = \sum_{k \in \mathbb{Z}} h_k \left(1 - (-1)^k \right)$$

Now the even-indexed terms in the above series are zero and the odd-indexed terms are $2h_k$, so that $\sqrt{2} = \sum\limits_{k \in \mathbb{Z}} 2h_{2k+1}$ or

$$\frac{\sqrt{2}}{2} = \sum_{k \in \mathbb{Z}} h_{2k+1}$$

The proof that $\frac{\sqrt{2}}{2} = \sum\limits_{k \in \mathbb{Z}} h_{2k}$ is similar and left as Problem 5.41.   ∎

Earlier in the section we saw that the symbol $H(\omega)$ for the scaling filter **h** satisfied the relation

$$|H(\omega)|^2 + |H(\omega + \pi)|^2 = 1 \qquad (5.53)$$

The identity is true because the functions $\{\phi(t - k)\}_{k \in \mathbb{Z}}$ are orthonormal and $\phi(t)$ satisfied the dilation equation (5.6). Since $\psi(t)$ also satisfies a dilation equation and the set $\{\psi(t - k)\}_{k \in \mathbb{Z}}$ is orthonormal, it is natural to ask if the symbol $G(\omega)$ for the wavelet filter **g** also satisfies a relation like (5.53). Let's first look at an example.

Example 5.6 (The Symbol $G(\omega)$ for the Haar Wavelet Function) *Find $G(\omega)$ for the Haar multiresolution analysis and determine whether*

$$|G(\omega)|^2 + |G(\omega + \pi)|^2 = 1$$

Solution
First we find $G(\omega)$:

$$\begin{aligned}
G(\omega) &= -e^{-i\omega} \overline{H(\omega + \pi)} \\
&= -e^{-i\omega} \left(\frac{1}{2} \overline{\left(1 + e^{-i(\omega + \pi)} \right)} \right) \\
&= -e^{-i\omega} \frac{1}{2} (1 - e^{i\omega}) \\
&= \frac{1}{2} (1 - e^{-i\omega})
\end{aligned}$$

so

$$|G(\omega)|^2 + |G(\omega + \pi)|^2 = G(\omega)\overline{G(\omega)} + G(\omega + \pi)\overline{G(\omega + \pi)}$$

$$= \frac{1}{4}(1 - e^{-i\omega})(1 - e^{i\omega})$$

$$+ \frac{1}{4}\left(1 - e^{-i(\omega + \pi)}\right)\left(1 - e^{i(\omega + \pi)}\right)$$

$$= \frac{1}{4}(2 - e^{-i\omega} - e^{i\omega}) + \frac{1}{4}(2 + e^{-i\omega} + e^{i\omega})$$

$$= 1$$

So the analogy works! We also know from Problem 3.25(c) that $\phi(t)$ and $\psi(t)$ are orthogonal, so an equation relating polynomials $H(\omega)$ and $G(\omega)$ is worth the search. We use (5.39) and after some experimenting and computation, we find that

$$H(\omega)\overline{G(\omega)} + H(\omega + \pi)\overline{G(\omega + \pi)}$$

$$= \frac{1}{4}(1 + e^{-i\omega})(1 - e^{i\omega}) + \frac{1}{4}\left(1 + e^{-i(\omega + \pi)}\right)\left(1 - e^{i(\omega + \pi)}\right) \qquad (5.54)$$

$$= 0$$

where we leave the computational details for Problem 5.42. ∎

It turns out that the equations we investigated in Example 5.6 are valid in general for any multiresolution analysis, as stated in the following theorem.

Theorem 5.8 (Wavelet Symbol Relations) *Suppose that $\phi(t)$ is a scaling function generating a multiresolution analysis $\{V_j\}_{j \in \mathbb{Z}}$ of $L^2(\mathbb{R})$ and wavelet function $\psi(t)$ is defined by (5.17). Then the following are true of their associated polynomials $H(\omega)$ and $G(\omega)$:*

$$|G(\omega)|^2 + |G(\omega + \pi)|^2 = 1 \qquad (5.55)$$

$$H(\omega)\overline{G(\omega)} + H(\omega + \pi)\overline{G(\omega + \pi)} = 0 \qquad (5.56)$$

for all $\omega \in \mathbb{R}$. ∎

Proof: This is a computational proof using part (b) of Proposition 5.11 along with complex arithmetic. We will leave the first identity for Problem 5.43 and prove (5.56).

$$H(\omega)\overline{G(\omega)} + H(\omega + \pi)\overline{G(\omega + \pi)}$$

$$= -H(\omega)e^{i\omega}\overline{H(\omega + \pi)} - H(\omega + \pi)e^{i(\omega + \pi)}\overline{H(\omega + 2\pi)}$$

$$= -e^{i\omega}H(\omega)H(\omega + \pi) + e^{i\omega}H(\omega + \pi)H(\omega + 2\pi)$$

Now recall that $H(\omega)$ must be a 2π-periodic function, so $H(\omega + 2\pi) = H(\omega)$. Thus

$$-e^{i\omega}H(\omega)H(\omega + \pi) + e^{i\omega}H(\omega + \pi)H(\omega + 2\pi) = 0$$

which proves the identity (5.56). ∎

We conclude this section with an important fact about wavelet functions.

Proposition 5.13 (The Wavelet Function Has a Zero Integral) *Suppose that $\psi(t)$ is a wavelet function. Then*

$$\hat{\psi}(0) = \int_{\mathbb{R}} \psi(t)\, dt = 0 \tag{5.57}$$

∎

Proof: From Proposition 5.11(a) we know that $\hat{\psi}(\omega) = G\left(\frac{\omega}{2}\right)\hat{\phi}\left(\frac{\omega}{2}\right)$ and since part (c) of the same proposition gives $G(0) = 0$, we have

$$0 = \hat{\psi}(0) = \frac{1}{\sqrt{2\pi}} \int_{\mathbb{R}} \psi(t) e^{-it\cdot 0}\, dt$$

∎

Proposition 5.13 indicates that the graph of $\psi(t)$ must vary equally above and below the t-axis; roughly speaking, the graph of wavelet $\psi(t)$ will have a *wavy* appearance. Since most wavelet functions have short support, the term "wave*let*" was adopted.

PROBLEMS

5.19   Use the formulas for the Haar $\phi(t)$, $\hat{\phi}(\omega)$, and $H(\omega)$ to verify the claims of Proposition 5.9 directly for the Haar multiresolution analysis.

5.20   In a general multiresolution analysis, what is the value of $\|\hat{\phi}_{j,k}\|$? Prove your claim.

5.21   In this problem you will complete the proof of Theorem 5.4.

(a) In the proof of Theorem 5.4, show that $B \geq N$. (*Hint:* Use a proof by contradiction and consider y values, $\frac{N+B}{2} > y$.)

(b) Show that $A = 0$ to complete the proof of the theorem.

5.22   The substitution $z = e^{-i\omega}$ is often useful. Perform the following conversions.

(a) Convert the following to z notation: $e^{-2i\omega}$, $e^{3i(\omega+\pi)}$, $e^{-i\omega/2}$

(b) Convert the following to $e^{-i\omega}$ notation: $-z$, z^7, $1/z^2$

5.23   Prove from its definition (5.40) that $A(\omega)$ is a 2π-periodic function.

5.24   Confirm the condition (5.46) on symbol $H(\omega)$ for the Haar case. Also look at the plots of $|H(\omega)|^2$, $|H(\omega + \pi)|^2$, and 1 to confirm this orthogonality condition visually.

5.25   Prove the converse of Theorem 5.5 using the following steps.

(a) Use the given half of the proof to show that

$$\int_{\mathbb{R}} \phi(t)\phi(t-k)\, dt = \int_0^{2\pi} \mathcal{A}(\omega)e^{ik\omega}\, d\omega$$

(b) Use converse hypothesis and (2.14) from Section 2.1 to show that

$$\int_{\mathbb{R}} \phi(t)\phi(t-k)\, dt = \delta_{k,0}$$

which shows that $\{\phi(t-k)\}_{k\in\mathbb{R}}$ is an orthonormal set.

★**5.26**   Confirm the identity $\widehat{\phi(t-k)} = e^{i\omega k}\hat{\phi}(\omega)$.

5.27   Investigate the symbol H with variable $z = e^{-iw}$.

(a) Prove Corollary 5.3, the symbol condition (5.46), to a form involving $z = e^{-iw}$ and $H(z)$.

(b) Explain why $H(z)$ must always have factor $1+z$.

(c) Write the symbol for Haar using $z = e^{-iw}$ and verify the symbol condition in part (a) for the Haar $H(z)$.

5.28   What is the exact value of c in Example 5.5, and how close is it to 0.16?

5.29   Prove Corollary 5.2. (*Hint:* Use (5.41) to write

$$|\hat{\phi}(\omega)|^2 + \sum_{\ell\neq 0}|\hat{\phi}(\omega+2\pi\ell)|^2 = \frac{1}{2\pi}$$

Then plug in $\omega = 0$ and use Proposition 5.9(b).)

5.30   Prove Proposition 5.10. The following steps will help you organize your work.

(a) Show that $F(t) = \sum_{k\in\mathbb{Z}} \phi(t-k)$ is a one-periodic function. We can then deduce from Problem 2.28 in Section 2.2 (with $L = \frac{1}{2}$) that $F(t)$ has a Fourier series representation in $L^2\left(\left[-\frac{1}{2},\frac{1}{2}\right]\right)$:

$$F(t) = \sum_{m\in\mathbb{Z}} c_m e^{2\pi imt}$$

where

$$c_m = \int_{-\frac{1}{2}}^{\frac{1}{2}} F(t)e^{-2\pi imt}\, dt = \int_0^1 F(t)e^{-2\pi imt}\, dt$$

(b) Use part (a) and the definition of $F(t)$ to show that

$$c_m = \sum_{k \in \mathbb{Z}} \int_0^1 \phi(t-k)e^{-2\pi i m t}\, dt$$

(c) Make the substitution $u = t - k$ in the integral in part (b) to show that

$$c_m = \int_{\mathbb{R}} \phi(u)e^{-2\pi i m u}\, du$$

(d) Use the definition of the Fourier transform (Definition 2.4) to show that

$$c_m = \sqrt{2\pi}\,\hat{\phi}(2\pi m)$$

(d) Use Proposition 5.9(b) and Corollary 5.2 to infer that $c_m = \delta_{0,m}$.

(e) Substitute your result from part (d) into the Fourier series for $F(t)$ in part (a) to complete the proof.

5.31   Recall from Example 5.4 that the symbol for the Haar scaling function $\phi(t) = \sqcap(t)$ is $H(\omega) = \frac{1+e^{-i\omega}}{2}$. Show that this symbol satisfies (5.46) from Theorem 5.6.

5.32   Let $H(\omega)$ be the symbol associated with scaling function $\phi(t)$. Show that $H(\pi) = 0$. (*Hint:* From Proposition 5.9(c) we know that $H(0) = 1$. Use this in conjunction with (5.46).)

5.33   Prove Proposition 5.11(a). (*Hint:* Start with the dilation equation (5.17) for $\psi(t)$ and take the Fourier transformations of both sides.)

5.34   Prove Proposition 5.11(b). The following steps will help you organize your work.

(a) Plug $\omega + \pi$ in for ω in (5.36) and simplify to show that

$$H(\omega + \pi) = \frac{1}{\sqrt{2}}\sum_{k \in \mathbb{Z}}(-1)^k h_k e^{-ik\omega}$$

(b) Use part (a) to show that $\overline{H(\omega + \pi)} = \frac{1}{\sqrt{2}}\sum_{k \in \mathbb{Z}}(-1)^k h_k e^{ik\omega}$.

(c) Change the summation index by the rule $\ell = 1 - k$ and simplify to obtain the desired result.

5.35   Using $z = e^{-i\omega}$, show that $G(z) = -z\overline{H(-z)}$.

5.36   If $\phi(t)$ is a scaling function with $\operatorname{supp}(\phi) = [0, N]$, then what is the support of the corresponding wavelet function $\psi(t)$?

5.37   If a scaling function is translated so that its compact support shifts from $[0, N]$ to $[L, N + L]$ for $L \in \mathbb{Z}$, what will be the corresponding change in the symbol $H(z)$?

5.38   In Theorem 5.4, N is assumed to be an integer. It can actually be proven that N is an integer. That is, in the proof of the theorem, assume only that N is a positive real number, and prove that N is an integer.

5.39   For the symbol $H(\omega) = \left(\frac{1+z}{2}\right)^2 \left(\frac{2+z}{3}\right)$, can we use Theorem 5.7 to determine whether the associated $\phi(t)$ is truly a scaling function generating a multiresolution analysis?

5.40   Use Proposition 5.11(c) along with Problem 5.43 to show that $|G(\pi)| = 1$.

5.41   Using an argument similar to that given in the proof of Proposition 5.12(ii) for odd-indexed terms, complete the proof of Proposition 5.12(ii) by showing that $\frac{\sqrt{2}}{2} = \sum_{k \in \mathbb{Z}} h_{2k}$.

5.42   Carry out the computations to confirm (5.54).

5.43   Prove the identity $|G(\omega)|^2 + |G(\omega + \pi)|^2 = 1$ in Theorem 5.8.

★5.44   Suppose that $H(\omega)$ and $G(\omega)$ satisfy (5.56) from Proposition 5.8. Show that for $\ell \in \mathbb{Z}$ we have

$$\sum_{k \in \mathbb{Z}} h_k g_{k-2\ell} = 0 \tag{5.58}$$

The following steps will help you organize your work.

(a) Write down the identity (5.56) and substitute in the associated polynomials.

(b) Simplify part (a) using the fact that $e^{ik(\omega + \pi)} = (-1)^k e^{ik\omega}$ to obtain

$$0 = \sum_{k \in \mathbb{Z}} \sum_{j \in \mathbb{Z}} h_k g_j e^{-i(k-j)\omega} + \sum_{k \in \mathbb{Z}} \sum_{j \in \mathbb{Z}} h_k g_j (-1)^{j+k} e^{-i(k-j)\omega}$$

(c) Combine terms in part (b) and simplify to obtain

$$0 = \sum_{k \in \mathbb{Z}} \sum_{j \in \mathbb{Z}} h_k g_j \left(1 + (-1)^{j+k}\right) e^{-i(k-j)\omega}$$

(d) Make the change of variable $\ell = k - j$ on the j sum and rewrite part (c) as

$$0 = \sum_{k \in \mathbb{Z}} \sum_{\ell \in \mathbb{Z}} h_k g_{k-\ell} \left(1 + (-1)^{\ell}\right) e^{-i\ell\omega}$$

(e) Simplify $\left(1 + (-1)^{\ell}\right)$ and interchange sums to obtain

$$0 = \sum_{\ell \in \mathbb{Z}} \left(\sum_{k \in \mathbb{Z}} h_k g_{k-2\ell}\right) e^{-2i\ell\omega}$$

(f) Use part (e) to prove (5.58).

★**5.45**   Suppose that L is an odd positive integer and define $\tilde{G}(\omega) = -e^{-iL\omega}\overline{H(\omega + \pi)}$, where $H(\omega)$ is the symbol for scaling function $\phi(t)$.

   (a) Write $\tilde{G}(\omega)$ in terms of the symbol $G(\omega)$ for the wavelet function $\psi(t)$.

   (b) Find the wavelet function $\tilde{\psi}(t)$, in terms of $\psi(t)$, that has $\tilde{G}(\omega)$ as its symbol. (*Hint:* Use part (a) and an appropriate rule from Section 2.2.)

   (c) Does the set $\{\tilde{\psi}(t - k)\}_{k\in\mathbb{Z}}$ form an orthonormal basis for W_j? Explain your answer.

   (d) Show that $\tilde{G}(\omega)$ satisfies (5.55) and (5.56) from Proposition 5.8.

★**5.46**   Show that the wavelet coefficients $\tilde{g}_k$ in terms of the scaling filter coefficients h_k for the symbol $\tilde{G}(\omega)$ from Problem 5.45 are

$$\tilde{g}_k = (-1)^k h_{L-k} \qquad\qquad (5.59)$$

5.3   EXAMPLES OF MULTIRESOLUTION ANALYSES

Now that we have seen the structure of multiresolution analyses and some properties in both the time and transform domains, it would be helpful to look some multiresolution analyses other than the Haar spaces. In this section we examine two more examples of multiresolution analyses. These examples illustrate the general results of the preceding section, as well as the need to generalize orthogonal multiresolution analyses. First we develop the Shannon/bandlimited multiresolution analysis, which can be viewed as the transform-domain version of the Haar multiresolution analysis: The Shannon $\hat{\phi}(\omega)$ is piecewise constant with short support in the transform domain. We also look at a simple application of projection onto V_0 for the Shannon system.

   Next we consider the linear B-spline in some detail, then examine higher-degree B-splines. The B-splines have huge advantages in many real-world applications and satisfy many of the multiresolution analysis properties. Unfortunately, they do not form an orthogonal basis, which leads us to wonder if the notion of a multiresolution analysis can be effectively generalized. We will see that the answer is fortunately "yes" and investigate this generalization in Chapter 8. But for now, let's look at the orthogonal Shannon multiresolution analysis.

Example 5.7 (Shannon Multiresolution Analysis)   *Define the Shannon scaling function $\phi(t)$ implicitly by defining its Fourier transform $\hat{\phi}(\omega)$ as the piecewise constant function*

$$\hat{\phi}(\omega) = \begin{cases} \dfrac{1}{\sqrt{2\pi}}, & -\pi \leq \omega < \pi \\ 0, & otherwise \end{cases}$$

Define the V_j spaces in the standard way: $V_j = \text{span}\left\{\phi(2^j t - k)_{k \in \mathbb{Z}}\right\}$.

We first show that $\phi(t)$ generates a multiresolution analysis by working in the transform domain to show that the function $\phi(t)$ and its integer translates are orthonormal and to establish a dilation equation. Recall from Theorem 5.5 that we can prove that $\{\phi(t-k)\}_{k \in \mathbb{Z}}$ forms an orthonormal set by showing that $A(\omega) = \frac{1}{2\pi}$ for all $\omega \in \mathbb{R}$. To see that this is true, first note that $\text{supp}\left(\hat{\phi}(\omega + 2\pi\ell)\right) = [-\pi - 2\pi\ell, \pi - 2\pi\ell)$ for each ℓ. Thus

$$A(\omega) = \sum_{\ell \in \mathbb{Z}} \left|\hat{\phi}(\omega + 2\pi\ell)\right|^2 = \left(\frac{1}{\sqrt{2\pi}}\right)^2 = \frac{1}{2\pi}$$

for all ω in the union of intervals $[-\pi - 2\pi\ell, \pi - 2\pi\ell)$, $\ell \in \mathbb{Z}$, which is $(-\infty, \infty)$.

That is, $A(\omega) = \frac{1}{2\pi}$ for all $\omega \in \mathbb{R}$, which proves that the set $\{\phi(t-k)\}_{k \in \mathbb{Z}}$ is orthonormal.

To show that $\phi(t)$ satisfies a dilation equation, we continue to work in the transform domain, using Theorem 5.3. We now prove that the symbol for $\phi(t)$ is

$$H(\omega) = \sqrt{2\pi} \sum_{k \in \mathbb{Z}} \hat{\phi}(2(\omega - 2\pi k)) \qquad (5.60)$$

To see that (5.60) is indeed a valid symbol, first note that the support of $\hat{\phi}(\omega - 4\pi k)$ is the interval $[-\pi + 4\pi k, \pi + 4\pi k)$ for each $k \in \mathbb{Z}$, so $\hat{\phi}(\omega - 4\pi k) \cdot \hat{\phi}\left(\frac{\omega}{2}\right) = 0$ for all $\omega \in \mathbb{R}$ unless $k = 0$. Thus

$$
\begin{aligned}
H\left(\frac{\omega}{2}\right)\hat{\phi}\left(\frac{\omega}{2}\right) &= \sqrt{2\pi} \sum_{k \in \mathbb{Z}} \hat{\phi}\left(2\left(\frac{\omega}{2} - 2\pi k\right)\right) \cdot \hat{\phi}\left(\frac{\omega}{2}\right) \\
&= \sqrt{2\pi}\,\hat{\phi}(\omega - 4\pi \cdot 0)\hat{\phi}\left(\frac{\omega}{2}\right) \\
&= \begin{cases} \sqrt{2\pi}\left(\frac{1}{\sqrt{2\pi}}\right)^2, & -\pi \le \omega < \pi \\ 0, & \text{otherwise} \end{cases}
\end{aligned}
$$

and this last expression simplifies to $\hat{\phi}(\omega)$, so we have shown that $H\left(\frac{\omega}{2}\right)\hat{\phi}\left(\frac{\omega}{2}\right) = \hat{\phi}(\omega)$, and the dilation equation is proved by Theorem 5.3.

Now what is the actual scaling function $\phi(t)$? Using the Fourier transform for $\sqcap(t)$ we see that $\phi(t)$ involves the sinc function. This is easily shown by direct computation using Theorem 2.4:

$$\phi(t) = \frac{1}{\sqrt{2\pi}} \int_{\mathbb{R}} \hat{\phi}(\omega)e^{it\omega}\,\mathrm{d}\omega = \frac{1}{\sqrt{2\pi}} \int_{-\pi}^{\pi} \frac{1}{\sqrt{2\pi}} e^{it\omega}\,\mathrm{d}\omega = \text{sinc}(\pi t) \qquad (5.61)$$

The intermediate details of this computation are left for Problem 5.47.

To sum up the analysis to this point, we have shown that $\{\phi(t-k)\}_{k \in \mathbb{Z}}$ forms an orthonormal basis for V_0, satisfies a dilation equation, and we have found an explicit formula for the scaling function $\phi(t)$. Since $\phi(t)$ satisfies a dilation equation, Proposition 5.4 tells us that the scaling and nested properties (see (5.1)) of a multiresolution *analysis are satisfied. The density and separation multiresolution analysis properties (5.1) are also valid, but the proofs are beyond the scope of this book.*

So what does the dilation equation of this multiresolution analysis look like in the time domain? This is not obvious! Some work with the Fourier series representation of $H(\omega)$ in Problem 5.48 will show that $\phi(t)$ satisfies the dilation equation

$$\phi(t) = \phi(2t) + \sum_{k \in \mathbb{Z}} \frac{2(-1)^{p(k)}}{(2k+1)\pi} \phi\big((2t - (2k+1))\big) \tag{5.62}$$

where

$$p(k) = \frac{\text{mod}\,(k,4) + 1}{2} \tag{5.63}$$

This infinite sum makes the dilation equation harder to visualize in the time domain. A plot of $\phi(t)$ and a partial sum of the right-hand side of (5.62) are shown in Figure 5.4. The two curves would be identical if we used the full infinite sum. Graphical representations of the dilation equation are explored more in Problem 5.50.

What can we say about the wavelet function $\psi(t) = \sum_{k \in \mathbb{Z}} (-1)^k h_{1-k} \phi(2t - k)$ for the Shannon multiresolution analysis? Like $\phi(t)$, $\psi(t)$ appears to have infinite support. Even calculating its value at a given t is tedious since $\psi(t)$ is defined by an infinite series. In Problem 5.55 you can use a CAS to generate a graph of $\psi(t)$. These problems of infinite support and infinite series computations suggest that the Shannon system will have only a short list of appropriate applications. ∎

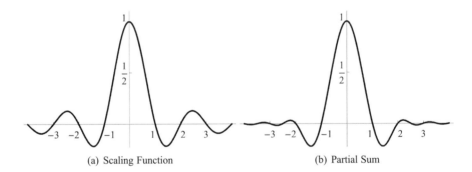

(a) Scaling Function   (b) Partial Sum

Figure 5.4   Plots of $\phi(t) = \text{sinc}(\pi t)$ and $\phi(2t) + \sum_{k=-2}^{1} \frac{2(-1)^{p(k)}}{(2k+1)\pi} \phi(2t - 2k - 1)$.

Approximation with the Shannon System

To familiarize ourselves with projections onto V_j spaces and to better understand the problems incurred using Shannon's scaling function with infinite support, consider the projection of

$$Q(t) = \begin{cases} 4 - t^2, & \text{if } |t| \leq 2 \\ 0, & \text{otherwise} \end{cases}$$

onto V_0. The projection definition from (5.13) is

$$P_{Q,0}(t) = \sum_{k \in \mathbb{Z}} \langle \phi(t-k), Q(t) \rangle \phi(t-k)$$

where

$$\langle \phi(t-k), Q(t) \rangle = \int_{\mathbb{R}} \mathrm{sinc}(\pi(t-k))Q(t)\,\mathrm{d}t$$
$$= \int_{-2}^{2} \mathrm{sinc}(\pi(t-k))(4 - t^2)\,\mathrm{d}t$$

In general, these integrals will be nonzero for each $k \in \mathbb{Z}$, so the projection $P_{Q,0}(t)$ will be an infinite sum. If we simply truncate the projection sum to a finite sum, we can find approximations to the projection, but the precision of the approximation is difficult to measure. The functions $Q(t)$ and $P_{Q,0}(t)$ are plotted in Figure 5.5. Notice the nice approximation for $|t| < 2$, but also the poor approximation outside the interval $(-2, 2)$.

We did not have the infinite sum complication for the compactly supported Haar scaling function $\sqcap(t)$. Hopefully, this example helps you appreciate the advantages of having a *compactly* supported scaling function. In the next multiresolution analysis example, we consider the compactly supported B-splines as scaling functions.

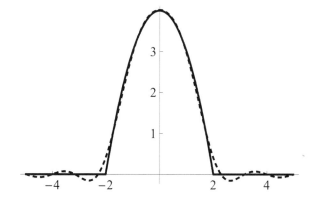

Figure 5.5   Graphs of $Q(t)$ (solid) and approximate projection (dashed) $P_{Q,0}(t) \approx \sum_{k=-3}^{3} \langle \phi(t-k), Q(t) \rangle \phi(t-k)$.

B-splines

The box function $B_0(t) = \sqcap(t)$ generates the Haar multiresolution analysis, so it makes sense to ask whether other B-splines generate multiresolution analyses. We first investigate the linear B-spline $B_1(t)$.

Example 5.8 (Linear B–Spline) *In Problem 2.60 we learned how to write a particular piecewise linear function with possible breakpoints at the integers in terms of linear B-splines. It turns out that any such piecewise linear function $f(t)$ with possible breakpoints at the integers, as long as it is continuous, can be written as*

$$f(t) = \sum_{k \in \mathbb{Z}} f(k) B_1(t - k) \tag{5.64}$$

The sum (5.64) converges pointwise for all $t \in \mathbb{R}$, and if $f(t) \in L^2(\mathbb{R})$, then the representation (5.64) is valid in $L^2(\mathbb{R})$ as well. Rigorous arguments justifying these statements appear in Walnut [62].

Let's first look at the crucial multiresolution analysis scaling and orthogonality properties (5.1). It is instructive to view these properties in both the time and frequency domains.

Time Domain. *A plot of $B_1(t)$ and some dilated translates of $B_1(2t)$ are given in Figure 5.6. We can use the graph to determine the values h_k for which the scaling relation $B_1(t) = \sum_{k \in \mathbb{Z}} h_k B_1(2t - k)$ is satisfied. See Problem 5.56 for details. As for orthogonality of $B_1(t)$ and its integer translates, we can see from Figure 5.6(b) that*

$$\int_{\mathbb{R}} B_1(2t) B_1(2t - 1)) \, dt > 0$$

so the orthogonality property is not satisfied. This of course ends the multiresolution analysis story for the linear B-spline, but fortunately the B-splines satisfy a generalized multiresolution analysis structure (see Chapter 8 for details), so we will examine them further.

Transform Domain. *The problem of satisfying the scaling property of (5.1) in the transform domain reduces to finding a symbol $H(\omega)$ for which*

$$\widehat{B_1}(\omega) = H\left(\frac{\omega}{2}\right) \widehat{B_1}\left(\frac{\omega}{2}\right) \tag{5.65}$$

Recall from Definition 2.6 that the piecewise linear B-spline $B_1(t)$ is given by

$$B_1(t) = B_0(t) * B_0(t)$$

where $B_0(t) = \sqcap(t)$ is the box function (1.4). Using the convolution theorem (Theorem 2.1) we have

$$\widehat{B_1}(\omega) = \widehat{B_0}(\omega) \cdot \widehat{B_0}(\omega) \tag{5.66}$$

From (5.39), we know that the symbol associated with the Haar scaling function is $H(\omega) = \frac{1+e^{-i\omega}}{2}$ and plugging this into (5.35) gives

$$\widehat{B_0}(\omega) = \frac{1 + e^{-i\omega}}{2} \widehat{B_0}\left(\frac{\omega}{2}\right) \tag{5.67}$$

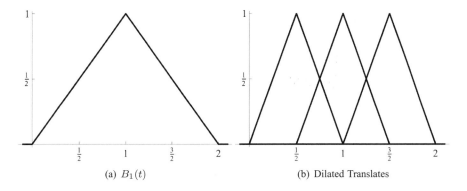

(a) $B_1(t)$

(b) Dilated Translates

Figure 5.6   Plots of $B_1(t)$ and the dilated translates $B_1(2t - k)$, $k = 0,1,2$.

If we plug (5.67) into (5.66), we obtain

$$\widehat{B_1}(\omega) = \widehat{B_0}(\omega) \cdot \widehat{B_0}(\omega)$$

$$= \left(\frac{1 + e^{-i\omega}}{2}\right) \widehat{B_0}\left(\frac{\omega}{2}\right) \cdot \left(\frac{1 + e^{-i\omega}}{2}\right) \widehat{B_0}\left(\frac{\omega}{2}\right)$$

$$= \left(\frac{1 + e^{-i\omega}}{2}\right)^2 \cdot \left(\widehat{B_0}\left(\frac{\omega}{2}\right) \cdot \widehat{B_0}\left(\frac{\omega}{2}\right)\right)$$

$$= \left(\frac{1 + e^{-i\omega}}{2}\right)^2 \cdot \widehat{B_1}\left(\frac{\omega}{2}\right) \qquad (5.68)$$

So we see that $B_1(t)$ has symbol $H(\omega) = \left(\frac{1 + e^{i\omega}}{2}\right)^2$. One advantage of the transform domain analysis is that it generalizes well to the B-splines of higher order, as we shall see later in the section.

Creating an Orthogonal Scaling Function. *We saw earlier that $B_1(t)$ does not generate an orthogonal multiresolution analysis. In the transform domain, this is equivalent to the fact that $A(\omega)$ for the linear B-spline is not constant (see Theorem 5.5). One way to circumvent this problem is to adjust $B_1(t)$ to create a new orthogonal scaling function $\phi^b(t)$ that generates the same V_j spaces. We can do this by working first in the transform domain and defining*

$$\widehat{\phi^b}(\omega) = \frac{1}{\sqrt{2\pi}} \frac{\widehat{B_1}(\omega)}{\sqrt{A(\omega)}} \qquad (5.69)$$

Using this "orthogonalization trick" and substantial work, it can be shown that

$$A^b(\omega) = \sum_{k \in \mathbb{Z}} \left|\widehat{\phi^b}(\omega + 2\pi k)\right|^2 = \frac{1}{2\pi} \sum_{k \in \mathbb{Z}} \frac{\left|\widehat{B_1}(\omega + 2\pi)\right|^2}{\sum_{\ell \in \mathbb{Z}} \left|\widehat{B_1}((\omega + 2\pi k) + 2\pi\ell)\right|^2} = \frac{1}{2\pi}$$

$$(5.70)$$

so $\left\{\phi^{b}(t-k)\right\}_{k\in\mathbb{Z}}$ *is an orthogonal set. Moreover, it generates a multiresolution analysis with the same V_{j} spaces as those of the original B-spline $B_{1}(t)$! See Problem 5.60 for details. However, it turns out that the new orthogonal scaling function $\phi^{b}(t)$ has infinite support, which makes it much less desirable for applications. For this reason we do not pursue the "orthogonalization trick" further. Interested readers can read more about this technique in Daubechies [20] or Walnut [62].* ∎

Dilation Equations and Support for General B–splines $B_{n}(t)$

B-splines of general degree n also satisfy dilation equations. In fact, $B_{n}(t)$ satisfies a dilation equation with the symbol

$$H(\omega) = \left(\frac{1+e^{-i\omega}}{2}\right)^{n+1} \tag{5.71}$$

as shown in Problem 5.57 using the same approach as used above (see 5.68) for $B_{1}(t)$. So, for example,

$$B_3(t) = \frac{\sqrt{2}}{16}B_3(2t) + \frac{4\sqrt{2}}{16}B_3(2t-1) + \frac{6\sqrt{2}}{16}B_3(2t-2)$$
$$+ \frac{4\sqrt{2}}{16}B_3(2t-3) + \frac{\sqrt{2}}{16}B_3(2t-4) \tag{5.72}$$

As with the linear B-spline, the orthogonality condition for multiresolution analyses will fail for each $B_{n}(t)$, but the orthogonality trick (5.69) can be used to build an orthogonal scaling function with infinite support. The interested reader is advised to consult Daubechies [20].

What about the support for the B-splines? Recall that we can apply one direction of Theorem 5.4 to determine the support of a function $\phi(t)$ satisfying a dilation equation, even if $\left\{\phi(t-k)\right\}_{k\in\mathbb{Z}}$ is not an orthogonal set. For example, the B-spline $B_3(t)$ has compact support $[0,4]$. See Problem 5.59 to compute the support interval for $B_2(t)$. Readers interested in B-splines are encouraged to see de Boor [21].

Approximation with B–splines

To better understand the advantages and issues with using B-splines for approximation, consider the representation of the piecewise linear function $f(t) \in V_0$ shown in Figure 5.7. The figure also contains plots of relevant translates of $B_1(t)$. We see that $f(t)$ can be written as

$$f(t) = a_{-1}B_1(t+1) + a_0 B_1(t) + a_1 B_1(t-1) + a_2 B_1(t-2) \tag{5.73}$$

for constants a_k, $k = -1, 0, 1, 2$. Now if the translates $B_1(t-k)$ are orthogonal, we can easily calculate the a_k by using Corollary 5.1 and computing $\langle B_1(t-k), f(t)\rangle$.

Since the translates are not orthogonal, instead we try multiplying (5.73) through by each relevant $B_1(t - k)$ and integrating. This step produces four equations:

$$\int_{\mathbb{R}} f(t)B_1(t - k)\, dt$$

$$= a_{-1} \int_{\mathbb{R}} B_1(t + 1)B_1(t - k)\, dt + a_0 \int_{\mathbb{R}} B_1(t)B_1(t - k)\, dt \qquad (5.74)$$

$$+ a_1 \int_{\mathbb{R}} B_1(t - 1)B_1(t - k)\, dt + a_2 \int_{\mathbb{R}} B_1(t - 2)B_1(t - k)\, dt$$

for $k = -1, 0, 1, 2$.

Note that the compact support of $B_1(t - k)$ makes each of these integrals easy to calculate, and we end up with a system of four equations and four variables a_k, which can be solved to obtain

$$a_{-1} = a_0 = 3, \quad a_1 = 1, \quad a_2 = 2$$

See Problem 5.62 for more information on this system. We emphasize two points here for approximation: the advantages of orthogonal translates of a scaling function, and the advantages of compact support of a scaling function. The alert reader may observe that the a_k coefficients can be found quickly by taking the values of $f(t)$ at integer t values. This is a desirable property enjoyed by linear B-splines.

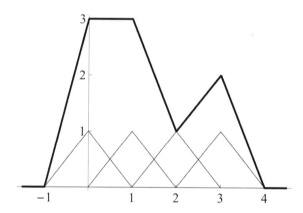

Figure 5.7   Graphs of piecewise linear $f(t) \in V_0$ and relevant translates $B_1(t - k)$.

In this chapter we have seen several functions satisfying a dilation equitation that have illustrated these desirable properties: continuity, compact support, and orthogonal translates. Unfortunately, none of them satisfied all three properties. The challenge of building a scaling function with all these properties was taken up successfully by Ingrid Daubechies [17] in the 1980s, and is discussed in Chapter 6.

PROBLEMS

Note: *A computer algebra system and the software on the course Web site will be useful for some of these problems. See the Preface for more details.*

5.47   Fill in the computational details in (5.61) to show that $\phi(t) = \text{sinc}(\pi t)$ for the Shannon multiresolution analysis.

5.48   In this problem you will establish the dilation equation for the Shannon scaling function $\phi(t) = \text{sinc}(\pi t)$. The following steps will help you organize your work.

(a) Use (2.54) in Problem 2.48 with $L = 2\pi$ to show that

$$\text{sinc}(\pi t) = \sum_{k \in \mathbb{Z}} \text{sinc}\left(\frac{k\pi}{2}\right) \text{sinc}(\pi(2t - k))$$

(b) Show that

$$\text{sinc}\left(\frac{k\pi}{2}\right) = \begin{cases} 1, & k = 0 \\ \dfrac{2}{(2k+1)\pi}, & k = \ldots, -5, -1, 3, 7, \ldots \\ \dfrac{-2}{(2k+1)\pi}, & k = \ldots, -7, -3, 1, 5, \ldots \\ 0, & \text{otherwise} \end{cases}$$

(c) Use part (b) to verify that $\text{sinc}\left(\frac{k\pi}{2}\right) = \frac{2(-1)^{p(k)}}{(2k+1)\pi}$ for $k \neq 0$. Here $p(k)$ is given by (5.63).

5.49   Show for the Shannon scaling function that $|H(\omega)|^2 + |H(\omega + \pi)|^2 = 1$ for all $\omega \in \mathbb{R}$ using the definition of the symbol $H(\omega)$.

5.50   A computer algebra system is needed for this problem. Plot some partial sums $\phi_n(t)$ for the Shannon dilation equation (5.62). How many terms are needed so that the error $|\phi(t) - \phi_n(t)|$ is less than 0.1 for all $|t| < 5$?

5.51   Comment on the degree of the Shannon symbol $H(z)$ in light of its support as detailed in Theorem 5.4.

5.52   Confirm Corollary 5.2 directly from the definition of the Shannon $\hat{\phi}(\omega)$.

5.53   Use a computer algebra system to calculate the values $\langle \phi(t - k), Q(t) \rangle$, for $k = -3, \ldots, 3$ correct to four decimal places, and then replicate Figure 5.5.

5.54   Use a computer algebra system to create a plot to visualize Proposition 5.10 for the Shannon scaling function. What complications are there?

5.55   Use a computer algebra system to create a plot of the Shannon wavelet function $\psi(t)$. What complications are there?

5.56   In this problem you will investigate the linear B-spline dilation equation.

(a) Use Figure 5.6 to work out the h_k values for the time-domain dilation equation coefficients h_k for the linear B-spline $B_1(t)$.

(b) Use the formula for B-spline $B_1(t)$ given in Example 2.8 and a computer algebra system to confirm this dilation equation graphically in the time domain.

5.57   Prove by induction that each B-spline B_n satisfies the transform-domain dilation equation (5.34) with symbol $H(\omega) = \left(\frac{1+e^{-i\omega}}{2}\right)^{n+1}$. (*Hint:* Mimic the method of the B_1 case.)

5.58   Show that none of the B-splines $B_n(t)$, $n \geq 1$, satisfy the orthogonality condition (5.48).

5.59   In this problem you will investigate the quadratic B-spline dilation equation.

(a) Use the symbol information in (5.71) to write down the time-domain dilation equation coefficients h_k for the quadratic B-spline $B_2(t)$.

(b) What does Theorem 5.4 tell you about the support of $B_2(t)$?

(c) Use the formula for the B-spline B_2 given in (2.60) and a computer algebra system to confirm this dilation equation graphically in the time domain.

5.60   Work out the details in (5.70).

★**5.61**   Explore the identity

$$\hat{\phi}(\omega) = \frac{1}{\sqrt{2\pi}} \prod_{j=1}^{\infty} H\left(e^{-i\omega/2^j}\right)$$

by iterating (5.34).

5.62   Use a computer algebra system to carry out the computational details of the B-spline approximation example to solve the system (5.74) and confirm the solution $a_{-1} = a_0 = 3$, $a_1 = 1$, and $a_2 = 2$.

5.4 SUMMARY

Before proceeding, it is important that we summarize the fundamental ideas presented in this chapter.

Multiresolution Analysis and Scaling Function

The fundamental idea of this chapter is that of a multiresolution analysis of $L^2(\mathbb{R})$. That is, we are interested in a nested list $V_j \subset V_{j+1}$, $j \in \mathbb{Z}$ of linear subspaces of

$L^2(\mathbb{R})$, where the spaces satisfy a separation property

$$\cdots \cap V_{-1} \cap V_0 \cap V_1 \cap \cdots = \{0\},$$

and a scaling property $f(t) \in V_0 \iff f(2^j t) \in V_j$. Furthermore, these spaces are dense in $L^2(\mathbb{R})$ in the sense that

$$\overline{\cup_{j \in \mathbb{Z}} V_j} = L^2(\mathbb{R})$$

Finally, there exists a *scaling function* $\phi(t) \in V_0$, with $\int_{\mathbb{R}} \phi(t)\, dt = 1$, that along with its integer translates form an orthonormal basis for V_0.

Scaling Function and Dilation Equation

The scaling function is paramount to the construction of a multiresolution analysis. Indeed, we learned that for $\phi_{j,k}(t) = 2^{j/2}\phi(2^j t - k)$, the set $\{\phi_{j,k}(t)\}_{k \in \mathbb{Z}}$ forms an orthonormal basis for $L^2(\mathbb{R})$.

Since the V_j spaces are nested, we know that $\phi(t) \in V_1$. Since $\{\phi_{1,k}(t)\}_{k \in \mathbb{Z}}$ forms an orthonormal basis for V_1, we know that $\phi(t)$ can be written as

$$\phi(t) = \sum_{k \in \mathbb{Z}} h_k \phi_{1,k}(t) = \sqrt{2} \sum_{k \in \mathbb{Z}} \phi(2t - k) \tag{5.75}$$

where $h_k = \langle \phi(t), \phi_{1,k}(t) \rangle$, $k \in \mathbb{Z}$. Equation (5.75) is called the *dilation equation* and the vector $\mathbf{h} = [\ldots, h_{-1}, h_0, h_1, \ldots]^T$ is called the *scaling filter*.

The scaling function $\phi_{j,k}(t)$ also satisfies a dilation equation:

$$\phi_{j,\ell}(t) = \sum_{k \in \mathbb{Z}} h_{k-2\ell} \phi_{j+1,k}(t)$$

The scaling filter satisfies the following properties:

$$\sum_{k \in \mathbb{Z}} h_k = \sqrt{2}, \qquad \sum_{k \in \mathbb{Z}} h_k h_{k-2\ell} = \delta_{0,\ell}, \ell \in \mathbb{Z}, \qquad \sum_{k \in \mathbb{Z}} h_k^2 = 1 \tag{5.76}$$

Projecting from V_{j+1} into V_j

The scaling filter is also useful for projecting function $f_{j+1}(t) \in V_{j+1}$ into the coarser space V_j. If $f_j(t) \in V_j$ is the orthogonal projection of

$$f_{j+1}(t) = \sum_{k \in \mathbb{Z}} a_k \phi_{j+1,k}(t) = \sum_{k \in \mathbb{Z}} \langle f_{j+1}(t), \phi_{j+1,k}(t) \rangle \phi_{j+1,k}(t) \tag{5.77}$$

into V_j, then

$$f_j(t) = \sum_{\ell \in \mathbb{Z}} b_\ell \phi_{j,\ell}(t) = \sum_{\ell \in \mathbb{Z}} \left(\sum_{k \in \mathbb{Z}} a_k h_{k-2\ell} \right) \phi_{j,\ell}(t) \tag{5.78}$$

Wavelet Function and Wavelet Spaces

Given a multiresolution analysis $\{V_j\}_{j\in\mathbb{Z}}$ of $L^2(\mathbb{R})$ with scaling function $\phi(t)$, we define the *wavelet function*

$$\psi(t) = \sum_{k\in\mathbb{Z}} g_k \phi_{1,k}(t) = \sqrt{2}\sum_{k\in\mathbb{Z}}(-1)^k h_{1-k}\phi(2t-k) \qquad (5.79)$$

The vector $\mathbf{g} = [\ldots, g_{-1}, g_0, g_1, \ldots]^T$ is called the *wavelet filter*. Note that $\psi(t) \in V_1$ satisfies a dilation equation and if we define $\psi_{j,k}(t) = 2^{j/2}\psi(2^j t - k)$, then we have the general dilation equation

$$\phi_{j,\ell}(t) = \sum_{k\in\mathbb{Z}} g_{k-2\ell}\phi_{j+1,k}(t) = \sum_{k\in\mathbb{Z}}(-1)^k h_{1+2\ell-k}\phi_{j+1,k}(t)$$

If we define the linear spaces $W_j = \mathrm{span}\{\psi_{j,k}(t)\}_{k\in\mathbb{Z}}$, then W_j is the orthogonal complement of V_j in V_{j+1}. That is,

$$V_{j+1} = V_j \oplus W_j \qquad (5.80)$$

Moreover, the set $\{\psi_{j,k}(t)\}_{k\in\mathbb{Z}}$ forms an orthonormal basis for W_j. For $m, M \in \mathbb{Z}$ with $m < M$, we can apply (5.80) iteratively to write

$$\begin{aligned}
V_M &= V_{M-1} \oplus W_{M-1} \\
&= V_{M-2} \oplus W_{M-2} \oplus W_{M-1} \\
&= V_{M-3} \oplus W_{M-3} \oplus W_{M-2} \oplus W_{M-1} \\
&\;\;\vdots \\
&= V_m \oplus W_m \oplus W_{m+1} \oplus \cdots \oplus W_{M-1}
\end{aligned}$$

so we can decompose functions in the fine space V_M into an approximation in coarse space V_m plus the sum from detail spaces $V_m, \ldots, V_{M-1}$. It can also be shown that

$$L^2(\mathbb{R}) = \cdots \oplus W_{-2} \oplus W_{-1} \oplus W_0 \oplus W_1 \oplus W_2 \oplus \cdots$$

Projecting from V_{j+1} into V_j and W_j

The wavelet filter is useful when we project function $f_{j+1}(t) \in V_{j+1}$ into W_j. If $f_{j+1}(t)$ is given by (5.77) and $w_j(t) \in W_j$ is the orthogonal projection of $f_{j+1}(t)$ into W_j, then

$$\begin{aligned}
w_j(t) &= \sum_{\ell\in\mathbb{Z}} c_\ell \psi_{j,\ell}(t) = \sum_{\ell\in\mathbb{Z}}\left(\sum_{k\in\mathbb{Z}} a_k g_{k-2\ell}\right)\psi_{j,\ell}(t) \\
&= \sum_{\ell\in\mathbb{Z}}\left(\sum_{k\in\mathbb{Z}} a_k(-1)^k h_{1+2\ell-k}\right)\psi_{j,\ell}(t) \qquad (5.81)
\end{aligned}$$

Thus we can decompose the function $f_{j+1}(t) \in V_{j+1}$ into the sum $f_{j+1}(t) = f_j(t) + w_j(t)$, where $f_j(t) \in V_j$ is given by (5.78) and $w_j(t) \in W_j$ is given by (5.81). Moreover, the functions $f_j(t)$ and $w_j(t)$ are orthogonal to each other.

Reconstruction

It is a straightforward process to reconstruct $f_{j+1}(t) \in V_{j+1}$ given $f_j(t) \in V_j$ and $w_j(t) \in W_j$. Since

$$f_{j+1}(t) = \sum_{\ell \in \mathbb{Z}} b_\ell \phi_{j,\ell}(t) + c_\ell \psi_{j,\ell}(t)$$

and $a_k = \langle f_{j+1}(t), \phi_{j+1,k}(t) \rangle$, we can write

$$
\begin{aligned}
a_k &= \langle f_{j+1}(t), \phi_{j+1,k}(t) \rangle \\
&= \left\langle \sum_{\ell \in \mathbb{Z}} b_\ell \phi_{j,\ell}(t) + c_\ell \psi_{j,\ell}(t), \phi_{j+1,k}(t) \right\rangle \\
&= \sum_{\ell \in \mathbb{Z}} b_\ell \langle \phi_{j,\ell}(t), \phi_{j+1,k}(t) \rangle + c_\ell \langle \psi_{j,\ell}(t), \phi_{j+1,k}(t) \rangle
\end{aligned}
$$

and use the fact that $\langle \phi_{j,\ell}(t), \phi_{j+1,k}(t) \rangle = h_{k-2\ell}$ and $\langle \psi_{j,\ell}(t), \phi_{j+1,k}(t) \rangle = g_{k-2\ell}$ to write

$$a_k = \sum_{\ell \in \mathbb{Z}} b_\ell h_{k-2\ell} + c_\ell g_{k-2\ell} = \sum_{\ell \in \mathbb{Z}} b_\ell h_{k-2\ell} + c_\ell (-1)^k h_{1+2\ell-k}$$

The Transform Domain

It is often easier to analyze or even construct the scaling function, its dilation equation, and other properties in the Fourier transformation domain. If we take Fourier transforms of both sides of the dilation equation (5.75), we obtain

$$\hat{\phi}(\omega) = H\left(\frac{\omega}{2}\right) \hat{\phi}\left(\frac{\omega}{2}\right) \tag{5.82}$$

where $H(\omega) = \frac{1}{\sqrt{2}} \sum_{k \in \mathbb{Z}} h_k e^{-ik\omega}$ is called the *symbol* of $\phi(t)$. We often set $z = e^{-i\omega}$ and write the symbol as a polynomial in z:

$$H(z) = \frac{1}{\sqrt{2}} \sum_{k \in \mathbb{Z}} h_k z^k$$

The symbol contains information about the support properties of scaling function $\phi(t)$. We can infer that $\overline{\text{supp}(\phi)} = [0, N]$ if and only if the symbol is the finite-length sum $H(z) = \frac{1}{\sqrt{2}} \sum_{k=0}^{N} h_k z^k$. Moreover, the symbol $H(z)$ must contain a factor of $1 + z$.

We can also use the symbol to show that the set $S = \{\phi(t - k)\}_{k\in\mathbb{Z}}$ forms an orthonormal basis for V_0 if and only if the *stability function*

$$A(\omega) = \sum_{\ell\in\mathbb{Z}} |\hat{\phi}(\omega + 2\pi\ell)|^2$$

is constant. In particular, S is an orthonormal set of functions if and only if

$$A(\omega) = \frac{1}{\sqrt{2\pi}}$$

for all $\omega \in \mathbb{R}$. The conditions (5.76) on the scaling filter can also be written in terms of the symbol:

$$|H(\omega)|^2 + |H(\omega + \pi)|^2 = 1 \tag{5.83}$$

for $\omega \in \mathbb{R}$.

We can write down the symbol for the wavelet function $\psi(t)$ as well:

$$G(\omega) = \frac{1}{2}\sum_{k\in\mathbb{Z}} g_k e^{-ik\omega} = \frac{1}{2}\sum_{k\in\mathbb{Z}} (-1)^k h_{1-k} e^{-ik\omega}$$

Since $\phi(t)$ can be written as a linear combination of $\phi_{1,k}(t)$, $k \in \mathbb{Z}$, it is not surprising that the symbol $G(\omega)$ can be expressed in terms of the symbol for $\phi(t)$. Indeed, we have $G(\omega) = -e^{-i\omega}\overline{H(\omega + \pi)}$. The symbol for $\psi(t)$ also satisfies the following properties:

$$|G(\omega)|^2 + |G(\omega + \pi)|^2 = 1, \qquad H(\omega)\overline{G(\omega)} + H(\omega + \pi)\overline{G(\omega + \pi)} = 0 \tag{5.84}$$

Moreover, $G(z)$ must contain a factor of $1 - z$. Finally, the symbol satisfies $G(0) = 0$, from which we can infer that $\int_{\mathbb{R}} \psi(t)\,\mathrm{d}t = 0$.

Constructing a Multiresolution Analysis

We have listed many properties regarding multiresolution analyses, scaling and wavelet functions, and symbols. We have looked at three examples of multiresolution analyses for $L^2(\mathbb{R})$ in Chapter 3 and Section 5.3. What we have *not* done is discuss how one would go about constructing a multiresolution analysis for $L^2(\mathbb{R})$.

The most common way that people proceed when building multiresolution analyses for $L^2(\mathbb{R})$ is to find a function $\phi(t) \in L^2(\mathbb{R})$ that satisfies the dilation equation (5.6). In addition, $\phi(t)$ is often selected based on smoothness and support criteria — for a particular application, we may insist that our scaling function is adequately smooth or that $\phi(t)$ has compact support.

We also need $\phi(t)$ to be orthogonal to its integer translates. We can either formulate this condition directly as $\langle\phi(t), \phi(t - k)\rangle = \delta_{0,k}$, $k \in \mathbb{Z}$, or we can use the relation (5.83) and characterize orthogonality in terms of the symbol $H(\omega)$. In fact, we can perform the entire construction in the Fourier transform domain — the dilation equation in the Fourier transform domain is given by (5.82), and if we insist that $\phi(t)$ is compactly supported on say $[0, N]$, then Theorem 5.4 tells us that the symbol $H(z)$ is a finite polynomial in z: $H(z) = \frac{1}{\sqrt{2}} \sum_{k=0}^{N} h_k z^k$.

Suppose that we have constructed $\phi(t)$ (either directly or in the Fourier transform domain) that satisfies the dilation equation (5.75) and $\langle \phi(t), \phi(t-k) \rangle = \delta_{0,k}$, $k \in \mathbb{Z}$. Then we can form the spaces $V_j = \text{span}\{\phi_{jk}(t)\}_{k \in \mathbb{Z}}$. Proposition 5.4 tells us that the scaling and nested properties of Definition 5.1 are satisfied. All that is left in order to show that the set $\{V_j\}_{j \in \mathbb{Z}}$ is a multiresolution analysis for $L^2(\mathbb{R})$ are the separation property and the density property from Definition 5.1. In [17], Daubechies shows that if we insist that $\hat{\phi}(\omega)$ is bounded and continuous near zero [21], then the density property in Definition 5.1 is satisfied. In addition, she shows that satisfying Theorem 5.5 is sufficient for the separation property in Definition 5.1 to hold. [22]

Connection to Discrete Wavelet Transformations

Once we have scaling function $\phi(t)$ that satisfies the dilation equation (5.75) and generates, along with its integer translates, an orthonormal basis for V_0, we have the scaling filter $\mathbf{h} = [\ldots, h_{-1}, h_0, h_1, \ldots]^T$ and hence the symbol $H(\omega)$. We can use $H(\omega)$ to form the symbol $G(\omega)$ for the wavelet function $\psi(t)$ and construct the wavelet filter $\mathbf{g}$ via the rule $g_k = (-1)^k h_{1-k}$, $k \in \mathbb{Z}$.

We learned in Section 5.1 that these filters are all we need to project $f_{j+1}(t) \in V_{j+1}$ into the complementary spaces V_j and W_j. For the projection (5.78) into V_j we have

$$
\begin{bmatrix} \vdots \\ b_{-2} \\ b_{-1} \\ b_0 \\ b_1 \\ b_2 \\ \vdots \end{bmatrix}
\begin{bmatrix} \ddots & & & & & & & & \ddots \\ \cdots & h_1 & \mathbf{h_2} & h_3 & h_4 & h_5 & h_6 & h_7 & h_8 & \cdots \\ \cdots & h_{-1} & h_0 & \mathbf{h_1} & h_2 & h_3 & h_4 & h_5 & h_6 & \cdots \\ \cdots & h_{-3} & h_{-2} & h_{-1} & \mathbf{h_0} & h_1 & h_2 & h_3 & h_4 & \cdots \\ \cdots & h_{-5} & h_{-4} & h_{-3} & h_{-2} & \mathbf{h_{-1}} & h_0 & h_1 & h_2 & \cdots \\ \cdots & h_{-7} & h_{-6} & h_{-5} & h_{-4} & h_{-3} & \mathbf{h_{-2}} & h_{-1} & h_0 & \cdots \\ & & & & \ddots & & & & \ddots \end{bmatrix}
\begin{bmatrix} \vdots \\ a_{-2} \\ a_{-1} \\ a_0 \\ a_1 \\ a_2 \\ \vdots \end{bmatrix}
$$

[21] Daubechies [17] also requires that $\hat{\phi}(0) \neq 0$, but we have adopted the convention that $\hat{\phi}(0) = \sqrt{2\pi}$ (see Convention 5.1) so that the condition is trivially satisfied.

[22] Actually, Daubechies gives a weaker condition than that given for $\mathcal{A}(\omega)$ in Theorem 5.5. We will consider this weaker condition in Chapter 8.

and for the projection (5.81), we have

$$
\begin{bmatrix} \vdots \\ c_{-2} \\ c_{-1} \\ c_0 \\ c_1 \\ c_2 \\ \vdots \end{bmatrix}
\begin{bmatrix} \ddots & & \ddots \\ \cdots & g_1 & \mathbf{g_2} & g_3 & g_4 & g_5 & g_6 & g_7 & g_8 & \cdots \\ \cdots & g_{-1} & g_0 & \mathbf{g_1} & g_2 & g_3 & g_4 & g_5 & g_6 & \cdots \\ \cdots & g_{-3} & g_{-2} & g_{-1} & \mathbf{g_0} & g_1 & g_2 & g_3 & g_4 & \cdots \\ \cdots & g_{-5} & g_{-4} & g_{-3} & g_{-2} & \mathbf{g_{-1}} & g_0 & g_1 & g_2 & \cdots \\ \cdots & g_{-7} & g_{-6} & g_{-5} & g_{-4} & g_{-3} & \mathbf{g_{-2}} & g_{-1} & g_0 & \cdots \\ & & & & \ddots & & \ddots & & \ddots \end{bmatrix}
\cdot
\begin{bmatrix} \vdots \\ a_{-2} \\ a_{-1} \\ a_0 \\ a_1 \\ a_2 \\ \vdots \end{bmatrix}
$$

In certain cases (see Problem 5.9) we can reverse the elements in each row and then the matrix product looks more like a downsampled convolution product (see Problem 2.22).

In subsequent chapters we will learn how to construct scaling functions that are compactly supported; in this case the scaling and wavelet filters are finite-length and the matrices in the products above are bandlimited.

If only a finite number of the input a_k are nonzero, then we can truncate the matrix products to form a finite-dimensional discrete wavelet transformation. For example, if we assume that $a_0, \ldots, a_{N-1}$, N even, are the only nonzero elements of $\mathbf{a}$, then the Haar scaling and wavelet filters can be used to form the discrete Haar wavelet transformation

$$
W_N = \begin{bmatrix} H_{N/2} \\ \hline G_{N/2} \end{bmatrix} =
\left[\begin{array}{cccccc}
\frac{\sqrt{2}}{2} & \frac{\sqrt{2}}{2} & 0 & 0 & 0 & 0 \\
0 & 0 & \frac{\sqrt{2}}{2} & \frac{\sqrt{2}}{2} & 0 & 0 \\
\vdots & & & & \ddots & \vdots \\
0 & 0 & 0 & 0 & \cdots & \frac{\sqrt{2}}{2} & \frac{\sqrt{2}}{2} \\
\hline
\frac{\sqrt{2}}{2} & -\frac{\sqrt{2}}{2} & 0 & 0 & 0 & 0 \\
0 & 0 & \frac{\sqrt{2}}{2} & -\frac{\sqrt{2}}{2} & 0 & 0 \\
\vdots & & & & \ddots & \vdots \\
0 & 0 & 0 & 0 & \cdots & \frac{\sqrt{2}}{2} & -\frac{\sqrt{2}}{2}
\end{array} \right]
$$

Filters of length longer than 2 present some challenges when forming finite-dimensional discrete wavelet transformations. These issues are addressed in Chapters 6 and 8.

CHAPTER 6

DAUBECHIES SCALING FUNCTIONS AND WAVELETS

A major problem in the development of wavelets during the 1980s was the search for a multiresolution analysis where the scaling function was compactly supported *and* continuous. As we saw in Chapter 3, the Haar multiresolution analysis is generated by a compactly supported scaling function that is not continuous. The B-splines are continuous and compactly supported but fail to form an orthonormal basis. In this chapter we build a family of multiresolution analyses generated by scaling functions both compactly supported and continuous. These multiresolution analyses were first constructed by Ingrid Daubechies, described in a 1988 paper [17] that created great excitement among mathematicians and scientists performing research in the area of wavelets. As you might guess, the construction was not an easy one!

In this chapter we first use the multiresolution analysis properties developed in Chapter 5 to build the family of *Daubechies scaling functions* and the accompanying wavelet functions. Daubechies' construction leads to a family of scaling functions that are compactly supported and smooth. Unfortunately, there are no explicit formulas for members of this family. In Section 6.2 we discuss a method, called the *cascade algorithm*, for approximating Daubechies scaling functions to arbitrary precision. In the final section of the chapter we investigate an efficient implementation of the cascade algorithm and use it to project functions $f(t) \in L^2(\mathbb{R})$ into V_j spaces.

Wavelet Theory: An Elementary Approach with Applications. By D. K. Ruch and P. J. Van Fleet
Copyright © 2009 John Wiley & Sons, Inc.

6.1   CONSTRUCTING THE DAUBECHIES SCALING FUNCTIONS

Ingrid Daubechies' inspired construction of scaling functions that are both continuous and compactly supported relies heavily on transform-domain methods. Her method focuses on building the scaling function's symbol, which in turn yields the dilation equation and the scaling function itself.

A Wish List of Properties

To start the process of developing these scaling functions, let's first list some desirable properties that we want our scaling function $\phi(t)$ to obey:

- **Compact Support.** As we saw with the B-spline and Haar scaling functions, the scaling filter $\mathbf{h}$ has finitely many nonzero entries for compactly supported scaling functions. Consequently, the dilation equation in the transform domain must be of the form

$$\hat{\phi}(\omega) = H\left(\frac{\omega}{2}\right)\hat{\phi}\left(\frac{\omega}{2}\right) \quad \text{where} \quad H(\omega) = \frac{1}{\sqrt{2}}\sum_{k=0}^{L}h_k e^{-ik\omega} \qquad (6.1)$$

  Here $H(\omega)$ is a trigonometric polynomial with degree L. Thus, we can use Theorem 5.4 to infer that the scaling function $\phi(t)$ has compact support with $\overline{\text{supp}(\phi)} = [0, L]$.

- **Orthonormality.** As we saw in Section 5.2, the symbol $H(\omega)$ for a multiresolution analysis must satisfy

$$|H(\omega)|^2 + |H(\omega + \pi)|^2 = 1 \qquad (6.2)$$

  for all $\omega \in \mathbb{R}$. We learned in Section 5.3 that B-splines do not satisfy (6.2).

- **Continuity.** For many applications, it is desirable that the scaling function $\phi(t)$ be sufficiently smooth. That is, we seek $\phi(t) \in C^{N-1}(\mathbb{R})$ for some positive integer N. Daubechies knew (see [20]) that in order for $\phi(t) \in C^{N-1}(\mathbb{R})$, it is *necessary* that the symbol $H(\omega)$ has $\left(\frac{1+e^{-i\omega}}{2}\right)^N$ as a factor. (Note that symbols for B-splines satisfy this fact – see Problem 5.57.) That is, the symbol $H(\omega)$ must be of the form

$$H(\omega) = \left(\frac{1 + e^{-i\omega}}{2}\right)^N S(\omega) \qquad (6.3)$$

  where $S(\omega)$ is the trigonometric polynomial

$$S(\omega) = \sum_{k=0}^{A} a_k e^{-ik\omega} \qquad (6.4)$$

We need to choose the coefficients of $S(\omega)$ carefully in order that the symbol $H(\omega)$ satisfies the orthonormality condition (6.2)! Also, we want the degree A of $S(\omega)$ to be small in order (via Theorem 5.4) to construct scaling functions $\phi(t)$ with short, compact support length. In terms of applications, a small A means the scaling filter $\mathbf{h}$ is short so that computation of the discrete wavelet transformation will be more efficient. Theorem 5.7 in Section 5.2 guarantees our scaling function will be valid as long as $\max|S(z)| \leq 2^{N-1}$. This is another reason to keep the degree A of $S(\omega)$ reasonably small.

Finding S(ω)

Now we fix a value for N and conduct our search for an appropriate $S(\omega)$. It turns out that this is best done indirectly: we begin with a sequence of clever transformations of the information given so far. First, we substitute symbol $H(\omega)$ from (6.3) into (6.2) to obtain the requirement

$$\left|\left(\frac{1+e^{-i\omega}}{2}\right)^N S(\omega)\right|^2 + \left|\left(\frac{1+e^{-i(\omega+\pi)}}{2}\right)^N S(\omega+\pi)\right|^2 = 1 \qquad (6.5)$$

Using some trigonometry and complex algebra facts (see Problem 6.2), we can rewrite (6.5) as

$$\left(\cos^2\left(\frac{\omega}{2}\right)\right)^N |S(\omega)|^2 + \left(\sin^2\left(\frac{\omega}{2}\right)\right)^N |S(\omega+\pi)|^2 = 1 \qquad (6.6)$$

We first focus on finding $|S(\omega)|^2$. For convenience, we let $L(\omega) = |S(\omega)|^2$. Our goal is to transform $L(\omega)$ into a polynomial $P(\omega)$ that is much easier to actually find and compute. Towards this end, we first observe that $|S(\omega)|^2 = S(\omega)\overline{S(\omega)} = S(\omega)S(-\omega)$ so that $L(\omega)$ is the product of trigonometric polynomials $S(\omega)$ and $S(-\omega)$. Thus $L(\omega)$ is itself a trigonometric polynomial with $L(\omega) = L(-\omega)$. From this even symmetry and Problem 2.11, we know that $L(\omega)$ can be written as a linear combination of the functions $\cos(j\omega)$ and, in particular, Problem 6.3 shows that $L(\omega)$ has the form

$$L(\omega) = c_0 + 2\sum_{j=1}^{A} c_j \cos(j\omega) \qquad (6.7)$$

Using elementary trigonometry (see Problem 6.4), we can rewrite (6.7) as a polynomial in $\cos(\omega)$. That is, $L(\omega)$ has the form

$$L(\omega) = \sum_{k=0}^{A} d_k \cos^k(\omega) \qquad (6.8)$$

Using the half-angle formula $\cos(\omega) = 1 - 2\sin^2\left(\frac{\omega}{2}\right)$ we can write $L(\omega)$ as

$$L(\omega) = \sum_{k=0}^{A} d_k \left(1 - 2\sin^2\left(\frac{\omega}{2}\right)\right)^k \qquad (6.9)$$

If we let

$$y = \sin^2\left(\frac{\omega}{2}\right) \tag{6.10}$$

we can write (6.9) as the polynomial

$$P(y) = \sum_{k=0}^{A} d_k(1 - 2y)^k \tag{6.11}$$

We next substitute (6.11) into (6.6) to obtain

$$(1 - y)^N P(y) + y^N P(1 - y) = 1 \tag{6.12}$$

The details of this step are left as Problem 6.5.

Summarizing the last few computations, we have converted requirement (6.6) and our search for $S(\omega)$ into requirement (6.12) and a search for polynomial $P(y)$. The problem of finding polynomial $P(y)$ turns out to be reasonably straightforward, certainly easier than finding $S(\omega)$ directly. We will describe its solution in a moment. Once we have $P(y)$ in hand, we can work backward to get $|S(\omega)|^2$. Then we can extract the "square root" $S(\omega)$ using a procedure called the *spectral factorization* and write down the symbol $H(\omega) = \left(\frac{1+e^{-i\omega}}{2}\right)^N S(\omega)$ for our fixed value of N. By varying the values of N, we actually obtain a family of scaling functions of various support lengths and smoothness, all generating orthonormal multiresolution analyses. You can see why this construction of Daubechies was lauded as quite an achievement!

Solving for $P(y)$

If you have studied greatest common divisors in number theory or abstract algebra, the equation $(1 - y)^N \cdot P(y) + y^N \cdot P(1 - y) = 1$ may remind you of this fact about relatively prime numbers:

> If integers a and b have no nontrivial common factors, then there exist unique integers p and q for which $a \cdot p + b \cdot q = 1$.

For example, if $a = 10$ and $b = 63$, then $p = 19$ and $q = -3$. This fact about relatively prime integers, proved using the Euclidean algorithm, generalizes nicely to polynomials, and it applies to our situation since $(1 - y)^N$ and y^N have no nontrivial common factors. The result, customized as a special case of Bézout's theorem for polynomials, is stated at this time.

Theorem 6.1 (A Formula for P(y)) *There exist unique polynomials $P(y)$ and $Q(y)$ of degree $N - 1$ that satisfy the equation*

$$(1 - y)^N \cdot P(y) + y^N \cdot Q(y) = 1 \tag{6.13}$$

The polynomial $P(y)$ is

$$P(y) = \sum_{k=0}^{N-1} \binom{2N-1}{k} y^k (1-y)^{N-1-k} \tag{6.14}$$

and $Q(y) = P(1 - y)$. ∎

Proof: The details of the proof are left as Problem 6.7. ∎

Theorem 6.1 gives us the solution for our $P(y)$ in (6.12) and we can compute $P(y)$ for any positive integer N. For example, when $N = 2$, $P(y) = 1 + 2y$, and when $N = 3$, $P(y) = 1 + 3y + 6y^2$.

Another desired feature of $P(y)$ is its nonnegativity: we require $P(y) \geq 0$ since we want $P(y) = L(\omega) = |S(\omega)|^2 \geq 0$. Fortunately it is not difficult to show that $P(y)$ in (6.14) is nonnegative, at least for $0 \leq y \leq 1$ (see Problem 6.9). This is the only set of y values that concerns us, since the values of $y = \sin^2\left(\frac{\omega}{2}\right)$ are in the interval $[0, 1]$.

Extracting S(ω) from P(y)

Our next task is to work back from $P(y)$ to determine $L(\omega) = |S(\omega)|^2$. We want to find the coefficients d_k in $L(\omega) = \sum_k d_k \cos^k(\omega)$, and we know that $L(\omega)$ can be found from $P(y)$ using our substitution (6.10) in reverse:

$$y = \sin^2\left(\frac{\omega}{2}\right) = \frac{1 - \cos\omega}{2}$$

We can use (6.14) from Theorem 6.1 in conjunction with (6.9) to write $L(\omega)$ in the following form, which will be more convenient for extracting the "square root" $S(\omega)$:

$$L(\omega) = \sum_{k=0}^{N-1} \binom{2N-1}{k} \left(\frac{1 - \cos\omega}{2}\right)^k \left(1 - \frac{1 - \cos\omega}{2}\right)^{N-1-k}$$

$$= \sum_{k=0}^{N-1} \binom{2N-1}{k} \left(\frac{1 - \cos\omega}{2}\right)^k \left(\frac{1 + \cos\omega}{2}\right)^{N-1-k} \tag{6.15}$$

Spectral Factorization

Now that we have $L(\omega)$, we want to find its "square root." That is, we seek a trigonometric polynomial $S(\omega)$ for which

$$S(\omega)S(-\omega) = |S(\omega)|^2 = L(\omega)$$

Our basic plan will be to factor $L(\omega)$ in such a way that if we take half of the factors to construct $S(\omega)$, we obtain $|S(\omega)|^2 = L(\omega)$. How to perform this task is not obvious. There are a couple of immediate facts that we need to keep in mind throughout the process:

$$L(\omega) \geq 0 \Rightarrow L(\omega) = |L(\omega)| \tag{6.16}$$

and

$$L(\omega) = L(-\omega) \tag{6.17}$$

Since $L(\omega)$ is even, each zero ω_0 we find of $L(\omega)$ gives a second zero $-\omega_0$. To construct the factors of $L(\omega)$, we can equivalently find its zeros.

Finding these zeros will be easier if we rewrite $L(\omega)$ in terms of $z = e^{-i\omega}$. We will need the identities

$$\frac{1 + \cos\omega}{2} = \left(\frac{1+z}{2}\right)\left(\frac{1+1/z}{2}\right) \tag{6.18}$$

$$\frac{1 - \cos\omega}{2} = \left(\frac{1-z}{2}\right)\left(\frac{1-1/z}{2}\right) \tag{6.19}$$

You are asked to verify these identities in Problem 6.10.

Using the identities (6.18) and (6.19), the formula (6.15) can be rewritten in terms of z. We name the corresponding function $T(z)$:

$$T(z) = \sum_{k=0}^{N-1} \binom{2N-1}{k} \left(\left(\frac{1-z}{2}\right)\left(\frac{1-1/z}{2}\right)\right)^k$$
$$\cdot \left(\left(\frac{1+z}{2}\right)\left(\frac{1+1/z}{2}\right)\right)^{N-1-k} \tag{6.20}$$

and equivalently (see Problem 6.12),

$$T(z) = \frac{1}{4^{N-1}} \sum_{k=0}^{N-1} \binom{2N-1}{k} \left(-z+2-z^{-1}\right)^k \left(z+2+z^{-1}\right)^{N-1-k} \tag{6.21}$$

From (6.20) we see that

$$T(z) = T\left(\frac{1}{z}\right) \tag{6.22}$$

$T(z)$ Properties

The following facts about $T(z)$ are easily verified.

Proposition 6.1 (Properties Obeyed by T(z)) *Let $T(z)$ be as defined by (6.20) or equivalently by (6.21). Then*

(a) If $z_0 \neq 0$ is a root of $T(z)$, then so is $1/z_0$.

(b) If z_0 is a root of $T(z)$, then so is $\overline{z_0}$.

(c) $T(z)$ can be written in the factored form

$$T(z) = \alpha z^{-(N-1)} \prod_{k=1}^{2N-2} (z - z_k) \tag{6.23}$$

where α is the leading coefficient in (6.20) and z_k denotes the roots of $T(z)$.

(d) $z = 1$ *and* $z = -1$ *are not roots of* $T(z)$.

(e) $T(z) = |T(z)| \geq 0$.

■

Proof: Part (a) is (6.22) while part (b) follows from basic complex algebra since $T(z)$ is a polynomial in z. Part (c) requires a fair amount of algebra and you are led through this work in Problem 6.12. The fact that $z = \pm 1$ are not roots of $T(z)$ follows by direct evaluation of (6.20). The final part of the proposition follows from the fact (6.16) that $L(\omega) = |L(\omega)| \geq 0$. ■

Using Proposition 6.1, we can categorize the roots according to where they lie in the complex plane.

$T(z)$ Root Categories and Analysis

We partition the complex plane into three sets: the real line, the unit circle excepting $z = \pm 1$, and the rest of the complex plane. Here are the three corresponding categories of $T(z)$ roots:

Case I: Real Roots. We have pairs of real roots r_k and $1/r_k$ by Proposition 6.1(a) and (b), since $z = \bar{z}$ when z is real. Without loss of generality we refer to the roots inside the unit circle by r_k, so $|r_k| < 1$ by Proposition 6.1(d). Denote the number of these pairs by $\mathcal{K}$.

Case II: Complex Roots on the Unit Circle.   From Proposition 6.1(b) we know that if z_j^U is a root of $T(z)$, then so is $\overline{z_j^U}$. But $|z_j^U| = 1$, and in this case $\overline{z_j^U} = 1/z_j^U$ (see Problem 1.7). In this way we obtain pairs of roots. From Problem 6.11 we know that the multiplicity of these roots is even. We form pairs of multiplicity 2 (there might be some redundancy with this formulation) and denote the number of these pairs by $\mathcal{J}$.

Case III: Complex Roots off the Unit Circle. We denote these roots by z_i^C. We obtain four distinct roots of $T(z)$ at a time since Proposition 6.1(a) and (b) imply that $1/z_i^C, \overline{z_i^C}$, and $1/\overline{z_i^C}$ are roots as well. We know that one of z_i^C and $1/z_i^C$ has magnitude less than 1 and the other has magnitude greater than 1. Without loss of generality, we refer to the roots inside the unit circle by z_i^C and also note that $\left|\overline{z_i^C}\right| = |z_i^C| < 1$. Denote the number of these quartets by $\mathcal{L}$.

Rewriting $T(z)$ as a product of its factors, grouped into categories and using Proposition 6.1(c), we have

$$T(z) = \alpha z^{-(N-1)} \prod_{i=1}^{\mathcal{L}} \left(z - z_i^C\right) \left(z - 1/z_i^C\right) \left(z - \overline{z_i^C}\right) \left(z - 1/\overline{z_i^C}\right)$$

$$\cdot \prod_{j=1}^{\mathcal{J}} \left(z - z_j^U\right)^2 \left(z - 1/z_j^U\right)^2 \cdot \prod_{k=1}^{\mathcal{K}} \left(z - r_k\right) \left(z - 1/r_k\right) \quad (6.24)$$

Recall that we want to build the square root $F(z)$ of $T(z)$ using the idea of taking half of the factors of $T(z)$ while ensuring that the square root $F(z)$ has real coefficients when expanded as a polynomial in z. To see how this can be done, we need to rewrite $T(z)$ again, appealing to (6.16) and regrouping the Case II and Case III roots to obtain

$$T(z) = |T(z)|$$

$$= |\alpha| \prod_{i=1}^{\mathcal{L}} \left| \left(z - z_i^C\right) \left(z - 1/\overline{z_i^C}\right) \right| \cdot \left| \left(z - \overline{z_i^C}\right) \left(z - 1/z_i^C\right) \right|$$

$$\cdot \prod_{j=1}^{\mathcal{J}} \left| \left(z - z_j^U\right) \left(z - 1/z_j^U\right) \right|^2 \cdot \prod_{k=1}^{\mathcal{K}} \left| \left(z - r_k\right) \left(z - 1/r_k\right) \right| \quad (6.25)$$

Next we use Problem 1.10 in Section 1.1 which states if $z, w \in \mathbb{C}$ with $|z| = 1$, then

$$|(z - w)(z - 1/\overline{w})| = |w|^{-1} |z - w|^2 \quad (6.26)$$

We apply (6.26) to each of the grouped pairs in the first and third products in (6.25) in addition to Proposition 6.1(e) to obtain

$$T(z) = |T(z)|$$

$$= |\alpha| \prod_{i=1}^{\mathcal{L}} |z_i^C|^{-1} |z - z_i^C|^2 \left| \overline{z_i^C} \right|^{-1} \left| z - \overline{z_i^C} \right|^2$$

$$\cdot \prod_{j=1}^{\mathcal{J}} \left| \left(z - z_j^U\right) \left(z - 1/z_j^U\right) \right|^2 \cdot \prod_{k=1}^{\mathcal{K}} |r_k|^{-1} |z - r_k|^2$$

$$= |\alpha| \prod_{i=1}^{\mathcal{L}} |z_i^C|^{-2} |z - z_i^C|^2 \left| z - \overline{z_i^C} \right|^2$$

$$\cdot \prod_{j=1}^{\mathcal{J}} \left| \left(z - z_j^U\right) \left(z - 1/z_j^U\right) \right|^2 \cdot \prod_{k=1}^{\mathcal{K}} |r_k|^{-1} |z - r_k|^2$$

Now we can see how to create the square root $F(z)$ by collecting all the factors with roots inside the unit circle, and one from each double root z_j^U on the unit circle.

Define $F(z)$ by

$$F(z) = \sqrt{|\alpha|} \left[\prod_{i=1}^{\mathcal{L}} |z_i^C|^{-1} \prod_{k=1}^{\mathcal{K}} |r_k|^{-1/2} \right] \prod_{i=1}^{\mathcal{L}} \left(z - z_i^C \right) \left(z - \overline{z_i^C} \right)$$

$$\cdot \prod_{j=1}^{\mathcal{J}} \left(z - z_j^U \right) \left(z - 1/z_j^U \right) \cdot \prod_{k=1}^{\mathcal{K}} \left(z - r_k \right) \tag{6.27}$$

and observe that

$$|F(z)|^2 = |T(z)| = T(z)$$

Note that since $T(z)$ has degree $2N - 2$ by (6.23), $F(z)$ has degree $N - 1$. In Problem 6.13 you will show that $F(z)$ has real coefficients. Summing up, we have outlined a proof that $T(z)$ can indeed be factored as desired and that a constructive method for carrying out the factorization exists. Formally, we have the following theorem.

Theorem 6.2 (Spectral Factorization) *The trigonometric polynomial $F(z)$ defined by (6.27) has degree $N - 1$, real coefficients and satisfies $T(z) = |F(z)|^2$ for $T(z)$ given by (6.20).* ■

Building the Symbol

The final step in obtaining a "square root" $S(\omega)$ is simply to make the substitution $z = e^{-i\omega}$ in $F(z)$, which gives us $S(\omega)$. Expanding (6.27) and making the aforementioned substitution gives (6.4):

$$S(\omega) = F(e^{-i\omega}) = \sum_{k=0}^{N-1} a_k e^{-ik\omega}$$

from which we can use (6.3) to write

$$H(\omega) = \left(\frac{1 + e^{-i\omega}}{2} \right)^N S(\omega) = \left(\frac{1 + e^{-i\omega}}{2} \right)^N \cdot \sum_{k=0}^{N-1} a_k e^{-ik\omega}$$

Now if we expand $H(\omega)$, the product of a polynomial of degree N and a polynomial of degree $N - 1$, we obtain the degree $2N - 1$ trigonometric polynomial from (6.1)

$$H(\omega) = \frac{1}{\sqrt{2}} \sum_{k=0}^{2N-1} h_k e^{-ik\omega} \tag{6.28}$$

for real numbers $h_0, \ldots, h_{2N-1}$. Moreover by construction, $H(\omega)$ satisfies (6.2). This solution is by no means unique. We could have constructed $F(z)$ using the roots outside the unit circle (see Problem 6.16). In most books and in applications, the coefficients are used in reverse order. The corresponding symbol from (6.1) is

$$H^*(\omega) = \frac{1}{\sqrt{2}} \sum_{k=0}^{2N-1} h_{2N-1-k} e^{-ik\omega} \qquad (6.29)$$

Does this formulation satisfy (6.2) and (6.3)? In Problem 6.15, you will show that

$$H^*(\omega) = e^{-(2N-1)i\omega} \overline{H(\omega)} \qquad (6.30)$$

It is trivial to see from (6.30) that $H^*(\omega)$ satisfies (6.2) and note that $\omega = \pm\pi$ are both values that make $\left(\frac{1+e^{-i\omega}}{2}\right)^N$ equal to zero, and this is essentially the condition Daubechies needed to ensure that $\phi(t) \in C^{N-1}(\mathbb{R})$. Thus our final step in constructing the symbol is to apply the map

$$h_k \mapsto h_{2N-1-k} \qquad (6.31)$$

to the coefficients in (6.28).

So, given positive integer N, we now know how to find the symbol and thus the scaling filter for the corresponding Daubechies scaling function. Let's summarize the algorithm and then look at an example.

Algorithm 6.1 (Algorithm for Building Daubechies Scaling Function Symbols)
This algorithm takes as input a positive integer N and returns the Daubechies length $2N$ scaling filter.

  1. *For given N, construct $P(y)$ using (6.14).*

  2. *Make the substitution*

$$y = \sin^2\left(\frac{\omega}{2}\right) = \frac{1-\cos\omega}{2} = \left(\frac{1-z}{2}\right)\left(\frac{1-1/z}{2}\right)$$

  to convert $P(y)$ to $L(\omega)$ and then to $T(z)$.

  3. *Find the leading coefficient α and zeros of $T(z)$. Make sure to choose zeros inside the unit circle. This can be done using a CAS for small N. For larger N, more sophisticated and specialized algorithms may be used (see [57]).*

  4. *Create $F(z)$ as stated in (6.27). Recall that the desired $S(\omega)$ is $F(z)$ using substitution $z = e^{-i\omega}$.*

  5. *Expand $\left(\frac{1+z}{2}\right)^N F(z)$ with substitution $z = e^{-i\omega}$ to obtain the $2N$ real numbers $\frac{1}{\sqrt{2}}h_0, \ldots, \frac{1}{\sqrt{2}}h_{2N-1}$.*

  6. *As a final step, reverse the order of the filter coefficients as described by (6.31) and multiply the resulting values by $\sqrt{2}$ to put the symbol in the form given by (6.1).*

■

The symbol $H(z)$ is the product of a degree N trigonometric polynomial $\left(\frac{1+z}{2}\right)^N$ and the trigonometric polynomial $F(z)$ of degree $N - 1$. The total degree of $H(z)$ is $N + N - 1 = 2N - 1$, so that $H(z)$ consists of $2N$ terms. Thus the Daubechies scaling filter will be of length $2N$.

Terminology Note. The Daubechies scaling function corresponding to the symbol $H(\omega)$ for a given N is often called the "D2N" scaling function, and the corresponding filter **h**, which has $2N$ nonzero components, is commonly referred to as the "$2N$-tap" or D2N Daubechies scaling filter. For example, when $N = 3$, we have the D6 scaling function and the six-tap Daubechies filter or the D6 scaling filter.

Let's look at an example of the implementation of Algorithm 6.1.

Example 6.1 (D6 Filter Coefficients) *Construct the coefficients of the Daubechies six-tap scaling filter.*
Solution
  Here $N = 3$ and first we find $P(y)$ using (6.14). We have

$$P(y) = 1 + 3y + 6y^2$$

Next construct $T(z)$:

$$
\begin{aligned}
T(z) &= 1 + 3\left(\frac{1-z}{2}\right)\left(\frac{1-1/z}{2}\right) + 6\left(\left(\frac{1-z}{2}\right)\left(\frac{1-1/z}{2}\right)\right)^2 \\
&= \frac{3}{8}z^2 + \frac{9}{4}z + \frac{19}{4} + \frac{9}{4}z^{-1} + \frac{3}{8}z^{-2} \\
&= \frac{3}{8}z^{-2}\left(z^4 + 6z^3 + \frac{38}{3}z^2 + 6z + 1\right)
\end{aligned}
$$

and note that $\alpha = \frac{3}{8}$.
  Third, we solve $T(z) = 0$ using a CAS to get the four approximate roots

$$2.71277 \pm 1.44389i \quad and \quad 0.28725 \pm 0.15289i \tag{6.32}$$

Clearly, this is a quartet of complex roots as detailed in Case III on page 239. There are no zeros of the other categories. We denote one of the exact zeros inside the unit circle as $z_1^C \approx 0.28725 + 0.15289i$. Our next step is to construct $F(z)$. We have

$$F(z) = \sqrt{\frac{3}{8}}\frac{1}{|z_1^C|}\left(z - z_1^C\right)\left(z - \overline{z_1^C}\right)$$

Now we build $H(\omega)$ using (6.3) and expand it to obtain the trigonometric polynomial. We have

$$H(\omega) = \left(\frac{1 + e^{-i\omega}}{2}\right)^3 F\left(e^{-i\omega}\right)$$

$$= 0.0249086 - 0.0604169e^{-i\omega} - 0.095467e^{-2i\omega}$$
$$+ 0.325186e^{-3i\omega} + 0.570563e^{-4i\omega} + 0.235235e^{-5i\omega} \quad (6.33)$$

Our final step is to reverse the order of the coefficients in the above polynomial and multiply each coefficient by $\sqrt{2}$. Thus the six nonzero scaling coefficients (rounded to six decimal digits) are

$$
\begin{array}{ll}
h_0 = 0.332671 & h_3 = -0.135011 \\
h_1 = 0.806892 & h_4 = -0.085441 \\
h_2 = 0.459878 & h_5 = 0.035226
\end{array}
\quad (6.34)
$$

These six coefficients form the D6 scaling filter[23] $\mathbf{h}$. In Problem 6.17 you will construct the coefficients of the famous "four-tap" D4 scaling filter, and in Problem 6.18 you will construct the D8 scaling filter. ∎

At this point, let's make sure that the Daubechies symbols really do lead to valid scaling functions and multiresolution analyses. We can perform the verification using Theorem 5.7. We have the following result.

Proposition 6.2 (Daubechies Scaling Functions and Multiresolution Analyses)
The Daubechies scaling functions with symbols defined by Algorithm 6.1 generate multiresolution analyses. ∎

Proof: For the case $N = 1$ we have $H(z) = \left(\frac{1+z}{2}\right)$ and the Haar system, which was discussed in Chapter 3. For $N \geq 2$ we use Theorem 5.7 to establish the result. From the construction of Algorithm 6.1, each Daubechies symbol satisfies (5.46), $H(0) = 1$, and has the form

$$H(z) = \left(\frac{1 + z}{2}\right)^N S(z)$$

so the proof will be complete once we verify that $\max |S(z)| < 2^{N-1}$. To establish this bound, first recall that $|S(z)|^2 = P(y)$ as given in (6.14), where $0 \leq y \leq 1$. Thus the proof is complete if we can show that

$$P(y) \leq \left(2^{N-1}\right)^2 = 2^{2N-2} \quad (6.35)$$

---

[23]It is possible to construct these filter coefficients exactly. See Van Fleet [60] Problem 7.14 for more details.

We leave the elementary case $N = 2$ as Problem 6.22. Whenever $N \geq 3$, we use the fact that $y \leq 1$ to show that

$$P(y) = \sum_{k=0}^{N-1} \binom{2N-1}{k} y^k (1-y)^{N-1-k}$$

$$\leq \sum_{k=0}^{N-1} \binom{2N-1}{k}$$

Next we use the fact that

$$\sum_{k=0}^{N-1} \binom{2N-1}{k} = \frac{1}{2} 2^{2N-1} = 2^{2N-2} \tag{6.36}$$

(see Problem 6.22(b) for details of this identity), so that $P(y) \leq 2^{2N-2}$. Thus

$$\max |S(z)| \leq \max \sqrt{P(y)} \leq 2^{(2N-2)/2}$$

and the proof is complete.   ∎

Daubechies Wavelet Filters

Now that we know how to construct the Daubechies scaling filters, construction of the wavelet filters is straightforward with the aid of (5.19) from Definition 5.3. We have

$$g_k = (-1)^k h_{1-k}, \quad k \in \mathbb{Z} \tag{6.37}$$

As stated above, the scaling filter coefficients are often listed in reverse order — in this case, the wavelet filter coefficients are reversed as well. Here is an example.

Example 6.2 (Wavelet Filter for the D6 Scaling Filter) *Find the wavelet filter associated with the Daubechies D6 scaling filter.*
Solution
   From Example 6.1 we know only six h_{1-k} coefficients are nonzero, $k = -4, \ldots, 1$, and we use (6.34) and (6.37) to obtain

$$
\begin{array}{lll}
g_{-4} = & h_5 = & 0.035226 \\
g_{-3} = -h_4 = & 0.085441 \\
g_{-2} = & h_3 = -0.135011
\end{array}
\qquad
\begin{array}{lll}
g_{-1} = -h_2 = -0.459878 \\
g_0 = & h_1 = & 0.806892 \\
g_1 = & -h_0 = -0.332671
\end{array}
\tag{6.38}
$$

∎

We are almost ready to look at applications such as the projection of functions onto the spaces V_j and W_j. We do not, at this point, have formulas for the Daubechies scaling functions and associated wavelet functions. Since these functions are only defined in terms of dilation equations, it is not obvious how to create the function graphs. This challenge is discussed in the next section.

PROBLEMS

Note: *A computer algebra system and the software on the course Web site will be useful for some of these problems. See the Preface for more details.*

6.1   What is the D2 scaling filter?

6.2   In this problem you will show (6.5) implies (6.6).

(a) Expand $\left|1 + e^{-i\omega}\right|^2 = \left(1 + e^{-i\omega}\right)\left(1 + e^{i\omega}\right)$ and use (2.8) in Section 2.1 to write this expansion in terms of $\cos\omega$.

(b) Use part (a) and an appropriate trigonometric identity to show that $\left|\frac{1+e^{-i\omega}}{2}\right|^2 = \cos^2\left(\frac{\omega}{2}\right)$.

(c) Use part (b) with the substitution $\omega \to \omega + \pi$ and an appropriate trigonometric identity to show that $\left|\frac{1+e^{-i(\omega+\pi)}}{2}\right|^2 = \sin^2\left(\frac{\omega}{2}\right)$.

(d) Use parts (b) and (c) to complete the proof that (6.5) implies (6.6).

6.3   In this problem you will show that $L(\omega)$ can be written as a linear combination of $\cos(j\omega)$, $j \in \mathbb{Z}$.

(a) Since $L(\omega) = S(\omega)S(-\omega) = \left(\sum\limits_{k=0}^{A} a_k e^{-ik\omega}\right)\left(\sum\limits_{k=0}^{A} a_k e^{ik\omega}\right)$, explain why

we can write $L(\omega) = \sum\limits_{j=-A}^{A} c_j e^{-ij\omega}$ and $c_j = c_{-j}$.

(b) Use the new expression for $L(\omega)$ in part (a) and regroup its terms to show that

$$L(\omega) = c_0 + 2\sum_{j=1}^{A} c_j \cos(j\omega)$$

6.4   In this problem you will show that each $\cos(k\omega)$ term can be written as a polynomial in $\cos(\omega)$.

(a) Use Problem 6.2 with Euler's formula (2.4) and two binomial expansions to show that

$$\cos(k\omega) = \frac{1}{2}\sum_{\ell=0}^{k} \binom{k}{\ell} \cos^{\ell}(\omega) \sin^{k-\ell}(\omega)\left(i^{k-\ell} + (-i)^{k-\ell}\right)$$

(b) Regroup the terms in the sum from part (a) according to whether $k - \ell$ is even or odd. Simplify and show that

$$\cos(k\omega) = \sum_{\substack{\ell=0 \\ k-\ell \text{ even}}}^{k} \binom{k}{\ell} \cos^{\ell}(\omega) \left(\sin^2(\omega)\right)^{(k-\ell)/2} i^{k-\ell}$$

(c) Use a basic trigonometric identity to show that $\cos(k\omega)$ can be written as a polynomial in $\cos(\omega)$.

6.5   In this problem you will show that (6.6) implies (6.13). The following steps will help you organize your work.

(a) Use the trigonometric identity $\cos^2\left(\frac{\omega}{2}\right) = 1 - \sin^2\left(\frac{\omega}{2}\right)$ to convert (6.6) to

$$(1 - y)^N P(y) + y^N \cdot \sum_k d_k \cos^k(\omega + \pi) = 1$$

(b) Now use trigonometry and the definition $y = \sin^2\left(\frac{\omega}{2}\right)$ to show that

$$\cos(\omega + \pi) = 1 - y$$

(c) Use the relations in parts (a) and (b) to obtain

$$(1 - y)^N P(y) + y^N P(1 - y) = 1$$

6.6   In this problem you will give an algebraic proof of the first part of Theorem 6.1.

(a) Use the Euclidean algorithm with $N = 3$ to find the polynomials $P(y)$ and $Q(y)$ for Theorem 6.1.

(b) Generalize your argument from part (a) to show the existence of polynomials $P(y)$ and $Q(y)$ for Theorem 6.1 with arbitrary N.

6.7   Derive the specific form of $P(y)$ in Theorem 6.1. The following steps will help you organize your work.

(a) Solve (6.12) for $P(y)$ in terms of $P(1 - y)$ and use a Taylor expansion for $(1 - y)^{-N}$ to obtain

$$P(y) = \sum_{k=0}^{N-1} \binom{2N-1}{k} y^k (1 - y)^{N-1-k} + T(N)$$

where $T(N)$ is the sum of all terms having degree N or higher.

(b) Use a degree argument to derive (6.14).

6.8   The crucial polynomial $P(y)$ is defined in equation (6.14).

(a) Find $P(y)$ for $n = 4$.

(b) Confirm that $P(y)$ from part (a) satisfies (6.13) from Theorem 6.1.

6.9   Prove that $P(y) \geq 0$ for $0 \leq y \leq 1$. (*Hint:* The binomial coefficients are positive.)

6.10   Use

$$\cos \omega = \frac{e^{-i\omega} + e^{i\omega}}{2} = \frac{z + 1/z}{2}$$

$$\frac{1 - \cos \omega}{2} = \frac{1}{2} \cdot \frac{2 - z - 1/z}{2} = \left(\frac{1-z}{2}\right)\left(\frac{1 - 1/z}{2}\right)$$

to prove (6.18) and (6.19). (*Hint:* Start by expanding the right–hand sides of both identities.)

6.11   This problem outlines a proof that roots of $T(z)$ on the unit circle (Case II on page 239) have even multiplicity. Suppose that z_0 is a root of multiplicity $n \geq 1$ on the unit circle for $T(z)$. Then z_0 is of the form $z_0 = e^{-i\omega_0}$ and we can write $T(z)$ as a function of ω:

$$L(\omega) = \left(e^{-i\omega} - z_0\right)^n Q(\omega)$$

where

$$Q(\omega) = \prod_k \left(e^{-i\omega} - z_k\right)^{p_k}$$

with $z_k \neq z_0$. The following steps will help you show that n is even.

(a) Let $P(\omega) = \left(e^{-i\omega} - z_0\right)^n$. Use mathematical induction to prove that for $k < n$,

$$P^{(k)}(\omega) = \sum_{j=1}^{k-1} c_j^k \left(e^{-i\omega} - z_0\right)^{n-j} e^{-ij\omega}$$

$$+ \frac{n!}{(n-k)!} (-i)^k e^{-ik\omega} \left(e^{-i\omega} - z_0\right)^{n-k}$$

for some constants c_j^k.

(b) Use part (a) to show that $P^{(k)}(\omega_0) = 0$ for $0 \leq k < n$ and $P^{(n)}(\omega_0) \neq 0$.

(c) The Leibnitz derivative formula gives

$$L^{(n)}(\omega) = \sum_{k=0}^{n} \binom{n}{k} P^{(k)}(\omega) Q^{(n-k)}(\omega)$$

Show that

$$L^{(n)}(\omega_0) = P^{(n)}(\omega_0) Q(\omega_0) \neq 0$$

(d) By Taylor's theorem, there exists a c for which $L^{(n)}(c) \neq 0$ and

$$L(\omega) = \sum_{k=0}^{n-1} L^{(k)}(\omega_0) + \frac{L^{(n)}(c)}{n!} (\omega - \omega_0)^n$$

Use this identity to show that $L(\omega) = \frac{L^{(n)}(c)}{n!}(\omega - \omega_0)^n$, and then use this formula and the nonnegativity of $L(\omega)$ to prove that n is even.

6.12   In this problem you will show that $T(z)$ has $2N - 2$ zeros and can be factored into the form (6.23).

(a) Extract the powers of 2 from the appropriate factors and then expand $(1 - z)(1 - 1/z)$ and $(1 + z)(1 + 1/z)$ to show that (6.21) follows from (6.20).

(b) From each term $(-z + 2 - z^{-1})^k$ in part (a), factor out $(-z)^{-k}$. Perform a similar factoring for the term $(z + 2 - z^{-1})^{-(N-1-k)}$ and show that

$$T(z) = z^{-(N-1)} \sum_{k=0}^{N-1} \binom{2N-1}{k} \frac{(-1)^k}{4^{N-1}}(z-1)^{2k}(z+1)^{2N-2-2k}$$

(c) Explain why the polynomial

$$p(z) = \sum_{k=0}^{N-1} \binom{2N-1}{k} \frac{(-1)^k}{4^{N-1}}(z-1)^{2k}(z+1)^{2N-2-2k}$$

in $T(z)$ has degree $2N - 2$. Then explain why $T(z)$ can be factored into the form (6.23) and why the leading coefficient of $p(z)$ is the constant α in (6.23).

6.13   This problem considers the possible values of $F(z)$ given in (6.27).

(a) Show that each term $\left(z - z_j^U\right)\left(z - \frac{1}{z_j}^U\right)$ in (6.27) has real coefficients when expanded.

(b) Show that each term $\left(z - z_i^C\right)\left(z - \overline{z_i^C}\right)$ in (6.27) has real coefficients when expanded.

(c) What can you conclude about $F(z)$ in (6.27)?

6.14   Show that the roots in (6.32) are not on the unit circle and are reciprocals.

6.15   Let $H(\omega)$, $H^*(\omega)$ be given by (6.1), (6.29), respectively. Prove the identity (6.30). (*Hint:* Start with the right-hand side of (6.30) and work to obtain (6.29).)

6.16   Show that we can use the outside zeros to build $F(z)$, except that this reverses the order of the coefficients. Start by showing that

$$z^{2N}\left(\frac{1 + \frac{1}{z}}{2}\right)^N F_{\text{inside}}\left(\frac{1}{z}\right) = F_{\text{outside}}(z)$$

6.17   In this problem you will derive the famous D4 scaling filter due to Ingrid Daubechies. Use Algorithm 6.1 to show that the Daubechies scaling filter coefficients (the nonzero terms) for $N = 2$ are

$$
\boxed{
\begin{aligned}
h_0 &= \frac{1 + \sqrt{3}}{4\sqrt{2}} & h_2 &= \frac{3 - \sqrt{3}}{4\sqrt{2}} \\
h_1 &= \frac{3 + \sqrt{3}}{4\sqrt{2}} & h_3 &= \frac{1 - \sqrt{3}}{4\sqrt{2}}
\end{aligned}
}
\qquad (6.39)
$$

(*Hint:* First verify that the roots of $T(z)$ are $2 \pm \sqrt{3}$ – use $r_1 = 2 - \sqrt{3}$. Then confirm the identity

$$
\frac{1 + \sqrt{3}}{\sqrt{2}} = \frac{1}{\sqrt{2 - \sqrt{3}}}
$$

and use it to get the desired coefficients.)

★**6.18**   Suppose your friend knows that for $N = 4$ the function $T(z)$ given in (6.20) has the following two roots:

$$
z = 0.328876, \qquad z = 2.03114 - 1.73895i
$$

(a) Using only the properties of $T(z)$ and without computing $P(y)$ or $T(z)$ directly, find the remaining roots of $T(z)$.

(b) Given that $T(z)$ has leading coefficient $-\frac{5}{16}$, use part (a) to find $F(z)$ and thus the D8 scaling filter (rounded to six decimal digits):

$$
\begin{array}{llll}
h_0 = & 0.230378 & h_4 = & -0.187035 \\
h_1 = & 0.714847 & h_5 = & 0.030841 \\
h_2 = & 0.630881 & h_6 = & 0.032883 \\
h_3 = & -0.027984 & h_7 = & -0.010597
\end{array}
\qquad (6.40)
$$

★**6.19**   Find the wavelet filter coefficients for the D4 scaling filter (6.39) and the D8 scaling filter (6.40).

★**6.20**   Use Algorithm 6.1 to find the D10 Daubechies scaling filter and then use these values to find the corresponding wavelet filter. A CAS is required for this problem.

6.21   In this problem you will verify that $P(y)$ and $Q(y)$ in Theorem 6.1 are unique.

(a) Use a contradiction argument to prove the uniqueness of $P(y)$ and $Q(y)$ in (6.12).

(b) Use uniqueness to show that $Q(y) = P(1 - y)$.

6.22 In this problem you will provide some details of the proof of Proposition 6.2.

(a) Prove the case $N = 2$ for Proposition 6.2 without using (6.36). That is, show $P(y) \leq 2^{2 \cdot 2 - 2}$ directly.

(b) Prove the combinatorial identity $\frac{1}{2} \sum_{k=0}^{2N-1} \binom{2N-1}{k} = \frac{1}{2} 2^{2N-1}$. (*Hint:* The binomial coefficients $\binom{2N-1}{k}$, $k = 0, \ldots, 2N - 1$ are the coefficients you get when you use the binomial theorem to expand $(x + y)^{2N-1}$. Now set $x = y = 1$.)

6.2 THE CASCADE ALGORITHM

What do the Daubechies scaling function graphs actually look like? From Section 6.1 we have the Daubechies scaling function symbols and the scaling filters but no closed-form formula for the scaling functions. To cope with this problem, an iterative routine called the *cascade algorithm* was introduced by Ingrid Daubechies and Jeffrey Lagarias in [19]. Like Newton's method or Euler's method in calculus, we start with a first guess $\phi_0(t)$ and iterate through an equation and get better approximations $\phi_n(t)$ until we converge sufficiently close to the true scaling function $\phi(t)$. The cascade algorithm uses the dilation equation (5.3) for its defining iteration:

$$\phi_{n+1}(t) = \sqrt{2} \sum_{k=0}^{M} h_k \phi_n(2t - k) \tag{6.41}$$

putting the nth approximation into the right-hand side and then "turning the crank" to produce the next approximation. We start with a safe first guess, the Haar scaling function, $\phi_0(t) = B_0(t)$. Under reasonable conditions, we can guarantee the sequence of approximations $\phi_0(t), \phi_1(t), \ldots, \phi_n(t), \ldots$ will converge to the scaling function associated with the symbol $H(\omega)$.

Before embarking upon the proof of convergence, let's look at some examples to appreciate how the algorithm works. Fortunately, the cascade algorithm converges for the B–splines and the Daubechies scaling functions. The next example illustrates cascading with these important wavelet families.

Example 6.3 (The Quadratic B-spline and D4 Scaling Functions) *Use the cascade algorithm to approximate the quadratic B-spline scaling function and the D4.*
Solution
We start by performing some iterations of the cascade algorithm for the B-spline $B_2(t)$, which is a piecewise quadratic polynomial and has symbol $H(z) = \left(\frac{1+z}{2}\right)^3$. Although this function, like all other B-splines, has an explicit representation (2.60), it is nonetheless a useful function for introducing the cascade algorithm.

Each h_k value is the coefficient of z^k in $\sqrt{2}H(z) = \frac{\sqrt{2}}{8}\left(1 + 3z + 3z^2 + z^3\right)$, so we find that

$$h_0 = h_3 = \frac{\sqrt{2}}{8} \quad and \quad h_1 = h_2 = \frac{3\sqrt{2}}{8} \tag{6.42}$$

Putting these values into (6.41) gives

$$\phi_{n+1}(t) = \sqrt{2}\sum_{k=0}^{3}h_k\phi_n(2t-k)$$

$$= \frac{1}{4}\phi_n(2t) + \frac{3}{4}\phi_n(2t-1) + \frac{3}{4}\phi_n(2t-2) + \frac{1}{4}\phi_n(2t-3)$$

Thus

$$\phi_1(t) = \frac{1}{4}\phi_0(2t) + \frac{3}{4}\phi_0(2t-1) + \frac{3}{4}\phi_0(2t-2) + \frac{1}{4}\phi_0(2t-3)$$

The first iteration $\phi_1(t)$ is plotted in Figure 6.1.

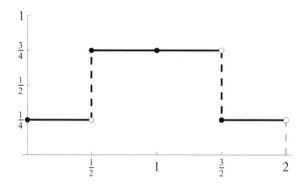

Figure 6.1   The first iteration of the cascade algorithm using the dilation equation for the quadratic B-spline $B_2(t)$ and the initial guess $\phi_0(t) = B_0(t)$.

We iterate again, this time using our approximate $\phi_1(t)$ in (6.41) to produce $\phi_2(t)$. We can also write $\phi_2(t)$ in terms of the approximate $\phi_0(t)$ if we so desire:

$$\phi_2(t) = \frac{1}{4}\phi_1(2t) + \frac{3}{4}\phi_1(2t-1) + \frac{3}{4}\phi_1(2t-2) + \frac{1}{4}\phi_1(2t-3)$$

$$= \frac{1}{16}\phi_0(4t) + \frac{3}{16}\phi_0(4t-1) + \cdots + \frac{3}{16}\phi_0(4t-8) + \frac{1}{16}\phi_0(4t-9)$$

$$(6.43)$$

You will find all of the coefficients for $\phi_0(4t-k)$, $k = 2,\ldots,7$ in Problem 6.24. Plots of $\phi_2(t)$ and $\phi_5(t)$ are shown in Figure 6.2. Note that $\phi_5(t)$ is beginning to look like the quadratic $B_2(t)$ B-spline displayed in Figure 2.14(c).

Next we carry out some iterations of the cascade algorithm for the D4 scaling function. Unlike the B-splines, there exists no explicit representation for the Daubechies scaling function, so the cascade algorithm is our only means of providing approximations to $\phi(t)$. Note from Figure 6.3 that the scaling function graph seems to have sharp cusps at $t = 1, 2$. It turns out that this scaling function is continuous but does not have derivatives everywhere.

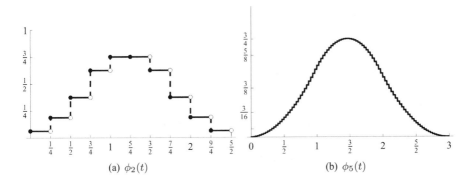

Figure 6.2   Plots of $\phi_2(t)$ and $\phi_5(t)$ approximates to the quadratic B-spline $B_2(t)$ using the cascade algorithm with initial guess $\phi_0(t) = B_0(t)$.

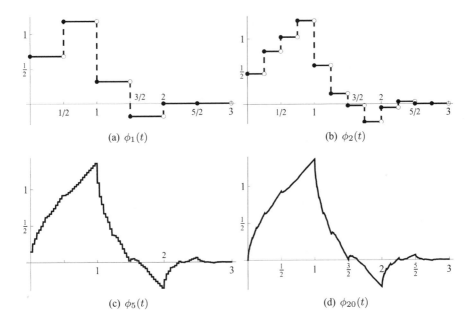

Figure 6.3   Plot of the approximates $\phi_n(t)$, $n = 1, 2, 5, 20$ to the D4 scaling function using $\phi_0(t) = B_0(t)$ as the initial guess.

What can we say about the function supports? For both $B_2(t)$ and the D4 scaling function, the support of $\phi_2(t)$ appears to be $\left[0, \frac{5}{2}\right]$. The support of each $\phi_5(t)$ appears to be contained in the interval $[0, 3]$. We investigate the support properties further in Problem 6.25 and later in the section. As expected from Theorem 5.4, the support of both scaling functions is $[0,3]$ since each symbol has degree 3. ∎

To gain a more analytical understanding of the convergence properties of the cascade algorithm, we need to move to the transform domain.

The Cascade Algorithm in the Transform Domain

Looking at the cascade algorithm in the transform domain provides some insight into the convergence question. If we apply the Fourier transform to both sides of (6.41), we obtain

$$\hat{\phi}_{n+1}(\omega) = H\left(\frac{\omega}{2}\right)\hat{\phi}_n\left(\frac{\omega}{2}\right) \tag{6.44}$$

Iterating from the initial guess

$$\hat{\phi}_0(\omega) = \widehat{B_0}(\omega) = \frac{1}{\sqrt{2\pi}}e^{-i\omega/2}\mathrm{sinc}\left(\frac{\omega}{2}\right)$$

(see Equation (1.34)), we obtain

$$\hat{\phi}_1(\omega) = H\left(\frac{\omega}{2}\right)\frac{1}{\sqrt{2\pi}}e^{-i\omega/4}\mathrm{sinc}\left(\frac{\omega}{4}\right)$$

$$\hat{\phi}_2(\omega) = H\left(\frac{\omega}{2}\right)\hat{\phi}_1\left(\frac{\omega}{2}\right)$$

$$= H\left(\frac{\omega}{2}\right)H\left(\frac{\omega}{4}\right)\frac{1}{\sqrt{2\pi}}e^{-i\omega/8}\mathrm{sinc}\left(\frac{\omega}{8}\right)$$

and in general we can write

$$\hat{\phi}_n(\omega) = \prod_{k=1}^{n} H\left(\frac{\omega}{2^k}\right) \cdot \frac{1}{\sqrt{2\pi}}e^{-i\omega/2^{n+1}}\mathrm{sinc}\left(\frac{\omega}{2^{n+1}}\right) \tag{6.45}$$

Thus if the cascade algorithm converges, the limit in the transform domain would have to be a function defined by

$$\lim_{n\to\infty}\prod_{k=1}^{n} H\left(\frac{\omega}{2^k}\right) \cdot \frac{1}{\sqrt{2\pi}}e^{-i\omega/2^{n+1}}\mathrm{sinc}\left(\frac{\omega}{2^{n+1}}\right)$$

Now for any given ω,

$$\lim_{n\to\infty} e^{-i\omega/2^{n+1}}\mathrm{sinc}\left(\frac{\omega}{2^{n+1}}\right) = e^0 \cdot \mathrm{sinc}(0) = 1$$

so we expect the limit function to be

$$g(\omega) = \frac{1}{\sqrt{2\pi}}\prod_{k=1}^{\infty} H\left(\frac{\omega}{2^k}\right) \tag{6.46}$$

where the notation $\prod_{k=1}^{\infty} H\left(\frac{\omega}{2^k}\right)$ is short for $\lim_{n\to\infty}\prod_{k=1}^{n} H\left(\frac{\omega}{2^k}\right)$. So, when the cascade algorithm converges, $g(\omega)$ should be $\hat{\phi}(\omega)$, the Fourier transform of the original $\phi(t)$. Before looking at a general proof, let's examine some important special cases.

Example 6.4 (The Quadratic B–Spline in the Transform Domain) *For the B-spline $B_2(t)$ discussed above, we know from Proposition 2.17 that*

$$\widehat{B_2}(\omega) = (2\pi)^{-1/2} e^{-3i\omega/2} \operatorname{sinc}^3\left(\frac{\omega}{2}\right)$$

and the symbol is $H(\omega) = \left(\frac{1+e^{-i\omega}}{2}\right)^3$. Let's confirm that $g(\omega)$ defined by (6.46) is indeed $\widehat{B_2}(\omega)$. This argument involves some clever algebra:

$$\prod_{k=1}^{n} H\left(\frac{\omega}{2^k}\right) = \prod_{k=1}^{n} \left(\frac{1+e^{-i\omega/2^k}}{2}\right)^3$$

$$= \prod_{k=1}^{n} \left(\frac{1+e^{-i\omega/2^k}}{2} \cdot \frac{1-e^{-i\omega/2^k}}{1-e^{-i\omega/2^k}}\right)^3$$

$$= \frac{1}{2^{3n}} \prod_{k=1}^{n} \left(\frac{1-e^{-i\omega/2^{k-1}}}{1-e^{-i\omega/2^k}}\right)^3$$

Expanding out this product and then canceling terms gives

$$\prod_{k=1}^{n} H\left(\frac{\omega}{2^k}\right) = \frac{1}{2^{3n}} \left(\frac{1-e^{-i\omega/2^{n-1}}}{1-e^{-i\omega/2^n}} \cdot \frac{1-e^{-i\omega/2^{n-2}}}{1-e^{-i\omega/2^{n-1}}} \cdots \frac{1-e^{-i\omega/2^0}}{1-e^{-i\omega/2^1}}\right)^3$$

$$= \frac{1}{2^{3n}} \left(\frac{1-e^{-i\omega}}{1-e^{-i\omega/2^n}}\right)^3 \qquad (6.47)$$

In Problem 6.32 you will use (6.47) to show that the Fourier transforms of the approximates $\hat{\phi}_n(\omega)$ converge to

$$\lim_{n\to\infty} \hat{\phi}_n(\omega) = (2\pi)^{-1/2} e^{-3i\omega/2} \operatorname{sinc}^3\left(\frac{\omega}{2}\right) \qquad (6.48)$$

which is indeed $\widehat{B_3}(\omega)$. ∎

What about the cascade algorithm in the transform domain for the Daubechies scaling functions? Let's look at the D4 case.

Example 6.5 (D4 Scaling Function in the Transform Domain) *For the Daubechies symbols, the algebra is much tougher than the B-spline case. Using a CAS (see Problem 6.34), we plot $\left|\hat{\phi}_n(\omega)\right|$ defined by (6.45) for $n = 1, 4, 7, 10$ in Figure 6.4. The graphs of $\hat{\phi}_n(\omega)$ for $n = 4, 7, 10$ are visually indistinguishable. It appears that the $\{\hat{\phi}_n(\omega)\}$ are converging quickly to a continuous function which decays rapidly to zero as $\omega \to \infty$.* ∎

The convergence of the $\{\hat{\phi}_n(\omega)\}$ to a continuous function, illustrated for the D4 scaling function, actually works for a large set of important symbols. The following

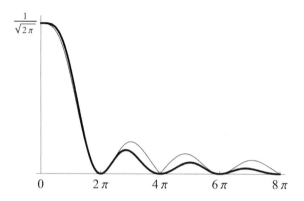

Figure 6.4 $\left|\hat{\phi}_n(\omega)\right|$, $n = 1,4,7,10$, for the D4 scaling function. The graph for $n = 1$ appears in the lighter color while the graphs for $n = 4,7,10$ are visually identical.

result formalizes a convergence criterion in the transform domain for the cascade algorithm.

Proposition 6.3 (Convergence in the Transform Domain) *Suppose that the symbol*

$$H(\omega) = \frac{1}{\sqrt{2}} \sum_{k=0}^{M} h_k e^{-ik\omega} \tag{6.49}$$

satisfies $H(0) = 1$. *Then the cascade algorithm iterates* $\{\hat{\phi}_n(\omega)\}$ *defined in (6.45) converge for each* ω *in the transform domain to a continuous function* $g(\omega)$. ∎

Proof: The proof is long, so we split it into three major steps.

Step 1. We will show that $\sum_{k=1}^{\infty} \left|H\left(\frac{\omega}{2^k}\right) - 1\right|$ converges to a finite value. Towards this end, we first rewrite $H(\omega)$ as follows, using $H(0) = 1$:

$$H(\omega) = \frac{1}{\sqrt{2}} \sum_{k=0}^{M} h_k \left(e^{-ik\omega} - 1\right) + \frac{1}{\sqrt{2}} \sum_{k=0}^{M} h_k$$

$$= \frac{1}{\sqrt{2}} \sum_{k=0}^{M} h_k \left(e^{-ik\omega} - 1\right) + H(0)$$

$$= \frac{1}{\sqrt{2}} \sum_{k=0}^{M} h_k \left(e^{-ik\omega} - 1\right) + 1 \tag{6.50}$$

Next we appeal to the useful identity (see Problem 2.8 in Section 2.1)

$$1 - e^{-i\omega} = 2ie^{-i\omega/2} \sin\left(\frac{\omega}{2}\right) \tag{6.51}$$

to obtain

$$
\left| \frac{1}{\sqrt{2}} \sum_{k=0}^{M} h_k \left(e^{-ik\omega} - 1 \right) \right| = \left| \frac{1}{\sqrt{2}} \sum_{k=0}^{M} h_k \left(2ie^{-ik\omega/2} \sin\left(-\frac{k\omega}{2} \right) \right) \right|
$$

$$
\leq \sqrt{2} \sum_{k=0}^{M} |h_k| \cdot \left| \sin\left(\frac{k\omega}{2} \right) \right| \tag{6.52}
$$

From calculus we know that $|\sin \omega| \leq |\omega|$ for all $\omega \in \mathbb{R}$. Set $\beta = \frac{\sqrt{2}}{2} \sum_{k=0}^{M} |h_k| \, k$ and use (6.50) and (6.52) to show that

$$
|H(\omega) - 1| \leq \beta |\omega| \tag{6.53}
$$

Then we use a geometric series to obtain

$$
\sum_{k=1}^{\infty} \left| H\left(\frac{\omega}{2^k} \right) - 1 \right| \leq \sum_{k=1}^{\infty} \beta \left| \frac{\omega}{2^k} \right| = \beta |\omega| \tag{6.54}
$$

so $\sum_{k=1}^{\infty} \left| H\left(\frac{\omega}{2^k} \right) - 1 \right|$ indeed converges for each $\omega \in \mathbb{R}$, and step 1 of the proof is complete.

Step 2. We next show $\prod_{k=1}^{\infty} H\left(\frac{\omega}{2^k} \right)$ converges. Let $c_k = H\left(\frac{\omega}{2^k} \right)$ for ease of notation. We need the following calculus fact (see Problem 6.28):

$$
\lim_{k \to \infty} \frac{\ln(c_k)}{c_k - 1} = 1 \tag{6.55}
$$

From step 1 we know that $\sum_{k=1}^{\infty} (c_k - 1)$ converges, so from the limit comparison test of calculus we conclude that $\sum_{k=1}^{\infty} \ln(c_k)$ converges. Next we use log properties to write

$$
\prod_{k=1}^{n} c_k = e^{\sum_{k=1}^{n} \ln(c_k)} \tag{6.56}
$$

Thus $\prod_{k=1}^{\infty} c_k$ must converge, completing step 2 of the proof. This justifies the existence of the function

$$
g(\omega) = \frac{1}{\sqrt{2\pi}} \prod_{k=1}^{\infty} H\left(\frac{\omega}{2^k} \right)
$$

defined in (6.46).

Step 3. Next we show that the sequence of cascade algorithm iterates $\left\{ \hat{\phi}_n(\omega) \right\}$ converges to $g(\omega)$. Using the definition of $\hat{\phi}_n(\omega)$ given in (6.45) and the convergence proved in step 2, we have

$$\lim_{n\to\infty} \hat{\phi}_n(\omega) = \lim_{n\to\infty} \prod_{k=1}^{n} H\left(\frac{\omega}{2^k}\right) \cdot \frac{1}{\sqrt{2\pi}} e^{-i\omega/2^{n+1}} \operatorname{sinc}\left(\frac{\omega}{2^{n+1}}\right)$$

$$= \prod_{k=1}^{\infty} H\left(\frac{\omega}{2^k}\right) \cdot \frac{1}{\sqrt{2\pi}} e^0 \operatorname{sinc}(0)$$

$$= g(\omega)$$

Now each $\hat{\phi}_n(\omega)$ is continuous, so it seems plausible that their limit $g(\omega)$ is continuous. And indeed, an advanced calculus (see Rudin [48]) result beyond the scope of our text can be used to show that $g(\omega)$ is continuous. ∎

Next we prove two more key properties of $g(\omega)$, which is our candidate for the scaling function's Fourier transformation. We have to insist on more requirements of a general symbol $H(z)$, but the following proposition suffices for symbols for scaling functions such as B-splines and the Daubechies family.

Proposition 6.4 (The Limit Function Satisfies the Dilation Equation) *If the symbol $H(z)$ given in (6.49) can be factored as*

$$H(z) = \left(\frac{1+z}{2}\right)^N S(z) \tag{6.57}$$

where $S(z)$ satisfies $S(1) = 1$ and

$$\max_{|z|=1} |S(z)| \le 2^{N-1}, \tag{6.58}$$

then $g(\omega)$ satisfies the transform-domain dilation equation

$$g(\omega) = H\left(\frac{\omega}{2}\right) g\left(\frac{\omega}{2}\right) \tag{6.59}$$

and

$$|g(\omega)| \le \frac{C}{1+|\omega|} \tag{6.60}$$

for some constant $C > 0$. ∎

Proof: We begin with the right side of (6.59) and compute

$$H\left(\frac{\omega}{2}\right) g\left(\frac{\omega}{2}\right) = H\left(\frac{\omega}{2}\right) \cdot \frac{1}{\sqrt{2\pi}} \lim_{n\to\infty} \prod_{k=1}^{n} H\left(\left(\frac{\omega}{2}\right)/2^k\right)$$

$$= \frac{1}{\sqrt{2\pi}} \lim_{n\to\infty} \left[H\left(\frac{\omega}{2}\right) \cdot H\left(\frac{\omega}{4}\right) \cdots H\left(\frac{\omega}{2^{n+1}}\right) \right] \tag{6.61}$$

where passing $H\left(\frac{\omega}{2}\right)$ through the limit is justifiable by an advanced calculus theorem (see Rudin [48]). Then reindexing $\ell = n + 1$ in (6.61), we have

$$H\left(\frac{\omega}{2}\right) g\left(\frac{\omega}{2}\right) = \frac{1}{\sqrt{2\pi}} \lim_{\ell \to \infty} \prod_{k=1}^{\ell} H\left(\frac{\omega}{2^k}\right)$$

$$= \frac{1}{\sqrt{2\pi}} \prod_{k=1}^{\infty} H\left(\frac{\omega}{2^k}\right)$$

$$= g(\omega)$$

which shows that $g(\omega)$ satisfies (6.59). To verify (6.60), we need the following identity, reminiscent of (6.47), with details given in Problem 6.27.

$$\prod_{k=1}^{\infty} \left(\frac{1 + e^{-i\omega/2^k}}{2}\right)^N = \left(\frac{1 - e^{-i\omega}}{i\omega}\right)^N \tag{6.62}$$

Using hypothesis (6.58) we have

$$g(\omega) = \frac{1}{\sqrt{2\pi}} \prod_{k=1}^{\infty} H\left(\frac{\omega}{2^k}\right)$$

$$= \frac{1}{\sqrt{2\pi}} \prod_{k=1}^{\infty} \left(\frac{1 + e^{-i\omega/2^k}}{2}\right)^N \cdot \prod_{k=1}^{\infty} S\left(\frac{\omega}{2^k}\right)$$

$$= \frac{1}{\sqrt{2\pi}} \left(\frac{1 - e^{-i\omega}}{i\omega}\right)^N \cdot \prod_{k=1}^{\infty} S\left(\frac{\omega}{2^k}\right) \tag{6.63}$$

It is not difficult to show that

$$\left|\frac{1 - e^{-i\omega}}{i\omega}\right|^N \leq 2^N \min(1, |\omega|^{-N}) \tag{6.64}$$

You will provide a proof of this identity in Problem 6.29. The other factor in (6.63) is a challenge.

Define $T(\omega) = \prod_{k=1}^{\infty} S\left(\frac{\omega}{2^k}\right)$. In order to verify (6.60), we need to ensure that $T(\omega)$ does not grow too fast as $|\omega| \to \infty$. Toward this end, fix an integer M and ω satisfying

$$2^{M-1} \leq |\omega| \leq 2^M \tag{6.65}$$

For this ω and M,

$$
\begin{aligned}
T(\omega) &= \prod_{k=1}^{\infty} S\left(\frac{\omega}{2^k}\right) \\
&= \prod_{k=1}^{M} S\left(\frac{\omega}{2^k}\right) \cdot \prod_{k=M+1}^{\infty} S\left(\frac{\omega}{2^k}\right) \\
&= \prod_{k=1}^{M} S\left(\frac{\omega}{2^k}\right) \cdot \prod_{\ell=1}^{\infty} S\left(\frac{2^{-M}\omega}{2^\ell}\right) \\
&= \prod_{k=1}^{M} S\left(\frac{\omega}{2^k}\right) \cdot T\left(2^{-M}\omega\right)
\end{aligned}
$$

Now set

$$
U = \max_{|\omega| \le 1} |T(\omega)|
$$

and observe that

$$
\left|T\left(2^{-M}\omega\right)\right| \le U \tag{6.66}
$$

We combine these facts with hypothesis (6.58) to see that

$$
\begin{aligned}
|T(\omega)| &= \left|\prod_{k=1}^{M} S\left(\frac{\omega}{2^k}\right)\right| \cdot \left|T\left(2^{-M}\omega\right)\right| \\
&\le \left(2^{N-1}\right)^M \cdot \left|T\left(2^{-M}\omega\right)\right| \\
&\le \left(2 \cdot 2^{M-1}\right)^{N-1} \cdot U \tag{6.67}
\end{aligned}
$$

Since $2^{M-1} \le |\omega|$ by (6.65), the bound (6.67) yields

$$
|T(\omega)| \le \left(2\,|\omega|\right)^{N-1} U \tag{6.68}
$$

Notice that (6.68) is independent of the integer M, so in fact (6.68) holds for *any* ω. We combine this fact with (6.64) to find, for any ω,

$$
\begin{aligned}
|g(\omega)| &\le \frac{1}{\sqrt{2\pi}} \left|\frac{1-e^{-i\omega}}{i\omega}\right|^N \cdot \left|\prod_{k=1}^{\infty} S\left(\frac{\omega}{2^k}\right)\right| \\
&\le \frac{1}{\sqrt{2\pi}} 2^N \min(1, |\omega|^{-N}) \cdot 2^{N-1} |\omega|^{N-1} U \\
&\le \frac{2^{2N-1}U}{\sqrt{2\pi}} \frac{2}{1+|\omega|} \tag{6.69}
\end{aligned}
$$

where the details of the last inequality (6.69) are explored in Problem 6.30. Putting $C = 2^{2N}U/\sqrt{2\pi}$ completes the proof of (6.60) and this proposition. ∎

As noted in Theorem 6.2, the Daubechies symbols all satisfy (6.58). The B-splines also satisfy this condition easily with the symbol $H(\omega) = \left(\frac{1+e^{-i\omega}}{2}\right)^{n+1}$, as given in (5.71). This gives us two important families of functions for which the cascade algorithm converges, at least in the transform domain.

Next we look at the cascade algorithm back in the time domain.

Theorem 6.3 (Convergence in the Time Domain) *Suppose that symbol $H(z)$ satisfies the hypotheses of Proposition 6.4. Then the sequence of functions $\{\phi_n(t)\}$ defined by (6.41), with $\phi_0(t) = B_0(t)$, converges to a function $\phi(t)$ that satisfies the dilation equation*

$$\phi(t) = \sqrt{2}\sum_{k=0}^{M} h_k \phi(2t - k) \tag{6.70}$$

Moreover, $\overline{\operatorname{supp}(\phi_n)} = [0, M - 2^{-n}(M-1)]$ *for $n \geq 1$, and* $\overline{\operatorname{supp}(\phi)} = [0, M]$. ∎

Proof: The function $g(\omega)$ discussed above in Proposition 6.4 is our candidate for $\hat{\phi}(\omega)$. But for this to translate sensibly to the time domain, we first need to show that $g(\omega)$ is a member of $L^2(\mathbb{R})$. To see this fact, we use (6.60) and observe (see Problem 6.31) that

$$\int_{\mathbb{R}} |g(\omega)|^2 \, d\omega \leq \int_{\mathbb{R}} \left(\frac{C}{1+|\omega|}\right)^2 \, d\omega < \infty \tag{6.71}$$

Thus $g(\omega)$ has Fourier transform $\hat{g}(t) \in L^2(\mathbb{R})$ by Proposition 2.4. We can thus justify defining $\phi(t)$ by $\phi(t) = \hat{g}(t)$. Now $\hat{\phi}(\omega) = g(\omega)$ satisfies the dilation equation (6.59) in the transform domain, so by Theorem 5.3, $\phi(t)$ satisfies dilation equation (6.70) in the time domain.

We have seen that the $\{\hat{\phi}_n(\omega)\}$ converges to $\hat{\phi}(\omega)$, so one would expect that the $\phi_n(t)$ would converge to $\phi(t)$ as well. The proof of this convergence, in the $L^2(\mathbb{R})$ sense, is quite technical and beyond the scope of this book. The interested reader may consult Walnut [62].

Next we consider the support claim $\overline{\operatorname{supp}(\phi_n)} = [0, M - 2^{-n}(M-1)]$ by first noting that

$$\phi_1(t) = \sqrt{2}\sum_{k=0}^{M} h_k \phi_0(2t - k)$$

$$= \sqrt{2}\sum_{k=0}^{M} h_k B_0(2t - k)$$

Since $B_0(t)$ has support $[0,1)$, we know that the support of each $B_0(2t - k)$ is $\left[\frac{k}{2}, \frac{k+1}{2}\right)$. These support intervals do not overlap, so $\operatorname{supp}(\phi_1)$ must be

$$\bigcup_{k=0}^{M}\left[\frac{k}{2}, \frac{k+1}{2}\right) = \left[0, \frac{M+1}{2}\right] = [0, M - 2^{-1}(M-1)]$$

The proof of the general case for $\overline{\text{supp}\,(\phi_n)}$ can be developed using mathematical induction (see Problem 6.35). The claim $\text{supp}(\phi) = [0, M]$ makes sense by taking the limit of the iterates' supports:

$$\lim_{n \to \infty} \overline{\text{supp}\,(\phi_n)} = \lim_{n \to \infty} \left[0, M - 2^{-n}(M - 1)\right] = [0, M]$$

A more formal proof comes from the argument of Theorem 5.4.   ∎

The following corollary is immediate.

Corollary 6.1 (Support Interval for Daubechies Scaling Functions) *Suppose that* $\phi(t)$ *is the Daubechies D2N scaling function, where* N *is a positive integer. Then*

$$\overline{\text{supp}(\phi)} = [0, 2N - 1]$$

∎

Proof: We know that $\phi(t)$ satisfies the dilation equation

$$\phi(t) = \sqrt{2} \sum_{k=0}^{2N-1} h_k \phi(2t - k)$$

for real numbers $h_0, \ldots, h_{2N-1}$. Using Theorem 6.3, we see that $\phi(t)$ is compactly supported on the interval $[0, 2N - 1]$.   ∎

To illustrate the results of Theorem 6.3, we consider the following example.

Example 6.6 (The D6 and D8 Scaling Functions) *Use the length 6 (6.34) and length 8 (6.40) Daubechies scaling filters to provide approximations to the D6 and D8 scaling functions, respectively.*
Solution
   The symbol for the D6 scaling function is given by (6.33) from Example 6.1 and in Problem 6.18 you found the symbol for the D8 scaling function. In Problem 6.26 you are asked to verify that both of these symbols satisfy the hypotheses of Proposition 6.4. Thus by Theorem 6.3, we know that the cascade algorithm converges for each scaling function with initial guess $\phi_0(t) = B_0(t)$. In Figure 6.5 we have plotted the twentieth iteration for each scaling function. We have used the "reverse" version of each scaling filter.
   Note from Corollary 6.1 that the support for the D6 scaling function is $[0, 5]$ and the support for the D8 scaling function is $[0, 7]$. It can also be shown (although it is not obvious from the graphs!) that the D6 scaling function is continuously differentiable and the D8 scaling function is twice continuously differentiable. See Daubechies [20] for more details.   ∎

Theorem 6.3 assures us that the cascade algorithm will converge in the time domain under reasonable conditions on the symbol. Note that orthogonality was not required, as evidenced by the splines. The issue of orthogonality and the cascade algorithm is examined in the next section.

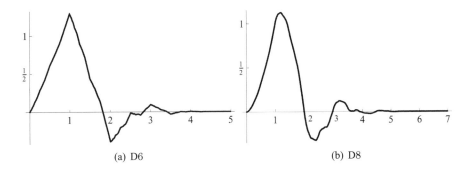

(a) D6   (b) D8

Figure 6.5   Plots of $\phi_{20}(t)$ for both the D6 and D8 scaling functions using the cascade algorithm with initial guess $\phi_0(t) = B_0(t)$.

Wavelet Functions

Now that we know how to produce approximations of the Daubechies DN scaling functions, we can use them to produce approximates of the corresponding wavelet functions. Let's look at the following example.

Example 6.7 (Daubechies Wavelet Functions) *Plot approximations of the Daubechies wavelet functions for the D4, D6, and D8 scaling functions.*
Solution
We start with the D6 scaling function $\phi(t)$. The corresponding wavelet $\psi_{D6}(t)$ satisfies the dilation equation (5.17)

$$\psi_{D6}(t) = \sqrt{2} \sum_{k\in\mathbb{Z}} g_k\phi(2t - k)$$

where the coefficients $g_k = (-1)^k h_{1-k}$. Moreover, from Example 6.2 we know the values of these coefficients and that they are nonzero for $-4 \le k \le 1$. Thus the dilation equation can be reduced to

$$\psi_{D6}(t) = \sqrt{2} \sum_{k=-4}^{1} g_k\phi(2t - k)$$

We simply plug these values into the dilation equation and use the approximate to the D6 scaling function obtained via the cascade algorithm (see Example 6.6) to obtain the approximate of $\psi_{D6}(t)$. The result is plotted in Figure 6.6.
We can use exactly the same ideas to construct wavelet functions (see Problem 6.37 for the dilation equation for Daubechies wavelet functions) for the D4 and D8 scaling functions. The approximates to the D4 and D8 scaling functions are given in Examples 6.3 and 6.6, respectively. The wavelet filter coefficients were found in Problem 6.19 in Section 6.1, and the dilation equations for the D6 and D8 wavelet

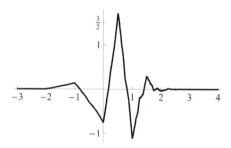

Figure 6.6   An approximation of the wavelet function $\psi_{D6}(t)$ for the D6 scaling function. Here 20 iterations where used to produce the approximate $\phi_{20}(t)$ to the scaling function.

functions are

$$\psi_{D4}(t) = \sqrt{2} \sum_{k=-2}^{1} g_k\phi(2t - k) \quad \text{and} \quad \psi_{D8}(t) = \sqrt{2} \sum_{k=-6}^{1} g_k\phi(2t - k),$$

respectively. We plot the approximates in Figure 6.7.   ∎

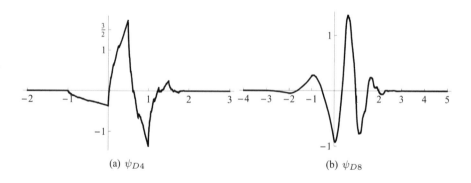

(a) ψ_{D4} (b) ψ_{D8}

Figure 6.7   Plots of approximates of the wavelet functions for the D4 and D8 scaling functions. In both cases we used $\phi_{20}(t)$ as an approximation to the scaling function.

Example 6.7 gives us some insight into the support for Daubechies wavelet functions. The result is given in Corollary 6.2.

Corollary 6.2 (Support Interval for Daubechies Wavelet Functions) *Let N be a positive integer and suppose that $\psi_{2N}(t)$ is the wavelet function associated with the Daubechies D2N scaling function. Then*

$$\overline{\text{supp}\,(\psi_{2N})} = [1 - N, N]$$

∎

Proof: Let $\phi(t)$ be the Daubechies D2N scaling function and $\psi_{2N}(t)$ the associated wavelet. From Problem 6.36, we know that for $k \in \mathbb{Z}$, $\phi(2t - k)$ is compactly supported on the interval $\left[\frac{k}{2}, N + \frac{k-1}{2}\right]$. The dilation equation obeyed by $\psi_{2N}(t)$ is given in Problem 6.37. Thus we see that $\psi(t)$ is built from functions

$$\phi\left(2t - (2 - 2N)\right), \phi\left(2t - (2 - 2N + 1)\right), \ldots, \phi(2t + 1), \phi(2t), \phi(2t - 1)$$

Again using Problem 6.36, the support intervals for each of these functions are given in Table 6.1. The union of these support intervals is $[1 - N, N]$, and the proof is complete. ∎

Table 6.1   The support intervals for the proof of Corollary 6.2.

Function	Support Interval
$\phi(2t - (2 - 2N))$	$\left[1 - N, \frac{1}{2}\right]$
$\phi(2t - (2 - 2N + 1))$	$\left[\frac{3}{2} - N, 1\right]$
$\vdots$	$\vdots$
$\phi(2t + 1)$	$\left[-\frac{1}{2}, N - 1\right]$
$\phi(2t)$	$\left[0, N - \frac{1}{2}\right]$
$\phi(2t - 1)$	$\left[\frac{1}{2}, N\right]$

PROBLEMS

Note: *A computer algebra system and the software on the course Web site will be useful for some of these problems. See the Preface for more details.*

★**6.23**   Suppose that you are given the symbol $H(z) = \left(\frac{1+z}{2}\right)^2 \left(\frac{2 + 4z - z^2}{5}\right)$.

(a) Show that the cascade algorithm will converge for this $H(z)$.

(b) Using a CAS, plot $\hat{\phi}_n(\omega)$ for $n = 1, 4, 5$. Is the convergence evident?

(c) Repeat part (b) but use $\phi_n(t)$.

6.24   Determine the coefficients of $\phi_0(4t - k)$, $k = 2, \ldots, 7$ in (6.43). Then use them to confirm the graph in Figure 6.2.

6.25   The support of $\phi_2(t)$ appears to be $\left[0, \frac{5}{2}\right]$ for both $B_2(t)$ and D4(t) in Figures 6.2 and 6.3.

(a) Verify this support claim.

(b) The plots of $\phi_5(t)$ are piecewise continuous. What are the segment lengths?

(c) What is the support of $\phi_5(t)$?

6.26   Use Example 6.1 and Problem 6.18 to verify that the symbols for the D6 and D8 scaling functions satisfy the hypotheses of Proposition 6.4.

6.27   In this problem you will prove (6.62). The following steps will help you organize your work.

(a) First use the ideas of Example 6.4 and Problem 6.27 to show that

$$\prod_{k=1}^{n}\left(\frac{1+e^{-i\omega/2^k}}{2}\right)^N = \left(\frac{1}{2^n} \cdot \frac{1-e^{-i\omega}}{2ie^{-i\omega/2^{n+1}}\sin\left(\omega/2^{n+1}\right)}\right)^N$$

(b) Now multiply the right-hand side of the identity in part (a) by $\left(\frac{\omega}{2^{n+1}}\right)^N$ top and bottom, rearrange terms, then extract the factor

$$\left(\frac{\omega/2^{n+1}}{\sin\left(\omega/2^{n+1}\right)}\right)^N$$

whose limit is known from calculus.

(c) Take the limit as $n \to \infty$ to finish the proof.

6.28   Assume that $\lim_{k\to\infty} c_k = 1$. Use this and calculus to prove (6.55).

6.29   In this problem you will prove (6.64). Note that we need to show that both

$$\left|\frac{1-e^{-i\omega}}{i\omega}\right|^N \le 2^N \cdot 1 \qquad \text{and} \qquad \left|\frac{1-e^{-i\omega}}{i\omega}\right|^N \le 2^N \cdot |\omega|^{-N}$$

to satisfy the "min" requirement. The following steps will help you organize your work:

(a) First prove the following helpful inequalities, using identity (6.51) and the triangle inequality, respectively:

$$\left|1-e^{-i\omega}\right| \le |\omega| \qquad \text{and} \qquad \left|1-e^{-i\omega}\right| \le 2$$

(b) Use part (a) to prove both required statements.

6.30   In this problem you will prove (6.69). *Hint:* The heart of the problem is to show that

$$\min\left(1, |\omega|^{-N}\right)|\omega|^{N-1} \le \frac{2}{1+|\omega|}$$

which is best done with two cases $|\omega| \le 1$ and when $|\omega| > 1$.)

6.31 Prove (6.71). *Hint:* Split up the integral

$$\int_{\mathbb{R}} \left(\frac{C}{1+|\omega|}\right)^2 d\omega = \int_{|\omega|\leq 1} \left(\frac{C}{1+|\omega|}\right)^2 d\omega + \int_{|\omega|>1} \left(\frac{C}{1+|\omega|}\right)^2 d\omega$$

Then use a different upper bound on $\frac{C}{1+|\omega|}$ for each integral.

6.32 Use Problem 6.27 to prove the limit (6.48).

6.33 Generalize Example 6.4 and Problem 6.32 from the quadratic B-spline $\widehat{B_2}(\omega)$ to a general B-spline $\widehat{B_m}(\omega)$.

6.34 This problem explores a graphical illustration of Proposition 6.3 using the D4 scaling function of Example 6.5.

(a) First use a CAS to confirm the plot of $\left|\hat{\phi}_{10}(\omega)\right|$ given in Figure 6.4.

(b) Use your plot in part (a) to find a C value, $C < 1.4$, such that the graph of $\frac{C}{1+|\omega|}$ lies above the graph of $\left|\hat{\phi}_{10}(\omega)\right|$ for $0 \leq \omega \leq 10\pi$.

6.35 Use mathematical induction to prove the support claim

$$\overline{\text{supp}\,(\phi_n)} = \left[0, M - 2^{-n}(M-1)\right]$$

in Theorem 6.3.

6.36 Let j, k be integers and $\phi(t)$ the Daubechies D2N scaling function. Define $\phi_{j,k}(t)$ using (5.3). Use Corollary 6.1 to show that

$$\overline{\text{supp}(\phi_{j,k})} = \left[\frac{k}{2^j}, \frac{2N-1+k}{2^j}\right]$$

6.37 Use Corollary 6.1 along with the fact that $g_k = (-1)^k h_{1-k}$ to show that the wavelet function $\psi_{2N}(t)$ associated with the Daubechies D2N scaling function $\phi(t)$ satisfies the dilation equation

$$\psi(t) = \sqrt{2} \sum_{k=2-2N}^{1} g_k \phi(2t - k)$$

6.38 Let $\psi_{2N}(t)$ be the wavelet function associated with the Daubechies D2N scaling function. Use Problem 6.37 to find $\overline{\text{supp}(\psi_{j,k})}$ where $\psi_{j,k}(t)$ is given by (5.18).

6.39 Use Problem 6.20 along with the cascade algorithm and the ideas from Example 6.7 to plot $\phi_{20}(t)$ for the D10 scaling function. Then use the ideas from Example 6.7 to plot an approximate of $\psi_{D10}(t)$. Verify the support of each function using Corollaries 6.1 and 6.2, respectively.

6.3   ORTHOGONAL TRANSLATES, CODING, AND PROJECTIONS

We explore three topics involving the cascade algorithm in this section, including a faster coding method in vector form suited for computer applications, and an application to projecting $L^2(\mathbb{R})$ functions into V_j spaces. But first we look at the orthogonality of translates $\phi(t - k)$ for a scaling function $\phi(t)$ generated by the cascade algorithm.

Theorem 6.3 in Section 6.2 promises convergence of the cascade algorithm to some function $\phi(t)$ satisfying a dilation equation. Do we have a true scaling function $\phi(t)$ satisfying the orthonormal basis condition in Definition 5.1 as well as the dilation equation? It depends on the symbol $H(z)$. If the symbol satisfies the orthogonality condition of Corollary 5.3, then the limit function $\phi(t)$ really is a scaling function, as seen in the next theorem.

Theorem 6.4 (Orthogonality and Cascading) *Suppose that the symbol $H(z)$ satisfies $H(1) = 1$, has the form $H(z) = \left(\frac{1+z}{2}\right)^N S(z)$, where $S(z)$ is a finite-degree trigonometric polynomial satisfying (6.58), and obeys the orthogonality criterion*

$$|H(z)|^2 + |H(-z)|^2 = 1$$

for all $|z| = 1$. Then the sequence of functions $\{\phi_n(t)\}$ defined by (6.41), with $\phi_0(t) = B_0(t)$, converges to a scaling function $\phi(t)$ that generates a multiresolution analysis of $L^2(\mathbb{R})$, and each $\phi_n(t)$ has orthogonal integer translates.   ∎

Proof: We show by mathematical induction that each $\phi_n(t)$ in the cascade algorithm has orthogonal integer translates. Thus it is quite reasonable to believe that the limit function $\phi(t)$ will also generate orthogonal translates. Unfortunately, the proof that $\phi(t) = \lim_{n\to\infty} \phi_n(t)$ also has orthogonal integer translates is beyond the scope of our book. The interested reader may find it in an appendix of Boggess and Narcowich [4].

To prove that each $\phi_n(t)$ has orthogonal integer translates, we use the transform domain characterization from Theorem 5.5. Define

$$\mathcal{A}_n(\omega) = \sum_{\ell \in \mathbb{Z}} \left| \hat{\phi}_n(\omega + 2\pi\ell) \right|^2$$

for each $n \in \mathbb{Z}$. We need to show that each $\mathcal{A}_n(\omega) = \frac{1}{2\pi}$. The induction base case $n = 0$ is clear since $\phi_0(t) = B_0(t)$, the Haar scaling function, which we know has orthogonal integer translates. We now assume as the induction hypothesis that $\mathcal{A}_n(\omega) = \frac{1}{2\pi}$ and prove that $\mathcal{A}_{n+1}(\omega) = \frac{1}{2\pi}$. We begin the argument by considering (6.44) evaluated at $\omega + 2\pi k$:

$$\hat{\phi}_{n+1}(\omega + 2\pi k) = H\left(\frac{\omega}{2} + \pi k\right) \hat{\phi}_n\left(\frac{\omega}{2} + \pi k\right)$$

Now take the modulus squared of each side and sum over all $k \in \mathbb{Z}$, and then separate the right-hand side into sums over odd and even integers:

$$\mathcal{A}_{n+1}(\omega) = \sum_{k \in \mathbb{Z}} \left| \hat{\phi}_{n+1}(\omega + 2\pi k) \right|^2$$

$$= \sum_{k \in \mathbb{Z}} \left| H\left(\frac{\omega}{2} + \pi k\right) \right|^2 \left| \hat{\phi}_n\left(\frac{\omega}{2} + \pi k\right) \right|^2$$

$$= \sum_{j \in \mathbb{Z}} \left| H\left(\frac{\omega}{2} + \pi(2j+1)\right) \right|^2 \left| \hat{\phi}_n\left(\frac{\omega}{2} + \pi(2j+1)\right) \right|^2$$

$$+ \sum_{\ell \in \mathbb{Z}} \left| H\left(\frac{\omega}{2} + 2\pi\ell\right) \right|^2 \left| \hat{\phi}_n\left(\frac{\omega}{2} + 2\pi\ell\right) \right|^2$$

Using the fact that $H(\omega)$ is a 2π-periodic function, we can write $H\left(\frac{\omega}{2} + \pi(2j+1)\right) = H\left(\frac{\omega}{2} + \pi\right)$ and $H\left(\frac{\omega}{2} + 2\pi\ell\right) = H\left(\frac{\omega}{2}\right)$. We can then write

$$\mathcal{A}_{n+1}(\omega) = \sum_{j \in \mathbb{Z}} \left| H\left(\frac{\omega}{2} + \pi\right) \right|^2 \left| \hat{\phi}_n\left(\frac{\omega}{2} + \pi(2j+1)\right) \right|^2$$

$$+ \sum_{\ell \in \mathbb{Z}} \left| H\left(\frac{\omega}{2}\right) \right|^2 \left| \hat{\phi}_n\left(\frac{\omega}{2} + 2\pi\ell\right) \right|^2$$

$$= \left| H\left(\frac{\omega}{2} + \pi\right) \right|^2 \sum_{j \in \mathbb{Z}} \left| \hat{\phi}_n\left(\left(\frac{\omega}{2} + \pi\right) + 2\pi j\right) \right|^2$$

$$+ \left| H\left(\frac{\omega}{2}\right) \right|^2 \sum_{\ell \in \mathbb{Z}} \left| \hat{\phi}_n\left(\frac{\omega}{2} + 2\pi\ell\right) \right|^2$$

$$= \left| H\left(\frac{\omega}{2} + \pi\right) \right|^2 \mathcal{A}_n\left(\frac{\omega}{2} + \pi\right) + \left| H\left(\frac{\omega}{2}\right) \right|^2 \mathcal{A}_n\left(\frac{\omega}{2}\right)$$

Now apply the induction hypothesis $\mathcal{A}_n(\omega) = \frac{1}{2\pi}$ for all $\omega \in \mathbb{R}$ to obtain

$$\mathcal{A}_{n+1}(\omega) = \sum_{k \in \mathbb{Z}} \left| \hat{\phi}_{n+1}(\omega + 2\pi k) \right|^2$$

$$= \left| H\left(\frac{\omega}{2} + \pi\right) \right|^2 \frac{1}{2\pi} + \left| H\left(\frac{\omega}{2}\right) \right|^2 \frac{1}{2\pi}$$

$$= \frac{1}{2\pi} \left(\left| H\left(\frac{\omega}{2} + \pi\right) \right|^2 + \left| H\left(\frac{\omega}{2}\right) \right|^2 \right)$$

$$= \frac{1}{2\pi}$$

where the last step uses the orthogonality hypothesis on the symbol. We conclude that $\mathcal{A}_{n+1}(\omega) = \frac{1}{2\pi}$, which finishes the induction proof that each $\phi_n(t)$ in the cascade algorithm has orthogonal integer translates. ∎

Of course, the B-spline symbols do not satisfy the orthogonality hypothesis of Theorem 6.4, but the cascade algorithm works for the B-splines by Theorem 6.3. We will see in Chapter 8 that the B-splines satisfy a *biorthogonal* condition and generate a useful structure very similar to a multiresolution analysis.

Coding the Cascade Algorithm in Vector Form

Suppose that we want to get a nice plot of $\phi(t)$ using the cascade algorithm. The methods of the previous Section 6.2 show how to get approximate piecewise constant plots, but the computation time becomes enormous if we use a large number of iterates in order to achieve high accuracy. Fortunately, there is a better way to implement the cascade algorithm.

In practice we perform the cascade algorithm in *vector form* for a suitable number of sample points in the support of $\phi(t)$. To simplify the process, we use *dyadic numbers* as our sample points. Dyadic numbers are rational numbers whose denominator is a power of 2. Examples include $\frac{9}{16}$, 3.25, and $-\frac{47}{32}$.

To illustrate the cascade algorithm in vector form, suppose that the scaling filter of interest is $\mathbf{h} = [h_0, h_1, h_2, h_3]^T$, with $\phi(t)$ having support on $[0, 3]$, and we want to approximate $\phi(t)$ at the points $t = \frac{j}{16}$ for $j = 0, \ldots, 47$. We start with a vector representing the Haar function evaluated at the given sample points, so we use an initial vector of length 48 with 16 ones and 32 zeros:

$$\mathbf{v}^0 = [1, \ldots, 1, 0, \ldots, 0]^T \tag{6.72}$$

Our challenge is this: How can we code the process of getting from the nth iterate vector $\mathbf{v}^n$, which represents the values of $\phi_n\left(\frac{j}{16}\right)$, to the $(n+1)$st iterate $\mathbf{v}^{n+1}$ vector, which represents the values of $\phi_{n+1}\left(\frac{j}{16}\right)$? We will use the notation

$$\mathbf{v}_j^n = \phi_n\left(\frac{j}{16}\right) \tag{6.73}$$

to denote the entry j of vector $\mathbf{v}^n$. In anticipation of using the iteration (6.41), we find for integer k that

$$\phi_n\left(\frac{2j}{16} - k\right) = \phi_n\left(\frac{2j - 16k}{16}\right) = \mathbf{v}_{2j-16k}^n$$

Then the cascade algorithm gives

$$\mathbf{v}_j^{n+1} = \phi_{n+1}\left(\frac{j}{16}\right) = \sqrt{2}\sum_{k=0}^{3} h_k \phi_n\left(\frac{2j}{16} - k\right) = \sqrt{2}\sum_{k=0}^{3} h_k \mathbf{v}_{2j-16k}^n \tag{6.74}$$

or

$$\mathbf{v}_j^{n+1} = \sqrt{2}\,\mathbf{h} \cdot \left[v_{2j}^n, v_{2j-16}^n, v_{2j-32}^n, v_{2j-48}^n\right]^T \tag{6.75}$$

for $j = 0, \ldots, 47$, where we set $\mathbf{v}_{2j-16k}^n = 0$ whenever $2j - 16k < 0$ or $2j - 16k > 47$. Note this is straightforward to code and very fast to calculate on a computer!

Example 6.8 (Vector Form of Cascade Algorithm for a B–spline) *Find* $\mathbf{v}^1$, $\mathbf{v}^2$, *and* $\mathbf{v}^3$ *sampling at* $t = \frac{j}{16}$ *for* $j = 0, \ldots, 16$ *for the quadratic B-spline* $B_2(t)$.
Solution
  The scaling filter coefficients for the quadratic B-spline $B_2(t)$ are given in (6.42) in Example 6.3. For the dilation equation (6.41), we need to multiply these coefficients by $\sqrt{2}$. Thus we use the coefficients

$$\sqrt{2}\,\mathbf{h} = \sqrt{2}\,[h_0, h_1, h_2, h_3]^T = \sqrt{2}\,\left[\frac{\sqrt{2}}{8}, \frac{3\sqrt{2}}{8}, \frac{3\sqrt{2}}{8}, \frac{\sqrt{2}}{8}\right]^T = \left[\frac{1}{4}, \frac{3}{4}, \frac{3}{4}, \frac{1}{4}\right]^T$$

Using initial vector (6.72) and iteration (6.75), we first find $\mathbf{v}^1$. Then we find $\mathbf{v}^2$ from $\mathbf{v}^1$. Some of the values for $\mathbf{v}_j^1$ and $\mathbf{v}_j^2$ with $j = 0, \ldots, 18$ are:

$$\mathbf{v}^1 = \left[\frac{1}{4}, \frac{1}{4}, \frac{1}{4}, \frac{1}{4}, \frac{1}{4}, \frac{1}{4}, \frac{1}{4}, \frac{1}{4}, \frac{3}{4}, \frac{3}{4}, \frac{3}{4}, \frac{3}{4}, \frac{3}{4}, \frac{3}{4}, \frac{3}{4}, \frac{3}{4}, \frac{3}{4}, \frac{3}{4}, \ldots\right]^T$$

$$\mathbf{v}^2 = \left[\frac{1}{16}, \frac{1}{16}, \frac{1}{16}, \frac{1}{16}, \frac{3}{16}, \frac{3}{16}, \frac{3}{16}, \frac{3}{16}, \frac{3}{8}, \frac{3}{8}, \frac{3}{8}, \frac{3}{8}, \frac{5}{8}, \frac{5}{8}, \frac{5}{8}, \frac{5}{8}, \frac{3}{4}, \frac{3}{4}, \frac{3}{4}, \ldots\right]^T$$

Both $\mathbf{v}^1$ and $\mathbf{v}^2$ are plotted in Figure 6.8.

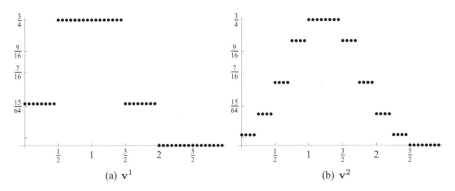

(a) $\mathbf{v}^1$ (b) $\mathbf{v}^2$

Figure 6.8   Plots of the vectors $\mathbf{v}^1$ and $\mathbf{v}^2$, which represent samples of the cascade algorithm iterates $\phi_1(t)$ and $\phi_2(t)$, where $\phi(t)$ is the quadratic B-spline $B_2(t)$.

  To illustrate the process (6.75) for the components of the third iterate $\mathbf{v}_j^3$, let $j = 9$ and calculate

$$\mathbf{v}_9^3 = \sqrt{2}\,\mathbf{h} \cdot \left[v_{18}^2, v_2^2, 0, 0\right]^T$$

$$= \left[\frac{1}{4}, \frac{3}{4}, \frac{3}{4}, \frac{1}{4}\right]^T \cdot \left[\frac{3}{4}, \frac{1}{16}, 0, 0\right]^T$$

$$= \frac{15}{64}$$

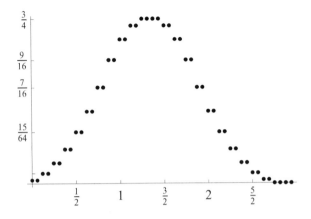

Figure 6.9   The third iteration $\mathbf{v}^3$ of the cascade algorithm for spline $B_2(t)$ using (6.75).

A plot of $\mathbf{v}^3$ is given in Figure 6.9. See Problem 6.42 for details.

∎

In Example 6.8 we used iterates at sampling points $\frac{j}{2^4}$ for $j = 0, \ldots, 3 \cdot 2^4 - 1$. We can gain better resolution of the scaling function by sampling at $\frac{j}{2^n}$ for $n > 4$. We illustrate this idea in the next example.

Example 6.9 (Vector Form for the D4 Scaling Function) *We now show an approximation of the D4 scaling function by implementing the cascade algorithm in vector form, using $n = 6$ and sampling points $\frac{j}{2^6}$ for $j = 0, \ldots, 3 \cdot 2^6 - 1 = 191$. We need the Daubechies four-tap scaling filter constructed in Problem 6.17 in Section 6.1:*

$$\sqrt{2}\,\mathbf{h} = \sqrt{2}\,[h_0, h_1, h_2, h_3]^T = \left[\frac{1+\sqrt{3}}{4}, \frac{3+\sqrt{3}}{4}, \frac{3-\sqrt{3}}{4}, \frac{1-\sqrt{3}}{4}\right]^T$$

Adjusting initial vector (6.72) and iteration (6.75) for $3 \cdots 2^6$ sampling points, we first find $\mathbf{v}^1$. Some values for $\mathbf{v}_j^1$ are

$$\mathbf{v}_0^1 = \frac{1+\sqrt{3}}{4}, \quad \mathbf{v}_8^1 = \frac{3+\sqrt{3}}{4}, \quad \mathbf{v}_{16}^1 = \frac{3-\sqrt{3}}{4}, \quad \mathbf{v}_{24}^1 = \frac{1-\sqrt{3}}{4}$$

The entire vector $\mathbf{v}^1$ is plotted in Figure 6.10(a). The iterates $\mathbf{v}^2$ and $\mathbf{v}^3$ are plotted in Figure 6.10 as well.

    In an actual computer implementation these values as well as the scaling filter would be in decimal form. After 20 iterations the iterations are quite close together, with $\|\mathbf{v}^{20} - \mathbf{v}^{19}\| \approx 0.0007$. The plot of $\mathbf{v}^{20}$ is given in Figure 6.10(d). Even with 192 points being plotted, the sharp corners of the D4 scaling function graph lead to clear gaps between the points plotted near $t = 1$ and $t = 2$.

∎

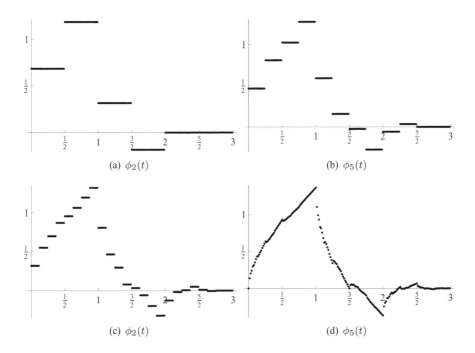

Figure 6.10   Plots of the vectors $\mathbf{v}^n$ that represent samples of the cascade algorithm iterates $\phi_n(t)$, $n = 1, 2, 3, 20$, for the D4 scaling function.

Projecting from $L^2(\mathbb{R})$ into V_j with Daubechies Scaling Functions

To project an $L^2(\mathbb{R})$ function into a V_j space generated by the D4 scaling function $\phi(t)$, we start with the projection definition (1.31) with basis functions $\phi_{j,k}(t)$:

$$f_j(t) = P_{f,j}(t) = \sum_{k \in \mathbb{Z}} a_{j,k} \phi_{j,k}(t)$$

Now to actually look at a plot of $f_j(t)$, we can use the results of Example 6.9 with the cascade algorithm to get plots of each $\phi_{j,k}(t)$.

We also need to calculate the integrals

$$a_{j,k} = \langle f(t), \phi_{j,k}(t) \rangle = \int_{\mathbb{R}} f(t), \phi_{j,k}(t) \, dt \qquad (6.76)$$

In the case of the multiresolution analysis generated by the D4 scaling function, the calculation of $a_{j,k}$ is complicated by the fact that we do not have a closed-form formula for the scaling function $\phi(t)$ for doing integration symbolically, so we must use a numerical integration method such as the trapezoidal rule to approximate $a_{j,k}$. Recalling that all numerical integration methods start with a set of sample function values, we use the cascade algorithm to compute sample values of $\phi(t)$ and then combine them with sample values of $f(t)$ to estimate the $a_{j,k}$ values. Let's illustrate the process with an example.

Example 6.10 (Projecting From $L^2(\mathbb{R})$ into V_j) *Consider the example function*

$$f(t) = \begin{cases} \frac{1}{2}t^3 - \frac{1}{2}t + 3, & -2 < t < 1 \\ 2t - 4, & 1 \le t < 2 \\ 0, & otherwise \end{cases} \tag{6.77}$$

which by Problem 1.16 in Section 1.2 is in $L^2(\mathbb{R})$. The function is plotted in Figure 6.11. Project $f(t)$ into the V_0 and V_4 spaces generated by the D4 scaling function.

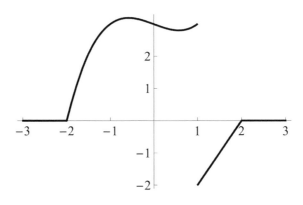

Figure 6.11   The function $f(t)$ given by (6.77).

Solution

Since $f(t)$ is compactly supported, for the V_0 projection we only need to find the $a_{0,k}$ values for $k = -4, \ldots, 1$, since $a_{0,k} = 0$ otherwise (see Problem 6.46). To keep the numerical integration simple, use the familiar trapezoidal rule from calculus with the results of Example 6.9.

$$a_{0,k} = \int_{-2}^{2} f(t)\phi(t - k)\,\mathrm{d}t = \int_{0}^{3} f(t + k)\phi(t)\,\mathrm{d}t$$
$$\approx \frac{\Delta t}{2}\left(f(0 + k)\phi(0) + 2\sum_{i=1}^{191} f(t_i + k)\phi(t_i) + f(3 + k)\phi(3) \right) \tag{6.78}$$

where we use $\Delta t = 2^{-6}$, $t_i = i\,\Delta t$, $i = 1, \ldots, 191$. With the aid of a computer, we find the following nonzero values of $a_{k,0}$, rounded to three decimal digits:

$$a_{0,-4} = 0.006 \qquad a_{0,-3} = -0.056 \qquad a_{0,-2} = 2.347$$
$$a_{0,-1} = 3.192 \qquad a_{0,0} = 2.009 \qquad a_{0,1} = -0.663$$

Of course, we can obtain more accurate approximations of the coefficients $a_{0,k}$ if we use a more sophisticated numerical integration method and sample at more than 2^6 points. In any case, we plot the projection

$$f_0(t) = \sum_{k=-4}^{1} a_{0,k}\phi(t-k) \tag{6.79}$$

into V_0, which is shown together with $f(t)$ in Figure 6.12(a). We can see from the plot that $f_0(t)$ is a poor approximation of $f(t)$, although it has the virtue of being stored using only six nonzero coefficients. In particular, the projection $f_0(t)$ has an especially difficult time near the discontinuity at $t=1$. To get a better approximation, we project $f(t)$ into a much finer space, say V_4, where we compute the $a_{4,k}$ values using numerical integration with samples of $\phi(16t-k)$ found from the cascade algorithm. The $a_{4,k}$ values are used to plot

$$f_4(t) = \sum_{k\in\mathbb{Z}} a_{4,k}\phi_{4,k}(t) \tag{6.80}$$

which is displayed in Figure 6.12(b). In Problem 6.46(b) you will verify that the nonzero values are $a_{4,-40}, \ldots, a_{4,39}$. Note the excellent approximation everywhere but the discontinuity at $t=1$. The "overshoot" just past $t=1$ is known as the Gibbs phenomenon, which we saw with the Fourier series in Example 2.1. It turns out that some overshoot is inevitable when there is a discontinuity in $f(t)$ and orthogonal wavelets or Fourier series are used for approximation (see Ruch and Van Fleet [47]). ∎

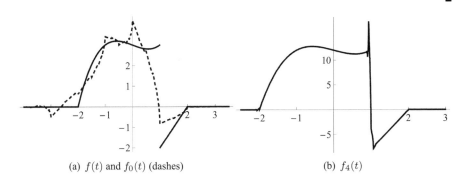

(a) $f(t)$ and $f_0(t)$ (dashes)   (b) $f_4(t)$

Figure 6.12 $f(t)$ and $f_0(t)$ projection into V_0, and projection $f_4(t)$ into V_4

It turns out that the projection coefficients $\{a_{j,k}\}$ can give more insight into the properties of function $f(t)$, especially after we decompose $f_4(t)$ into its components in

$$V_4 = V_0 \oplus W_0 \oplus W_1 \oplus W_2 \oplus W_3 \oplus W_4$$

We explore this decomposition and the accompanying insights in Section 7.1 as we see how to develop and use the discrete Daubechies wavelet transformation.

PROBLEMS

Note: *A computer algebra system and the software on the course Web site will be useful for some of these problems. See the Preface for more details.*

6.40   In Problem 6.23, you showed that there is a $\phi(t)$ satisfying the dilation equation for symbol $H(z) = \left(\frac{1+z}{2}\right)^2 \left(\frac{2+4z-z^2}{5}\right)$. Determine whether $\phi(t)$ is a scaling function that generates a multiresolution analysis of $L^2(\mathbb{R})$.

6.41   Write the first cascade iterate $\phi_1(t)$ for the D4 scaling function in terms of $B_0(t)$. Then prove directly in the time domain that $\phi_1(t)$ has orthogonal integer translates.

6.42   Use (6.75) to calculate the rest of the elements in $\mathbf{v}^1$, $\mathbf{v}^2$, and $\mathbf{v}^3$ for the spline $B_2(t)$ in Example 6.8.

6.43   Adjust (6.75) for $3 \cdot 2^6$ sampling points to calculate the rest of the elements in $\mathbf{v}^1$ for the D4 scaling function in Example 6.9.

6.44   Write code to compute the general $\mathbf{v}^n$ vectors for Example 6.8, and plot $\mathbf{v}^{10}$.

6.45   Write code to compute the general $\mathbf{v}^n$ vectors for the D6 scaling function with sample points at $\frac{j}{2^6}$, and plot $\mathbf{v}^{10}$.

6.46   This problem refers to Example 6.10. Recall that $\phi(t) = 0$ when $t \leq 0$ and $t \geq 3$, and $f(t) = 0$ when $t \leq -2$ and $t \geq 2$.

(a) Use these facts and integral properties to show that $a_{0,k} = 0$ if $k \leq -5$ or $k \geq 2$.

(b) Determine the support of $\phi_{4,k}(t)$ and use this fact to show that $a_{4,k} = 0$ if $k \leq -41$ or $k \geq 40$.

(c) More generally, use the support properties for $\phi_{j,k}(t)$ and $f(t)$ to find the j, k values for which $a_{j,k} = 0$.

6.47   Repeat the projections of Example 6.10 using in place of $f(t)$, the heavisine function of Example 4.2.

CHAPTER 7

THE DISCRETE DAUBECHIES TRANSFORMATION AND APPLICATIONS

In this chapter we develop the one- and two-dimensional *discrete wavelet transformation* in Section 7.1 and use it in Section 7.2 to decompose functions $f_{j+1}(t) \in V_{j+1}$ into component functions $f_0(t) \in V_0$ and $g_m \in W_m, m = 0, \ldots, j$. We also consider applications of the discrete wavelet transformation to imaging applications. We use the discrete wavelet transformation to perform image compression in Section 7.2 and image segmentation in Section 7.3.

Another important concept that is emphasized in this chapter is that of the difficulties encountered when the decomposition (5.15), (5.26) and reconstruction formula (5.32) are truncated. We present a procedure for dealing with the truncation issue in Section 7.2. Given a vector $\mathbf{v} \in \mathbb{R}^N$, we avoid issues that arise when the projection formulas are truncated by padding $\mathbf{v}$ with an appropriate number of zeros. As we see during derivation of the process, the number of zeros needed for padding can grow large in size (relative to N) and lead to slow computation speeds. We see that although zero padding is a solution to the truncation problem, it is not optimal and another approach is desirable. Thus the material in Section 7.2 serves as motivation for the development of biorthogonal scaling functions in Chapter 8. We conclude the chapter with an application of the discrete Daubechies wavelet transformation to the image segmentation problem.

Wavelet Theory: An Elementary Approach with Applications. By D. K. Ruch and P. J. Van Fleet **277**
Copyright © 2009 John Wiley & Sons, Inc.

7.1   THE DISCRETE DAUBECHIES WAVELET TRANSFORM

In this section we develop a discrete Daubechies wavelet transform with finite matrices for use in applications, much as we did with the discrete Haar wavelet transform from Chapter 4. We anticipate some benefits for applications, since the Daubechies scaling functions and wavelets are smoother than their Haar counterparts. Our development follows the ideas from the discrete Haar wavelet transform in Chapter 4, using the general multiresolution analysis material about projections as matrices given in Section 5.1, the $L^2(\mathbb{R})$ projections discussed in Section 6.3, and the Daubechies filters constructed in Section 6.1.

In practice we process *finite-length* signals or images, so our general setting is the projection of a function $f_{j+1}(t) = \sum_{k \in \mathbb{Z}} a_{j+1,k}\phi_{j+1,k}(t)$ from V_{j+1} into V_j, where only a *finite* number of the coefficients $a_{j+1,k}$ are nonzero. For ease of notation we fix a positive even integer N and assume that $a_{j+1,k} = 0$ for $k < 0$ and $k > N - 1$:

$$f_{j+1}(t) = \sum_{k=0}^{N-1} a_{j+1,k}\phi_{j+1,k}(t)$$

and write our general length $(L + 1)$ Daubechies scaling filter, where L is an odd positive integer, as

$$\mathbf{h} = [h_0, h_1, \ldots, h_L]^T$$

Truncating the Projection Matrix from V_{j+1} to V_j

We can use (5.16) to write the projection of $f_{j+1}(t)$ into V_j as an infinite matrix product

$$
\begin{bmatrix}
\ddots & \ddots & & & & & & & \\
\cdots & h_1 & \mathbf{h_2} & h_3 & h_4 & h_5 & h_6 & h_7 & h_8 & \cdots \\
\cdots & h_{-1} & h_0 & \mathbf{h_1} & h_2 & h_3 & h_4 & h_5 & h_6 & \cdots \\
\cdots & h_{-3} & h_{-2} & h_{-1} & \mathbf{h_0} & h_1 & h_2 & h_3 & h_4 & \cdots \\
\cdots & h_{-5} & h_{-4} & h_{-3} & h_{-2} & \mathbf{h_{-1}} & h_0 & h_1 & h_2 & \cdots \\
\cdots & h_{-7} & h_{-6} & h_{-5} & h_{-4} & h_{-3} & \mathbf{h_{-2}} & h_{-1} & h_0 & \cdots \\
& & & \ddots & & \ddots & & \ddots &
\end{bmatrix}
\cdot
\begin{bmatrix}
\vdots \\
0 \\
0 \\
a_{j+1,0} \\
a_{j+1,1} \\
a_{j+1,2} \\
\vdots \\
a_{j+1,N-1} \\
0 \\
0 \\
\vdots
\end{bmatrix}
\quad (7.1)
$$

The boldface elements in the matrix above are elements on the main diagonal. Our goal is to truncate this matrix product. Since $a_{j+1,k} = 0$ for $k < 0$ and $k > N - 1$, it is natural to form the vector $\mathbf{a} = [a_{j+1,0}, a_{j+1,1}, \ldots, a_{j+1,N-1}]^T \in \mathbb{R}^N$. The question, then, is how do we truncate the matrix in (7.1)? Let's do the same as we did for the discrete Haar wavelet transform and make the h_0 diagonal element in (7.1) the element in the $(1,1)$ position of our truncated matrix. We then retain the subsequent $N/2 - 1$ rows and truncate each of these to be of length N. The resulting product is

$$
\begin{bmatrix}
h_0 & h_1 & h_2 & h_3 & \cdots & h_{L-1} & h_L & 0 & 0 & \cdots \\
0 & 0 & h_0 & h_1 & \cdots & h_{L-3} & h_{L-2} & h_{L-1} & h_L & \cdots \\
0 & 0 & 0 & 0 & \cdots & h_{L-5} & h_{L-4} & h_{L-3} & h_{L-2} & \cdots \\
& & & \ddots & & & & & & \ddots
\end{bmatrix}
\cdot
\begin{bmatrix}
a_0 \\
a_1 \\
\vdots \\
a_{N-1}
\end{bmatrix}
\qquad (7.2)
$$

We denote the product as $\tilde{H}_{N/2} \cdot \mathbf{a}$, where we assume that N is much greater than L. This is a very safe assumption in practical applications since $\mathbf{a} = [a_0, \ldots, a_N]^T$ might be a signal or a column of a digital image and L is typically an odd number $3, 5, \ldots, 15$.

We have purposely left the last few rows of $\tilde{H}_{N/2}$ without entries. The problem is that there is not enough room for the filter coefficients in the last few rows of $\tilde{H}_{N/2}$. Let's illustrate the problem with an example.

Example 7.1 (The Projection Matrix $\tilde{H}_{N/2}$ for N $=$ 12 and L $=$ 5) *Write $\tilde{H}_6$ for the D6 scaling filter.*
Solution
For the values of $\mathbf{h}$ *given in (6.33), we have the* 6×12 *matrix*

$$
\tilde{H}_6 =
\begin{bmatrix}
h_0 & h_1 & h_2 & h_3 & h_4 & h_5 & 0 & 0 & 0 & 0 & 0 & 0 \\
0 & 0 & h_0 & h_1 & h_2 & h_3 & h_4 & h_5 & 0 & 0 & 0 & 0 \\
0 & 0 & 0 & 0 & h_0 & h_1 & h_2 & h_3 & h_4 & h_5 & 0 & 0 \\
0 & 0 & 0 & 0 & 0 & 0 & h_0 & h_1 & h_2 & h_3 & h_4 & h_5 \\
0 & 0 & 0 & 0 & 0 & 0 & 0 & 0 & h_0 & h_1 & h_2 & h_3 \\
0 & 0 & 0 & 0 & 0 & 0 & 0 & 0 & 0 & 0 & h_0 & h_1
\end{bmatrix}
\qquad (7.3)
$$

Note that the first four rows of $\tilde{H}_6$ contain all the scaling filter coefficients, but the last two rows contain only four and two filter coefficients, respectively – we have simply truncated these rows so that they have length 12. While the corresponding projection coefficients

$$
b_{j,4} = h_0 a_{j+1,8} + h_1 a_{j+1,9} + h_2 a_{j+1,10} + h_3 a_{j+1,11} + h_4 \cdot 0 + h_5 \cdot 0
$$
$$
b_{j,5} = h_0 a_{j+1,10} + h_1 a_{j+1,11} + h_2 \cdot 0 + h_3 \cdot 0 + h_4 \cdot 0 + h_5 \cdot 0
$$

of $\mathbf{b} = \tilde{H}_6 \mathbf{a}$ *are correct, we can check directly (see Problem 7.2) that the rows of $\tilde{H}_6$ are not orthonormal. Since our goal is to construct an orthogonal transformation matrix, the formulation (7.3) for $\tilde{H}_6$ is unacceptable.* ∎

Modifying $\tilde{H}_{N/2}$

There are a couple of ways to modify $\tilde{H}_6$ (and in general $\tilde{H}_{N/2}$) so that it is orthogonal. The first is to replace the bottom rows (and in some cases a few of the top rows) of

$\tilde{H}_{N/2}$ with rows that are structurally similar (i.e., sparse with values at either the beginning or end of the row for new top or bottom rows, respectively) so that $\tilde{H}_{N/2}$ is an orthogonal matrix. We do not pursue this *matrix completion* approach in this book; the interested reader is referred to Keinert [39] or Van Fleet [60], Problem 7.19. Recall from the scaling filter property (ii) given in Proposition 5.5 that

$$\sum_{k \in \mathbb{Z}} h_k h_{k-2\ell} = \delta_{0,\ell} \qquad (7.4)$$

We can easily check (see Problem 7.1) in the case of the D6 scaling filter that (7.4) yields

$$h_0^2 + h_1^2 + h_2^2 + h_3^2 + h_4^2 + h_5^2 = 1$$
$$h_0 h_2 + h_1 h_3 + h_2 h_4 + h_3 h_5 = 0 \qquad (7.5)$$
$$h_0 h_4 + h_1 h_5 = 0$$

We can use (7.5) to make $\tilde{H}_6$ in (7.3) orthogonal. Indeed, if we *wrap* the filter coefficients in the last two rows rather than truncating them, we have

$$H_6 = \begin{bmatrix} h_0 & h_1 & h_2 & h_3 & h_4 & h_5 & 0 & 0 & 0 & 0 & 0 & 0 \\ 0 & 0 & h_0 & h_1 & h_2 & h_3 & h_4 & h_5 & 0 & 0 & 0 & 0 \\ 0 & 0 & 0 & 0 & h_0 & h_1 & h_2 & h_3 & h_4 & h_5 & 0 & 0 \\ 0 & 0 & 0 & 0 & 0 & 0 & h_0 & h_1 & h_2 & h_3 & h_4 & h_5 \\ h_4 & h_5 & 0 & 0 & 0 & 0 & 0 & 0 & h_0 & h_1 & h_2 & h_3 \\ h_2 & h_3 & h_4 & h_5 & 0 & 0 & 0 & 0 & 0 & 0 & h_0 & h_1 \end{bmatrix} \qquad (7.6)$$

It is easy to check that the rows of H_6 satisfy (7.5), so that $H_6 H_6^T = I_6$, where I_6 is the 6×6 identity matrix.

Defining $H_{N/2}$ — and Some Issues

Some comments are in order about the effects of wrapping rows on the projection coefficients $b_{j,0}, \ldots, b_{j,N-1}$. If $a_{j,0}, \ldots, a_{j,N-1}$ are samples of a periodic functions (with the $a_{j,k}$ equally spaced across the entire period), then $H_6 \mathbf{a}$ will produce the correct $\mathbf{b}$. If the data are not taken from periodic input, the last two projection coefficients will not be correct.

The boundary rows (i.e., top or bottom rows) are the bane of all discrete wavelet transformations constructed from orthogonal scaling filters. As we will see in Chapter 8, the ability to construct *symmetric* scaling filters (e.g., $h_k = h_{-k}$ or $h_k = h_{1-k}$) gives us a way to handle boundary rows adequately in the discrete transformation matrix. As Daubechies proved in [18], the only member of her family of scaling filters that possesses any symmetry is the Haar filter. But this filter comes from a scaling function that is piecewise continuous only and thus not desirable for many applications. For now, we are faced with the dilemma of producing accurate projection coefficients via truncation and losing orthogonality or sacrificing a few accurate projection coefficients in order to maintain orthogonality. In this book we choose the latter and make the following definition:

Definition 7.1 (The Truncated Projection Matrix $H_{N/2}$) *For L an odd positive integer, let* $\mathbf{h} = [h_0, \ldots, h_{L-1}]^T$ *be the length $(L+1)$ Daubechies scaling filter. Suppose that N is an even integer with $N > L$. Then we define the $\frac{N}{2} \times N$ truncated projection matrix $H_{N/2}$ as*

$$H_{N/2} = \begin{bmatrix} h_0 & h_1 & h_2 & h_3 & \cdots & h_{L-1} & h_L & \cdots & 0 & 0 \\ 0 & 0 & h_0 & h_1 & \cdots & h_{L-3} & h_{L-2} & \cdots & 0 & 0 \\ 0 & 0 & 0 & 0 & \cdots & h_{L-5} & h_{L-4} & \cdots & 0 & 0 \\ & & & \ddots & & & \ddots & & & \\ 0 & 0 & 0 & 0 & \cdots & 0 & 0 & \cdots & h_{L-1} & h_L \\ h_{L-1} & h_L & 0 & 0 & \cdots & 0 & 0 & \cdots & h_{L-3} & h_{L-2} \\ & & & \ddots & & & & \ddots & & \\ h_2 & h_3 & h_4 & h_5 & \cdots & 0 & 0 & \cdots & h_0 & h_1 \end{bmatrix} \qquad (7.7)$$

∎

The following proposition is immediate.

Proposition 7.1 (The Rows of $H_{N/2}$ are Orthogonal) *Let $H_{N/2}$ be the matrix from Definition 7.1. Then*

$$H_{N/2}H_{N/2}^T = I_{N/2}$$

where $I_{N/2}$ is the $\frac{N}{2} \times \frac{N}{2}$ identity matrix. In other words, the rows of $H_{N/2}$ are orthogonal to each other, with each row having unit norm.

∎

Proof: The proof is left as Problem 7.5.

∎

Constructing $G_{N/2}$

We next work on a truncated version of the matrix used to project $f_{j+1}(t) \in V_{j+1}$ into the wavelet space W_j. From (5.27) we have

$$\begin{bmatrix} \ddots & \ddots & & & & & & & \\ \cdots & g_1 & \mathbf{g_2} & g_3 & g_4 & g_5 & g_6 & g_7 & g_8 & \cdots \\ \cdots & g_{-1} & g_0 & \mathbf{g_1} & g_2 & g_3 & g_4 & g_5 & g_6 & \cdots \\ \cdots & g_{-3} & g_{-2} & g_{-1} & \mathbf{g_0} & g_1 & g_2 & g_3 & g_4 & \cdots \\ \cdots & g_{-5} & g_{-4} & g_{-3} & g_{-2} & \mathbf{g_{-1}} & g_0 & g_1 & g_2 & \cdots \\ \cdots & g_{-7} & g_{-6} & g_{-5} & g_{-4} & g_{-3} & \mathbf{g_{-2}} & g_{-1} & g_0 & \cdots \\ & & \ddots & & & & \ddots & & \ddots \end{bmatrix} \begin{bmatrix} \vdots \\ 0 \\ 0 \\ a_0 \\ a_1 \\ \vdots \\ a_{N-1} \\ 0 \\ 0 \\ \vdots \end{bmatrix} \qquad (7.8)$$

The boldface elements in (7.8) are elements on the main diagonal. The g_k are the wavelet coefficients defined by

$$g_k = (-1)^k h_{1-k} \tag{7.9}$$

for $k = 1 - L, \ldots, 1$. In this case, the nonzero wavelet filter coefficients are

$$g_{1-L}, \cdots, g_{-2}, g_{-1}, g_0, g_1$$

Because of these negative indices, the truncation process for $G_{N/2}$ is a bit different than it is for $H_{N/2}$. Instead of using a wavelet filter with negative indices, we appeal instead to Problem 5.45 in Section 5.2. Since L is an odd positive integer, Problem 5.45 tells us that we can use the symbol $\tilde{G}(\omega) = -e^{-iL\omega}\overline{H(\omega + \pi)}$ instead of $G(\omega)$. Problem 5.46 provides a formula for the associated wavelet filter coefficients:

$$\tilde{g}_k = (-1)^k h_{L-k} \tag{7.10}$$

Since the nonzero values for h_k occur when $k = 0, \ldots, L$, we see that the nonzero wavelet filter coefficients are $\tilde{g}_L, \tilde{g}_{L-1}, \ldots, \tilde{g}_0$ — exactly the same range of indices as those for nonzero h_k! Thus if we use this filter, the truncation process for $G_{N/2}$ is exactly the same as it was for $H_{N/2}$. We have the following definition:

Definition 7.2 (The Truncated Projection Matrix $G_{N/2}$) *For L an odd positive integer, let* $\mathbf{h} = [h_0, \ldots, h_{L-1}]^T$ *be the $(L+1)$–tap Daubechies scaling filter. Let* $\mathbf{g}$ *be the wavelet filter whose elements are given by the formula* $g_k = (-1)^k h_{L-k}$. *Suppose that N is an even integer with $N > L$. Then we define the* $\frac{N}{2} \times N$ *truncated projection matrix $G_{N/2}$ as*

$$G_{N/2} = \begin{bmatrix} g_0 & g_1 & g_2 & g_3 & \cdots & g_{L-1} & g_L & \cdots & 0 & 0 \\ 0 & 0 & g_0 & g_1 & \cdots & g_{L-3} & g_{L-2} & \cdots & 0 & 0 \\ 0 & 0 & 0 & 0 & \cdots & g_{L-5} & g_{L-4} & \cdots & 0 & 0 \\ & & & & \ddots & & & \ddots & & \\ 0 & 0 & 0 & 0 & \cdots & 0 & 0 & \cdots & g_{L-1} & g_L \\ g_{L-1} & g_L & 0 & 0 & \cdots & 0 & 0 & \cdots & g_{L-3} & g_{L-2} \\ & & & & \ddots & & & \ddots & & \\ g_2 & g_3 & g_4 & g_5 & \cdots & 0 & 0 & \cdots & g_0 & g_1 \end{bmatrix} \tag{7.11}$$

■

There is an added benefit of using wavelet filter coefficients that have the same index range as that of the scaling filter coefficients. When we compute $H_{N/2}\mathbf{a}$, we are not only computing projection coefficients for $f_{j+1}(t)$ in the coarser space V_j, but we are also computing in some sense the *weighted averages* of $L + 1$ elements from $\mathbf{a}$. Choosing the wavelet filter coefficients index range to be identical to that of the scaling filter coefficients means that the elements of $G_{N/2}\mathbf{a}$ each use the same $L + 1$ elements of $\mathbf{a}$ as do the corresponding elements of $H_{N/2}\mathbf{a}$. In applications where we quantize the detail portion $G_{N/2}\mathbf{a}$ of the transformation or look for large values in $G_{N/2}\mathbf{a}$, then it is desirable that the index range of the two filters be the same. We need to verify that the rows of $G_{N/2}$ are orthogonal to each other with unit length and that the rows of $G_{N/2}$ are orthogonal to the rows of $H_{N/2}$. We have the following proposition.

Proposition 7.2 (Orthogonality and $G_{N/2}$) *Let $G_{N/2}$ be the matrix from Definition 7.2. Then*

$$G_{N/2}G_{N/2}^T = I_{N/2} \qquad (7.12)$$

where $I_{N/2}$ is the $\frac{N}{2} \times \frac{N}{2}$ identity matrix. In other words, the rows of $G_{N/2}$ are orthogonal to each other, with each row having unit norm. Moreover, for $H_{N/2}$ defined by (7.7), we have

$$H_{N/2}G_{N/2}^T = 0_{N/2} \qquad (7.13)$$

where $0_{N/2}$ is the $\frac{N}{2} \times \frac{N}{2}$ zero matrix. In other words, the rows of $H_{N/2}$ are orthogonal to the rows of $G_{N/2}$. ∎

Proof: Suppose that $1 \le i < j \le N$. Set $\ell = j - i$. Then row j is a cyclic shift of row i by 2ℓ units. Thus the inner product of rows i and j takes the form

$$\sum_{k=0}^{L} g_k g_{k+2\ell} = \sum_{k \in \mathbb{Z}} g_k g_{k+2\ell}$$

if we view $g_k = 0$ for $k < 0$ and $k > L$. By Problem 5.11 in Section 5.1 (replacing ℓ by $-\ell$), this inner product is 0. Thus the rows of $G_{N/2}$ are orthogonal. If we compute the inner product of row i with itself, we obtain

$$\sum_{k=0}^{L} g_k^2 = \sum_{k \in \mathbb{Z}} g_k^2$$

We again appeal to Problem 5.11 (with $\ell = 0$) to infer that this inner product is 1. This establishes (7.12). The proof of (7.13) is left as Problem 7.11. ∎

The Discrete Daubechies Wavelet Transformation Matrix

We are now ready to define the Daubechies wavelet transformation matrix W_N. We mimic the ideas of the discrete Haar wavelet transformation and build W_N in block form using $H_{N/2}$ and $G_{N/2}$. We have the following definition.

Definition 7.3 (Daubechies Discrete Wavelet Transformation) *Suppose that $\mathbf{h}$ is the length $(L+1)$ Daubechies scaling filter and assume that N is an even positive integer with $N > L$. Form $H_{N/2}$ and $G_{N/2}$ as prescribed in Definitions 7.1 and 7.2, respectively. We define the* discrete Daubechies wavelet transformation matrix W_N

as

$$W_N = \begin{bmatrix} H_{N/2} \\ \hline G_{N/2} \end{bmatrix} = \left[\begin{array}{ccccccccccc} h_0 & h_1 & h_2 & h_3 & \cdots & h_{L-1} & h_L & \cdots & 0 & 0 \\ 0 & 0 & h_0 & h_1 & \cdots & h_{L-3} & h_{L-2} & \cdots & 0 & 0 \\ 0 & 0 & 0 & 0 & \cdots & h_{L-5} & h_{L-4} & \cdots & 0 & 0 \\ & & & \ddots & & & \ddots & & & \\ 0 & 0 & 0 & 0 & \cdots & 0 & 0 & \cdots & h_{L-1} & h_L \\ h_{L-1} & h_L & 0 & 0 & \cdots & 0 & 0 & \cdots & h_{L-3} & h_{L-2} \\ & & & \ddots & & & \ddots & & & \\ h_2 & h_3 & h_4 & h_5 & \cdots & 0 & 0 & \cdots & h_0 & h_1 \\ \hline g_0 & g_1 & g_2 & g_3 & \cdots & g_{L-1} & g_L & \cdots & 0 & 0 \\ 0 & 0 & g_0 & g_1 & \cdots & g_{L-3} & g_{L-2} & \cdots & 0 & 0 \\ 0 & 0 & 0 & 0 & \cdots & g_{L-5} & g_{L-4} & \cdots & 0 & 0 \\ & & & \ddots & & & \ddots & & & \\ 0 & 0 & 0 & 0 & \cdots & 0 & 0 & \cdots & g_{L-1} & g_L \\ g_{L-1} & g_L & 0 & 0 & \cdots & 0 & 0 & \cdots & g_{L-3} & g_{L-2} \\ & & & \ddots & & & \ddots & & & \\ g_2 & g_3 & g_4 & g_5 & \cdots & 0 & 0 & \cdots & g_0 & g_1 \end{array} \right] \tag{7.14}$$

∎

The following proposition follows immediately from Propositions 7.1 and 7.2.

Proposition 7.3 (The Matrix W_N is Orthogonal) *The matrix W_N given in Definition 7.3 is orthogonal.* ∎

Proof: Using (7.14) we have

$$W_N W_N^T = \begin{bmatrix} H_{N/2} \\ \hline G_{N/2} \end{bmatrix} \begin{bmatrix} H_{N/2} \\ \hline G_{N/2} \end{bmatrix}^T$$

$$= \begin{bmatrix} H_{N/2} \\ \hline G_{N/2} \end{bmatrix} \left[H_{N/2}^T \middle| G_{N/2}^T \right]$$

$$= \begin{bmatrix} H_{N/2} H_{N/2}^T & H_{N/2} G_{N/2}^T \\ \hline G_{N/2} H_{N/2}^T & G_{N/2} G_{N/2}^T \end{bmatrix}$$

From Proposition 7.1 we know that $H_{N/2} H_{N/2}^T = I_{N/2}$, and from Proposition 7.2 we have $G_{N/2} G_{N/2}^T = I_{N/2}$ and

$$0 = H_{N/2} G_{N/2}^T = \left(G_{N/2} H_{N/2}^T \right)^T$$

so that $W_N W_N^T = \begin{bmatrix} I_{N/2} & 0 \\ \hline 0 & I_{N/2} \end{bmatrix} = I_N.$ ∎

Let's look at an example of a Daubechies discrete wavelet transform matrix.

Example 7.2 (The Matrix W_8 for the D4 Scaling Filter) *Construct W_8 for the D4 scaling filter.*

Solution

The scaling filter is given by (6.39):

$$h_0 = \frac{1 + \sqrt{3}}{4\sqrt{2}}, \quad h_1 = \frac{3 + \sqrt{3}}{4\sqrt{2}}, \quad h_2 = \frac{3 - \sqrt{3}}{4\sqrt{2}}, \quad h_3 = \frac{1 - \sqrt{3}}{4\sqrt{2}}$$

Since $L = 3$, we have $g_k = (-1)^k h_{3-k}$, $k = 0,1,2,3$:

$$g_0 = h_3 = \frac{1 - \sqrt{3}}{4\sqrt{2}}, \qquad\qquad g_1 = -h_2 = \frac{\sqrt{3} - 3}{4\sqrt{2}}$$

$$g_2 = h_1 = \frac{3 + \sqrt{3}}{4\sqrt{2}}, \qquad\qquad g_3 = -h_0 = -\frac{1 + \sqrt{3}}{4\sqrt{2}}$$

We use (7.14) to write

$$W_8 = \left[\begin{array}{cccccccc}
h_0 & h_1 & h_2 & h_3 & 0 & 0 & 0 & 0 \\
0 & 0 & h_0 & h_1 & h_2 & h_3 & 0 & 0 \\
0 & 0 & 0 & 0 & h_0 & h_1 & h_2 & h_3 \\
h_2 & h_3 & 0 & 0 & 0 & 0 & h_0 & h_1 \\
\hline
g_0 & g_1 & g_2 & g_3 & 0 & 0 & 0 & 0 \\
0 & 0 & g_0 & g_1 & g_2 & g_3 & 0 & 0 \\
0 & 0 & 0 & 0 & g_0 & g_1 & g_2 & g_3 \\
g_2 & g_3 & 0 & 0 & 0 & 0 & g_0 & g_1
\end{array}\right]$$

$$= \left[\begin{array}{cccccccc}
\frac{1+\sqrt{3}}{4\sqrt{2}} & \frac{3+\sqrt{3}}{4\sqrt{2}} & \frac{3-\sqrt{3}}{4\sqrt{2}} & \frac{1-\sqrt{3}}{4\sqrt{2}} & 0 & 0 & 0 & 0 \\
0 & 0 & \frac{1+\sqrt{3}}{4\sqrt{2}} & \frac{3+\sqrt{3}}{4\sqrt{2}} & \frac{3-\sqrt{3}}{4\sqrt{2}} & \frac{1-\sqrt{3}}{4\sqrt{2}} & 0 & 0 \\
0 & 0 & 0 & 0 & \frac{1+\sqrt{3}}{4\sqrt{2}} & \frac{3+\sqrt{3}}{4\sqrt{2}} & \frac{3-\sqrt{3}}{4\sqrt{2}} & \frac{1-\sqrt{3}}{4\sqrt{2}} \\
\frac{3-\sqrt{3}}{4\sqrt{2}} & \frac{1-\sqrt{3}}{4\sqrt{2}} & 0 & 0 & 0 & 0 & \frac{1+\sqrt{3}}{4\sqrt{2}} & \frac{3+\sqrt{3}}{4\sqrt{2}} \\
\frac{1-\sqrt{3}}{4\sqrt{2}} & \frac{\sqrt{3}-3}{4\sqrt{2}} & \frac{3+\sqrt{3}}{4\sqrt{2}} & -\frac{1+\sqrt{3}}{4\sqrt{2}} & 0 & 0 & 0 & 0 \\
0 & 0 & \frac{1-\sqrt{3}}{4\sqrt{2}} & \frac{\sqrt{3}-3}{4\sqrt{2}} & \frac{3+\sqrt{3}}{4\sqrt{2}} & -\frac{1+\sqrt{3}}{4\sqrt{2}} & 0 & 0 \\
0 & 0 & 0 & 0 & \frac{1-\sqrt{3}}{4\sqrt{2}} & \frac{\sqrt{3}-3}{4\sqrt{2}} & \frac{3+\sqrt{3}}{4\sqrt{2}} & -\frac{1+\sqrt{3}}{4\sqrt{2}} \\
\frac{3+\sqrt{3}}{4\sqrt{2}} & -\frac{1+\sqrt{3}}{4\sqrt{2}} & 0 & 0 & 0 & 0 & \frac{1-\sqrt{3}}{4\sqrt{2}} & \frac{\sqrt{3}-3}{4\sqrt{2}}
\end{array}\right]$$

■

The next example shows the results of applying W_N to a vector **a**.

Example 7.3 (Applying W_N to a Vector) *We consider the function $f(t)$ given by (6.77) and plotted in Figure 6.12. We first project $f(t)$ into V_4. The projection, plotted in Figure 7.1, has the form*

$$f_4(t) = \sum_{k=-40}^{39} c_k \phi_{4,k}(t)$$

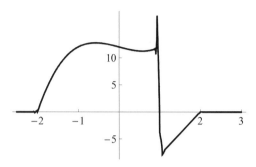

Figure 7.1   The projection $f_4(t)$ of $f(t)$ into V_4.

We create the vector $\mathbf{a} \in \mathbf{R}^{80}$ *where the elements are given by* $a_k = c_{k-41}$, $k = 1, \ldots, 80$. *The* 80 *coefficients are plotted in Figure 7.2(a) and form a decent approximation to* $f(t)$. *We next apply the matrix* W_{80} *using the D4 scaling filter to* $\mathbf{a}$. *The result is plotted in Figure 7.2(b).*

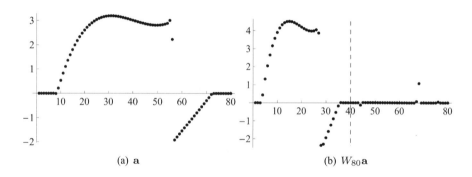

(a) $\mathbf{a}$ (b) $W_{80}\mathbf{a}$

Figure 7.2   The projection coefficients $\mathbf{a}$ and the product $\mathbf{y} = W_{80}\mathbf{a}$.

The first 40 *elements of* $\mathbf{y} = W_{80}\mathbf{a}$ *are the coefficients of the projection* $f_3(t)$ *of* $f_4(t)$ *into* V_3. *This function is plotted in Figure 7.3(a). The matrix* W_{80} *has two wrapping rows (see Problem 7.4(a)) and since the vector* $\mathbf{a}$ *has zeros at the beginning and the end, the transform treats it as periodic data and the projection coefficients are correct. The second* 40 *values of* $W_{80}\mathbf{a}$ *are the coefficients of the projection* $g_3(t)$ *of* $f_4(t)$ *into* W_3. *This function is plotted in Figure 7.3(b).*

Since the function is suitably smooth over most of its domain, the wavelet coefficients are near zero. But also note that both the jumps in the first derivative of $f(t)$ *and the discontinuity of* $f(t)$ *are detected in the wavelet coefficients — these values are larger (in absolute value) relative to the other wavelet coefficients.* ∎

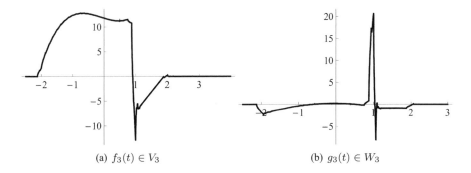

(a) $f_3(t) \in V_3$ (b) $g_3(t) \in W_3$

Figure 7.3   The projections of $f_4(t)$ into V_3 and W_3.

Our next example considers the effects of the wrapping rows of a Daubechies discrete wavelet transformation when the input vector cannot be interpreted as "periodic."

Example 7.4 (Applying W_N to a Vector) *Let* $\mathbf{v} = [2, 4, 6, \ldots, 46, 48]^T \in \mathbb{R}^{24}$. *Using the D4 scaling filter, apply* W_{24} *to* $\mathbf{v}$ *and analyze the results.*
Solution
   The output $\mathbf{y} = W_{24}\mathbf{v}$, *rounded to two decimal digits, is*

$$\mathbf{y} = [9.52, 15.18, 20.83, 26.49, 32.15, 37.80, 43.46, 49.12, 54.78,$$
$$60.43, 66.09, 8.41 \mid 0, 0, 0, 0, 0, 0, 0, 0, 0, 0, 0, 16.97]^T$$

and is plotted in Figure 7.4.

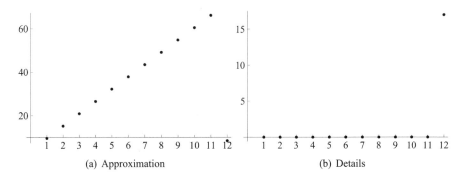

(a) Approximation   (b) Details

Figure 7.4   The output of the wavelet transformation applied to $\mathbf{v}$.

   We can see the effect of the wrapping row in the transformed data. The first eleven values of the approximation portion of the transform are linear but the last element is the inner product $[h_0, h_1, h_2, h_3] \cdot [22, 24, 2, 4]^T$. *In a similar manner, the last element of the details portion of the transform is the inner product* $[g_0, g_1, g_2, g_3]^T \cdot$ $[22, 24, 2, 4]^T$.

It is interesting to note that the details portion of the transform consists of zeros wherever wrapping rows do not figure into the computation. In Problem 7.10 you will show that if vector $\mathbf{v} \in \mathbb{R}^4$ is composed of linear data, then $\mathbf{v} \cdot \mathbf{g} = 0$. ∎

Iterating the Transformation

The iterative process for the Haar wavelet transform described in Section 4.1 works exactly the same way for the discrete Daubechies wavelet transform; we just have different entries for the W_N matrices! Suppose that $\mathbf{v} \in \mathbb{R}^N$, where N is divisible by 2^j. By analogy, from (4.15) and (4.16) we say that the jth iteration of the discrete Daubechies wavelet transform applied to $\mathbf{v}$ is

$$\mathbf{y}^j = W_{N,j-1} \cdots W_{N,0} \mathbf{v} \tag{7.15}$$

where W_N is defined by the matrix (7.14), $W_{N,0} = W_N$, and for $j \geq 1$,

$$W_{N,j} = \mathrm{diag}\left[W_{N/2^j}, I_{N/2^j}, I_{N/2^{j-1}}, \ldots, I_{N/2}\right] \tag{7.16}$$

Let's look at the following example.

Example 7.5 (The Iterated Wavelet Transformation) *Apply three iterations of the Daubechies wavelet transformation using the D4 scaling filter to the vector* $\mathbf{v}$ *in Example 7.4.*
Solution
We compute the product

$$\mathbf{y}^3 = W_{24,2} W_{24,1} W_{24,0} \mathbf{v}$$

where

$$\mathbf{y}^3 = \begin{bmatrix} \mathbf{y}^{3,a} \\ \mathbf{y}^{3,d} \\ \mathbf{y}^{2,d} \\ \mathbf{y}^{1,d} \end{bmatrix}$$

and rounded to one decimal digit, the elements of each component are

$$\mathbf{y}^{3,a} = [99.3, 68.9, 43.9]^T$$
$$\mathbf{y}^{3,d} = [0, 20.3, -38]^T$$
$$\mathbf{y}^{2,d} = [0, 0, 0, 0, 8.2, -29]^T$$
$$\mathbf{y}^{1,d} = [0, 0, 0, 0, 0, 0, 0, 0, 0, 0, 0, 17]^T$$

These vectors are plotted in Figure 7.5. ∎

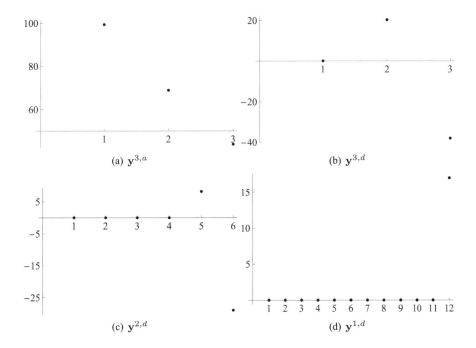

Figure 7.5   The component vectors of three iterates of the discrete wavelet transformation.

Iterating the Inverse Transformation

Recall from Section 4.1 that we iterate the inverse Haar wavelet transform in a straight-forward manner. Exactly the same process works for the discrete Daubechies wavelet transform because we have only changed the elements of the W_N matrices but have kept intact the crucial orthogonality property $W_{N,k}^{-1} = W_{N,k}^T$. To be precise, to recover a vector $\mathbf{v}$ from $\mathbf{y}^j$, we compute

$$\mathbf{v} = W_{N,0}^T W_{N,1}^T \cdots W_{N,j-1}^T \mathbf{y}^j \qquad (7.17)$$

The Two-Dimensional Discrete Daubechies Wavelet Transform

The basic principles of the two-dimensional discrete Daubechies wavelet transform and the two-dimensional Haar wavelet transform are the same. For an $M \times N$ matrix A, the *two-dimensional discrete Daubechies wavelet transform* of A is defined as

$$B = W_M A W_N^T \qquad (7.18)$$

where M and N are even positive integers and the matrices W_M and W_N are defined by (7.14). As in the Haar case, we can write B in 2×2 block form:

$$
\begin{aligned}
B &= W_M A W_N^T \\[4pt]
&= \left[\dfrac{H_{M/2}}{G_{M/2}} \right] A \left[\dfrac{H_{N/2}}{G_{N/2}} \right]^T \\[4pt]
&= \left[\dfrac{H_{M/2}}{G_{M/2}} \right] A \left[H_{N/2}^T \middle| G_{N/2}^T \right] \\[4pt]
&= \left[\begin{array}{c|c} H_{M/2} A H_{N/2}^T & H_{M/2} A G_{N/2}^T \\ \hline G_{M/2} A H_{N/2}^T & G_{M/2} A G_{N/2}^T \end{array} \right] \\[4pt]
&= \left[\begin{array}{c|c} \mathcal{A} & \mathcal{V} \\ \hline \mathcal{H} & \mathcal{D} \end{array} \right]
\end{aligned}
$$

where we think of

$$
B = \left[\begin{array}{c|c} \mathcal{A} & \mathcal{V} \\ \hline \mathcal{H} & \mathcal{D} \end{array} \right] = \left[\begin{array}{c|c} \text{averages} & \begin{array}{c}\text{vertical}\\\text{differences}\end{array} \\ \hline \begin{array}{c}\text{horizontal}\\\text{differences}\end{array} & \begin{array}{c}\text{diagonal}\\\text{differences}\end{array} \end{array} \right]
$$

We can iterate the two-dimensional transformation in exactly the same manner as we did with the Haar transformation. In Figure 7.6 we have plotted several iterations of a discrete Daubechies wavelet transformation using the D4 scaling filter.

Although the structure for the discrete Daubechies transformation is the same as that of the Haar wavelet transformation, the fact that we can use different scaling filters gives us more flexibility in applications.

Inverting the two-dimensional wavelet transformation is straightforward. We have $B = W_M A W_N^T$, and since W_M and W_N are orthogonal, we can write

$$
W_M^T B W_N = \left(W_M^T W_M \right) A \left(W_N^T W_N \right) = A
$$

This process can be applied iteratively as well.

PROBLEMS

Note: *A computer algebra system and the software on the course Web site will be useful for some of these problems. See the Preface for more details.*

7.1   Use the D6 scaling filter in (7.4) to produce the three equations in (7.5).

7.2   Show by direct computation that H_6 in (7.3) is not an orthogonal matrix. (*Hint:* You can reduce the number of computations by using (7.5).)

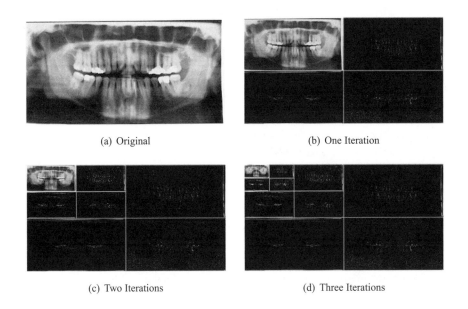

(a) Original

(b) One Iteration

(c) Two Iterations

(d) Three Iterations

Figure 7.6   An image and three iterations of the discrete Daubechies wavelet transformation.

7.3   Write H_4 for the D4 scaling filter and then verify that the rows of H_4 are orthogonal to each other.

7.4   Refer to (7.7). How many wrapping rows are at the bottom of $H_{N/2}$ for the

(a) The D4 scaling filter?

(b) The D8 scaling filter?

(c) The D10 scaling filter?

(d) The length $(L+1)$ scaling filter where L is an odd positive integer?

7.5   Prove Proposition 7.1.

7.6   In this problem we compare the projection matrix $\mathcal{H}$ (5.16) to $H_{N/2}$ (7.2) when $\mathcal{H}$ is applied to a bi–infinite sequence and $H_{N/2}$ is applied to a related vector. Suppose that $N \geq 6$ is an even positive integer and let $\mathbf{a} = (\ldots, a_0, \ldots, a_{N-1}, \ldots)$ with $a_k = 0$ for $k < 0$ and $k \geq N$. Construct vector $\mathbf{a}^* = [a_0, \ldots, a_{N-1}]^T$. Finally, let $\mathbf{b} = \mathcal{H}\mathbf{a}$ and $\mathbf{b}^* = H_{N/2}\mathbf{a}^*$.

(a) For the D4 scaling filter, show that $b_k = b_k^*$ for $0 \leq k < \frac{N}{2} - 1$, but that $b_{N/2-1} \neq b_{N/2-1}^*$ in general.

(b) Prove a similar result for the D6 filter. What happens to $b_{N/2-2}$?

(c) Generalize your results for the length $(L+1)$ Daubechies filter where $L < N$ is an odd positive integer.

7.7   In this problem we compare the projection matrix $\mathcal{H}$ (5.16) to $H_{N/2}$ (7.2) when $H_{N/2}$ is applied to vector and $\mathcal{H}$ is applied to a bi-infinite sequence that is created by periodizing the vector.

Suppose that $N \geq 6$ is an even positive integer, and let $\mathbf{a}^* = [a_0, \ldots, a_{N-1}]^T$. Now construct the N-periodic bi-infinite sequence $\mathbf{a}$ using the rule $a_k = a_{N+k}$, $k \in \mathbb{Z}$. Finally, let $\mathbf{b} = \mathcal{H}\mathbf{a}$ and $\mathbf{b}^* = H_{N/2}\mathbf{a}^*$.

(a) For the D4 scaling filter, show that $b_k = b_k^*$ for $0 \leq k < \frac{N}{2}$.

(b) Show that $\mathbf{b}$ is N-periodic.

(c) Generalize your results for the length $(L+1)$ Daubechies scaling filter. Here $L < N$ is an odd positive integer.

7.8   Write the matrix G_8 in terms of the D4 scaling filter and show that the rows are orthogonal to each other.

★**7.9**   In Corollary 6.2 in Section 6.2 we learned that the support of the wavelet function $\psi(t)$ associated with the Daubechies scaling function $\phi_{2L}(t)$ is $[1 - L, L]$. Suppose that we use the wavelet filter given by (7.10) to generate $\psi(t)$ instead of the filter given by (7.9). Show that the support of the resulting wavelet function $\psi(t)$ is supp$(\psi) = [0, 2L]$.

7.10   Suppose that $\mathbf{h}$ is the D4 scaling filter. Form the wavelet filter $\mathbf{g}$ using the rule $g_k = (-1)^k h_{3-k}$, $k = 0,1,2,3$. Suppose that $\mathbf{v} \in \mathbb{R}^4$ with $v_j = mj + b$, where $m, b \in \mathbb{R}$ with $m \neq 0$. Show that $\mathbf{v} \cdot \mathbf{g} = 0$. What happens if we repeat the exercise with the D6 filter? The D8 filter?

7.11   Verify the identity (7.13) to complete the proof of Proposition 7.2. (*Hint:* Use the ideas from the proof of (7.12) along with Problem 5.44 from Section 5.2.)

7.12   Write W_{12} for both the D6 and D8 scaling filters. Use a CAS to verify that these matrices are orthogonal.

7.13   Wrapping rows in a wavelet transform can have boundary effects. To illustrate, let $\mathbf{v}$ be a vector whose components are given by

$$v_k = \sin\left(\frac{k-6}{5}\right)$$

for $k = 0, \ldots, 47$.

(a) Compute $H_{24}\mathbf{v}$ using the D6 filter and plot $\mathbf{v}$ and $H_{24}\mathbf{v}$.

(b) In general we expect $H_{24}\mathbf{v}$ to be an approximation of $\mathbf{v}$ with half its length. Is this true for our example? Where do you notice boundary effects? Why did wrapping rows cause these effects?

7.14   Consider the Heavisine function $h(t)$ given by (4.7) in Section 4.1. Using a CAS, construct vector **a** with components $a_k = h\left(\frac{k}{100}\right)$, $k = 0, \ldots, 127$. Use the D6 scaling filter (6.33) and compute $\mathbf{y} = W_{128}\mathbf{a}$. Plot both **a** and **y**.

7.15   For the vector defined in Problem 7.14, compute three iterations of the discrete wavelet transformation using the D6 scaling filter. Plot both **a** and the transformed vector.

7.16   We can see problems caused by the wrapping rows in the two-dimensional wavelet transform via this simple example. Construct four 16×16 matrices A_k, each of which is composed of constant elements $64k$, $k = 0, 1, 2, 3$. Now construct the 32×32 matrix

$$A = \left[\begin{array}{c|c} A_0 & A_1 \\ \hline A_2 & A_3 \end{array}\right]$$

Using a CAS and the D4 scaling filter, compute the wavelet transform $B = W_{32}AW_{32}^T$. Look at the upper left corner of B. It is supposed to be an approximation of A. Describe the discrepancies and what caused them.

7.2   PROJECTIONS AND SIGNAL AND IMAGE COMPRESSION

In this section we consider two applications of the discrete Daubechies wavelet transformation. We will see how it can be used to create the coefficients necessary to project a function $f_{j+1}(t) \in V_{j+1}$ into V_j and W_j. This application leads to the *pyramid algorithm*. We also use Daubechies scaling filters to perform compression on functions in $L^2(\mathbb{R})$ and digital images.

Projections and the Discrete Wavelet Transformation

It is interesting to see how the discrete wavelet transformation relates to the projection and decomposition of a function in V_{j+1} into components in V_j and W_j discussed in Section 5.1. Recall that the projection matrices (5.16) and (5.27) were both infinite-dimensional, so we expect some issues when we use the discrete wavelet transform matrix (7.14) given in Definition 7.3. Let's look at an example.

Example 7.6 (Decomposition and Iteration)   *Use an iterated discrete Daubechies wavelet transform to decompose the function*

$$f_3(t) = 2\phi_{3,0}(t) + 4\phi_{3,1}(t) + 6\phi_{3,2}(t) + \cdots + 48\phi_{3,11}(t) \qquad (7.19)$$

into its components in V_0, W_0, W_1, and W_2, where $\phi(t)$ is the D4 scaling function.
Solution
    The function $f_3(t)$ is plotted in Figure 7.7.

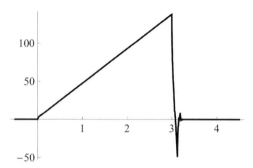

Figure 7.7   The function $f_3(t) \in V_3$ defined by (7.19).

We have $a_{3,k} = 2(k+1)$, $k = 0, \ldots, 23$ with $a_{3,k} = 0$ for $k < 0$ and $k > 11$. To project $f_3(t) \in V_3$ into V_2, we use the relation (5.15) to write

$$a_{2,\ell} = \sum_{k \in \mathbb{Z}} a_{3,k} h_{k-2\ell} = \sum_{k=0}^{23} a_{3,k} h_{k-2\ell}$$

Now only finitely many of the $a_{2,\ell}$ are nonzero. In particular, we consider only those $a_{2,\ell}$ for which $0 \leq k - 2\ell \leq 3$ since these are the indices that correspond to nonzero filter coefficients. Solving the inequality for ℓ gives

$$\frac{k-3}{2} \leq \ell \leq \frac{k}{2}$$

Since $0 \leq k \leq 23$, we can write

$$-\frac{3}{2} \leq \ell \leq \frac{23}{2}$$

and in this way we see that the only nonzero $a_{2,\ell}$ occur when $\ell = -1, \ldots, 11$.
  If we naively apply H_{12} to $[2, 4, \ldots, 44, 48]^T$, we obtain only 12 projection coefficients. We miss completely $a_{2,-1}$ completely since the first element of the product is $a_{2,0}$. Moreover, due to wrapping (see Example 7.4), the value for $a_{2,11}$ is incorrect. We can still use matrix multiplication to find the projection coefficients – we just need to account for the wrapping rows. The easiest solution is to pad the vector $[2, 4, \ldots, 48]^T$ with two zeros to create the vector

$$\mathbf{a}^3 = [0, 0, 2, 4, \ldots, 46, 48]^T \in \mathbb{R}^{26} \tag{7.20}$$

The padding step allows the matrix multiplication $H_{13}\mathbf{a}^3$ to correctly compute

$$a_{2,-1} = 0 \cdot h_0 + 0 \cdot h_1 + 2h_2 + 4h_3$$
$$a_{2,11} = 46h_0 + 48h_1 + 0 \cdot h_2 + 0 \cdot h_3$$

Since the wavelet filter components $g_k = (-1)^k h_{3-k}$, $k = 0, \ldots, 3$, *have the same range of nonzero indices as the scaling filter components, the same ideas can be used to project* $f_3(t)$ *into* W_2 *to obtain projection coefficients (see (5.26)).*

$$c_{2,\ell} = \sum_{k=-1}^{11} a_{3,k} g_{k-2\ell}$$

for $\ell = -1, \ldots, 11$.

We apply W_{26} *to the vector* $\mathbf{a}^3$ *in (7.20). Using matrix/vector notation, we have*

$$\mathbf{y}^1 = \begin{bmatrix} \mathbf{y}^{1,a} \\ \mathbf{y}^{1,d} \end{bmatrix} = W_{26} \mathbf{a}^3$$

The results, rounded to one decimal digit, are

$$\mathbf{y}^{1,a} = \begin{bmatrix} y_{-1}^{1,a}, \ldots, y_{11}^{1,a} \end{bmatrix}^T$$
$$= [-0.1, 4.6, 10.3, 15.9, 21.6, 27.2, 32.9, 38.6, 44.2, 49.9, 55.5, 61.2, 62.4]^T$$
$$\mathbf{y}^{1,d} = \begin{bmatrix} y_{-1}^{1,d}, \ldots, y_{11}^{1,d} \end{bmatrix}^T$$
$$= [0.3, 0, 0, 0, 0, 0, 0, 0, 0, 0, 0, 0, 16.7]^T$$

The values $\mathbf{y}_\ell^{1,a}$ *are the coefficients* $\{a_{2,\ell}\}$ *of the* $\phi_{2,\ell}(t)$ *functions in the projection of* $f_3(t)$ *into* V_2. *We have*

$$f_2(t) = \sum_{\ell=-1}^{11} a_{2,\ell} \phi_{2,\ell}(t) = \sum_{\ell=-1}^{11} \mathbf{y}_\ell^{1,a} \phi_{2,\ell}(t)$$

This projection is plotted in Figure 7.8(a). The projection into W_2 *is obtained from* $\mathbf{y}^{1,d}$:

$$g_2(t) = \sum_{\ell=-1}^{11} c_{2,\ell} \psi_{2,\ell}(t) = 0.3\psi_{2,-1}(t) + 16.7\psi_{2,11}(t) \tag{7.21}$$

This projection is plotted in Figure 7.9(d).

The next step is to project $f_2(t)$ *into* V_1 *or, equivalently, multiply* $\mathbf{y}^{1,a}$ *by a wavelet transform matrix. We seek the projection coefficients*

$$a_{1,\ell} = \sum_{k=-1}^{11} a_{2,\ell} h_{k-2\ell}$$

An argument similar to that used to find the nonzero values $a_{2,\ell}$ *shows that* $a_{1,\ell} = 0$ *for* $\ell < -2$ *and* $\ell > 5$. *Now the first value in* $\mathbf{y}^{1,a}$ *is* $a_{2,-1}$ *and the length of* $\mathbf{y}^{1,a}$ *is 13. If we pad* $\mathbf{y}^{1,a}$ *with one zero, we obtain an even-length vector and we have an appropriate setting so that matrix multiplication will return* $a_{1,-2}$. *But we*

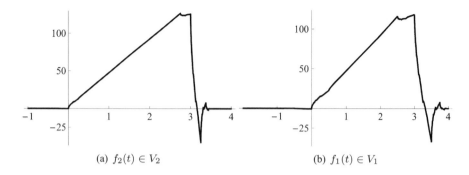

(a) $f_2(t) \in V_2$ (b) $f_1(t) \in V_1$

Figure 7.8   The projections $f_2(t)$ and $f_1(t)$ of $f_3(t) \in V_3$ into V_2 and V_1, respectively.

need another zero to take into account the wrapping that occurs when we use matrix multiplication to compute $a_{1,5}$. To maintain an even-length vector, we pad $\mathbf{y}^{1,a}$ with three zeros:

$$\mathbf{y}^{1,a} = \left[0, 0, 0, y_{-1}^{1,a}, \ldots, y_{11}^{1,a}\right]^T$$

Since $\mathbf{y}^{1,a}$ is now a vector of length 16, we apply W_{16} to obtain $\mathbf{y}^2$:

$$\mathbf{y}^2 = W_{16}\mathbf{y}^1 = \begin{bmatrix} \mathbf{y}^{2,a} \\ \mathbf{y}^{2,d} \\ \mathbf{y}^{1,d} \end{bmatrix}$$

These values, rounded to one decimal digit, are

$$\mathbf{y}^{2,a} = \left[y_{-2}^{2,a}, \ldots, y_5^{2,a}\right]^T$$
$$= [0, -0.4, 11.6, 27.6, 43.6, 59.6, 76.2, 81.7]^T$$
$$\mathbf{y}^{2,d} = \left[y_{-2}^{2,d}, \ldots, y_5^{2,d}\right]^T$$
$$= [0, 1.1, 0, 0, 0, 0, -2.2, 21.9]^T$$

The $\mathbf{y}_\ell^{2,a}$ are the coefficients $\{a_{1,\ell}\}$ of the $\phi_{1,\ell}(t)$ functions in the projection of $f_2(t)$ into V_1:

$$f_1(t) = \sum_{\ell=-2}^5 a_{1,\ell}\phi_{1,\ell}(t) = \sum_{\ell=-2}^5 \mathbf{y}_\ell^{2,a}\phi_{1,\ell}(t)$$

This projection is plotted in Figure 7.8(b). The projection into W_1 is constructed using $\mathbf{y}^{2,d}$. We have

$$g_1(t) = \sum_{\ell=-2}^5 c_{1,\ell}\psi_{1,\ell}(t)$$
$$= 1.1\psi_{1,-1}(t) - 2.2\psi_{1,4}(t) + 21.9\psi_{1,5}(t) \tag{7.22}$$

This projection is plotted in Figure 7.9(c).

The final step is to find $\mathbf{y}^3$. *We pad* $\mathbf{y}^2$ *with two zeros to create a vector of length 10. We have*

$$\mathbf{y}^3 = W_{10}\mathbf{y}^{2,a} = \begin{bmatrix} \mathbf{y}^{3,a} \\ \mathbf{y}^{3,d} \end{bmatrix}$$

with numerical values

$$\mathbf{y}^{3,a} = \begin{bmatrix} y_{-2}^{3,a}, \ldots, y_2^{3,a} \end{bmatrix}^T$$
$$= [0, -1.3, 30.8, 77.4, 105.2]^T$$
$$\mathbf{y}^{3,d} = \begin{bmatrix} y_{-2}^{3,d}, \ldots, y_2^{3,d} \end{bmatrix}^T$$
$$= [-0.2, 3.5, 0, -5.3, 28.2]^T$$

The $\mathbf{y}_{\ell}^{3,a}$ *are the coefficients* $\{a_{0,\ell}\}$ *of the* $\phi_{0,\ell}(t)$ *functions in the projection of* $f_1(t)$ *into* V_0:

$$f_0(t) = -1.3\phi(t+1) + 30.8\phi(t) + 77.4\phi(t-1) + 105.2\phi(t-2) \quad (7.23)$$

The projection into W_0 *is obtained from* $\mathbf{y}^{3,d}$:

$$g_0(t) = -0.2\psi(t+2) + 3.5\psi(t+1) - 5.3\psi(t-1) + 28.2\psi(t-2) \quad (7.24)$$

These projections are plotted in Figure 7.9(a) and 7.9(b), respectively. So the component functions (plotted in Figure 7.9) of $f_3(t)$ *in* V_0, W_0, W_1, *and* W_2 *are given by equations (7.23), (7.24), (7.22), and (7.21), respectively, and*

$$f_3(t) = f_0(t) + g_0(t) + g_1(t) + g_2(t)$$

All the information can be stored concisely on a computer with the vector

$$\begin{bmatrix} \mathbf{y}^{3,a} \\ \hline \mathbf{y}^{3,d} \\ \hline \mathbf{y}^{2,d} \\ \hline \mathbf{y}^{1,d} \end{bmatrix}$$

If you compare the values of $\mathbf{y}^{3,a}$, $\mathbf{y}^{3,d}$, $\mathbf{y}^{2,d}$, *and* $\mathbf{y}^{1,d}$ *with those obtained in Example 7.4, you will see the effects of ignoring the wrapping rows when using* W_N *to compute projection coefficients.* ∎

The General Procedure for Computing Projection Coefficients

We now develop a general procedure for using the discrete wavelet transformation to compute projection coefficients. The process for padding zeros at each iterative

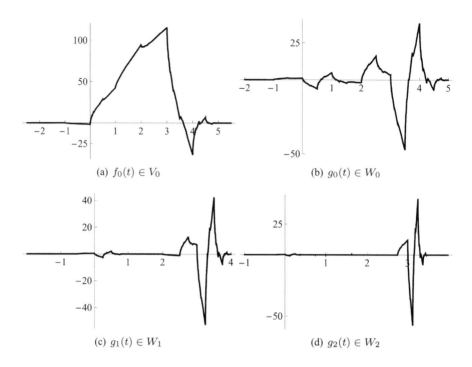

(a) $f_0(t) \in V_0$ (b) $g_0(t) \in W_0$

(c) $g_1(t) \in W_1$ (d) $g_2(t) \in W_2$

Figure 7.9   The component functions used to write $f_3(t) \in V_3$ in terms of functions in V_0, W_0, W_1, and W_2.

step in Example 7.6 was somewhat ad hoc. We first describe a way to perform zero padding for any length $(L+1)$ Daubechies scaling filter so that the discrete wavelet transform can be used to compute the coefficients when projecting $f_{j+1}(t) \in V_{j+1}$ into V_j and W_j.

To keep the notation simple, we assume that $f_{j+1}(t) \in V_{j+1}$ with $\overline{\text{supp}\,(f_{j+1})} = [0, A]$ for some positive real number A. We assume the multiresolution analysis $\{V_j\}_{j \in \mathbb{Z}}$ is generated by the scaling function $\phi(t)$ associated with the length $(L+1)$ Daubechies scaling filter $\mathbf{h} = [h_0, \ldots, h_L]^T$ and the wavelet filter $\mathbf{g}$ has components $g_k = (-1)^k h_{L-k}$. For these choices, both $\phi(t)$ and the wavelet function $\psi(t)$ are supported on the interval $[0, L]$ (see Theorem 6.3 and the discussion preceding 7.10). We have

$$f_{j+1}(t) = \sum_{k=0}^{N-1} a_{j+1,k}\phi_{j+1,k}(t) \tag{7.25}$$

with $a_{j+1,k} = 0$ for $k < 0$ and $k \geq N$. Using Propositions 5.6 and 5.8, we can decompose $f_{j+1}(t)$ as a sum of functions in V_j and W_j

$$f_{j+1}(t) = \sum_{\ell \in \mathbb{Z}} a_{j,\ell}\phi_{j,\ell}(t) + \sum_{\ell \in \mathbb{Z}} c_{j,\ell}\psi_{j,\ell}(t)$$

where

$$a_{j,\ell} = \sum_{k \in \mathbb{Z}} a_{j+1,k} h_{k-2\ell} \quad \text{and} \quad c_{j,\ell} = \sum_{k \in \mathbb{Z}} a_{j+1,k} g_{k-2\ell} \qquad (7.26)$$

Since $a_{j+1,k} = 0$ for $k < 0$ and $k \geq N$, we can rewrite (7.26) using the finite sums

$$a_{j,\ell} = \sum_{k=0}^{N-1} a_{j+1,k} h_{k-2\ell} \quad \text{and} \quad c_{j,\ell} = \sum_{k=0}^{N-1} a_{j+1,k} g_{k-2\ell}$$

Moreover $h_m, g_m = 0$ for $m < 0$ and $m > L$ so it follows that $a_{j,\ell}, c_{j,\ell} = 0$ whenever $k - 2\ell < 0$ and $k - 2\ell > L$. Thus we are interested in those indices $k - 2\ell$ for which

$$0 \leq k - 2\ell \leq L$$

or, solving for ℓ,

$$\frac{k - L}{2} \leq \ell \leq \frac{k}{2} \qquad (7.27)$$

The smallest and biggest values we can choose for k are 0 and $N - 1$, respectively. Inserting $k = 0$ into the leftmost term and $N - 1$ in the rightmost term of (7.27) gives

$$-\frac{L}{2} \leq \ell \leq \frac{N - 1}{2} \qquad (7.28)$$

Thus the only possible nonzero values for $a_{j,\ell}$ and $c_{j,\ell}$ in (7.26) occur when

$$\ell = \frac{1 - L}{2}, \dots, 0, \dots, \frac{N}{2} - 1$$

Here, we use the fact that N is even to realize that the last index for which $a_{j,\ell}$ is possibly nonzero is $\frac{N}{2} - 1$. We also use the fact that L is odd to see that $\ell = \frac{1-L}{2}$ is the first index for which $a_{j,\ell}$ is possibly nonzero.

Let's consider for the moment the product $H_{N/2}\tilde{\mathbf{a}}$, where

$$\tilde{\mathbf{a}} = [a_{j+1,0}, \dots, a_{j+1,N-1}]^T$$

The first element in the product is

$$a_{j+1,0} h_0 + \cdots + a_{j+1,L} h_L$$

and using (7.26), we see that the value is exactly $a_{j,0}$. In fact, before we encounter wrapping rows when computing the $H_{N/2}\tilde{\mathbf{a}}$, the output will be the values $a_{j,\ell}$. From Problem 7.4(d) we know that there are $\frac{L-1}{2}$ wrapping rows in $H_{N/2}\tilde{\mathbf{a}}$. Thus for $k = \frac{L-1}{2} + 1, \dots, N$, the elements in $H_{N/2}\tilde{\mathbf{a}}$ will errantly dot the scaling filter $\mathbf{h}$ with elements from both the beginning and end of $\tilde{\mathbf{a}}$. Moreover, the product $H_{N/2}\tilde{\mathbf{a}}$ completely omits the computation of $a_{j,\ell}$ for $\ell = \frac{1-L}{2}, \dots, -1$.

How do we solve this problem? The answer is quite simple. Since $a_{j+1,k} = 0$ for $k < 0$ and $k \geq N$, we need to add some zeros to $\tilde{\mathbf{a}}$. Consider the last row of $H_{N/2}$:

$$h_2 \quad h_3 \quad \cdots \quad h_{L-1} \quad h_L \quad 0 \quad 0 \quad \cdots \quad 0 \quad 0 \quad h_0 \quad h_1$$

There are $L - 1$ elements of the scaling filter at the beginning of this last wrapping row. If we pad $\tilde{\mathbf{a}}$ with $L - 1$ zeros to create

$$
\mathbf{a} = \left[\underbrace{0 \quad \cdots \quad 0}_{L-1} \quad a_{j+1,0} \quad \cdots \quad a_{j+1,N-1} \right]^{T} \tag{7.29}
$$

then $\mathbf{a} \in \mathbb{R}^{\tilde{N}}$ where $\tilde{N} = N + L - 1$, and not only will $H_{\tilde{N}/2}\mathbf{a}$ produce the correct values for $a_{j,\ell}$ for the wrapping row indices $\ell = \frac{L-1}{2} + 1, \ldots, N - 1$, but it will also compute the values $a_{j,\ell}$ for $\ell = \frac{1-L}{2}, \ldots, -1$. The analysis for $G_{N/2}$ is exactly the same since the indices of nonzero wavelet filter elements are also $0, \ldots, L$. Thus we compute our projection coefficients using the matrix product $W_{\tilde{N}}\mathbf{a}$.

We can also use the vector notation (4.10)–(4.12) developed in Section 4.1 to represent the components of the matrix product $W_{\tilde{N}}\mathbf{a}$. We have

$$
W_{\tilde{N}} \left[\begin{array}{c} 0 \\ 0 \\ \vdots \\ 0 \\ \hline a_{j+1,0} \\ \vdots \\ a_{j+1,N-1} \end{array} \right] = \mathbf{y}^{1} = \left[\begin{array}{c} \mathbf{y}^{1,a} \\ \mathbf{y}^{1,d} \end{array} \right] = \left[\begin{array}{c} a_{j,(1-L)/2} \\ \vdots \\ a_{j,0} \\ \vdots \\ a_{j,N-1} \\ \hline c_{j,(1-L)/2} \\ \vdots \\ c_{j,0} \\ \vdots \\ c_{j,N-1} \end{array} \right]
$$

where $y_{\ell}^{1,a} = a_{j,\ell}$ and $y_{\ell}^{1,d} = c_{j,\ell}$.

Iteration and Computing Projection Coefficients

The process becomes a bit trickier should we wish to project $f_{j+1}(t)$ into, say, V_{j-i} and W_{j-m}, $m = 1, \ldots, i$. Ideally, we would like to employ the iterated form (7.15) of the discrete wavelet transform, but again we must concern ourselves with the wrapping effects caused by the transform matrix. For simplicity, assume now that the number of projection coefficients N in (7.25) can be written as $N = 2^{i}M$.

The natural solution to the wrapping problem is again to pad $\tilde{\mathbf{a}}$ with zeros — at first glance, the logical choice is $i(L-1)$ zeros. The problem here is that $\tilde{N} = N+i(L-1)$ may not be divisible by 2^{i}, making it impossible to use (7.15). Moreover, as we learned in constructing $\mathbf{a}$ (7.29), we need to have $L - 1$ leading zeros in $\mathbf{a}$ before applying the discrete wavelet transformation. Instead, we create $\mathbf{a}$ from $\tilde{\mathbf{a}}$ by padding $\tilde{\mathbf{a}}$ with $k(L - 1)$ zeros, where

1. $\tilde{N} = k(L - 1) + N$ is divisible by 2^{i}.

2. At each iteration, the vector to be processed still has at least $L - 1$ leading zeros.

We illustrate this construction with an example.

Example 7.7 (Decomposition and Discrete Iteration Revisited) *Consider the function $f_3(t)$ from Example 7.6. We use a fast implementation of (7.15) to compute all possible nonzero projection coefficients. We have*

$$\tilde{\mathbf{a}} = [2, 4, \ldots, 46, 48]^T \in \mathbb{R}^{24}$$

We use the D4 scaling filter, so that $L = 3$ and $L - 1 = 2$. We need $\tilde{N} = N + k(L - 1) = 24 + 2k$ to be divisible by 8, so k must be divisible by 4. Moreover, our choice of k must leave two leading zeros in $a_{2,\ell}$, $a_{1,\ell}$, and $a_{0,\ell}$. Our choices for k are $4, 8, 12, \ldots$ In Problem 7.21 you will learn why $k = 4$ does not work but any $k = 8, 12, \ldots$ will work. Since we want the smallest possible vector to process, we take $k = 8$ so that $\mathbf{a}$ is formed by padding 16 leading zeros to $\tilde{\mathbf{a}}$. We have

$$\mathbf{a} = [0, \cdots, 0, a_{3,0}, \cdots, a_{3,23}]^T \in \mathbb{R}^{40}$$

Application of the iterated wavelet transform routine WT1D from the Discrete-Wavelets *package described by Van Fleet [60] returns the values of the projection coefficients given in (7.21)–(7.24).* ∎

To summarize, if we initially form

$$\mathbf{a} = \left[\underbrace{0, \ldots, 0}_{k(L-1)}, a_{j+1,0}, \cdots, a_{j+1,N-1} \right]^T \in \mathbb{R}^{\tilde{N}} \tag{7.30}$$

where $\tilde{N} = N + k(L - 1)$, then each multiplication by a wavelet transform matrix $W_{\tilde{N}}$ (7.14) corresponds to a decomposition step of a function $f_{j+1} \in V_{j+1}$ into component functions $f_j(t) \in V_j$ and $g_j(t) \in W_j$.

The Pyramid Algorithm

The entire process described above, often called the *pyramid algorithm*, is represented in Figure 7.10, which should remind the reader of (5.25). Suppose that $f_j(t) \in V_j$ with projection coefficients $\{a_{j,k}\}$. We create $\mathbf{a} \in \mathbb{R}^{\tilde{N}}$ as prescribed by (7.30), where $\tilde{N} = N + k(L - 1)$. For some positive integer i we wish to write

$$f_{j+1}(t) = f_{j-i}(t) + g_{j-1}(t) + g_{j-2}(t) + \cdots + g_{j-i}(t)$$

where $f_{j-i}(t) \in V_{j-i}$ and $g_m \in V_{j-m}$ for $m = 1, \ldots, i$.

We pad the projection coefficients $\{a_{j+1,k}\}$ of $f_{j+1}(t)$ with $k(L-1)$ zeros and use length $(L + 1)$ Daubechies scaling filter and the iterated discrete wavelet transform

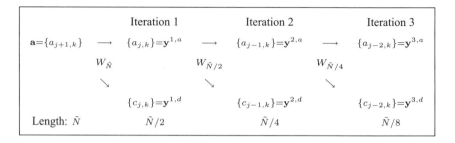

Figure 7.10   Three iterations of the pyramid algorithm.

to find the projection coefficients $\{a_{j-i,k}\}$ and $\{c_{j-m,k}\}$, $m = 1, \ldots, i$. Figure 7.10 shows the first three iterations of the pyramid algorithm.

We offer two observations on the material discussed above. First, how can we interpret the vectors $\mathbf{y}^{j,a}$ and $\mathbf{y}^{j,d}$ for the Daubechies filters? Although $\mathbf{y}^{j,a}$ is not literally the "averages" portion of the transformation as was the case with the discrete Haar transformation, we can think of the $\mathbf{y}^{j,a}$ from the Daubechies wavelet transform as *weighted averages* of the $\mathbf{y}^{j+1,a}$ elements. As we shall see in future examples, the $\mathbf{y}^{j,a}$ vector is a smoothed and blurred version of $\mathbf{y}^{j+1,a}$, with half the length (see Problem 7.25). Similarly, $\mathbf{y}^{j,d}$ holds the details portion of the transformation, produced by the wavelet filter coefficients in $G_{N/2}$.

We also note that when computing the vectors $\mathbf{y}^j$ from $\mathbf{a}$, in practice the matrix form is not used. The process would be too slow and use too much computer memory when N is large. Instead, the decomposition formulas (5.15) and (5.26) are coded, which speeds things up tremendously. In fact, when they are iterated it is commonly referred to as the *fast wavelet transformation*. See Strang and Nguyen [57] for more on this topic.

Reconstruction via the Discrete Wavelet Transformation

It is worth noting that each multiplication by $W_{\tilde{N},i}^T$ in (7.17) corresponds to a step in the reconstruction of a function in V_{j+1} from its components in coarser V_j and W_j spaces as given by formulas (5.32) in Section 5.1. Examples exploring these ideas are given in Problems 7.23 and 7.24.

It is also important to note that zero padding to avoid wrapping row effects when computing projection coefficients can significantly increase the length of a vector (see Problem 7.22). When the vector or image size is large, we typically do not use zero padding (see Example 7.9) in conjunction with Daubechies scaling filters to perform applications such as compression. In Chapter 8 we learn about different scaling filters that offer a better solution than zero padding for handling wrapping row issues.

Compression of Functions in $L^2(\mathbb{R})$

In this subsection we look at $L^2(\mathbb{R})$ function compression using Daubechies wavelets and in the process show how the wavelet decomposition of a function can lead to insights about function properties.

The compression of $f(t) \in L^2(\mathbb{R})$ is similar (no coding step is necessary) to the image compression scheme discussed for the Haar wavelet transform in Section 4.3. The idea here is to write $f(t)$ in some space V_j using relatively few projection coefficients and at the same time, produce a good approximate $f_j(t)$ of $f(t)$. Algorithm 7.1 describes the compression process:

Algorithm 7.1 (Compression of Functions in $L^2(\mathbb{R})$) *This algorithm takes a projection $f_j(t) \in V_j$ of compactly supported $f(t) \in L^2(\mathbb{R})$ and produces an approximate $\widetilde{f}_j(t) \in V_j$ that can be reconstructed from relatively few projection coefficients. For a prescribed tolerance $\epsilon > 0$, the approximation satisfies $\left\| \widetilde{f}_j(t) - f_j(t) \right\| < \epsilon$.*

1. *Choose a compression tolerance level $\epsilon > 0$.*

2. *Use (6.76) in Section 6.3 to project $f(t)$ into some V_j space.*

3. *Using the ideas of Example 7.6 and the ensuing discussion, apply i iterations of the discrete Daubechies wavelet transformation to obtain projection coefficients in $\{a_{j-i,k}\}$ for V_{j-i} and $\{c_{m,k}\}$ for W_{j-m}, where $m = 1, \ldots, i$.*

4. *Quantize the projection coefficients at each decomposition step. First form the vector $\mathbf{y}$ by concatenating the various sets of projection coefficients:*

$$\mathbf{y} = \left[\, \{a_{j-i,k}\} \,\middle|\, \{c_{j-i,k}\} \,\middle|\, \cdots \,\middle|\, \{c_{j-1,k}\} \,\right]$$

We use vector norms to perform our quantization. We form $\tilde{\mathbf{y}}$ from $\mathbf{y}$ by converting to zero the smallest u_k values, so that

$$\|\mathbf{y} - \tilde{\mathbf{y}}\| < \epsilon$$

We will use the cumulative energy function from Section 4.1 to perform the quantization.

∎

The process should produce a vector $\tilde{\mathbf{u}}$ that is sparse relative to the original projection coefficients $\{a_{j,k}\}$. Assuming that the resolution level j is fine enough, the function

$$\widetilde{f}_j(t) = \sum_k \tilde{a}_{j,k} \phi_{j,k}(t)$$

should give a good approximation to the original $f(t)$. A natural way to measure the quality of the approximation is the $L^2(\mathbb{R})$ difference $\left\| \widetilde{f}_j(t) - f_j(t) \right\|$, which can be interpreted in signal parlance as the *energy* of the difference. The orthogonality of the functions $\phi_{k,j}(t)$ tells us that the energy of contribution $a_{j,k}\phi_{k,j}(t)$ comes from $|a_{j,k}|$. Thus our tolerance level ϵ is also a bound on $\left\|\tilde{f}_j(t) - f_j(t)\right\|$ (see Problem 7.27).

To illustrate Algorithm 7.1, let's look at an example. We consider the $f(t)$ studied in Example 6.10. We saw there that projection into V_0 gave a poor approximation, so we perform compression at a finer level with $j = 4$.

Example 7.8 (Compression in $\mathbf{L}^2(\mathbb{R})$) *We compress the function $f(t)$ from Example 6.10 defined by (6.77). This function is plotted in Figure 6.12. We use Algorithm 7.1 with $i = 4$ iterations of the discrete wavelet transformation with the D4 scaling filter and compression tolerance level $\epsilon = 0.0001$.*

For step 1 of the algorithm, we project $f(t)$ into the finer space V_4, where we use the $a_{4,k}$ values computed in Example 6.10. This function is plotted in Figure 7.1. The 80 nonzero $a_{4,k}$, $k = -40, \ldots, 39$, are plotted in Figure 7.2(a). Note that these values are easily interpreted as decent approximations to the function values from $t = -2\frac{1}{2}$ to $2\frac{7}{16}$. All other $a_{4,k}$ values are clearly zero. You can also see from this example why sample function values are often used for the $a_{j,k}$ values when j is large and there is no explicit formula for $f(t)$.

We next create vector $\mathbf{a} \in \mathbb{R}^{112}$ by padding the value $a_{4,k}$, $k = -40, \ldots, 39$ with 32 zeros (see Problem 7.22). We now apply four iterations of the discrete wavelet transformation using the D4 scaling filter. The result is the vector

$$\mathbf{y} = \left[\mathbf{y}^{4,a} \middle| \mathbf{y}^{4,d} \middle| \mathbf{y}^{3,d} \middle| \mathbf{y}^{2,d} \middle| \mathbf{y}^{1,d} \right]$$

This vector is plotted in Figure 7.11(a).

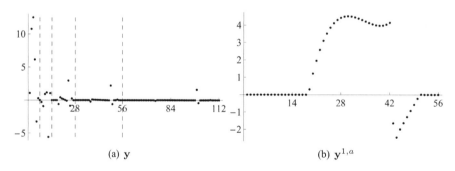

(a) y   (b) $\mathbf{y}^{1,a}$

Figure 7.11   At left is $\mathbf{y}$, which contains the four iterations of the discrete wavelet transformation applied to $\mathbf{a}$. The different iterations are separated by dashed lines. The blocks, left to right, are $\mathbf{y}^{4,a}$, $\mathbf{y}^{4,d}$, $\mathbf{y}^{3,d}$, $\mathbf{y}^{2,d}$, and $\mathbf{y}^{1,d}$. The image at the right is the first iteration, $\mathbf{y}^{1,a}$.

The lengths of $\mathbf{y}^{4,a}$ and $\mathbf{y}^{4,d}$ are each 7 and the lengths of $\mathbf{y}^{3,d}$, $\mathbf{y}^{2,d}$ and $\mathbf{y}^{1,d}$ are 14, 28, and 56, respectively. The detail coefficients $\mathbf{y}^{j,d}$, $j = 1, 2, 3, 4$ are plotted in Figure 7.12.

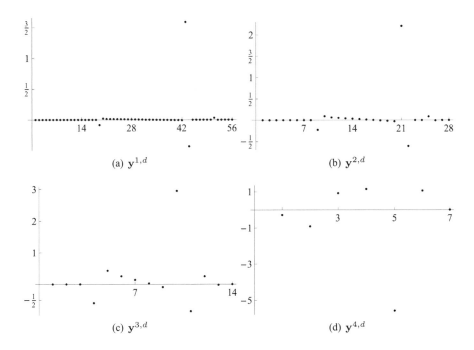

Figure 7.12   Wavelet coefficients for $f(t)$ in Example 7.8 and the compressed function $\widetilde{f}_4(t)$ The nonzero values $c_{j,k}$ values are stored in $\mathbf{y}^{4-j,d}$ for $j = 1,2,3,4$.

Let's digress for just a moment. The $\mathbf{y}^{1,a}$ and $\mathbf{y}^{1,d}$ (plotted in Figures 7.11(b) and 7.12(a), respectively) correspond to 56 values each of $a_{3,k}$ and $c_{3,k}$ for the spaces V_3 and W_3, respectively (see Problem 7.26). Recall that the $c_{3,k}$ values are estimates of weighted differences of function values nearby $t = \ell/2^4$. Insight into the original $f(t)$ can be gained by looking at these values. From Figure 7.12(a) we see that nearly all $c_{3,k}$ are close to zero, since $f(t)$ is continuous and differentiable nearly everywhere. The exceptions are at $t = -2,1,2$. In Figure 7.12(a) we see the huge spike at entry $k = 43$, which corresponds to the discontinuity in $f(t)$ at $t = 1$. Note there are also relatively significant $c_{3,k}$ values at $k = 19, 44, 51$. The first and last correspond to the change of $f(t)$ at $t = \pm 2$, where the derivative $f'(t)$ is undefined. This illustrates why it is often said that wavelet coefficients indicate the local "sharpness" of functions.

The $\mathbf{y}^{2,d}$ correspond to 28 projection coefficients $c_{2,k}$ for the space W_2. The $c_{k,2}$ values are plotted in Figure 7.12(b). If we remember that these are weighted differences of the weighted differences of the approximate function values, it makes sense to interpret these values in terms of the second derivative of $f(t)$. The plot scale is chosen to highlight the striking near-linearity of the $c_{2,k}$ values between $k = 3$ and 12. Why does this phenomenon occur? The $c_{2,k}$ linearity suggests that $f''(t)$ is nearly linear for t between -1.75 and -2.5. For $f''(t)$ to be linear, $f(t)$ must be *cubic on this interval. This fact is readily verified from the defining relation (6.77) for $f(t)$. This is another insight that we gain from analyzing the wavelet coefficients!*

We now quantize with $\epsilon = 0.0001$. Applying the cumulative energy function (see Definition 4.2 in Section 4.1) with this choice of ϵ, we convert to zero all but 25 of the original 80 coefficients. The quantized transform values are stored in $\tilde{\mathbf{y}}$. This vector is plotted in Figure 7.13(b). Note that most of the nonzero values in the differences portions of the transformation occur where the original function or its derivative have jump discontinuities. For convenience, we have plotted $\mathbf{y}$ in Figure 7.13(a).

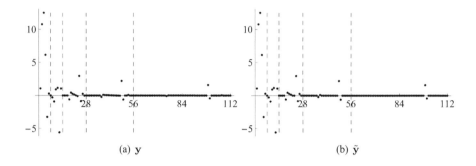

(a) $\mathbf{y}$ (b) $\tilde{\mathbf{y}}$

Figure 7.13   The vectors $\mathbf{y}$ and $\tilde{\mathbf{y}}$.

What have we gained by retaining only 25 projection coefficients? We can view the compressed version of $\tilde{f}_4(t)$ by applying four iterations of the inverse wavelet transformation to the vector $\tilde{\mathbf{y}}$. The result is vector $\tilde{\mathbf{v}}$ which holds the modified projection coefficients. The functions $f_4(t)$ and $\tilde{f}_4(t)$ are plotted side by side in Figure 7.14. The approximation is quite good considering that we constructed $\tilde{f}_4(t)$ from only 25 nonzero coefficients.

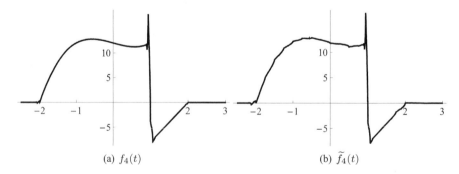

(a) $f_4(t)$ (b) $\tilde{f}_4(t)$

Figure 7.14   The function $f_4(t)$ and its approximate $\tilde{f}_4(t)$.

To the naked eye, the graphs of $f_4(t)$ and $\tilde{f}_4(t)$ are nearly identical. It is also important to remember the information that the wavelet coefficients provide about *the original function. Since each coefficient reflects the behavior of a function over a specific time interval, the coefficients should capture interesting behaviors of the function, such as sharp changes or very smooth portions of the function graph. The wavelet decomposition allows us to change scale and zoom in on a function at points of interest. The ability to provide such a localized view of a function illustrates one of the powerful features of wavelet analysis.* ∎

Image Compression

We perform image compression exactly as described in Algorithm 4.1 in Section 4.3, but we use Daubechies scaling filters to construct our wavelet transformation instead of the Haar scaling filter. The second step in Algorithm 4.1 uses a modified version of the Haar wavelet transformation matrix — multiplying the matrix by $\sqrt{2}$ produces integer-valued entries. The Huffman coding method is much more effective when the input is integer-valued rather than real-valued. For example, two copies of an irrational number expressed as double precision values on a computer can differ in the last position. In this case the Huffman coding method counts these as distinct values, and the resulting tree includes another node.

The output of a wavelet-transformed vector using a Daubechies scaling filter will typically be comprised primarily of irrational numbers. It is possible to modify the transformation process when using the D4 and D6 filters so that integers are mapped to integers (see [11]). The process is beyond the scope of this book so we do not consider it here. Since we are performing lossy compression, we will simply round transformed data to the nearest integer. We also perform no zero padding when we apply the wavelet transformation; Problem 7.28 gives an example where zero padding produces an input matrix substantially larger than the original.

Let's look at an example.

Example 7.9 (Image Compression with Daubechies Filters) *We again consider the digital grayscale image (stored in a 512×512 matrix A) plotted in Figure 4.11(a) in Section 4.3. We first subtract 128 from each element of A. For comparison purposes, we perform compression using the D4 scaling filter and the D6 scaling filter.*

We next compute four iterations of the discrete wavelet transformations. The results are plotted in Figure 7.15.

We use cumulative energy to quantize the transformations, and like Example 4.10, we seek to retain 99.5% of the energy in each transformation. The cumulative energy vectors are virtually indistinguishable but significantly more conservative than the cumulative energy vector for the Haar transformation (see Figure 4.14). The cumulative energy vector for the transformation using the D4 scaling filter is plotted in Figure 7.16. Table 7.1 gives information about the cumulative energy of the two transformations.

We now quantize the transformations keeping only the largest (in absolute value) elements, which contribute 99.5% of the energy of the transform. The results are plotted in Figure 7.17. While the transforms look much like their counterparts in Figure 7.15, the mean squared errors between the D4 and the D6 transforms and the

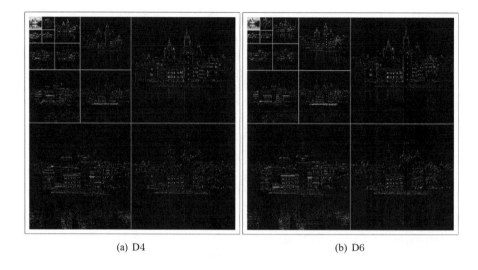

(a) D4   (b) D6

Figure 7.15 Four iterations of the discrete wavelet transformation using D4 and D6 scaling filters.

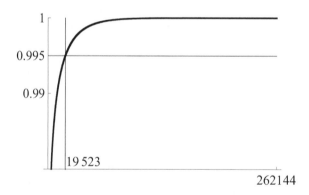

Figure 7.16 The cumulative energy vector for the wavelet transform using the D4 scaling filter.

quantized version are both approximately 128.57. *This is quite a difference from the MSE computed for the modified Haar transform in Example 4.10.*

We now perform Huffman coding (see Appendix A) on each of the quantized transformations. The results are summarized in Table 7.2. For this particular image, the D4 and D6 filter perform virtually the same and a bit better than the Haar filter. The original image consists of $512^2 \cdot 8 = 2{,}097{,}152$ *bits, so the savings even via our naive compression method are quite substantial. To view the compressed images, we decode the bitstream, apply four iterations of the inverse wavelet transformation, and*

Table 7.1   Cumulative energy information for the D4 and D6 transforms as well as the Haar transform from Example 4.10.

Transform	Nonzero Terms	% Zero
Haar	72996	72.15
D4	19523	92.56
D6	19708	92.48

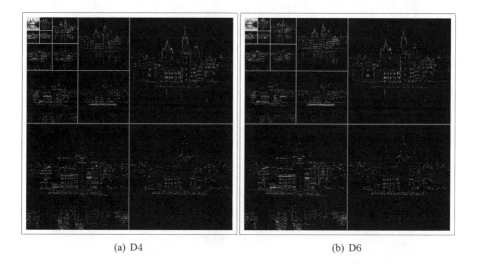

(a) D4   (b) D6

Figure 7.17   Four iterations of the discrete wavelet transformation using the D4 and D6 scaling filters.

add 128 *to each element of the transformed matrix. Figure 7.18 shows the original image and the compressed images.*

Table 7.2   Coding information for the D4 and D6 transforms as well as the Haar transform from Example 4.10.

Transform	Bitstream Length	bpp	Entropy	PSNR
Haar	705536	2.69	2.53	43.90
D4	426240	1.63	1.01	27.04
D6	427309	1.63	1.01	27.04

(a) Original   (b) Haar

(c) D4   (d) D6

Figure 7.18   The original and compressed images using the Haar, D4, and D6 scaling filters.

The results for the peak signal-to-noise ratios are somewhat surprising. Although the Haar–compressed image enjoys a higher PSNR, a closer look reveals some block artifacts. In Figure 7.19 we have enlarged a 128×128 portion of both the Haar- and D4-compressed image. Note the block artifacts that appear in the Haar-compressed image. Block artifacts exist in the D4-compressed image but are less substantial. ■

PROBLEMS

Note: *A computer algebra system and the software on the course Web site will be useful for some of these problems. See the Preface for more details.*

(a) Haar   (b) D4

Figure 7.19 Enlarged versions of the 128×128 portions of two of the compressed images. Note the block artifacts that appear in the Haar-compressed image.

7.17 Repeat the computations in Example 7.6 using, instead, the D6 scaling function.

7.18 Use ideas similar to those outlined in Example 7.6 and a CAS to decompose the function

$$f_4(t) = \sum_{k=-5}^{42} \sin\left(\frac{k}{5}\right) \phi_{4,k}(t)$$

into its component functions in V_0, W_0, W_1, W_2, and W_3 using the D8 scaling function. Plot $f_4(t)$ and each of the component functions.

7.19 In this problem you will follow the ideas of Example 7.8 to compress the heavisine function $f(t)$ with formula (4.7) on $[0,1]$ and $f(t) = 0$ elsewhere. Use the D4 scaling function and make an initial projection into space V_4.

(a) Carry out one iteration and plot the nontrivial coefficients $\{a_{3,k}\}$ and $\{c_{3,k}\}$. Comment on these coefficients.

(b) Carry out the pyramid algorithm with $i = 4$ iterations and compression tolerance level $\epsilon = 0.0001$. How many coefficients can you convert to zero in the quantization step?

(c) Apply four iterations of the inverse transform. View the compressed function and comment on how it compares with the original $f(t)$.

7.20 Suppose that we wish to project a function $f_3(t) \in V_3$ into the component function $f_0(t) \in V_0$, $g_0(t) \in W_0$, $g_1(t) \in W_1$, and $g_2(t) \in W_2$. We use the D8 scaling function and assume that

$$f_3(t) = \sum_{k=0}^{31} a_{3,k}\phi_{3,k}(t)$$

with $a_{3,k} = 0$ for $k < 0$ and $k \geq 32$.

(a) What is the range of indices k for which the projection coefficients $\{c_{2,k}\}$ of $g_2(t)$ are possibly nonzero?

(b) Find similar index ranges for possible nonzero projection coefficients $\{a_{0,k}\}$, $\{c_{0,k}\}$, and $\{c_{1,k}\}$.

7.21 In this problem you will investigate further the need to have $L - 1$ leading zeros in the input vector of each iteration of the discrete wavelet transformation in order to compute projection coefficients accurately.

Suppose that $\tilde{\mathbf{a}} = [a_{2,0}, \dots, a_{2,7}]^T$ are the only nonzero coefficients of a function $f_2(t) \in V_2$. We wish to use the D4 scaling function to project $f_2(t)$ into V_0, W_0, and W_1. Since $\tilde{\mathbf{a}} \in \mathbb{R}^8$, and $L - 1 = 2$, we decide to create $\mathbf{a}$ by padding eight zeros to the front of $\tilde{\mathbf{a}}$ to create a vector of length 16. Write down the possible nonzero coefficients $\{a_{1,\ell}\}$ and $\{c_{1,\ell}\}$ produced by $W_{16}\mathbf{a}$. What happens when you iterate again and apply W_8 to the coefficients $a_{1,\ell}$? What happens if you pad the vector $\tilde{\mathbf{a}}$ by $16, 24, \dots$ zeros?

7.22 Suppose that $\mathbf{v} = [v_0, \dots, v_{31}] \in \mathbb{R}^{32}$ are the nonzero coefficients of the function

$$f_4(t) = \sum_{k=0}^{31} v_k\phi_{4,k}(t)$$

where $\phi(t)$ is the D4 scaling function. We wish to project $f_4(t)$ into V_0 and W_j, $j = 0, 1, 2, 3$ using the iterated discrete wavelet transformation.

(a) How many zeros must we pad to $\mathbf{v}$ so that projection coefficients are computed correctly by the iterated transform?

(b) Repeat the entire exercise using the D6 scaling function.

7.23 Suppose that a function $f_3(t)$ has been decomposed into components

$$f_0(t) = \sum_{\ell} a_{0,\ell}\phi(t - \ell) \quad g_j(t) = \sum_{\ell} c_{j,\ell}\psi_{j,\ell}(t)$$

in V_0 and W_j for $j = 0,1,2$. Write an inverse pyramid algorithm schematic reversing the pyramid algorithm outline in Figure 7.10 that shows how to recover the original coefficients $\{a_{3,k}\}$ of $f_3(t)$ from the coefficients $\{a_{0,k}\}$, $\{c_{0,k}\}$, $\{c_{1,k}\}$, and $\{c_{2,k}\}$.

7.24 Use the inverse pyramid schematic of Problem 7.23 and a CAS to compute the coefficients of $f_3(t)$ defined in Example 7.6 from the coefficients $\{a_{0,k}\}$, $\{c_{2,k}\}$, $\{c_{k,1}\}$, and $\{c_{k,0}\}$.

7.25   We mention in the text that the transform coefficients $\mathbf{y}^{j,a}$ can be thought of as weighted averages, giving a smoother, blurred version of the original signal. Discuss this claim with regard to Example 7.8.

7.26   For Example 7.8, recall that the "trivial" values $a_{4,k} = 0$ for $k < -40$ and $k \geq 40$. Use this information and the decomposition formulas (7.26) to show that $a_{3,k} = c_{3,k} = 0$ for $k < -21$ and $k \geq 20$.

7.27   In this problem we investigate the continuous–discrete connection in the compression of functions in $L^2(\mathbb{R})$. Suppose that we use the Daubechies scaling function $\phi(t)$ associated with the length $(L+1)$ Daubechies scaling filter to perform function compression as described in Algorithm 7.1.

(a) Use the orthogonality of the $\phi_{j,k}(t)$ to show that

$$\left\| \tilde{f}_j(t) - f_j(t) \right\|^2 = \left\| \{\tilde{a}_{j,k}\} - \{a_{j,k}\} \right\|^2$$

(b) Show that

$$\left\| \{\tilde{a}_{j,k}\} - \{a_{j,k}\} \right\|^2 = \left\| \prod_{m=0}^{j-1} W_{N,m}^T \tilde{\mathbf{u}} - \prod_{m=0}^{j-1} W_{N,m}^T \mathbf{u} \right\|^2$$

(c) Prove that the matrices $W_{N,m}$ (7.16) are orthogonal and use this to show that

$$\left\| \prod_{i=0}^{j-1} W_{N,i}^T \tilde{\mathbf{u}} - \prod_{i=0}^{j-1} W_{N,i}^T \mathbf{u} \right\|^2 = \| \tilde{\mathbf{u}} - \mathbf{u} \|^2$$

(d) Conclude that

$$\left\| \tilde{f}_j(t) - f_j(t) \right\| = \| \tilde{\mathbf{u}} - \mathbf{u} \|$$

so if ϵ is a bound on $\| \tilde{\mathbf{u}} - \mathbf{u} \|$ then it is a bound on $\left\| \tilde{f}_j(t) - f_j(t) \right\|$ as well.

7.28   Suppose that we wish to apply four iterations of wavelet transformation constructed from the D6 scaling filter to the the vector $\mathbf{v} \in \mathbb{R}^{32}$.

(a) Show that we need to pad $\mathbf{v}$ with 64 zeros to compute all projection coefficients accurately.

(b) If we wish to pad a 32×32 matrix A with zeros, then we must pad the matrix with zero columns on the left and zero rows on the bottom. Use part (a) to determine the number of zero columns and rows that we must add to A before processing it with the wavelet transformation. What are the dimensions of this new matrix?

7.3 NAIVE IMAGE SEGMENTATION

Image segmentation is the process of splitting a digital image into multiple regions. A common image segmentation goal is to locate objects or boundaries in the image. Some applications include tumor identification in medical imaging, and determining regions of interest (e.g., rainforests, roads) in satellite images.

A common segmentation method is based on finding regions with similar pixel intensity. We use this method in our naive image segmentation. To illustrate the process, we consider grayscale images that consist of two different regions. To complicate matters, we add mild noise and texture to the images. The primary role of wavelets in image segmentation is denoising the image before the segmentation is performed. For a full treatment of image segmentation, see Gonzalez and Woods [31].

To demonstrate the naive image segmentation method, a test image is plotted in Figure 7.20. There are two basic regions with similar pixel intensity in addition to some complications in texture and noise.

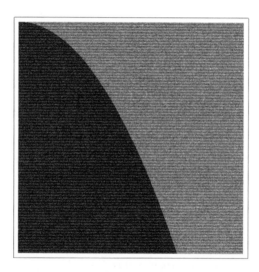

Figure 7.20   A test image consisting of two distinct region with mild noise and texture.

A good way to view the distribution of pixel values is via a *pixel intensity frequency histogram*. The histogram for the test image is plotted in Figure 7.21. The horizontal axis is pixel intensity, ranging from 0 (black) to 255 (white), and the vertical axis is the frequency of each intensity. Each "bump" in the pixel intensity histogram corresponds to a region in the image, with a dividing point near intensity value 100. We call this value the *dividing value T*.

In practice an automated method for finding the dividing value T in the pixel histogram is very desirable. We postpone a discussion of the automation, which computes the value $T = 101$. At this point we partition pixel intensities into sets

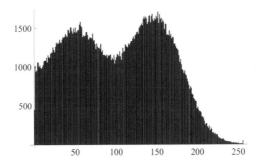

Figure 7.21   The pixel intensity histogram for the image plotted in Figure 7.20.

S_L and S_H, where S_L is the set of all intensities with value less than T, and S_H is the set of all intensities with value T or higher. We compute the average intensities $\mu_L = 51, \mu_H = 171$ for each set.

To reconstruct the image in segmented form, we set $s = \mu_L$ for all $s \in S_L$ and $t = \mu_H$ for all $t \in S_H$. The test image after segmentation is plotted in Figure 7.22. The result is better than the original, but there are "speckling" problems that arise from the noise and texture in original image. Thus we can improve the process by performing denoising before partitioning pixels. We use a wavelet–based method called *wavelet shrinkage* to perform the denoising. Wavelet shrinkage is an involved process, which is useful in its own right, so we leave our discussion of image segmentation for now and turn our attention to wavelet shrinkage.

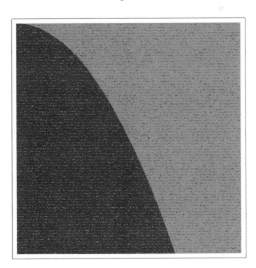

Figure 7.22   Test image segmented without denoising.

Wavelet Shrinkage and Denoising

Wavelet shrinkage is a method of denoising a signal or image. The method is due to Donoho and Johnstone [24]. The interested reader may also consult books by Van Fleet [60] or Vidakovic [61]. We provide a brief treatment of the subject.

For illustrative purposes, we start with the one-dimensional noisy signal $\mathbf{y}$ stored as an N-length vector where

$$\mathbf{y} = \mathbf{v} + \mathbf{e}$$

Here $\mathbf{v}$ is the true signal and $\mathbf{e}$ is a pure noise vector. As an example, recall Figure 4.1 from Section 4.1. As is the case in this example, we assume that the entries e_k of the noise vector $\mathbf{e}$ are independent samples from a normal distribution with mean zero and variance σ^2. Such noise is commonly called *Gaussian white noise*, and σ is referred to as the *noise level*. In practice, we know $\mathbf{y}$ only — our task is to approximate $\mathbf{v}$. The basic approach for estimating $\mathbf{v}$ begins with the application of the discrete wavelet transform W_N to $\mathbf{y}$. The result is

$$\mathbf{y}^1 = W_N \mathbf{v} + W_N \mathbf{e}$$

It can be shown (see, e. g., Problem 9.1 in Van Fleet [60]) that $W_N \mathbf{e}$ is also Gaussian white noise and the detail portion $\mathbf{y}^{1,d}$ of the transformation is primarily noise (see [24]). Thus if we convert to zero sufficiently small values in $\mathbf{y}^{1,d}$ and shrink the remaining entries toward zero to obtain $\tilde{\mathbf{y}}^1$, then we expect the inverse transform $W_N^T \tilde{\mathbf{y}}^1$ to be a vector close to $\mathbf{v}$.

There are a variety of rules for shrinking the entries of $\mathbf{y}^{1,d}$ toward zero. We will use a *shrinkage function* $s_\lambda(t)$ with threshold value $\lambda > 0$. We will use the piecewise linear function

$$s_\lambda(t) = \begin{cases} t - \lambda, & t > \lambda \\ 0, & -\lambda \le t \le \lambda \\ t + \lambda, & t < \lambda \end{cases} \tag{7.31}$$

plotted in Figure 7.23 to shrink entries of $\mathbf{y}^{1,d}$.

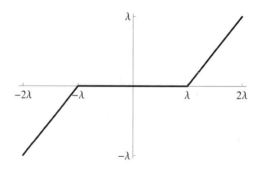

Figure 7.23   The shrinkage function $s_\lambda(t)$.

It is easy to see how the function works: If $|t| < \lambda$, it is converted to zero; otherwise, it is moved λ units closer to zero. For example, if we choose threshold value $\lambda = 3$ and apply $s_\lambda(t)$ component-wise to the vector $\mathbf{d} = [2, -3, 1, -5, 9, 4]^T$, we obtain $\tilde{\mathbf{d}} = [0, 0, 0, -2, 6, 1]^T$.

Here is the algorithm we use for denoising:

Algorithm 7.2 (The Wavelet Shrinkage Algorithm) *Given* $\mathbf{y} \in \mathbb{R}^N$*, where* $N = 2^p M$*, and*

$$\mathbf{y} = \mathbf{v} + \mathbf{e}$$

where $\mathbf{e}$ *is a Gaussian white noise vector, the following algorithm produces a "denoised" approximate of* $\mathbf{v}$:

1. *Apply* i *iterations,* $1 \le i \le p$*, of a wavelet transform to* $\mathbf{y}$ *to obtain*

$$\mathbf{y}^i = \left[\mathbf{y}^{i,a} \,\middle|\, \mathbf{y}^{i,d} \,\middle|\, \mathbf{y}^{i-1,d} \,\middle|\, \cdots \,\middle|\, \mathbf{y}^{1,d} \right]^T$$

2. *Apply the shrinkage function* $s_\lambda(t)$ *given by (7.31) to the detail portions*

$$\mathbf{d} = \left[\mathbf{y}^{i,d} \,\middle|\, \mathbf{y}^{i-1,d} \,\middle|\, \cdots \,\middle|\, \mathbf{y}^{1,d} \right]^T \tag{7.32}$$

   of $\mathbf{y}^i$*. Denote the result by* $\tilde{\mathbf{d}}$.

3. *Join* $\tilde{\mathbf{d}}$ *with* $\mathbf{y}^{i,a}$ *to create the modified transform vector*

$$\hat{\mathbf{z}} = \left[\mathbf{y}^{i,a} \,\middle|\, \tilde{\mathbf{d}} \right]^T$$

4. *Apply* i *iterations of the inverse wavelet transform to* $\hat{\mathbf{z}}$ *to obtain the estimate* $\hat{\mathbf{v}}$ *to* $\mathbf{v}$.

■

The algorithm can be applied to matrices as well as to vectors. In fact, the method can be enhanced by applying different shrinkage functions to different portions of the detail portion of the signal or image.

Choosing the Threshold Value λ

The big challenge when using Algorithm 7.2 is selecting a good threshold value λ. If λ is too large, then $\tilde{\mathbf{d}} \approx \mathbf{0}$. The inverse transformation applied to $\hat{\mathbf{z}}$ will consequently be too blurry. On the other hand, if λ is too small, then $\tilde{\mathbf{d}} \approx \mathbf{d}$ and $\tilde{\mathbf{v}}$ will still contain a significant amount of noise. Donoho and Johnstone developed the wavelet–based VisuShrink and SureShrink methods for choosing λ. We provide an overview of these methods at this time. For a more detailed treatment with derivations of threshold methods, see Donoho and Johnstone [24, 25].

The VisuShrink method uses the *universal threshold* λ^{univ}, which is defined by

$$\lambda^{\text{univ}} = \hat{\sigma}\sqrt{2\ln(M)} \tag{7.33}$$

where M is the length of the detail vector $\mathbf{d}$ and $\hat{\sigma}$ is an estimate of the noise level σ. In Section 4.1 we learned that

$$\hat{\sigma} = \text{MAD}\left(\mathbf{y}^{1,d}\right)/0.6745 \tag{7.34}$$

where the median absolute deviation was defined by (4.5) in Section 4.1. Let's apply Algorithm 7.2 in conjunction with VisuShrink to denoise the signal given in Example 4.2.

Example 7.10 (Signal Denoising) *Denoise the samples of the heavisine function given in Example 4.2.*
Solution
   We use Algorithm 7.2 and first perform five iterations of the wavelet transform constructed from the D6 scaling filter to the noisy vector $\mathbf{y} \in \mathbb{R}^{2048}$ *plotted in Figure 4.1(b). We use the detail portion of the first iterate* $\mathbf{y}^{1,d}$ *and (7.34) to estimate the noise level to be* $\hat{\sigma} = 0.47$. *The length of* $\mathbf{y}^{1,a}$ *is* $2048/2^5 = 64$, *so the the length of* $\mathbf{d}$ *is* $M = 2048 - 64 = 1984$. *Using (7.33) we see that*

$$\lambda^{\text{univ}} = 0.47\sqrt{2\ln(1984)} \approx 1.85$$

This value is used in conjunction with the shrinkage function $s_\lambda(t)$ *in (7.31) and applied component-wise to the detail vector* $\mathbf{d}$ *to obtain* $\tilde{\mathbf{d}}$. *The vectors* $\mathbf{d}$ *and* $\tilde{\mathbf{d}}$ *are plotted in Figure 7.24.*

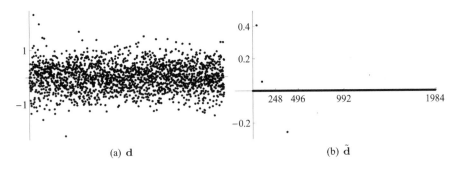

(a) $\mathbf{d}$ (b) $\tilde{\mathbf{d}}$

Figure 7.24   The detail portion portion $\mathbf{d}$ of the wavelet transform is at left. The modified detail $\tilde{\mathbf{d}}$ after application of the shrinkage function is plotted at right.

We next form the modified transform vector $\hat{\mathbf{z}} = \left[\mathbf{y}^{i,a}\middle|\tilde{\mathbf{d}}\right]^T$ *and apply five iterations of the inverse transform to obtain our estimate* $\hat{\mathbf{v}}$ *of the true signal. This estimate is plotted in Figure 7.25(b). The original signal* $\mathbf{v}$ *is plotted for convenience*

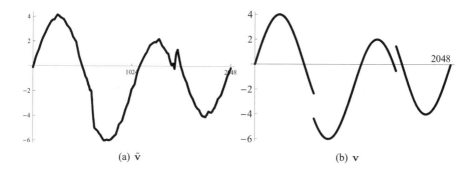

(a) ṽ   (b) v

Figure 7.25   The denoised approximate ṽ of v.

in Figure 7.25(a), and we can see that the VisuShrink method has worked well on this signal.

∎

The universal threshold λ^{univ} works well in practice if the detail portion **d** of the transformed vector is neither too long nor its norm $\|\mathbf{d}\|$ is too large; the vector **d** is termed *sparse* in this case (see Problem 7.32). Some insight can be gained by considering what happens to λ^{univ} in (7.33) if the length M of **d** is large. The threshold λ^{univ} increases proportionally to M and is also large. In this case, the detail portion **d** is shrunk too much. In the case where vector **d** is not sparse, Donoho and Johnstone recommend using the more complicated SureShrink method of wavelet shrinkage. SureShrink creates a different tolerance for each detail portion of the transformation and this value does not depend on the length of the input vector **y**. Unfortunately calculating these tolerances is much more involved than a simple formula, and the derivation goes beyond the scope of this book. The interested reader is encouraged to consult Donoho and Johnstone [25] or Van Fleet [60] for the derivation of λ^{sure}. In the case where **d** is not sparse, SureShrink vastly outperforms VisuShrink.

To gain some empirical perspective on these methods, both are used to denoise the original image segmentation test image from Figure 7.20. Three iterations of the wavelet transformation using the D6 scaling filter were used in the second step of Algorithm 7.2. As evident in Figure 7.26(b), the SureShrink method removes most of the noise and retains the textured pattern of oscillating row intensities. The VisuShrink method seems to blur excessively the image, losing the texture in Figure 7.26(a). However, for the purposes of image segmentation, this is actually convenient, since we are trying to identify just the two main regions.

Image Segmentation Continued

Now that we have discussed wavelet shrinkage, we return to the original problem of image segmentation and complete the segmentation of the test image in Figure 7.20.

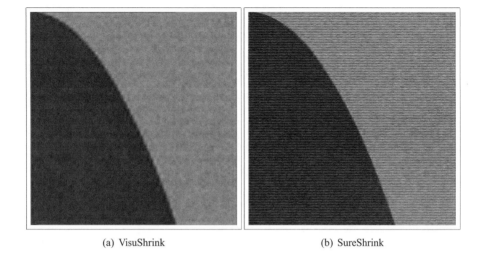

(a) VisuShrink   (b) SureShrink

Figure 7.26   Test image after denoising with three iterations of the wavelet transform.

Example 7.11 (Segmenting the Test Image) *We now segment the test image in Figure 7.20 using both VisuShrink and SureShrink wavelet shrinkage for comparison purposes.*

First the test image is denoised by each method, applying three iterations of the wavelet transformation constructed from the D6 scaling filter. The results are plotted in Figure 7.26. We next create a pixel intensity histogram for each image. These histograms are plotted in Figure 7.27.

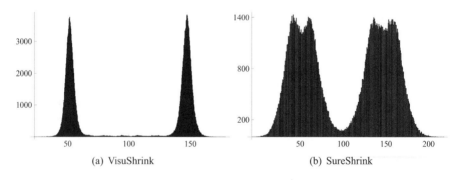

(a) VisuShrink   (b) SureShrink

Figure 7.27   The pixel intensity histograms of the denoised test image. Greater separation is obtained using VisuShrink.

Notice how much more defined and separated the bumps are in these histograms than the one in Figure 7.21 generated without denoising. The fact that the denoising methods produced separation between the bumps in the intensity histograms should *reduce the speckling problem evident in Figure 7.22. An automated method for finding the dividing value T is employed for each histogram, with the results being $T = 100$ for both the SureShrink and VisuShrink methods. To reconstruct the image in segmented form for each method, we set the grayscale values to the appropriate μ_L and μ_H values. The results of segmentation of the test image by each method are plotted in Figure 7.28.*

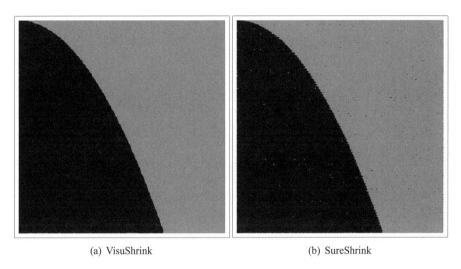

(a) VisuShrink   (b) SureShrink

Figure 7.28   Final segmentation of the test image using the pixel intensity histograms post-denoising.

The VisuShrink result shows no speckling and is clearly superior for this test image. The SureShrink result still contains some "speckling" problems from the texture preserved by this method of denoising. Notice also the large number of dots on the side edges of the images. These dots are the effects of the wrapping rows that occur in the wavelet transform matrix.   ■

To maintain the flow of the segmentation narrative, we have delayed describing an automated method for finding the dividing value T for a pixel intensity histogram. Gonzalez and Woods [31] describe an iterative method that is simple to implement on a computer. The idea is to create a sequence of dividing value estimates T_n that converge to an optimal value. To ensure that this process stops after a finite number of steps on the computer, a small stopping value $\epsilon > 0$ is prescribed and the iteration stops when consecutive estimates satisfy $|T_{n+1} - T_n| < \epsilon$. Since the pixel intensities are integers $0, \ldots, 255$, setting $\epsilon = 1$ is generally a safe choice.

Algorithm 7.3 (Calculating the Dividing Value) *Given a pixel intensity histogram and a stopping value $\epsilon > 0$, this algorithm estimates the dividing value T. The iterative scheme produces the estimates T_n and stops when $|T_{n+1} - T_n| < \epsilon$.*

1. *Select the initial estimate T_1. Typically, this is done by averaging the minimum and maximum frequency values.*

2. *Partition pixels into sets S_L and S_H, where S_L is the set of all frequency values less than T_1, and S_H is set of all frequency values greater than or equal to T_1.*

3. *Compute the average values μ_L and μ_H for the sets S_L and S_H.*

4. *Update the dividing value by averaging the values found in the step 3:*

$$T_2 = \frac{\mu_L + \mu_H}{2}$$

5. *Repeat steps 2–4, replacing the old dividing value T_n by the new value T_{n+1}. Continue until $|T_{n+1} - T_n| < \epsilon$.*

∎

This algorithm will generally work if the histogram data form two "bumps," as we have seen in Example 7.11. At each step of the algorithm, the average values μ_L and μ_H should correspond approximately to the centers of the two bumps. See Problem 7.30 for more on this algorithm.

PROBLEMS

Note: *A computer algebra system and the software on the course Web site will be useful for some of these problems. See the Preface for more details.*

7.29   Apply the wavelet shrinkage function $s_2(t)$ to vector

$$\mathbf{d} = [-1, 5, -4, 1.9, -2.4]^T$$

7.30   Use the dividing value algorithm with $\epsilon = 1$ on the set

$$[0, 0, 1, 2, 2, 2, 3, 4, 124, 130, 140, 146, 147, 147, 147, 149, 160, 250]^T$$

Comment on your results. Do your final values for T, μ_L and μ_H seem correct for the data set?

7.31   To see the effects of threshold λ being too large or too small, denoise the near-linear signal

$$L = [1, 2, 3.1, 3.9, 4.8, 6.1, 7, 7.8, 9.3, 10]^T$$

using the wavelet shrinkage algorithm with one iteration of the Haar wavelet transform for $\lambda = 3$. Repeat the exercise with $\lambda = 0.001$. Comment on your results.

7.32   Whether to use VisuShrink or SureShrink depends on the sparseness of the detail component $\mathbf{d}$ (7.32) of the transformed signal or image. More precisely, Donoho and Johnstone use statistical arguments [25] to recommend using λ^{univ} for shrinkage whenever

$$\frac{\|\mathbf{d}\|^2}{M} - 1 \le \frac{3 \log_2(M)}{2\sqrt{M}}$$

is satisfied. In such a case we say that the detail vector $\mathbf{d}$ is sparse.

(a) The details portion $\mathbf{d}$ in Example 7.10 is known to have norm $\|\mathbf{d}\| = 515.76$. Determine whether $\mathbf{d}$ is sparse and whether the VisuShrink method is indeed an appropriate method for denoising the image.

(b) The test image in Figure 7.20 has 512×512 pixels. In Example 7.11 we compute three iterations of the wavelet transform of the image. If we load the details portions of the transform into vector $\mathbf{d}$, we can compute $\|\mathbf{d}\| = 12{,}364.9$. Determine whether $\mathbf{d}$ is sparse and whether the VisuShrink method is an appropriate method for denoising the image.

CHAPTER 8

BIORTHOGONAL SCALING FUNCTIONS AND WAVELETS

Biorthogonality is useful generalization of orthogonality that we explore in this chapter. In Chapter 6 we noticed some limitations of the Daubechies wavelets, such as the asymmetry of the Daubechies filter coefficients. In fact, it can be proven that orthogonal multiresolution analysis, other than the one generated by the Haar scaling function, cannot produce symmetric scaling filter coefficients. In applications such as image processing, symmetric scaling filters can be used to handle the edge issues we encountered with the discrete wavelet transformations of Chapter 6. Moreover, the requirement that the scaling function generate an orthogonal basis for V_0 excludes popular approximating functions such as the B-splines.

In this chapter we build a biorthogonal structure called a *dual multiresolution analysis* that allows for the construction of symmetric scaling filters and that can incorporate spline functions. Some extra work is involved in this construction. For example, instead of just a scaling and wavelet filter pair $\mathbf{h}$ and $\mathbf{g}$, the new construct yields two pairs: a scaling and wavelet pair $\tilde{\mathbf{h}}, \tilde{\mathbf{g}}$ for decomposition and a pair $\mathbf{h}, \mathbf{g}$ for reconstruction. Instead of a single scaling function $\phi(t)$ and wavelet function $\psi(t)$, the dual multiresolution analysis requires a pair of scaling functions $\phi(t)$ and $\tilde{\phi}(t)$ related by a *duality condition*, and a pair of associated wavelet functions $\psi(t)$ and $\tilde{\psi}(t)$.

Wavelet Theory: An Elementary Approach with Applications. By D. K. Ruch and P. J. Van Fleet
Copyright © 2009 John Wiley & Sons, Inc.

In Section 8.1 we introduce the basic idea of duality via a linear spline example and motivate a few key results that are presented for the general theory of duality in Section 8.2. The development of this theory allows us to construction dual multiresolution analysis and the related scaling functions and wavelet functions.

In Section 8.3 we use the theoretical tools developed earlier in the chapter to design a family of symmetric, *biorthogonal spline filter pairs*. One of these pairs, the $(5,3)$ biorthogonal spline filter pair, is used to perform lossless image compression in the JPEG2000 image compression standard. The standard also performs lossy image compression and the *CDF97 biorthogonal filter pair*, developed by Cohen, Daubechies, and Feauveau, is used for this task. This important filter pair is also discussed in Section 8.3.

In Section 8.4 we examine how the pyramid algorithm for decomposition and reconstruction can be generalized for biorthogonal scaling and wavelet functions. We use these ideas to create a discrete biorthogonal wavelet transformation in Section 8.5.

Finally, the theoretically-minded student can learn how biorthogonality concepts utilize the notion of a *Riesz basis* in Section 8.6. These bases generalize an orthogonal bases in such a way as to allow for the dual multiresolution analysis structure.

8.1   A BIORTHOGONAL EXAMPLE AND DUALITY

We mentioned in the chapter introduction that biorthogonality is a concept that allows B-splines to be used within a multiresolution-like structure. In this section we examine a biorthogonal spline example and use it to motivate the properties essential to the construction of biorthogonal scaling functions and wavelets. The notion of *dual* functions is critical to biorthogonality.

To illustrate, consider the linear B-spline $B_1(t)$ and define $\tilde{\phi}(t)$ by $\tilde{\phi}(t) = B_1(t+1)$. It turns out that $\tilde{\phi}(t)$ has a dual function $\phi(t)$ associated with it that satisfies the remarkable *duality condition*

$$\boxed{\left\langle \tilde{\phi}(t-m), \phi(t-k) \right\rangle = \int_{\mathbb{R}} \tilde{\phi}(t-m)\phi(t-k)\,\mathrm{d}t = \delta_{m,k}} \qquad (8.1)$$

The linear B-spline and its *dual* are plotted in Figure 8.1. We shall see later how $\phi(t)$ is created from a dilation equation and why it satisfies (8.1). For now, we accept this duality as fact and illustrate an advantage of such dual functions via the following example.

Example 8.1 (Using Duality to Represent Functions) *Use the duality property (8.1) to solve for the coefficients* $\tilde{a}_j$ *such that*

$$f(t) = \tilde{a}_{-1}\tilde{\phi}(t+1) + \tilde{a}_0\tilde{\phi}(t) + \tilde{a}_1\tilde{\phi}(t-1) + \tilde{a}_2\tilde{\phi}(t-2) \qquad (8.2)$$

where $f(t)$ *is the piecewise linear function given by (5.73) in Section 5.3 and* $\tilde{\phi}(t) = B_1(t+1)$.

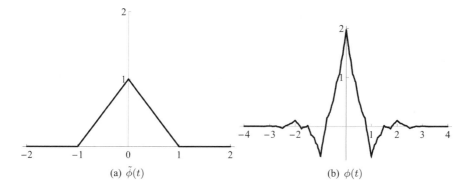

(a) $\tilde{\phi}(t)$ (b) $\phi(t)$

Figure 8.1 Dual functions $\tilde{\phi}(t) = B_1(t+1)$ and $\phi(t)$.

Solution

We first show that $\tilde{a}_2$ can be computed from an integral. Anticipating the use of the duality property (8.1) with $j = 2$, we multiply each side of (8.2) by $\phi(t-2)$ and then integrate to obtain

$$\int_{\mathbb{R}} f(t)\phi(t-2)\,\mathrm{d}t = \int_{\mathbb{R}} \tilde{a}_{-1}\tilde{\phi}(t+1)\phi(t-2)\,\mathrm{d}t + \int_{\mathbb{R}} \tilde{a}_0\tilde{\phi}(t)\phi(t-2)\,\mathrm{d}t$$
$$+ \int_{\mathbb{R}} \tilde{a}_1\tilde{\phi}(t-1)\phi(t-2)\,\mathrm{d}t + \int_{\mathbb{R}} \tilde{a}_2\tilde{\phi}(t-2)\phi(t-2)\,\mathrm{d}t$$

We then use (8.1) to simplify, yielding

$$\int_{\mathbb{R}} f(t)\phi(t-2)\,\mathrm{d}t = \tilde{a}_{-1}\cdot 0 + \tilde{a}_0 \cdot 0 + \tilde{a}_1 \cdot 0 + \tilde{a}_2 \cdot 1 = a_2$$

More generally, we see that each $\tilde{a}_k$ coefficient can be obtained using the integral

$$\tilde{a}_k = \langle f(t), \phi(t-k)\rangle = \int_{\mathbb{R}} f(t)\phi(t-k)\,\mathrm{d}t \qquad (8.3)$$

instead of solving the system of equations (5.74).

To compute this integral, we use the dilation equation for $\phi(t)$ and the cascade algorithm to obtain numerical values for $\phi(t)$, which can then be used to find the integrals (8.3). As we shall see in Section 8.3, the dilation equation for $\phi(t)$ is

$$\phi(t) = \sqrt{2}\sum_{k=-4}^{4} h_k\phi(2t-k)$$

with nonzero filter coefficients

$$h_0 = \frac{45\sqrt{2}}{64} \qquad h_{\pm 1} = \frac{19\sqrt{2}}{64} \qquad h_{\pm 2} = -\frac{\sqrt{2}}{8}$$
$$h_{\pm 3} = -\frac{3\sqrt{2}}{64} \qquad h_{\pm 4} = \frac{3\sqrt{2}}{128} \qquad (8.4)$$

The details of calculating the integrals are left as Problem 8.1. In this problem you will show that $\tilde{a}_{-1} = \tilde{a}_0 = 3$, $\tilde{a}_1 = 1$, and $\tilde{a}_2 = 2$. These values agree with the results satisfying (5.74). ∎

With the motivation given by this example, we now formalize some concepts.

Definition 8.1 (Biorthogonal Sets and Dual Spaces) *Let $\tilde{\phi}(t)$ and $\phi(t)$ be functions in $L^2(\mathbb{R})$ and define the spaces*

$$V_0 = \text{span}\{\phi(t-k)\}_{k\in\mathbb{Z}} \quad and \quad \tilde{V}_0 = \text{span}\{\tilde{\phi}(t-k)\}_{k\in\mathbb{Z}}$$

The set $\{\tilde{\phi}(t-k)\}_{k\in\mathbb{Z}}$ is biorthogonal to the set $\{\phi(t-k)\}_{k\in\mathbb{Z}}$ if for all $m, k \in \mathbb{Z}$, we have

$$\left\langle \tilde{\phi}(t-m), \phi(t-k) \right\rangle = \int_{\mathbb{R}} \tilde{\phi}(t-m)\phi(t-k)\,dt = \delta_{m,k} \tag{8.5}$$

We say that $\tilde{\phi}(t)$ and $\phi(t)$ are dual generators and V_0 and $\tilde{V}_0$ are dual spaces. ∎

One consequence of being dual generators is that the biorthogonal sets $\{\tilde{\phi}(t-k)\}_{k\in\mathbb{Z}}$ and $\{\phi(t-k)\}_{k\in\mathbb{Z}}$ are automatically linearly independent.

Proposition 8.1 (Biorthogonal Sets Are Linearly Independent) *Let $\tilde{\phi}(t)$ and $\phi(t)$ be dual generators in $L^2(\mathbb{R})$. Then the biorthogonal sets*

$$\{\tilde{\phi}(t-k)\}_{k\in\mathbb{Z}} \quad and \quad \{\phi(t-k)\}_{k\in\mathbb{Z}}$$

are linearly independent. ∎

Proof: The proof is a direct consequence of the definitions of biorthogonality and linear independence. See Problem 8.3. ∎

The sets $\{\tilde{\phi}(t-k)\}_{k\in\mathbb{Z}}$ and $\{\phi(t-k)\}_{k\in\mathbb{Z}}$ may not be orthogonal as we know for B-splines from Section 5.3. However, if these sets are nearly orthogonal in a certain sense, they are called *Riesz bases* of V_0 and $\tilde{V}_0$. All examples V_0 and $\tilde{V}_0$ in this chapter have Riesz bases, and the formal definition of Riesz bases and some theoretical results are given Section 8.6. The important fact that we will use is that Riesz bases allow us to construct a multiresolution analysis-like structure from V_0 and $\tilde{V}_0$.

An important feature of the dual generators plotted in Figure 8.1 is the symmetry about $t = 0$ of the graphs of $\phi(t)$, $\tilde{\phi}(t)$ and the symmetry of the scaling filter coefficients h_k, $\tilde{h}_k$ about $k = 0$. Recall (see Problem 2.58) that the scaling filter coefficients for the linear B-spline $B_1(t)$ are

$$\tilde{h}_{\pm 1} = \frac{\sqrt{2}}{4} \quad and \quad \tilde{h}_0 = \frac{\sqrt{2}}{2} \tag{8.6}$$

and

$$\tilde{\phi}(t) = \sqrt{2} \sum_{k=-1}^{1} \tilde{h}_k \tilde{\phi}(2t-k)$$

This type of symmetry is useful in applications and is not possible for orthogonal scaling functions. This symmetry theme is explored in Problem 8.8 and again later in the chapter.

As was the case of the development of multiresolution analysis in Chapter 5, we will consider translates and dilates of $\tilde{\phi}(t)$ and $\phi(t)$ and the resulting spaces. We have the following definition.

Definition 8.2 (The Functions $\tilde{\phi}_{j,k}(t)$ and $\phi_{j,k}(t)$) *Suppose that $\tilde{\phi}(t)$ and $\phi(t)$ are dual generators of $\tilde{V}_0$ and V_0. For all integers j and k, we define the functions*

$$\boxed{\tilde{\phi}_{j,k}(t) = 2^{j/2}\tilde{\phi}\left(2^j t - k\right) \qquad and \qquad \phi_{j,k}(t) = 2^{j/2}\phi\left(2^j t - k\right)} \qquad (8.7)$$

∎

The following result is immediate.

Proposition 8.2 (Biorthogonality for $\tilde{\phi}_{jk}$ and ϕ_{jk}) *Suppose j, k and m are integers. Then*

$$\left\langle \tilde{\phi}_{j,k}(t), \phi_{j,m}(t) \right\rangle = \delta_{k,m} \qquad (8.8)$$

∎

Proof: The proof is left as Problem 8.4.   ∎

We can now define the spaces $\tilde{V}_j$ and V_j.

Definition 8.3 (The Spaces $\tilde{V}_j$ and V_j) *Let $j, k \in \mathbb{Z}$. If $\tilde{\phi}_{j,k}(t)$ and $\phi_{j,k}(t)$ are given by (8.7), then we define the* dual spaces

$$\boxed{\tilde{V}_j = \mathrm{span}\{\tilde{\phi}_{j,k}(t)\}_{k\in\mathbb{Z}} \qquad and \qquad V_j = \mathrm{span}\left\{\phi_{j,k}(t)\right\}_{k\in\mathbb{Z}}} \qquad (8.9)$$

∎

We want to extend the ideas of projection to dual generators and spaces. The process used to find $\tilde{a}_k$ coefficients in Example 8.1 provides some motivation for the following definition.

Definition 8.4 (Biorthogonal Projections) *Let $f(t) \in L^2(\mathbb{R})$ and suppose that V_j and $\tilde{V}_j$ are dual spaces. The projections of $f(t)$ into V_j and $\tilde{V}_j$ are given by*

$$f_j(t) = P_{f,j}(t) = \sum_{k\in\mathbb{Z}} a_{j,k}\phi_{j,k}(t) \qquad and \qquad \tilde{f}_j(t) = \tilde{P}_{f,j}(t) = \sum_{k\in\mathbb{Z}} \tilde{a}_{j,k}\tilde{\phi}_{j,k}(t)$$

respectively, where

$$a_{j,k} = \langle f(t), \tilde{\phi}_{j,k}(t) \rangle \qquad and \qquad \tilde{a}_{j,k} = \langle f(t), \phi_{j,k}(t) \rangle$$

∎

Notice in particular that the $\tilde{a}_{j,k}$ coefficients come from $\phi(t)$ and not the dual $\tilde{\phi}(t)$, as illustrated in Example 8.1.

Example 8.2 (Projections into V_0 and $\tilde{V}_0$) *Using Definition 8.4, project the function*

$$f(t) = \begin{cases} t^3/2 - t/2 + 3, & -2 < t < 1 \\ 2t - 4, & 1 \le t < 2 \\ 0, & otherwise \end{cases} \tag{8.10}$$

from (6.77) into the V_0 and $\tilde{V}_0$ spaces generated by the linear B-spline and its dual given in Example 8.1.

Solution

To compute $f_0(t)$, we need to find $a_{0,k}$ for $k \in \mathbb{Z}$. Since $\overline{\text{supp}(\tilde{\phi})} = [-1,1]$ and $\overline{\text{supp}(f)} = [-2,2]$ we can deduce that $a_{0,k} = \int_{\mathbb{R}} f(t)\tilde{\phi}(t-k)\,dt = 0$ for $k \ge 3$ and $k \le -3$. The nontrivial $a_{0,k}$ values are easy to calculate since $\tilde{\phi}(t)$ is piecewise linear. On the other hand, plotting $\phi(t)$ requires the cascade algorithm.

We find the projection $\tilde{P}_{f,0}(t) = \tilde{f}(t)$ into $\tilde{V}_0$ in a similar way. Using the support properties of $f(t)$ and $\tilde{\phi}(t)$, we can show that all but a finite number of $\tilde{a}_{0,k}$ values are zero. Finding the nontrivial $\tilde{a}_{0,k}$ values requires numerical integration since the $\phi(t)$ is approximated using the cascade algorithm. The details for performing these computations are addressed in Problems 8.1 and 8.5. The two projections are plotted in Figure 8.2. ∎

One thing the two projections in Example 8.2 have in common is poor resolution. To obtain better resolution, we need to build two dual multiresolution analysis-like structures with finer resolution spaces V_j, $\tilde{V}_j$ and dual wavelet spaces W_j, $\tilde{W}_j$. Although the construction is a bit more involved than that of the multiresolution analyses from Chapter 5, the advantages of using symmetric biorthogonal scaling filters in applications will more than justify the work.

To carry out this program of building the dual multiresolution analysis-like structures with symmetric scaling filters, we need to further investigate the connections between the two dilation equations, the dual filters, and the scaling function symbols. To begin, we combine the biorthogonality and the dilation equation relationships by evaluating $\left\langle \tilde{\phi}(t), \phi(t-j) \right\rangle$ using the dilation equations. This should give some insight into the relationship between the filter coefficients $\tilde{h}_k$ and h_k. For now, let's assume that $\tilde{\phi}(t)$ and $\phi(t)$ satisfy dilation equations

$$\tilde{\phi}(t) = \sqrt{2}\sum_{k\in\mathbb{Z}} \tilde{h}_k\tilde{\phi}(2t-k) \quad \text{and} \quad \phi(t) = \sqrt{2}\sum_{k\in\mathbb{Z}} h_k\phi(2t-k) \tag{8.11}$$

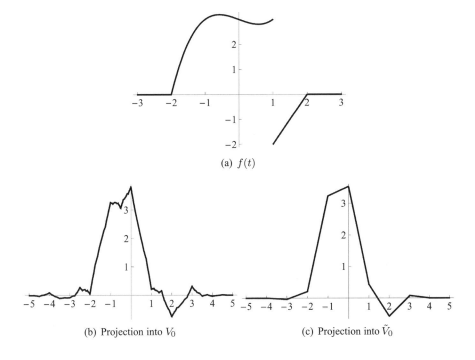

(a) $f(t)$

(b) Projection into V_0

(c) Projection into $\tilde{V}_0$

Figure 8.2   The projections of $f(t)$ into V_0 and $\tilde{V}_0$.

and use them to evaluate the inner product $\langle \tilde{\phi}(t), \phi(t-k) \rangle$ as follows:

$$
\begin{aligned}
\langle \tilde{\phi}(t), \phi(t-k) \rangle &= \left\langle \sqrt{2} \sum_{\ell \in \mathbb{Z}} \tilde{h}_\ell \tilde{\phi}(2t-\ell), \sqrt{2} \sum_{n \in \mathbb{Z}} h_n \phi(2(t-k)-n) \right\rangle \\
&= 2 \int_{\mathbb{R}} \sum_{\ell \in \mathbb{Z}} \tilde{h}_\ell \tilde{\phi}(2t-\ell) \cdot \sum_{n \in \mathbb{Z}} h_n \phi(2(t-k)-n) \, \mathrm{d}t \\
&= 2 \sum_{\ell \in \mathbb{Z}} \sum_{n \in \mathbb{Z}} \tilde{h}_\ell h_n \left(\frac{1}{2} \int_{\mathbb{R}} \tilde{\phi}(u-\ell) \phi(u-(2k+n)) \, \mathrm{d}u \right) \quad (8.12)
\end{aligned}
$$

where the last step uses the substitution $u = 2t$. Now remember that $\tilde{\phi}(t)$ and $\phi(t)$ are dual, so by (8.1), the integral in (8.12) is $\delta_{\ell, 2k+n}$.

For each $\ell \in \mathbb{Z}$, consider the term where $\ell = 2k + n$. Replacing n by $\ell - 2k$ from (8.12) yields

$$
\tilde{h}_\ell h_n \int_{\mathbb{R}} \tilde{\phi}(u-\ell) \phi(u-(2k+n)) \, \mathrm{d}u = \tilde{h}_\ell h_{\ell-2k} \int_{\mathbb{R}} \tilde{\phi}(u-\ell) \phi(u-\ell) \, \mathrm{d}u
$$

$$
= \tilde{h}_\ell h_{\ell-2k}
$$

Next consider the terms in (8.12), where $\ell \neq 2k + n$. By the duality property (8.5) we obtain

$$\tilde{h}_\ell h_n \int_{\mathbb{R}} \tilde{\phi}(u - \ell)\phi(u - (2k + n)) \, du = \tilde{h}_\ell h_n \cdot 0 = 0$$

Now combine these observations to obtain

$$\langle \tilde{\phi}(t), \phi(t - k) \rangle = 2\frac{1}{2} \sum_{\ell \in \mathbb{Z}} \tilde{h}_\ell h_{\ell - 2k}$$

We have found an important relationship between filter coefficients $\{\tilde{h}_k\}$ and $\{h_k\}$ when we have dual functions satisfying dilation equations. We record this observation formally in a theorem.

Theorem 8.1 (Filter Condition for Dual Generators) *Suppose that $\tilde{\phi}(t)$ and $\phi(t)$ satisfy dilation equations (8.11). These functions are dual generators if and only if for all $k \in \mathbb{Z}$,*

$$\sum_\ell \tilde{h}_\ell h_{\ell - 2k} = \delta_{k,0} \tag{8.13}$$

∎

Proof: If we assume that $\tilde{\phi}(t)$ and $\phi(t)$ are dual generators, the proof of (8.13) is given above for $k = 0$. The proof for nonzero integers k is straightforward and omitted. For the converse, see Problem 8.7.   ∎

Next we illustrate the previous result with the dual linear B-spline example.

Example 8.3 (Filter Condition for the Linear Spline and its Dual) *Recall the linear spline $B_1(t+1)$ from Example 8.1. Use Theorem 8.1 to confirm that the functions $\phi(t)$ and $\tilde{\phi}(t)$ are dual generators.*

Solution
   Recall that the scaling filter coefficients for $\tilde{\phi}(t)$ and $\phi(t)$ are given by (8.6) and (8.4), respectively. To verify (8.13) we check for $k = 0$ and $k = 2$ and leave rest for Problem 8.6. For $k = 0$ we have

$$\sum_{\ell \in \mathbb{Z}} \tilde{h}_\ell h_{\ell - 2 \cdot 0} = \left(\sqrt{2}\right)^2 \left(\frac{1}{4} \cdot \frac{19}{64} + \frac{1}{2} \cdot \frac{45}{64} + \frac{1}{4} \cdot \frac{19}{64} \right) = 1 = \delta_{0,0}$$

and when $k = 2$, (8.13) yields

$$\sum_{\ell \in \mathbb{Z}} \tilde{h}_\ell h_{\ell - 2 \cdot 2} = \left(\sqrt{2}\right)^2 \left(\frac{1}{4} \cdot 0 + \frac{1}{2} \cdot \frac{3}{128} + \frac{1}{4} \cdot -\frac{3}{64} \right) = 0 = \delta_{2,0}$$

∎

The biorthogonality condition (8.13) should remind you of a similar condition,

$$\sum_\ell h_\ell h_{\ell - 2k} = \delta_{k,0}$$

for a multiresolution analysis given in Proposition 5.5(ii). In the next section we examine the connection between this filter coefficient condition and scaling function symbols $H(\omega)$ and $\tilde{H}(\omega)$.

PROBLEMS

Note: *A computer algebra system and the software on the course Web site will be useful for some of these problems. See the Preface for more details.*

8.1   Use the cascade algorithm to generate numerical values for $\phi(t)$ in Example 8.1 and use them to reproduce the graph in Figure 8.1.

8.2   Use the results of Problem 8.1 to calculate the integrals, giving the $\tilde{a}_k$ values in Example 8.1.

8.3   Recall that a set of functions $\{f_k(t)\}_{k\in\mathbb{Z}}$ is linearly independent if $\sum_{k\in\mathbb{Z}} c_k f_k(t) = 0$ for all $t \in \mathbb{R}$ implies that $c_k = 0$ for all $k \in \mathbb{Z}$. Use this fact and Definition 8.1 to prove Proposition 8.1.

8.4   Make the substitution $u = 2^j t - k$ in the inner product (8.8) to prove Proposition 8.2.

8.5   In Example 3.2, recall that $f(t) = 0$ when $t \le -2$ and $t \ge 2$ and $\overline{\text{supp}(\tilde{\phi})} = [-1, 1]$. In Section 8.3 we will see that $\text{supp}(\phi) = [-4, 4]$.

  (a) Use these facts and integral properties to show that $a_{0,k} = 0$ if $|k| \ge 3$.

  (b) More generally, use the support properties for $\tilde{\phi}(t)$ and $f(t)$ to find the values for which $a_{j,k} = 0$.

  (c) Use the support properties for $\phi(t)$, $f(t)$ to find the values for which $\tilde{a}_{j,k} = 0$.

8.6   Prove that the linear B-spline $\tilde{\phi}(t) = B_1(t+1)$ and its dual $\phi(t)$ from Example 8.1 satisfy condition (8.13) for all $k \in \mathbb{Z}$.

8.7   Prove the converse of Theorem 8.1.

★**8.8**   Recall that a real function $f(t) \in L^2(\mathbb{R})$ is symmetric about $t = 0$ if $f(t) = f(-t)$ for all $t \in \mathbb{R}$ and a real filter **h** is symmetric about $k = 0$ if $h_k = h_{-k}$ for all $k \in \mathbb{Z}$.

  (a) Show that a real function $f(t) \in L^2(\mathbb{R})$ is symmetric about $t = 0$ if and only if $\hat{f}(\omega)$ is a real function.

  (b) Show that a real filter **h** is symmetric about $k = 0$ if and only if its symbol $H(\omega) = \frac{1}{\sqrt{2}} \sum_{k\in\mathbb{Z}} h_k e^{-ik\omega}$ is a real function.

  (c) Suppose that a real function $\phi(t)$ satisfies a dilation equation

$$\phi(t) = \sqrt{2} \sum_{k\in\mathbb{Z}} h_k \phi(2t - k)$$

and is symmetric about $t = 0$. Show that $\mathbf{h}$ is symmetric about $k = 0$. (*Hint:* Consider the dilation equation in the transform domain and use parts (a) and (b).)

(d) Suppose that a real function $\phi(t)$ satisfies a dilation equation and that filter $\mathbf{h}$ is symmetric about $k = 0$. Use the infinite product formulation

$$\hat{\phi}(\omega) = \frac{1}{\sqrt{2\pi}} \prod_{k=1}^{\infty} H\left(\frac{\omega}{2^k}\right)$$

from (6.46) and parts (a) and (b) to show $\phi(t)$ must be symmetric about $t = 0$.

8.2 BIORTHOGONALITY CONDITIONS FOR SYMBOLS AND WAVELET SPACES

We now wish to connect the filter condition (8.13) for dual scaling functions $\tilde{\phi}(t)$ and $\phi(t)$ to their symbols and then derive an analog to Theorem 5.6

$$|H(\omega)|^2 + |H(\omega + \pi)|^2 = 1$$

for all $\omega \in \mathbb{R}$, or equivalently,

$$H(\omega)\overline{H(\omega)} + H(\omega + \pi)\overline{H(\omega + \pi)} = 1$$

for all $\omega \in \mathbb{R}$. In the biorthogonal case we have two symbols,

$$\tilde{H}(\omega) = \frac{1}{\sqrt{2}} \sum_{k \in \mathbb{Z}} \tilde{h}_k e^{-ik\omega} \qquad \text{and} \qquad H(\omega) = \frac{1}{\sqrt{2}} \sum_{k \in \mathbb{Z}} h_k e^{-ik\omega} \qquad (8.14)$$

for the dual scaling functions $\phi(t)$ and $\tilde{\phi}(t)$, respectively.

Let's try using (8.13) to see if the orthogonality condition generalizes to

$$\tilde{H}(\omega)\overline{H(\omega)} + \tilde{H}(\omega + \pi)\overline{H(\omega + \pi)} = 1 \qquad (8.15)$$

under the duality assumption on $\phi(t)$ and $\tilde{\phi}(t)$.

First we check a concrete example before proceeding to the general case. For convenience of computation, it is often easier to use the notation $z = e^{-i\omega}$. In Problem 8.13 you will show that (8.15) is equivalent to

$$\tilde{H}(z)H(z^{-1}) + \tilde{H}(-z)H(-z^{-1}) = 1 \qquad (8.16)$$

Example 8.4 (Symbol Condition Computation) *Confirm that (8.16) holds for the linear B-spline and its dual from Example 8.1.*
Solution
We use the notation $z = e^{-i\omega}$, so that

$$\tilde{H}(z) = \frac{1}{4}z^{-1} + \frac{1}{2} + \frac{1}{4}z$$

and

$$H(z) = \frac{3}{128}z^{-4} + \frac{-3}{64}z^{-3} + \frac{-1}{8}z^{-2} + \frac{19}{64}z^{-1} + \frac{45}{64} + \frac{19}{64}z + \frac{-1}{8}z^2 + \frac{-3}{64}z^3 + \frac{3}{128}z^4$$

We first compute

$$\tilde{H}(z)H(z^{-1}) = \left(\frac{1}{4}z^{-1} + \frac{1}{2} + \frac{1}{4}z\right)$$

$$\cdot \left(\frac{3}{128}z^4 - \frac{3}{64}z^3 - \frac{1}{8}z^2 + \frac{19}{64}z + \frac{45}{64} + \frac{19}{64}z^{-1} - \frac{-1}{8}z^{-2} - \frac{3}{64}z^{-3} + \frac{3}{128}z^{-4}\right)$$

$$= \frac{3}{512}z^{-5} - \frac{25}{512}z^{-3} + \frac{75}{256}z^{-1} + \frac{1}{2} + \frac{75}{256}z - \frac{25}{512}z^3 + \frac{3}{512}z^5 \qquad (8.17)$$

and a similar computation for $\tilde{H}(-z)H(-z^{-1})$ yields

$$\tilde{H}(-z)H(-z^{-1}) = \left(-\frac{1}{4}z^{-1} + \frac{1}{2} - \frac{1}{4}z\right)$$

$$\cdot \left(\frac{3}{128}z^{-4} - \frac{3}{64}z^{-3} - \frac{1}{8}z^{-2} - \frac{19}{64}z^{-1} + \frac{45}{64} - \frac{19}{64}z - \frac{-1}{8}z^2 + \frac{3}{64}z^3 + \frac{3}{128}z^4\right)$$

$$= -\frac{3}{512}z^{-5} + \frac{25}{512}z^{-3} - \frac{75}{256}z^{-1} + \frac{1}{2} - \frac{75}{256}z + \frac{25}{512}z^3 - \frac{3}{512}z^5 \quad (8.18)$$

Combining (8.17) and (8.18) gives

$$\tilde{H}(z)H(z^{-1}) + \tilde{H}(-z)H(-z^{-1}) = 1$$

so the conjecture is valid for this example. ∎

The linear B-spline example above gives encouraging evidence that (8.15) is correct! Now to confirm that it works in general, we state and prove the following theorem.

Theorem 8.2 (Duality Symbol Condition) *Consider the symbols*

$$H(\omega) = \frac{1}{\sqrt{2}} \sum_{k \in \mathbb{Z}} h_k e^{-ik\omega} \qquad and \qquad \tilde{H}(\omega) = \frac{1}{\sqrt{2}} \sum_{k \in \mathbb{Z}} \tilde{h}_k e^{-ik\omega}$$

The symbols satisfy

$$\tilde{H}(\omega)\overline{H(\omega)} + \tilde{H}(\omega + \pi)\overline{H(\omega + \pi)} = 1 \qquad (8.19)$$

for all $\omega \in \mathbb{R}$ if and only if for all $k \in \mathbb{Z}$

$$\sum_{\ell \in \mathbb{Z}} \tilde{h}_\ell h_{\ell - 2k} = \delta_{k,0} \qquad (8.20)$$

∎

Proof: We give the proof of (8.19) assuming the filter coefficient condition (8.20). The proof in the other direction in considered in Problem 8.10.

Assume that (8.20) holds for all $k \in \mathbb{Z}$. To prove (8.19) we first rearrange the left-hand side as follows:

$$\tilde{H}(\omega)\overline{H(\omega)} + \tilde{H}(\omega + \pi)\overline{H(\omega + \pi)}$$

$$= \frac{1}{2}\left(\sum_{k\in\mathbb{Z}}\tilde{h}_k e^{-ik\omega}\right)\left(\sum_{j\in\mathbb{Z}}h_j e^{ij\omega}\right) + \frac{1}{2}\left(\sum_{k\in\mathbb{Z}}\tilde{h}_k e^{-ik(\omega+\pi)}\right)\left(\sum_{j\in\mathbb{Z}}h_j e^{ij(\omega+\pi)}\right)$$

$$= \frac{1}{2}\sum_{k\in\mathbb{Z}}\sum_{j\in\mathbb{Z}}\tilde{h}_k h_j e^{ij\omega}e^{-ik\omega} + \frac{1}{2}\sum_{k\in\mathbb{Z}}\sum_{j\in\mathbb{Z}}\tilde{h}_k h_j e^{ij\omega}e^{-ik\omega}(-1)^{k-j}$$

$$= \frac{1}{2}\sum_{k\in\mathbb{Z}}\sum_{j\in\mathbb{Z}}\tilde{h}_k h_j e^{-i(k-j)\omega}\left(1 + (-1)^{k-j}\right) \tag{8.21}$$

using $e^{-ik\pi} = (-1)^k$ in the next-to-last step. We now make the substitution $m = k-j$ in (8.21) and note the inner sum still runs over all integers m and $j = k - m$. We have

$$\tilde{H}(\omega)\overline{H(\omega)} + \tilde{H}(\omega + \pi)\overline{H(\omega + \pi)} = \frac{1}{2}\sum_{k\in\mathbb{Z}}\sum_{m\in\mathbb{Z}}\tilde{h}_k h_{k-m}e^{-im\omega}\left(1 + (-1)^m\right)$$

Now when m is odd, $1 + (-1)^m = 0$, and when m is even, $1 + (-1)^m = 2$, so we can discard the odd terms above and replace m by $2m$ to write

$$\tilde{H}(\omega)\overline{H(\omega)} + \tilde{H}(\omega + \pi)\overline{H(\omega + \pi)} = \sum_{k\in\mathbb{Z}}\sum_{m\in\mathbb{Z}}\tilde{h}_k h_{k-2m}e^{-2im\omega}$$

$$= \sum_{m\in\mathbb{Z}}\left(\sum_{k\in\mathbb{Z}}\tilde{h}_k h_{k-2m}\right)e^{-2im}$$

Using (8.20), we see that the inner sum above is $\delta_{m,0}$, so that

$$\tilde{H}(\omega)\overline{H(\omega)} + \tilde{H}(\omega + \pi)\overline{H(\omega + \pi)} = \sum_{m\in\mathbb{Z}}\delta_{m,0}e^{-2im} = 1$$

and the proof is complete. ∎

Combining the last two theorems for $\tilde{\phi}(t)$, $\phi(t)$ satisfying dilation equations gives us three equivalent conditions: the duality of the function pair $\tilde{\phi}(t)$, $\phi(t)$, a symbol condition (8.19), and a filter condition (8.13). Each will be useful in building a biorthogonal multiresolution analysis-like framework.

We have some major tasks ahead of us before examining applications. We next need to determine how to construct the wavelet functions and wavelet filters in the biorthogonal setting, and we need to learn how to design symbols $H(\omega)$ and $\tilde{H}(\omega)$ from (8.19) that are useful in applications.

Filters and Wavelets for Biorthogonal Scaling Functions

Our program is to build two dual multiresolution analysis-like structures, one gener-
ated by $\tilde{\phi}(t)$ and the other by $\phi(t)$ with associated wavelets $\psi(t)$ and $\tilde{\psi}(t)$. How are
these four functions related? What about the associated filters and symbols? These
are the issues we tackle in this subsection.

Before looking at the symbols, let's first consider the situation from the filtering
point of view. The idea is to take a vector $\mathbf{v} \in \mathbb{R}^N$ and decompose it using the scaling
filter $\tilde{\mathbf{h}}$ and an associated wavelet filter $\tilde{\mathbf{g}}$. We can combine these filters in a single
matrix much as we did in the orthogonal setting and define

$$\tilde{W}_N = \begin{bmatrix} \tilde{H} \\ \tilde{G} \end{bmatrix}$$

where $\tilde{H}$ and $\tilde{G}$ are $N/2 \times N$ matrices built from filters $\tilde{\mathbf{h}}$ and $\tilde{\mathbf{g}}$, respectively. The
transformed vector is thus $\tilde{W}_N \mathbf{v}$ in matrix form. We then want *exact* reconstruction
from the dual filters $\mathbf{h}$ and $\mathbf{g}$ and their associated $N/2 \times N$ matrices H and G, which
we combine into a single matrix as

$$W_N = \begin{bmatrix} H \\ G \end{bmatrix}$$

Reconstruction comes from the transpose of W_N, so we need W_N^T and $\tilde{W}_N$ to satisfy

$$W_N^T \left(\tilde{W}_N \mathbf{v} \right) = \mathbf{v}$$

for any vector $\mathbf{v} \in \mathbb{R}^N$. That is, we require the matrices W_N^T and $\tilde{W}_N$ to be inverses.
Equivalently, we write $\tilde{W}_N W_N^T = I_N$ in block matrix form as

$$\begin{bmatrix} \tilde{H} \\ \tilde{G} \end{bmatrix} \cdot \begin{bmatrix} H \\ G \end{bmatrix}^T = I_N$$

Expanding this block matrix product yields

$$\begin{bmatrix} \tilde{H} \\ \tilde{G} \end{bmatrix} \cdot \left[H^T \middle| G^T \right] = \left[\begin{array}{c|c} \tilde{H}H^T & \tilde{H}G^T \\ \hline \tilde{G}H^T & \tilde{G}G^T \end{array} \right] = \left[\begin{array}{c|c} I_{N/2} & 0_{N/2} \\ \hline 0_{N/2} & I_{N/2} \end{array} \right]$$

which we organize in tabular form as

$$\begin{aligned} \tilde{H}H^T &= I_{N/2} & \tilde{H}G^T &= 0_{N/2} \\ \tilde{G}H^T &= 0_{N/2} & \tilde{G}G^T &= I_{N/2} \end{aligned} \tag{8.22}$$

Summing up, for exact reconstruction we need two filter pairs $\mathbf{h}$, $\mathbf{g}$ and $\tilde{\mathbf{h}}$, $\tilde{\mathbf{g}}$,
whose corresponding matrices satisfy (8.22). How does this translate to correspond-
ing relations on the symbols? A clue comes from comparing the matrix formula

$\tilde{H}H^T = I_{N/2}$ in (8.22) with the symbol condition (8.19) from Theorem 8.2. Letting $G(\omega)$ and $\tilde{G}(\omega)$ denote the symbols associated with wavelet filters **g** and **g̃**, respectively, we make the following conjectures on symbol requirements corresponding to the four relations in (8.22), organized in the same tabular format.

$$
\begin{array}{|c|c|}
\hline
\tilde{H}(\omega)\overline{H(\omega)}+\tilde{H}(\omega+\pi)\overline{H(\omega+\pi)}=1 & \tilde{H}(\omega)\overline{G(\omega)}+\tilde{H}(\omega+\pi)\overline{G(\omega+\pi)}=0 \\
\hline
\tilde{G}(\omega)\overline{H(\omega)}+\tilde{G}(\omega+\pi)\overline{H(\omega+\pi)}=0 & \tilde{G}(\omega)\overline{G(\omega)}+\tilde{G}(\omega+\pi)\overline{G(\omega+\pi)}=1 \\
\hline
\end{array}
\tag{8.23}
$$

Reviewing Theorem 5.8 in the orthogonal setting should also remind you of these symbol requirements. We can also construct $\tilde{G}(\omega)$ and $G(\omega)$ from $H(\omega)$ and $\tilde{H}(\omega)$, respectively, if we have symbols $H(\omega)$ and $\tilde{H}(\omega)$ for which (8.19) holds. Proposition 5.11(b) from Section 5.2 provides some motivation here. Since each term of (8.19) contains both $\tilde{H}(\omega)$ and $H(\omega)$, it is natural to believe that $\tilde{G}(\omega)$ and $G(\omega)$ can be expressed in terms of $H(\omega)$ and $\tilde{H}(\omega)$, respectively. We have the following proposition.

Proposition 8.3 (Constructing $\tilde{G}(\omega)$ and $G(\omega)$) *Suppose that symbols $\tilde{H}(\omega)$ and $H(\omega)$ satisfy (8.19). If $\tilde{G}(\omega)$ and $G(\omega)$ are given by*

$$
\boxed{\tilde{G}(\omega) = -e^{-i\omega}\overline{H(\omega+\pi)} \qquad and \qquad G(\omega) = -e^{-i\omega}\overline{\tilde{H}(\omega+\pi)}}
\tag{8.24}
$$

then

$$
\tilde{G}(\omega)\overline{G(\omega)} + \tilde{G}(\omega+\pi)\overline{G(\omega+\pi)} = 1
\tag{8.25}
$$

$$
\tilde{H}(\omega)\overline{G(\omega)} + \tilde{H}(\omega+\pi)\overline{G(\omega+\pi)} = 0
\tag{8.26}
$$

$$
H(\omega)\overline{\tilde{G}(\omega)} + H(\omega+\pi)\overline{\tilde{G}(\omega+\pi)} = 0
\tag{8.27}
$$

∎

Proof: Using (8.24) makes the proof of (8.25) easy! We leave this proof as Problem 8.11. We prove (8.26) and leave the proof of (8.27) as Problem 8.12. Using (8.24) we write

$$
\tilde{H}(\omega)\overline{G(\omega)} = -\tilde{H}(\omega)\overline{e^{-i\omega}\overline{\tilde{H}(\omega+\pi)}} = -e^{i\omega}\tilde{H}(\omega)\tilde{H}(\omega+\pi)
\tag{8.28}
$$

Replacing ω by $\omega+\pi$ in (8.28) and using the fact that $\tilde{H}(\omega)$ is a 2π-periodic function gives

$$
\begin{aligned}
\tilde{H}(\omega+\pi)\overline{G(\omega+\pi)} &= -e^{i(\omega+\pi)}\tilde{H}(\omega+\pi)\tilde{H}(\omega+2\pi) \\
&= -e^{i\omega}e^{i\pi}\tilde{H}(\omega+\pi)\tilde{H}(\omega) \\
&= e^{i\omega}\tilde{H}(\omega)\tilde{H}(\omega+\pi)
\end{aligned}
$$

The last identity is the opposite of (8.28), so adding it to (8.28) establishes (8.26). ∎

The choices (8.24) are by no means unique. Other choices are described in Problem 8.15. Once we have the symbols $\tilde{G}(\omega)$ and $G(\omega)$ defined by (8.24), we can record the associated filter coefficients.

Proposition 8.4 (The Filters $\tilde{g}$ and g) *The filter coefficients $\tilde{g}$ and g associated with the symbols $\tilde{G}(\omega)$ and $G(\omega)$ given in (8.24) are given component-wise by*

$$\boxed{\tilde{g}_k = (-1)^k \tilde{h}_{1-k} \qquad \text{and} \qquad g_k = (-1)^k \tilde{h}_{1-k}} \tag{8.29}$$

■

Proof: We use (8.24) with $H(\omega) = \frac{1}{\sqrt{2}} \sum_{k \in \mathbb{Z}} h_k e^{-ik\omega}$ to write

$$\tilde{G}(\omega) = -e^{-i\omega} \overline{H(\omega + \pi)} = -\frac{1}{\sqrt{2}} e^{-i\omega} \sum_{k \in \mathbb{Z}} h_k e^{-ik(\omega + \pi)}$$

$$= -\frac{1}{\sqrt{2}} e^{-i\omega} \sum_{k \in \mathbb{Z}} (-1)^k h_k e^{ik\omega} = \frac{1}{\sqrt{2}} \sum_{k \in \mathbb{Z}} (-1)^{k+1} h_k e^{-i(1-k)\omega}$$

Make the substitution $m = 1 - k$ in the last identity and note that $k = 1 - m$ so that $k + 1 = 2 - m$. This gives $(-1)^{k+1} = (-1)^{2-m} = (-1)^m$. We have

$$\tilde{G}(\omega) = \frac{1}{\sqrt{2}} \sum_{m \in \mathbb{Z}} (-1)^m h_{1-m} e^{-im\omega}$$

so that $\tilde{g}_k = (-1)^k h_{1-k}$, as desired. The proof establishing the coefficients g_k is identical and thus omitted. ■

In Theorem 8.2 we learned of the interplay between the symbols $\tilde{H}(\omega)$, $H(\omega)$ and the scaling filters $\tilde{h}$ and h. Relations also exist for the symbols $\tilde{G}(\omega)$ and $G(\omega)$.

Proposition 8.5 (Relating Symbols to Filters) *Suppose that symbols $\tilde{G}(\omega)$ and $G(\omega)$ are given by (8.24). Then for all $k \in \mathbb{Z}$,*

$$\tilde{H}(\omega)\overline{G(\omega)} + \tilde{H}(\omega + \pi)\overline{G(\omega + \pi)} = 0 \iff \sum_{j \in \mathbb{Z}} \tilde{h}_{j-2k} g_j = 0 \tag{8.30}$$

and

$$\tilde{G}(\omega)\overline{H(\omega)} + \tilde{G}(\omega + \pi)\overline{H(\omega + \pi)} = 0 \iff \sum_{j \in \mathbb{Z}} \tilde{g}_{j-2k} h_j = 0 \tag{8.31}$$

Furthermore, if $\tilde{H}(\omega)$ and $H(\omega)$ satisfy Theorem 8.2, then for all $k \in \mathbb{Z}$,

$$\tilde{G}(\omega)\overline{G(\omega)} + \tilde{G}(\omega + \pi)\overline{G(\omega + \pi)} = 1 \iff \sum_{j \in \mathbb{Z}} \tilde{g}_j g_{j-2k} = \delta_{k,0}, k \in \mathbb{Z} \tag{8.32}$$

■

Proof: We prove (8.30). The proof for (8.31) is identical and omitted. The proof of (8.32) is left as Problem 8.16.

We start by using (8.24) and write

$$\tilde{H}(\omega)\overline{G(\omega)} = \frac{1}{\sqrt{2}}\sum_{k\in\mathbb{Z}}\tilde{h}_k e^{-ik\omega} \cdot \frac{1}{\sqrt{2}}\overline{\sum_{j\in\mathbb{Z}}g_j e^{-ij\omega}}$$

$$= \frac{1}{2}\sum_{k\in\mathbb{Z}}\tilde{h}_k e^{-ik\omega} \cdot \sum_{j\in\mathbb{Z}}g_j e^{ij\omega}$$

$$= \frac{1}{2}\sum_{k\in\mathbb{Z}}\sum_{j\in\mathbb{Z}}\tilde{h}_k g_j e^{-i(k-j)\omega} \tag{8.33}$$

Replacing ω by $\omega + \pi$ in (8.33) gives

$$\tilde{H}(\omega+\pi)\overline{G(\omega+\pi)} = \frac{1}{2}\sum_{k\in\mathbb{Z}}\sum_{j\in\mathbb{Z}}\tilde{h}_k g_j e^{-i(k-j)(\omega+\pi)}$$

$$= \frac{1}{2}\sum_{k\in\mathbb{Z}}\sum_{j\in\mathbb{Z}}\tilde{h}_k g_j (-1)^{k-j} e^{-i(k-j)\omega} \tag{8.34}$$

Adding (8.33) and (8.34) and combining and interchanging sums gives

$$\tilde{H}(\omega)\overline{G(\omega)} + \tilde{H}(\omega+\pi)\overline{G(\omega+\pi)} = \frac{1}{2}\sum_{j\in\mathbb{Z}}\sum_{k\in\mathbb{Z}}\tilde{h}_k g_j \left(1+(-1)^{k-j}\right) e^{-i(k-j)\omega}$$

Now let $m = j - k$, so that $k = j - m$. Replacing the inner sum variable k with this substitution yields

$$\tilde{H}(\omega)\overline{G(\omega)} + \tilde{H}(\omega+\pi)\overline{G(\omega+\pi)} = \frac{1}{2}\sum_{j\in\mathbb{Z}}\sum_{m\in\mathbb{Z}}\tilde{h}_{j-m} g_j \left(1+(-1)^{-m}\right) e^{im\omega}$$

Now if m is even, the factor $(1+(-1)^{-m})$ is 2; otherwise the factor is zero. Thus we need only even terms in the sum above. Replacing m by $2m$ and interchanging the order of the summations gives

$$\tilde{H}(\omega)\overline{G(\omega)} + \tilde{H}(\omega+\pi)\overline{G(\omega+\pi)} = \sum_{m\in\mathbb{Z}}\left(\sum_{j\in\mathbb{Z}}\tilde{h}_{j-2m} g_j\right) e^{2im\omega} \tag{8.35}$$

If $\sum_{j\in\mathbb{Z}}\tilde{h}_{j-2m}g_j = 0$, certainly the left-hand side of (8.35) is zero. Now assume that the left-hand size of (8.35) is zero. We view the right-hand side of (8.35) as a Fourier series with Fourier coefficients $\sum_{j\in\mathbb{Z}}\tilde{h}_{j-2m}g_j$. Since the Fourier series is zero, it must be that the coefficients are zero, and the result is established. ∎

We can now define the wavelet functions $\tilde{\psi}(t)$ and $\psi(t)$.

Definition 8.5 (The Wavelet Functions $\tilde{\psi}(t)$ and $\psi(t)$) *Suppose that $\tilde{\phi}(t)$ and $\phi(t)$ satisfy the dilation equations given by (8.11). We define the* wavelet functions *$\tilde{\psi}(t)$ and $\psi(t)$ by*

$$\boxed{\tilde{\psi}(t) = \sqrt{2} \sum_{k \in \mathbb{Z}} \tilde{g}_k \tilde{\phi}(2t - k) \qquad and \qquad \psi(t) = \sqrt{2} \sum_{k \in \mathbb{Z}} g_k \phi(2t - k)} \qquad (8.36)$$

where the coefficients $\tilde{g}_k$ and g_k are given by (8.29) from Proposition 8.4. ∎

Let's look at an example of wavelet symbols, coefficients, and functions.

Example 8.5 (Wavelet Symbols, Coefficients, and Functions) *For the linear B-spline $\tilde{\phi}(t) = B_1(t+1)$ and its dual $\phi(t)$ introduced in Example 8.1, find $\tilde{G}(\omega)$, $\tilde{\mathbf{g}}$, and $\tilde{\psi}(t)$.*

Solution

We start with the wavelet filter $\tilde{\mathbf{g}}$. It is constructed from $\mathbf{h}$ and the nonzero filter coefficients for $\mathbf{h}$ are given in (8.4). Using the formula $\tilde{g}_k = (-1)^k h_{1-k}$, we have

$$\tilde{\mathbf{g}} = (\tilde{g}_{-3}, \tilde{g}_{-2}, \tilde{g}_{-1}, \tilde{g}_0, \tilde{g}_1, \tilde{g}_2, \tilde{g}_3, \tilde{g}_4, \tilde{g}_5)$$

$$= \left(-\frac{3\sqrt{2}}{128}, -\frac{3\sqrt{2}}{64}, \frac{\sqrt{2}}{8}, \frac{19\sqrt{2}}{64}, -\frac{45\sqrt{2}}{64}, \frac{19\sqrt{2}}{64}, \frac{\sqrt{2}}{8}, -\frac{3\sqrt{2}}{64}, -\frac{3\sqrt{2}}{128} \right)$$

The symbol is given by

$$\tilde{G}(\omega) = \frac{1}{\sqrt{2}} \sum_{k=-3}^{5} \tilde{g}_k e^{-ik\omega}$$

$$= -\frac{3}{128} e^{-5i\omega} - \frac{3}{64} e^{-4i\omega} + \frac{1}{8} e^{-3i\omega} + \frac{19}{64} e^{-2i\omega} - \frac{45}{64} e^{-i\omega} + \frac{19}{64}$$

$$+ \frac{1}{8} e^{i\omega} - \frac{3}{64} e^{2i\omega} - \frac{3}{128} e^{3i\omega}$$

and $|\tilde{G}(\omega)|$ is plotted in Figure 8.3(a). The wavelet function is

$$\tilde{\psi}(t) = \sqrt{2} \sum_{k=-3}^{5} \tilde{g}_k \tilde{\phi}(2t - k)$$

$$= -\frac{3}{64} \tilde{\phi}(2t + 3) - \frac{3}{32} \tilde{\phi}(2t + 2) + \frac{1}{4} \tilde{\phi}(2t + 1) + \frac{19}{32} \tilde{\phi}(2t) - \frac{45}{32} \tilde{\phi}(2t - 1)$$

$$+ \frac{19}{32} \tilde{\phi}(2t - 2) + \frac{1}{4} \tilde{\phi}(2t - 3) - \frac{3}{32} \tilde{\phi}(2t - 4) - \frac{3}{64} \tilde{\phi}(2t - 5)$$

The wavelet function $\tilde{\psi}(t)$ is plotted in Figure 8.3(b).

The plot for $|\tilde{G}(\omega)|$ is somewhat consistent with what we expect for a wavelet filter (see Figure 5.2 in Section 5.1 and the discussion that precedes it). The maximum that occurs right of $\frac{\pi}{2}$ is undesirable, and in Section 8.3 we design new filters that alleviate

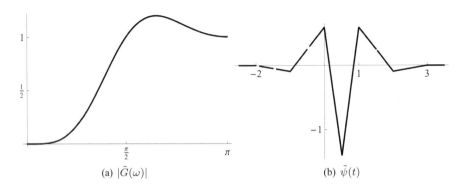

(a) $|\tilde{G}(\omega)|$ (b) $\tilde{\psi}(t)$

Figure 8.3   The modulus of $\tilde{G}(\omega)$ and the wavelet function $\tilde{\psi}(t)$.

this problem. Compare this construction with (5.69) in Example 5.8, where we used an "orthogonalization trick" to create a multiresolution analysis from linear B-splines, but lost compact support. This illustrates some benefits of the new biorthogonal approach! In Problem 8.19 you will find expressions for $G(\omega)$, $\mathbf{g}$, and $\psi(t)$.   ∎

Now that we have the wavelet function $\psi(t)$ from Example 8.5, let's look at one more example illustrating the relationship between the symbols.

Example 8.6 (Symbol Relationships) *Verify (8.26) for the scaling function $\tilde{\phi}(t)$ from Example 8.1 and the wavelet function $\psi(t)$ from Example 8.5.*
Solution
We need to show that $\tilde{H}(\omega)\overline{G(\omega)} + \tilde{H}(\omega + \pi)\overline{G(\omega + \pi)} = 1$. From Example 8.1 we can compute the symbol $\tilde{H}(\omega) = \frac{1}{4}e^{-i\omega} + \frac{1}{2} + \frac{1}{4}e^{i\omega}$. It is easily verified (see Problem 8.17) that $\tilde{H}(\omega) = \cos^2\left(\frac{\omega}{2}\right)$, and from Problem 8.18 we know that the symbol $G(\omega)$ satisfies

$$G(\omega) = \frac{1}{4} - \frac{1}{2}e^{i\omega} + \frac{1}{4}e^{2i\omega} = -e^{i\omega}\sin^2\left(\frac{\omega}{2}\right)$$

Thus

$$\tilde{H}(\omega)\overline{G(\omega)} = -e^{-i\omega}\cos^2\left(\frac{\omega}{2}\right)\sin^2\left(\frac{\omega}{2}\right) \tag{8.37}$$

and

$$\tilde{H}(\omega + \pi)\overline{G(\omega + \pi)} = -e^{-i(\omega+\pi)}\cos^2\left(\frac{\omega + \pi}{2}\right)\sin^2\left(\frac{\omega + \pi}{2}\right)$$

$$= e^{-i\omega}\cos^2\left(\frac{\omega}{2} + \frac{\pi}{2}\right) \cdot \sin^2\left(\frac{\omega}{2} + \frac{\pi}{2}\right)$$

Using the trigonometric identities $\cos\left(\theta + \frac{\pi}{2}\right) = -\sin(\theta)$ and $\sin\left(\theta + \frac{\pi}{2}\right) = \cos(\theta)$, the identity above becomes

$$\tilde{H}(\omega + \pi)\overline{G(\omega + \pi)} = e^{-i\omega}\sin^2\left(\frac{\omega}{2}\right)\cos^2\left(\frac{\omega}{2}\right)$$

This value is opposite the one obtained in (8.37) for $\tilde{H}(\omega)\overline{G(\omega)}$. Thus the sum is zero. ∎

We need notation for the translates and dilations of our wavelet functions.

Definition 8.6 (The Functions $\tilde{\psi}_{j,k}(t)$ and $\psi_{j,k}(t)$) *Suppose that $\tilde{\psi}(t)$ and $\psi(t)$ are the wavelet functions given by Definition 8.5. Then for integers j and k we define*

$$\tilde{\psi}_{j,k}(t) = 2^{j/2}\tilde{\psi}(2^j t - k) \quad and \quad \psi_{j,k}(t) = 2^{j/2}\psi(2^j t - k) \tag{8.38}$$

∎

We can now define the *wavelet spaces* $\tilde{W}_j$ and W_j.

Definition 8.7 (The Wavelet Spaces $\tilde{W}_j$ and W_j) *Let $j, k \in \mathbb{Z}$. If $\tilde{\psi}_{j,k}(t)$ and $\psi_{j,k}(t)$ are given by (8.38), then we define the* wavelet spaces

$$\tilde{W}_j = \text{span}\left\{\tilde{\psi}_{j,k}(t)\right\}_{k\in\mathbb{Z}} \quad and \quad W_j = \text{span}\left\{\psi_{j,k}(t)\right\}_{k\in\mathbb{Z}} \tag{8.39}$$

∎

Our goal now is to develop relations between the dual spaces and wavelet spaces analogous to those detailed in Chapter 5. In particular, we show that the dual spaces can be decomposed as

$$V_{j+1} = V_j + W_j \quad and \quad \tilde{V}_{j+1} = \tilde{V}_j + \tilde{W}_j$$

where $\tilde{V}_j$ and V_j are given in Definition 8.3. Note that the decompositions are not orthogonal: $V_{j+1} = V_j + W_j$ means that any function $f_{j+1}(t) \in V_{j+1}$ can be expressed uniquely as a sum $f_j(t) + g_j(t)$ with $f_j(t) \in V_j$ and $g_j(t) \in W_j$.

We will show that the wavelet functions satisfy (8.1) and are dual to each other:

$$\langle \psi(t-m), \tilde{\psi}(t-k) \rangle = \delta_{m,k}$$

and this will lead to the biorthogonality relations $W_j \perp \tilde{V}_j$ and $\tilde{W}_j \perp V_j$.

These relations allow the multiresolution hierarchies $\{V_j\}_{j\in\mathbb{Z}}$ and $\left\{\tilde{V}_j\right\}_{j\in\mathbb{Z}}$ and their sequences of complement spaces $\{W_j\}_{j\in\mathbb{Z}}$ and $\left\{\tilde{W}_j\right\}_{j\in\mathbb{Z}}$ to "fit together like a giant zipper," as Daubechies describes the structure [20]. Figure 8.4 is a diagram to facilitate a visualization of some of these space relations. This hierarchy of coarser to finer spaces will in turn allow us to create finer resolution projections than the ones in Example 8.2.

We now state several orthogonality relations obeyed by the scaling functions and wavelet functions.

$$
\begin{array}{ccccccc}
& & & \vdots & & & \\
\tilde{V}_3 & \supset & \tilde{W}_3 & \perp & V_3 & = & V_2 + W_2 \\
\cup & & & & \cup & & \\
\tilde{V}_2 & \supset & \tilde{W}_2 & \perp & V_2 & = & V_1 + W_1 \\
\cup & & & & \cup & & \\
\tilde{V}_1 & \supset & \tilde{W}_1 & \perp & V_1 & = & V_0 + W_0 \\
& & & \vdots & & &
\end{array}
$$

Figure 8.4 The space ladders: $\tilde{W}_j \perp V_j$, $V_{j+1} = V_j + W_j$, $V_j \subset V_{j+1}$, $\tilde{V}_j \subset \tilde{V}_{j+1}$.

Theorem 8.3 (Orthogonality Relations for Scaling and Wavelet Functions)

Suppose that $\tilde{\phi}(t)$ and $\phi(t)$ satisfy the duality condition (8.1) and the dilation equations (8.11). If $\tilde{\phi}_{j,k}(t)$, $\phi_{j,k}(t)$ are given by Definition 8.2 and $\tilde{\psi}_{j,k}(t)$, $\psi_{j,k}(t)$ are given by Definition 8.6, then following orthogonality relations hold for all integers j, k, ℓ, and m:

$$
\langle \phi_{j,k}(t), \tilde{\psi}_{j,\ell}(t) \rangle = 0 \tag{8.40}
$$

$$
\langle \tilde{\phi}_{j,k}(t), \psi_{j,\ell}(t) \rangle = 0 \tag{8.41}
$$

$$
\langle \tilde{\psi}_{j,k}(t), \psi_{m,\ell}(t) \rangle = \delta_{j,m} \cdot \delta_{k,\ell} \tag{8.42}
$$

∎

Proof: We first prove (8.40). We have

$$
\begin{aligned}
\langle \phi_{j,k}(t), \tilde{\psi}_{j,\ell}(t) \rangle &= \int_{\mathbb{R}} 2^{j/2} \phi(2^j t - k) 2^{j/2} \tilde{\psi}(2^j - \ell)\, dt \\
&= 2^j \int_{\mathbb{R}} \phi(2^j t - k) \tilde{\psi}(2^j t - \ell)\, dt \\
&= \int_{\mathbb{R}} \phi(u) \tilde{\psi}(u - r)\, du
\end{aligned}
$$

where we have used (1.21) from Proposition 1.7 and set $r = \ell - k$ for convenience. Now rewrite the right-hand side of the identity above using the dilation equations for $\phi(t)$ and $\tilde{\psi}(t)$:

$$
\begin{aligned}
\langle \phi_{j,k}(t), \tilde{\psi}_{j,\ell}(t) \rangle &= \int_{\mathbb{R}} \sqrt{2} \sum_{p \in \mathbb{Z}} h_p \phi(2u - p) \cdot \sqrt{2} \sum_{n \in \mathbb{Z}} \tilde{g}_n \tilde{\phi}(2(u - r) - n)\, du \\
&= 2 \sum_{p \in \mathbb{Z}} \sum_{n \in \mathbb{Z}} h_p \tilde{g}_n \int_{\mathbb{R}} \phi(2u - p) \tilde{\phi}(2u - (2r + n))\, du \tag{8.43}
\end{aligned}
$$

Let's consider the integral on the right-hand side of (8.43). Make the substitution $v = 2u$ to obtain

$$\int_{\mathbb{R}} \phi(2u - p)\tilde{\phi}(2u - (2r + n))\,du = \frac{1}{2}\int_{\mathbb{R}} \phi(v - p)\tilde{\phi}(v - (2r + n))\,dv$$

$$= \frac{1}{2}\delta_{p,2r+n}$$

by the duality relation (8.1). Inserting this value into (8.43) gives

$$\langle \phi_{j,k}(t), \tilde{\psi}_{j,\ell}(t)\rangle = 2\sum_{p\in\mathbb{Z}}\sum_{n\in\mathbb{Z}} h_p \tilde{g}_n \frac{1}{2}\delta_{p,2r+n}$$

$$= \sum_{n\in\mathbb{Z}} h_{2r+n}\tilde{g}_n$$

$$= \sum_{i\in\mathbb{Z}} h_i \tilde{g}_{i-2r}$$

where we have made the substitution $i = 2r + n$ to obtain the last identity. Since $\tilde{\phi}(t)$ and $\phi(t)$ satisfy the duality relation (8.1), it follows that (8.27) holds from Proposition 8.3. This in turn allows us to use (8.31) from Proposition 8.5 to infer that the right-hand side of the last identity is zero. This establishes (8.40).

The proof of (8.41) is nearly identical to the proof above and is thus omitted. To establish (8.42) we consider the two cases $j = m$ and $j \neq m$. The $j = m$ case is outlined in Problem 8.23. Without loss of generality, assume that $j < m$. Then $\tilde{\psi}_{j,k}(t) \in \tilde{W}_j \subset \tilde{V}_{j+1} \subset \tilde{V}_{j+2}$ by the dilation equation in Definition 8.5. Continuing in this manner, we see that

$$\tilde{\psi}_{j,k}(t) \in \tilde{V}_{j+1} \subset \tilde{V}_{j+2} \subset \cdots \subset \tilde{V}_m$$

So $\tilde{\psi}_{j,k}(t)$ is a linear combination of functions the $\tilde{\phi}_{m,\ell}, \ell \in \mathbb{Z}$. But from (8.40) we know that the inner product of $\tilde{\psi}_{j,k}(t)$ with each of the $\tilde{\phi}_{m,\ell}$ is zero.

Putting the cases together, we see that $\langle \tilde{\psi}_{j,k}(t), \psi_{m,\ell}(t)\rangle$ will be zero unless $m = j$, and in this case $\langle \tilde{\psi}_{j,k}(t), \psi_{m,\ell}(t)\rangle$ will be zero unless we also have $k = \ell$. This can be written as $\langle \tilde{\psi}_{j,k}(t), \psi_{m,\ell}(t)\rangle = \delta_{m,j} \cdot \delta_{n,k}$, which is (8.42). ∎

A remark is in order. We have not shown that

$$\langle \phi_{j,k}(t), \tilde{\psi}_{j,\ell}(t)\,dt = 0$$

In fact, using a CAS in Problem 8.22, you can show that

$$\int_{\mathbb{R}} \phi(t)\psi(t)\,dt \neq 0$$

Thus V_j and W_j may not be orthogonal in general.

We have the immediate corollary to Theorem 8.3.

Corollary 8.1 (Orthogonality of Spaces) *We have*

$$V_j \perp \tilde{W}_j \qquad and \qquad \tilde{V}_j \perp W_j \qquad (8.44)$$

∎

Proof: The proof is left as Problem 8.24. ∎

The last result we have yet to prove is the fact that $V_{j+1} = V_j + W_j$ and $\tilde{V}_{j+1} = \tilde{V}_j + \tilde{W}_j$. We have the following theorem.

Theorem 8.4 (Sums of Spaces) *For V_j, $\tilde{V}_j$ given in Definition 8.3 and W_j, $\tilde{W}_j$ given in Definition 8.7, we have*

$$\boxed{V_{j+1} = V_j + W_j \qquad and \qquad \tilde{V}_{j+1} = \tilde{V}_j + \tilde{W}_j} \qquad (8.45)$$

∎

Proof: The proof that $V_j + W_j \subset V_{j+1}$ is straightforward since the basis functions $\phi_{j,k}(t)$, $\psi_{j,k}(t)$ for V_j, W_j, respectively, can be expressed as linear combinations of $\phi_{j+1,k}(t) \in V_{j+1}$ via the dilation equations. The same is true for the dual spaces. The proof that $V_{j+1} \subset V_j + W_j$ is quite technical and beyond the scope of this course. The interested reader is referred to Walnut [62]. ∎

Suppose that we have functions $\tilde{\phi}(t)$ and $\phi(t)$ that satisfy the biorthogonality condition (8.1), dilation equations (8.11) and Theorem 8.2. Then we can create the wavelet functions $\tilde{\psi}(t)$ and $\psi(t)$ that satisfy Theorem 8.3 and nest spaces $V_j \subset V_{j+1}$ and $\tilde{V}_j \subset \tilde{V}_{j+1}$ and wavelet spaces W_j, $\tilde{W}_j$ that satisfy Theorem 8.4. We will also assume that $\{\phi(t - k)\}_{k \in \mathbb{R}}$ and $\{\tilde{\phi}(t - k)\}_{k \in \mathbb{Z}}$ form Riesz bases for V_0, $\tilde{V}_0$, respectively. We mentioned briefly Riesz bases in Section 8.1. You can view Riesz bases as orthogonal or "nearly orthogonal" in a certain sense; we discuss them in more detail in Section 8.6. Then we have all the structure we need to proceed in the new setting. We have the following two definitions.

Definition 8.8 (Biorthogonal Scaling Functions) *If two functions $\tilde{\phi}(t)$ and $\phi(t)$ satisfy the biorthogonality condition (8.1), can be constructed via dilation equations (8.11), and satisfy Theorem 8.2, then we say that $\tilde{\phi}(t)$ and $\phi(t)$ are biorthogonal scaling functions.* ∎

Here is a formal definition generalizing a multiresolution analysis to our biorthogonal setting.

Definition 8.9 (Dual (Biorthogonal) Muliresolution Analysis) *Suppose that $\{V_j\}$ and $\{\tilde{V}_j\}$ are sequences of nested subspaces of $L^2(\mathbb{R})$, $V_j \subset V_{j+1}$ and $\tilde{V}_j \subset \tilde{V}_{j+1}$, and that each satisfies the criterion (5.1) from Definition 5.1 of a multiresolution analysis of $L^2(\mathbb{R})$. If $\{\phi(t - k)\}_{k \in \mathbb{Z}}$ and $\{\tilde{\phi}(t - k)\}_{k \in \mathbb{Z}}$ are Riesz bases of V_0 and $\tilde{V}_0$, respectively, then we say $\{V_j\}$ and $\{\tilde{V}_j\}$ together form a dual multiresolution analysis or a biorthogonal multiresolution analysis of $L^2(\mathbb{R})$.* ∎

Function Decomposition

The decompositions possible for a dual multiresolution analysis are of interest in applications. We next illustrate the basic idea with the linear B-spline $\tilde{\phi}(t) = B_1(t + 1)$ and its dual from Example 8.1.

Example 8.7 (Function Decomposition) *Decompose*

$$f(t) = c_1 \tilde{\phi}(2t - 1) \tag{8.46}$$

in $\tilde{V}_1$ into its components in $\tilde{V}_0$ and $\tilde{W}_0$.
Solution
  By Theorem 8.4, $f(t) \in \tilde{V}_0 + \tilde{W}_0$, so we can write

$$f(t) = \sum_{k \in \mathbb{Z}} a_k \tilde{\phi}(t - k) + \sum_{\ell} b_\ell \tilde{\psi}(t - \ell) \tag{8.47}$$

for some a_k and b_ℓ. We can find the coefficients a_k using the duality of $\phi(t)$ and $\tilde{\phi}(t)$. We first multiply (8.47) by $\phi(t - n)$ and then integrate over $\mathbb{R}$ to obtain

$$\int_{\mathbb{R}} \left(\sum_{k \in \mathbb{Z}} a_k \tilde{\phi}(t - k)\phi(t - n) + \sum_{\ell \in \mathbb{Z}} b_\ell \tilde{\psi}(t - \ell)\phi(t - n) \right) dt$$

$$= \sum_{k \in \mathbb{Z}} a_k \int_{\mathbb{R}} \tilde{\phi}(t - k)\phi(t - n) \, dt + \sum_{\ell \in \mathbb{Z}} b_\ell \int_{\mathbb{R}} \tilde{\psi}(t - \ell)\phi(t - n) \, dt$$

$$= \sum_{k \in \mathbb{R}} a_k \delta_{k,n} + \sum_{\ell \in \mathbb{Z}} b_\ell \cdot 0$$

since the first integral in the identity above is the biorthogonality relation (8.1) and the second integral is the inner product of orthogonal functions (see Theorem 8.3). So the result of this computation is a_n, and it must be the same as multiplying (8.46) by $\phi(t - n)$ and integrating over $\mathbb{R}$. We have

$$a_n = \int_{\mathbb{R}} c_1 \tilde{\phi}(2t - 1)\phi(t - n) \, dt \tag{8.48}$$

Next we use the dilation equation (8.11) to write

$$\phi(t - n) = \sqrt{2} \sum_{\ell \in \mathbb{Z}} h_\ell \phi(2(t - n) - \ell) = \sqrt{2} \sum_{\ell \in \mathbb{Z}} h_\ell \phi(2t - (2n + \ell)) \, dt$$

and substitute this into (8.48) to obtain, with the aid of duality

$$a_n = \sqrt{2}\,c_1 \sum_{\ell \in \mathbb{Z}} h_\ell \int_{\mathbb{R}} \phi(2t - (2n + \ell))\tilde{\phi}(2t - 1)\,dt$$

$$= \frac{\sqrt{2}}{2} c_1 \sum_{\ell \in \mathbb{Z}} h_\ell \int_{\mathbb{R}} \phi(u - (2n + \ell))\tilde{\phi}(u - 1)\,du$$

$$= \frac{\sqrt{2}}{2} c_1 \sum_{\ell \in \mathbb{Z}} h_\ell \delta_{2n+\ell,1}$$

$$= \frac{\sqrt{2}}{2} c_1 h_{1-2n}$$

We use a similar approach to find the coefficients b_k using the duality of $\psi(t)$ and $\tilde{\psi}(t)$. We multiply (8.47) by $\psi(t - n)$ and then integrate over $\mathbb{R}$ to obtain

$$\int_{\mathbb{R}} \left(\sum_{k \in \mathbb{Z}} a_k \,\tilde{\phi}(t - k)\psi(t - n) + \sum_{\ell \in \mathbb{Z}} b_\ell \tilde{\psi}(t - \ell)\psi(t - n) \right) dt$$

$$= \sum_{k \in \mathbb{Z}} a_k \int_{\mathbb{R}} \tilde{\phi}(t - k)\psi(t - n)\,dt + \sum_{\ell \in \mathbb{Z}} b_\ell \int_{\mathbb{R}} \tilde{\psi}(t - \ell)\psi(t - n)\,dt$$

$$= \sum_{k \in \mathbb{R}} a_k \cdot 0 + \sum_{\ell \in \mathbb{Z}} b_\ell \delta_{\ell,n} = b_n$$

We set this equal to the integral over $\mathbf{R}$ of the product of $\psi(t - n)$ and (8.46) to see that

$$b_n = \int_{\mathbb{R}} c_1 \tilde{\phi}(2t - 1)\psi(t - n)\,dt$$

Next we use the dilation equation (8.36) to write

$$\psi(t - n) = \sqrt{2} \sum_{\ell \in \mathbb{Z}} g_\ell \phi(2t - (2n + \ell))$$

and with the aid of duality we have

$$b_n = \sqrt{2}\,c_1 \sum_{\ell \in \mathbb{Z}} g_\ell \int_{\mathbb{R}} \phi(2t - (2n + \ell))\tilde{\phi}(2t - 1)\,dt = \frac{\sqrt{2}}{2} c_1 g_{1-2n}$$

Thus we can write $f(t)$ as

$$f(t) = \frac{\sqrt{2}}{2} c_1 \left(\sum_{k \in \mathbb{Z}} h_{1-2k}\tilde{\phi}(t - k) + \sum_{\ell \in \mathbb{Z}} g_{1-2\ell}\tilde{\psi}(t - \ell) \right)$$

The only nonzero filter coefficients for the first sum occur when $k = -1, 0, 1, 2$ and the only nonzero filter coefficient for the second term is when $\ell = 0$. We have

$$f(t) = \frac{\sqrt{2}}{2} c_1 \left(\sum_{k=-1}^{2} h_{1-2k} \tilde{\phi}(t-k) + g_1 \tilde{\psi}(t) \right)$$

$$= \frac{\sqrt{2}}{2} c_1 \left(h_3 \tilde{\phi}(t+1) + h_1 \tilde{\phi}(t) + h_{-1} \tilde{\phi}(t-1) + h_{-3} \tilde{\phi}(t-2) + g_1 \tilde{\psi}(t) \right)$$

$$= c_1 \left(-\frac{3}{64} \tilde{\phi}(t-1) + \frac{19}{64} \tilde{\phi}(t) + \frac{19}{64} \tilde{\phi}(t-1) - \frac{3}{64} \tilde{\phi}(t-2) - \frac{1}{2} \tilde{\psi}(t) \right)$$

∎

The topic of biorthogonal decomposition is explored further in Section 8.4.

Biorthogonal Filters

We are now in a position to formally define biorthogonal filters. We are particularly interested in constructing pairs of scaling filters $\tilde{\mathbf{h}}$, $\mathbf{h}$ and wavelet filters $\tilde{\mathbf{g}}$, $\mathbf{g}$ for applications. As we saw in Section 5.1, we need the symbols $H(\omega)$, $\tilde{H}(\omega)$ of the scaling filters to satisfy $H(\pi) = \tilde{H}(\pi) = 0$. We naturally want these symbols to satisfy the condition

$$\tilde{H}(\omega)\overline{H(\omega)} + \tilde{H}(\omega + \pi)\overline{H(\omega + \pi)} = 1$$

Evaluating this identity at $\omega = 0$ gives $\tilde{H}(0) \cdot \overline{H(0)} = 1$. As a normalization, we will insist that $\tilde{H}(0) = H(0)$, which forces $\tilde{H}(0) = H(0) = \pm 1$. Since $H(0)$ and $\tilde{H}(0)$ are the sums of our scaling filters divided by $\sqrt{2}$, we will choose to have the sum be positive. Here is the formal definition.

Definition 8.10 (Biorthogonal Filter Pair) *Suppose that $\tilde{\mathbf{h}}$, $\mathbf{h}$ are filters whose symbols satisfy*

$$\tilde{H}(\omega)\overline{H(\omega)} + \tilde{H}(\omega + \pi)\overline{H(\omega + \pi)} = 1 \qquad (8.49)$$

with

$$H(\pi) = \tilde{H}(\pi) = 0 \qquad and \qquad \tilde{H}(0) = H(0) = 1$$

Then we say that $\tilde{\mathbf{h}}$ and $\mathbf{h}$ are a biorthogonal filter pair.

∎

Many authors refer to such filters as *quadrature mirror filters*, which has significance in the engineering literature.

In the following example we design a biorthogonal filter pair. We pick $\tilde{\mathbf{h}}$ to be the filter associated with the linear B-spline function $\tilde{\phi}(t) = B_1(t+1)$ and try to construct a shorter filter $\mathbf{h}$ than was used in Example 8.1. We insist that this new filter $\mathbf{h}$ is symmetric about $k = 0$.

Example 8.8 (A Biorthogonal Filter Pair) *Choose* $\tilde{\mathbf{h}}$ *to be the filter for the linear B-spline function* $\tilde{\phi}(t) = B_1(t+1)$ *and find a symmetric, length 5 filter* $\mathbf{h} = (h_2, h_1, h_0, h_1, h_2)$ *so that* $\tilde{\mathbf{h}}, \mathbf{h}$ *form a biorthogonal filter pair.*
Solution
We know from Example 8.1 that $\tilde{h}_{-1} = \tilde{h}_1 = \frac{\sqrt{2}}{4}$ *and* $\tilde{h}_0 = \frac{\sqrt{2}}{2}$. *The beauty of picking one of the filters is that the resulting system we must solve to find the other filter is linear!*
We use Theorem 8.2 and write

$$\delta_{k,0} = \sum_{\ell \in \mathbb{Z}} \tilde{h}_\ell h_{\ell - 2k}$$

$$= \frac{\sqrt{2}}{4} h_{-1-2k} + \frac{\sqrt{2}}{2} h_{-2k} + \frac{\sqrt{2}}{4} h_{1-2k}$$

$$= \frac{\sqrt{2}}{4} h_{1+2k} + \frac{\sqrt{2}}{2} h_{2k} + \frac{\sqrt{2}}{4} h_{1-2k}$$

Since $h_{-2} = h_2$, $h_{-1} = h_1$, *and* h_0 *are the only nonzero filter terms in* $\mathbf{h}$, *the identity above yields two equations:*

$$\sqrt{2}\left(\frac{1}{4}h_1 + \frac{1}{2}h_0 + \frac{1}{4}h_1\right) = 1$$

$$\sqrt{2}\left(\frac{1}{2}h_2 + \frac{1}{4}h_1\right) = 0$$

The condition $H(\pi) = 0$ *adds the equation*

$$h_{-2} - h_{-1} + h_0 - h_1 + h_2 = h_0 - 2h_1 + 2h_2 = 0$$

Solving this linear system gives $h_0 = \frac{3\sqrt{2}}{4}$, $h_1 = \frac{\sqrt{2}}{4}$ *and* $h_2 = -\frac{\sqrt{2}}{8}$. *You will verify this solution in Problem 8.26. Thus we have the symmetric biorthogonal filter pair*

$$\mathbf{h} = \sqrt{2}\left(-\frac{1}{8}, \frac{1}{4}, \frac{3}{4}, \frac{1}{4}, -\frac{1}{8}\right) \quad and \quad \tilde{\mathbf{h}} = \sqrt{2}\left(\frac{1}{4}, \frac{1}{2}, \frac{1}{4}\right) \tag{8.50}$$

which is often called the $(5,3)$ *biorthogonal spline filter pair.* ∎

Biorthogonal spline filter pairs as well as more general methods for constructing biorthogonal filter pairs are discussed in Section 8.3.

PROBLEMS

Note: *A computer algebra system and the software on the course Web site will be useful for some of these problems. See the Preface for more details.*

8.9 For $f(t) = c_{-5}\phi(2t+5)$, where $\phi(t)$ is as defined in Example 8.1, write out the decomposition of $f(t)$ into its components in V_0 and W_0.

8.10   Prove the converse of Theorem 8.2. (*Hint:* From the proof of Theorem 8.2, we know that

$$\tilde{H}(\omega)\overline{H(\omega)} + \tilde{H}(\omega + \pi)\overline{H(\omega + \pi)} = \sum_{m \in \mathbb{Z}} \left(\sum_{k \in \mathbb{Z}} \tilde{h}_k h_{k-2m} \right) e^{-2im}$$

By assumption, the sum on the left is 1 and this constant function can also be viewed as a Fourier series $P(\omega)$ with $p_0 = 1$ and $p_k = 0$, for $k \neq 0$. Compare these Fourier coefficients to those on the right-hand side of the identity above.)

8.11   Establish (8.25) from Proposition 8.3 using (8.24). (*Hint:* The proof is similar to that of Theorem 8.2.)

8.12   Prove (8.27) from Proposition 8.3.

8.13   Suppose that $\tilde{H}(\omega)$, $H(\omega)$ satisfy (8.19) and $\tilde{G}(\omega)$, $G(\omega)$ are given by (8.24). Express equations (8.19) and (8.25)–(8.27) in terms of $z = e^{-i\omega}$.

8.14   Use Problem 8.13 to rewrite $G(\omega)$ and $\tilde{H}(\omega)$ from Example 8.6 in terms of $z = e^{-i\omega}$. Use your results to verify the symbol relationship (8.26).

8.15   We can add some flexibility to the wavelet symbol definitions (8.24) by using the more general versions

$$G(\omega) = e^{-i(S\omega + b)}\overline{\tilde{H}(\omega + \pi)} \qquad \text{and} \qquad \tilde{G}(\omega) = e^{-i(S\omega + b)}\overline{H(\omega + \pi)}$$

for odd integer S.

(a) Show that if we use these symbols instead of those given by (8.24), the conclusions of Proposition 8.3 still hold.

(b) Show that the wavelet filter coefficients are

$$g_k = -e^{ib}(-1)^k \tilde{h}_{S-k} \qquad \text{and} \qquad \tilde{g}_k = -e^{ib}(-1)^k h_{S-k}$$

(c) Since we insist on real-valued filters in this book, show that the only possible values for b are 0 or π. What are the effects of choosing each value?

(c) What is the effect of choosing different values for S?

8.16   Prove (8.32) from Proposition 8.5.

8.17   Show that

$$\cos^2\left(\frac{\omega}{2}\right) = \frac{1}{4}e^{-i\omega} + \frac{1}{2} + \frac{1}{4}e^{i\omega}$$

8.18   Show that

$$-e^{i\omega}\sin^2\left(\frac{\omega}{2}\right) = \frac{1}{4} - \frac{1}{2}e^{i\omega} + \frac{1}{4}e^{2i\omega}$$

8.19   Find $G(\omega)$, **g**, and $\psi(t)$ for the scaling functions given in Example 8.5. Use a CAS to plot $|G(\omega)|$ and $\psi(t)$.

8.20   Show that the support of $\tilde{\psi}(t)$ from Problem 8.19 is $\overline{\text{supp}(\tilde{\psi})} = [-2,3]$.

8.21   Consider $\phi(t)$ and $\psi(t)$ from Example 8.5.

(a) Show that $\psi(t)$ is symmetric about $t = \frac{1}{2}$.

(b) Find the support of $\psi(t)$.

8.22   Consider $\phi(t)$ and $\psi(t)$ from Example 8.5.

(a) Use a support argument to show that $\int_{\mathbb{R}} \psi(t)\phi(t - k)\,dt$ is automatically zero for all but a finite number of integer values. For what k values do the supports of $\psi(t)$ and $\phi(t - k)$ overlap?

(b) Use a CAS to compute

$$\int_{\mathbb{R}} \psi(t)\phi(t - k)\,dt$$

for the relevant k values found in part (a). What can you conclude about V_0 and W_0 for this example?

8.23   Prove (8.42) when $j = m$ to complete the proof of Theorem 8.3. The following steps will help you organize your work.

(a) Use the dilation equations for $\tilde{\psi}(t)$ and $\psi(t)$ to write the inner product. Make the substitution $u = 2^j t$ in the integral to obtain

$$\langle \psi_{j,k}(t), \tilde{\psi}_{m,\ell}(t) \rangle = \sum_{k \in \mathbb{Z}} \sum_{\ell \in \mathbb{Z}} g_k \tilde{g}_\ell \int_{\mathbb{R}} \phi(u - k)\tilde{\phi}(u - \ell)\,du$$

(b) Use the duality relation (8.1) to replace the integral in the identity above with $\delta_{k,\ell}$.

(c) From part (b) we can infer that the inner product is zero when $k \neq \ell$. Use (8.32) from Proposition 8.5 to infer the inner product equals 1 when $k = \ell$. Thus the inner product is $\delta_{k,\ell}$ when $j = m$.

8.24   Prove Corollary 8.1.

8.25   In a dual multiresolution analysis, show that $W_m \subset V_j$ whenever $m < j$.

8.26   Solve the system of equations in Example 8.8 and confirm the biorthogonal filter pair (8.50).

8.27   For the filter $\mathbf{h}$ in Example 8.8, use a CAS to find and plot $|H(\omega)|$, $|G(\omega)|$, $\phi(t)$, and $\psi(t)$. Find the support of $\phi(t)$ and $\psi(t)$.

8.3   BIORTHOGONAL SPLINE FILTER PAIRS AND THE CDF97 FILTER PAIR

In Section 8.2 we constructed a symmetric biorthogonal filter pair in Example 8.8 by creating and then solving a system of equations. In this section we outline Daubechies' method for constructing a family of symmetric *biorthogonal spline filter pairs*. In this construction, the filter $\tilde{\mathbf{h}}$ comes from a B-spline function. The filter $\mathbf{h}$ comes from the symbol $H(\omega)$ that Daubechies constructed to satisfy the conditions of Definition 8.10. We conclude this section with a discussion of a general method for defining $\tilde{H}(\omega)$ and $H(\omega)$ that holds the biorthogonal spline filter pair as a special case. These symbols were derived by Cohen, Daubechies and Feauveau in [13] and are so named the *CDF biorthogonal filter pairs*.

Symmetric Filters and Spline Filters

We begin with a formal definition of symmetric filters.

Definition 8.11 (Symmetric Filter) *Let* $\mathbf{h} = (h_\ell, \ldots, h_L)$ *be a finite-length filter with length* $N = L - \ell + 1$ *and* $h_k = 0$ *for* $k < \ell$ *and* $k > L$. *We say that* $\mathbf{h}$ *is a symmetric filter if* $h_k = h_{-k}$ *when* N *is odd and* $h_k = h_{1-k}$ *if* N *is even.*   ∎

Note that if N is odd, the filter is symmetric about $k = 0$ and $\ell = -L$. If N is even, the filter is symmetric about $k = \frac{1}{2}$ and $\ell = -L + 1$.

The Haar filter $\mathbf{h} = (h_0, h_1) = \left(\frac{\sqrt{2}}{2}, \frac{\sqrt{2}}{2} \right)$ is symmetric about $k = \frac{1}{2}$ and the filters in Example 8.8 both have odd length and are symmetric about $k = 0$.

The Haar scaling filter comes from the constant scaling function $\phi(t) = \sqcap(t) = B_0(t)$, and the filter $\tilde{\mathbf{h}}$ from Example 8.8 is derived from the linear B-spline function $\tilde{\phi}(t) = B_1(t+1)$. We can find symmetric filters associated with the B-splines $B_n(t)$.

Recall from (5.71) in Section 5.3 that the B-spline $B_n(t)$ satisfies a dilation equation with symbol $S(\omega) = \left(\frac{1+e^{-i\omega}}{2} \right)^{n+1}$, and from Section 2.4 we know that $B_n(t)$ is symmetric about $t = (n+1)/2$. To create symmetric filters we will need to center the spline about $t = 0$ (n odd) or $t = 1/2$ (n even).

In the case $n = 2\tilde{\ell} - 1$ odd, we translate $B_n(t)$ by $t = \frac{n+1}{2} = \tilde{\ell}$ units to the left so that it will be symmetric about $t = 0$. In this case, we can use the translation rule from Section 2.2 to show that (see Problem 8.28) the resulting symbol is

$$\tilde{H}(\omega) = e^{i\tilde{\ell}\omega} S(\omega) = e^{i\tilde{\ell}\omega} \left(\frac{1 + e^{-i\omega}}{2} \right)^{n+1} \tag{8.51}$$

Since the translated B-spline is an even function, its Fourier series must be a real-valued function (see Problem 2.11 in Section 2.2) and in fact, a series involving $\cos(t)$. We can write

$$\frac{1 + e^{-i\omega}}{2} = e^{-i\omega/2} \left(\frac{e^{i\omega/2} + e^{-i\omega/2}}{2} \right) = e^{-i\omega/2} \cos\left(\frac{\omega}{2} \right) \tag{8.52}$$

Inserting this identity into (8.51) and noting that $n + 1 = 2\tilde{\ell}$ gives

$$
\begin{aligned}
\tilde{H}(\omega) &= e^{i\tilde{\ell}\omega}\left(e^{-i\omega/2}\cos\left(\frac{\omega}{2}\right)\right)^{2\tilde{\ell}} \\
&= e^{i\tilde{\ell}\omega} \cdot e^{-i\tilde{\ell}\omega}\cos^{n+1}\left(\frac{\omega}{2}\right) \\
&= \cos^{n+1}\left(\frac{\omega}{2}\right)
\end{aligned}
\tag{8.53}
$$

In the case $n = 2\tilde{\ell}$ even, we translate $B_n(t)$ by $t = \frac{n}{2} = \tilde{\ell}$ units left so that it is symmetric about $t = \frac{1}{2}$. You will show in Problem 8.29 that in this case, the symbol is

$$
\begin{aligned}
\tilde{H}(\omega) &= e^{i\tilde{\ell}\omega}\left(\frac{1 + e^{-i\omega}}{2}\right)^{n+1} \\
&= e^{-i\omega/2}\cos^{n+1}\left(\frac{\omega}{2}\right)
\end{aligned}
\tag{8.54}
$$

We summarize the preceding discussion with the following proposition.

Proposition 8.6 (The Symbol for the Centered B–spline) *Suppose that $B_n(t)$ is a B-spline. If we translate $B_n(t)$ by $\frac{n+1}{2}$ units left if n is odd or $\frac{n}{2}$ units left if n is even, then the symbol for the centered spline is*

$$
\tilde{H}(\omega) = \begin{cases} \cos^{n+1}\left(\frac{\omega}{2}\right), & n \text{ odd} \\ e^{-i\omega/2}\cos^{n+1}\left(\frac{\omega}{2}\right), & n \text{ even} \end{cases}
\tag{8.55}
$$

∎

Let's return to (8.52) and rewrite it as

$$
\cos\left(\frac{\omega}{2}\right) = \frac{1}{2}e^{i\omega/2}\left(1 + e^{-i\omega}\right)
$$

Suppose that $n = 2\tilde{\ell} - 1$ is odd. Then $n + 1 = 2\tilde{\ell}$ and we raise both sides of the previous identity to the power $n + 1 = 2\tilde{\ell}$ and use the binomial theorem to obtain the expanded version of $\tilde{H}(\omega)$:

$$\tilde{H}(\omega) = \cos^{2\tilde{\ell}}\left(\frac{\omega}{2}\right) = \frac{1}{2^{n+1}}e^{i\tilde{\ell}\omega}\left(1 + e^{-i\omega}\right)^{2\tilde{\ell}}$$

$$= \frac{1}{2^{n+1}}e^{i\tilde{\ell}\omega}\sum_{k=0}^{2\tilde{\ell}}\binom{2\tilde{\ell}}{k}e^{-ik\omega}$$

$$= \frac{1}{2^{n+1}}\sum_{k=0}^{2\tilde{\ell}}\binom{2\tilde{\ell}}{k}e^{i(\tilde{\ell}-k)\omega}$$

$$= \frac{1}{2^{n+1}}\sum_{m=-\tilde{\ell}}^{\tilde{\ell}}\binom{2\tilde{\ell}}{\tilde{\ell}+m}e^{-im\omega}$$

$$= \frac{1}{\sqrt{2}}\sum_{m=-\tilde{\ell}}^{\tilde{\ell}}\left(\frac{\sqrt{2}}{2^{n+1}}\cdot\binom{2\tilde{\ell}}{\tilde{\ell}+m}\right)e^{-im\omega}$$

We can read the filter coefficients directly from the last identity. Putting everything in terms of n, we see that the filter coefficients are

$$\tilde{h}_k = \frac{\sqrt{2}}{2^{n+1}}\binom{n+1}{\frac{n+1}{2}-k} \tag{8.56}$$

for $k = -\frac{n+1}{2},\ldots,\frac{n+1}{2}$. Since the binomial coefficients are symmetric, we can show (see Problem 8.31) that $\tilde{h}_k = \tilde{h}_{-k}$. Moreover the length of this filter is $n + 2$.

In Problem 8.30 you will find a formula for $\tilde{h}_k$ when n is even. We summarize our results in the following definition.

Definition 8.12 (Spline Filters) *Let n be a nonnegative integer. We define the spline filter $\tilde{\mathbf{h}}$ of length $n + 2$ as*

$$\tilde{h}_k = \frac{\sqrt{2}}{2^{n+1}}\begin{cases} \dbinom{n+1}{\frac{n+1}{2}-k}, & k = -\frac{n+1}{2},\ldots,\frac{n+1}{2}, \quad n \text{ odd} \\[4mm] \dbinom{n+1}{\frac{n}{2}+k}, & k = -\frac{n}{2},\ldots,\frac{n}{2}+1, \quad n \text{ even} \end{cases} \tag{8.57}$$

All other $\tilde{h}_k = 0$. ∎

In Problem 8.31 you will show that $\tilde{h}_k = \tilde{h}_{-k}$ if n is odd and $\tilde{h}_k = \tilde{h}_{1-k}$ if n is even. Let's look at an example.

Example 8.9 (Spline Filters) *Find the spline filters associated with $B_1(t+1)$, $B_2(t+1)$, and $B_3(t+2)$.*
Solution

For $B_1(t+1)$, we have $n = 1$, so that $\frac{n+1}{2} = 1$ and (8.57) gives

$$\tilde{h}_k = \frac{\sqrt{2}}{4}\binom{2}{1-k}$$

for $k = -1, 0, 1$. This agrees with the filter used for $B_1(t+1)$ in Example 8.1. For $B_2(t+1)$, we use (8.57) with $n = 2$:

$$\tilde{h}_k = \frac{\sqrt{2}}{8}\binom{3}{1+k}$$

where $k = -1, 0, 1, 2$. We have

$$\tilde{\mathbf{h}} = \left(\tilde{h}_{-1}, \tilde{h}_0, \tilde{h}_1, \tilde{h}_2\right) = \left(\frac{\sqrt{2}}{8}, \frac{3\sqrt{2}}{8}, \frac{3\sqrt{2}}{8}, \frac{\sqrt{2}}{8}\right) \tag{8.58}$$

Finally, for $B_3(t+2)$, we use (8.57) with $n = 3$ to write

$$\tilde{h}_k = \frac{\sqrt{2}}{16}\binom{4}{2-k}$$

where $k = -2, \ldots, 2$. We have

$$\tilde{\mathbf{h}} = \left(\tilde{h}_{-2}, \tilde{h}_{-1}, \tilde{h}_0, \tilde{h}_1, \tilde{h}_2\right) = \left(\frac{\sqrt{2}}{16}, \frac{\sqrt{2}}{4}, \frac{3\sqrt{2}}{8}, \frac{\sqrt{2}}{4}, \frac{\sqrt{2}}{16}\right) \tag{8.59}$$

∎

Constructing the Dual Symbol: An Example

With these preliminary notions in hand, we turn to the main problem of finding a symmetric filter **h** dual to the spline filter $\tilde{\mathbf{h}}$ from Definition 8.12. The basic idea is to follow the methods of the orthogonal setting from Section 6.1. We start with the symbol $\tilde{H}(\omega)$ for a given B-spline and construct the symbol $H(\omega)$ of the dual scaling function so that it satisfies

$$\tilde{H}(\omega)\overline{H(\omega)} + \tilde{H}(\omega + \pi)\overline{H(\omega + \pi)} = 1 \tag{8.60}$$

which we know is required from Definition 8.10 for lowpass biorthogonal filters.

We use an example to illustrate the process. Consider the translated linear B-spline $B_1(t+1)$ and its symbol

$$\tilde{H}(\omega) = \frac{1}{4}e^{-i\omega} + \frac{1}{2} + \frac{1}{4}e^{i\omega}$$

from Example 8.4. We wish to construct a dual filter **h** symmetric about $k = 0$ with corresponding symbol $H(\omega)$. Now the coefficient symmetry $h_{-k} = h_k$ induces even function symmetry on the symbol and guarantees that $H(\omega)$ is real-valued (see Problem 2.11 in Section 2.2):

$$H(\omega) = H(-\omega) \qquad \text{and} \qquad H(\omega) = \overline{H(\omega)} \tag{8.61}$$

Recall from (6.8) in the orthogonal Daubechies construction that we can write even function $H(\omega)$ as a polynomial in the variable $\cos(\omega)$, where we use the term polynomial in the broader sense that a finite number of negative powers are allowed. Thus we define

$$y = \sin^2\left(\frac{\omega}{2}\right) \tag{8.62}$$

and use the half-angle formula $\cos(\omega) = 1 - 2\sin^2\left(\frac{\omega}{2}\right) = 1 - 2y$ to see that we can write the symbol $H(\omega)$ as a polynomial H_1 in the variable y

$$H(\omega) = H_1(y) \tag{8.63}$$

Since we want to substitute (8.63) into (8.60), we need to write $H(\omega + \pi)$ using variable y. We therefore observe that

$$\sin^2\left(\frac{\omega + \pi}{2}\right) = \cos^2\left(\frac{\omega}{2}\right) = 1 - y \tag{8.64}$$

so that

$$H(\omega + \pi) = H_1(1 - y) \tag{8.65}$$

For the linear B-spline symbol $\tilde{H}(\omega)$, we can convert to variable y using (8.53)

$$\tilde{H}(\omega) = \cos^2\left(\frac{\omega}{2}\right) = 1 - y \tag{8.66}$$

Combining (8.66), (8.65), and (8.61) allows us to rewrite (8.60) as

$$(1 - y)H_1(y) + yH_1(1 - y) = 1 \tag{8.67}$$

Before solving (8.67) for $H_1(y)$, consider the form of $H(\omega)$ again. We need **h** to be a scaling filter, so from Definition 8.10 we must have $H(\pi) = 0$, or equivalently, $H_1(1) = 0$. Thus $(1 - y)^\ell$ must be a factor of $H_1(y)$ for some $\ell \geq 1$ and

$$H_1(y) = (1 - y)^\ell P(y) \tag{8.68}$$

where $P(y)$ is as yet unknown. Putting these ideas together, we rewrite (8.67) as

$$(1 - y)^{\ell+1}P(y) + y^{\ell+1}P(1 - y) = 1 \tag{8.69}$$

This equation should look familiar — we saw it in Section 6.1, where we found a formula for its solution $P(y)$ in Theorem 6.1. At this time we will set $\ell = 2$ to form a concrete example, and get an explicit formula for $P(y)$ from (6.14):

$$P(y) = \sum_{k=0}^{\ell} \binom{2\ell + 1}{k} y^k (1 - y)^{\ell-k} = 1 + 3y + 6y^2 \tag{8.70}$$

We now retrace our steps to obtain symbol $H(\omega)$. Recall that

$$y = \sin^2\left(\frac{\omega}{2}\right) = \left(\frac{e^{i\omega/2} - e^{-i\omega/2}}{2i}\right)^2 \tag{8.71}$$

which we can substitute into $P(y)$ to obtain the dual symbol from (8.68):

$$\begin{aligned}
H(\omega) &= H_1(y) \\
&= (1-y)^2 P(y) \\
&= (1-y)^2 (1 + 3y + 6y^2) \\
&= \left(1 - \sin^2\left(\frac{\omega}{2}\right)\right)^2 \left(1 + 3\sin^2\left(\frac{\omega}{2}\right) + 6\sin^4\left(\frac{\omega}{2}\right)\right) \\
&= \frac{3}{128} e^{-4i\omega} - \frac{3}{64} e^{-3i\omega} - \frac{1}{8} e^{-2i\omega} + \frac{19}{64} e^{-i\omega} + \frac{45}{64} \\
&\quad + \frac{19}{64} e^{i\omega} - \frac{1}{8} e^{2i\omega} - \frac{3}{64} e^{3i\omega} + \frac{3}{128} e^{4i\omega}
\end{aligned}$$

The corresponding symmetric filter is

$$\begin{aligned}
\mathbf{h} &= (h_{-4}, h_{-3}, h_{-2}, h_{-1}, h_0, h_1, h_2, h_3, h_4) \\
&= (h_4, h_3, h_2, h_1, h_0, h_1, h_2, h_3, h_4) \\
&= \left(\frac{3\sqrt{2}}{128}, -\frac{3\sqrt{2}}{64}, -\frac{\sqrt{2}}{8}, \frac{19\sqrt{2}}{64}, \frac{45\sqrt{2}}{64}, \frac{19\sqrt{2}}{64}, -\frac{\sqrt{2}}{8}, -\frac{3\sqrt{2}}{64}, \frac{3\sqrt{2}}{128}\right)
\end{aligned}$$

Note that this is the familiar filter (8.4) we introduced early in the chapter as the dual for the translated linear B-spline. Now we know how to build this filter systematically and many more! The power and beauty of this method of Daubechies for constructing dual filters relies on equation (8.69), which has a known solution $P(y)$ for any positive integer ℓ, complete with a formula from (8.70).

The Smoothness of the Scaling Function $\phi(t)$

We make one more connection to the orthogonal setting before giving a general algorithm for the Daubechies method. We know that $(1-y)^\ell$ is a factor of $H_1(y)$. What consequence does this have for $H(\omega)$? We recall (8.52) and perform the variable conversion

$$\begin{aligned}
(1-y)^\ell &= \cos^{2\ell}\left(\frac{\omega}{2}\right) \\
&= \left(\frac{e^{-i\omega/2} + e^{i\omega/2}}{2}\right)^{2\ell} \\
&= e^{i\ell\omega}\left(\frac{1 + e^{-i\omega}}{2}\right)^{2\ell}
\end{aligned}$$

which forces $\left(\frac{1+e^{-i\omega}}{2}\right)^{2\ell}$ to be a factor of $H(\omega)$, so we can write

$$H(\omega) = \left(\frac{1+e^{-i\omega}}{2}\right)^{2\ell} S(\omega)$$

for some $S(\omega)$, similar to (6.3) in the orthogonal setting. Recall that ℓ indicates the smoothness of the corresponding scaling function.

Algorithms for Computing the Symbol $\mathbf{H}(\omega)$

To recap the biorthogonal Daubechies construction method, we started with the linear B-spline with scaling filter $\tilde{\mathbf{h}}$, then shifted the spline to be symmetric about 0 and an unknown symmetric filter $\mathbf{h}$ with "smoothness" level ℓ. We constructed filter $\mathbf{h}$ using (8.70), where there are two free parameters: $\tilde{\ell}$ for the particular B-spline and the value of ℓ for $\mathbf{h}$. We now give a formal algorithm for this method when the B-spline has an odd number of nonzero filter coefficients (taps), as is the case for the linear B-spline.

Algorithm 8.1 (Dual Filters of Odd Length) *Let $n = 2\tilde{\ell} - 1$, so that $n + 1 = 2\tilde{\ell}$ or $\tilde{\ell} = \frac{n+1}{2}$. This algorithm computes the dual filter $\mathbf{h}$ to the spline filter $\tilde{\mathbf{h}}$ generated from the centered B-spline $B_n(t + \tilde{\ell})$. From (8.57) we know that $\tilde{\mathbf{h}}$ is length $n + 2 = 2\tilde{\ell} + 1$.*

1. *Let $y = \sin^2\left(\frac{\omega}{2}\right)$ and choose a smoothness level ℓ and symbol*

$$H(\omega) = H_1(y) = (1-y)^\ell P(y) \tag{8.72}$$

2. *Solve $(1-y)^{\ell+\tilde{\ell}} P(y) + y^{\ell+\tilde{\ell}} P(1-y) = 1$ for $P(y)$ of degree $\ell + \tilde{\ell} - 1$ and obtain*

$$P(y) = \sum_{k=0}^{\ell+\tilde{\ell}-1} \binom{2\ell + 2\tilde{\ell} - 1}{k} y^k (1-y)^{\ell+\tilde{\ell}-1-k}$$

3. *Convert $H_1(y)$ from a polynomial in $y = \sin^2(\omega/2)$ to a polynomial in $e^{-i\omega}$ using (8.72) and (8.71). The filter $\mathbf{h}$ is built from the coefficients of $H(\omega)$.*

∎

Theorem 8.5 gives more information on the length of filter $\mathbf{h}$ and support of the corresponding scaling function $\phi(t)$, as well as full justification of the validity of this algorithm. In Problem 8.35 you can use this algorithm to build systematically the $(5,3)$ biorthogonal spline filter pair of Example 8.8.

The process for constructing even-length filters $\mathbf{h}$ is nearly identical to the odd-length case, but now the symmetry for $\phi(t)$ is about $t = 1/2$ and the filter coefficients will have symmetry $h_k = h_{1-k}$. This shifting results in an extra factor of $e^{-i\omega/2}$ in the symbol. Moreover, the spline filters of even length have an odd power of $\cos\left(\frac{\omega}{2}\right)$

in their symbols, so the dual filters H will need an extra $\cos\left(\frac{\omega}{2}\right)$ factor. Here is the algorithm for constructing even-length filters dual to even-length spline filters.

Algorithm 8.2 (Dual Filters of Even Length) *Let* $n = 2\tilde{\ell}$, *so that* $\tilde{\ell} = \frac{n}{2}$. *This algorithm computes the dual filter* $\mathbf{h}$ *to the spline filter* $\tilde{\mathbf{h}}$ *generated from the centered B-spline* $B_n(t + \tilde{\ell})$. *From (8.57) we know that* $\mathbf{h}$ *is length* $n + 2 = 2\tilde{\ell} + 2$.

1. *Let* $y = \sin^2\left(\frac{\omega}{2}\right)$ *and choose a smoothness level* ℓ *and symbol*

$$H(\omega) = H_1(y) = e^{-i\omega/2}\cos^{2\ell+1}\left(\frac{\omega}{2}\right)P(y)$$

$$= e^{-i\omega/2}\cos\left(\frac{\omega}{2}\right)(1-y)^\ell P(y)$$

2. *Solve* $(1-y)^{\ell+\tilde{\ell}+1}P(y) + y^{\ell+\tilde{\ell}+1}P(1-y) = 1$ *for* $P(y)$ *of degree* $\ell + \tilde{\ell}$ *and obtain*

$$P(y) = \sum_{k=0}^{\ell+\tilde{\ell}}\binom{2\ell + 2\tilde{\ell} + 1}{k}y^k(1-y)^{\ell+\tilde{\ell}-k} \qquad (8.73)$$

3. *Convert* $H_1(y)$ *from a polynomial in* $y = \sin^2(\omega/2)$ *to a polynomial in* $e^{-i\omega}$ *using (8.71) and*

$$e^{-i\omega/2}\cos\left(\frac{\omega}{2}\right) = \frac{1 + e^{-i\omega}}{2} \qquad (8.74)$$

The filter $\mathbf{h}$ *is built from the coefficients of* $H(\omega)$.

■

We now demonstrate Algorithm 8.2 with an example.

Example 8.10 (An Even-Length Filter Pair) *Start with the quadratic B-spline and use Algorithm 8.2 to create a filter* $\mathbf{h}$ *with smoothness level* $\ell = 1$.
Solution
  With quadratic B-spline $B_2(t)$, we have $n = 2$, so $\tilde{\ell} = n/2 = 1$ and thus $\ell + \tilde{\ell} + 1 = 3$. We follow the algorithm steps.
  For $\ell = 1$ and $y = \sin^2\left(\frac{\omega}{2}\right)$, we construct the symbol

$$H(\omega) = H(y) = e^{-i\omega/2}\cos\left(\frac{\omega}{2}\right)(1-y)P(y)$$

We next solve $(1-y)^3P(y) + y^3P(1-y) = 1$ for $P(y)$ of degree 2 using (8.73)

$$P(y) = \sum_{k=0}^{2}\binom{5}{k}y^k(1-y)^{2-k} = 1 + 3y + 6y^2$$

We convert $H(\omega)$ from a polynomial in y to a polynomial in $e^{-i\omega}$. We use (8.74) to write

$$H(\omega) = e^{-i\omega/2}\cos\left(\frac{\omega}{2}\right)\left(1 - \sin^2\left(\frac{\omega}{2}\right)\right)\left(1 + 3\sin^2\left(\frac{\omega}{2}\right) + 6\sin^4\left(\frac{\omega}{2}\right)\right)$$

$$= \frac{1 + e^{-i\omega}}{2}\left(1 - \sin^2\left(\frac{\omega}{2}\right)\right)\left(1 + 3\sin^2\left(\frac{\omega}{2}\right) + 6\sin^4\left(\frac{\omega}{2}\right)\right)$$

Finally, use (8.71) and a CAS to expand the expression and obtain

$$H(\omega) = \frac{3}{64}e^{-3i\omega} - \frac{9}{64}e^{-2i\omega} - \frac{7}{64}e^{-i\omega}$$
$$+ \frac{45}{64} + \frac{45}{64}e^{i\omega} - \frac{7}{64}e^{2i\omega} - \frac{9}{64}e^{3i\omega} + \frac{3}{64}e^{4i\omega}$$

Using (8.58) from Example 8.9 and the previous identity, the biorthogonal filter pair is

$$\tilde{\mathbf{h}} = (\tilde{h}_{-1}, \tilde{h}_0, \tilde{h}_1, \tilde{h}_2)$$
$$= \left(\frac{\sqrt{2}}{8}, \frac{3\sqrt{2}}{8}, \frac{3\sqrt{2}}{8}, \frac{\sqrt{2}}{8}\right)$$

$$\mathbf{h} = (h_{-3}, h_{-2}, h_{-1}, h_0, h_1, h_2, h_3, h_4)$$
$$= \left(\frac{3\sqrt{2}}{64}, -\frac{9\sqrt{2}}{64}, -\frac{7\sqrt{2}}{64}, \frac{45\sqrt{2}}{64}, \frac{45\sqrt{2}}{64}, -\frac{7\sqrt{2}}{64}, -\frac{9\sqrt{2}}{64}, \frac{3\sqrt{2}}{64}\right)$$

This filter pair is typically called the $(8,4)$ biorthogonal spline filter pair.   ∎

Daubechies Symbol Construction

We now state a theorem due to Ingrid Daubechies [18] that formally confirms the validity of the two algorithms above and gives some information on the filter lengths and scaling function supports. In particular, we note that the lengths of both filters will have the same parity (i.e., lengths both even or lengths both odd), even with the two free parameters ℓ and $\tilde{\ell}$.

Theorem 8.5 (Daubechies Dual Symbol) *Suppose that $\tilde{\mathbf{h}}$ is a spline filter (8.57) of length $n + 2$ with $n = 2\tilde{\ell} - 1$ if $\tilde{n}$ is odd and $n = 2\tilde{\ell}$ if n is even. Suppose that $\mathbf{h}$ is the filter built from the symbol*

$$H(\omega) = \cos^{2\ell}\left(\frac{\omega}{2}\right)\sum_{k=0}^{\ell+\tilde{\ell}-1}\binom{2\ell + 2\tilde{\ell} - 1}{k}\sin^{2k}\left(\frac{\omega}{2}\right)\left(1 - \sin^2\left(\frac{\omega}{2}\right)\right)^{\ell+\tilde{\ell}-1-k}$$

$$(8.75)$$

when n is odd or

$$H(\omega) = e^{-i\omega/2} \cos^{2\ell+1}\left(\frac{\omega}{2}\right) \sum_{k=0}^{\ell+\tilde{\ell}} \binom{2\ell + 2\tilde{\ell} + 1}{k} \sin^{2k}\left(\frac{\omega}{2}\right) \left(1 - \sin^2\left(\frac{\omega}{2}\right)\right)^{\ell+\tilde{\ell}-k}$$

(8.76)

when n is even. Then $\tilde{\mathbf{h}}$, $\mathbf{h}$ *form a biorthogonal filter pair as defined in Definition 8.10.*

Moreover, if $\tilde{n}$ *is odd, then the length of* $\mathbf{h}$ *is also odd, with the number of nonzero terms equal to* $4\ell + 2\tilde{\ell} - 1$. *The scaling function* $\phi(t)$ *satisfies* $\mathrm{supp}(\phi) = [-2\ell - \tilde{\ell}, 2\ell + \tilde{\ell}]$. *If n is even, then the length of* $\mathbf{h}$ *is also even, with the number of nonzero terms equal to* $4\ell + 2\tilde{\ell} + 2$. *In this case,* $\mathrm{supp}(\phi) = [-2\ell - \tilde{\ell}, 2\ell + \tilde{\ell} + 1]$. ∎

Proof: Confirming that $H(0) = 1$ and $H(\pi) = 0$ is left to Problem 8.37. We now need to establish that (8.49) holds for the spline symbols $\tilde{H}(\omega)$ and $H(\omega)$. We first consider the case where n is even.

We let $y = \sin^2\left(\frac{\omega}{2}\right)$ and rewrite $H(\omega)$ from (8.76) as

$$H(\omega) = e^{-i\omega/2} \cos\left(\frac{\omega}{2}\right) (1 - y)^\ell \sum_{k=0}^{\ell+\tilde{\ell}} \binom{2\ell + 2\tilde{\ell} + 1}{k} y^k (1 - y)^{\ell+\tilde{\ell}-k}$$

(8.77)

$$= e^{-i\omega/2} \cos\left(\frac{\omega}{2}\right) (1 - y)^\ell P(y)$$

(8.78)

Here

$$P(y) = \sum_{k=0}^{\ell+\tilde{\ell}} \binom{2\ell + 2\tilde{\ell} + 1}{k} y^k (1 - y)^{\ell+\tilde{\ell}-k}$$

(8.79)

From Theorem 6.1 we know that $P(y)$ solves

$$(1 - y)^{\ell+\tilde{\ell}+1} P(y) + y^{\ell+\tilde{\ell}+1} P(1 - y) = 1$$

(8.80)

Let's consider the first term $\tilde{H}(\omega)\overline{H(\omega)}$ on the left-hand side of (8.49). We have

$$\tilde{H}(\omega)\overline{H(\omega)} = e^{-i\omega/2} \cos^{2\tilde{\ell}+1}\left(\frac{\omega}{2}\right) \cdot \overline{e^{-i\omega/2} \cos\left(\frac{\omega}{2}\right) (1 - y)^\ell P(y)}$$

$$= \cos^{2\tilde{\ell}+2}\left(\frac{\omega}{2}\right) (1 - y)^\ell P(y)$$

$$= \left(1 - \sin^2\left(\frac{\omega}{2}\right)\right)^{\tilde{\ell}+1} (1 - y)^\ell P(y)$$

$$= (1 - y)^{\tilde{\ell}+1} (1 - y)^\ell P(y)$$

$$= (1 - y)^{\tilde{\ell}+\ell+1} P(y)$$

(8.81)

We want to replace ω by $\omega + \pi$ in (8.81). Using (8.64) and (8.65), we see that replacing ω by $\omega + \pi$ is the same as replacing y by $1 - y$. We make this substitution in (8.81) to obtain

$$\tilde{H}(\omega + \pi)\overline{H(\omega + \pi)} = y^{\tilde{\ell}+\ell+1} P(1 - y)$$

(8.82)

Adding (8.81) and (8.82) gives

$$\tilde{H}(\omega)\overline{H(\omega)} + \tilde{H}(\omega + \pi)\overline{H(\omega + \pi)} = (1 - y)^{\tilde{\ell}+\ell+1}P(y) + y^{\tilde{\ell}+\ell+1}P(1 - y)$$

But the right hand-side here is 1 because of (8.79) and (8.80).

To prove the claims on the number of nonzero filter coefficients and the support of $\phi(t)$, we need to view $H(\omega)$ as a polynomial in $e^{-i\omega}$. First note from (8.77) that $(1 - y)^{\ell}P(y)$ has degree $2\ell + \tilde{\ell}$ in $y = \sin^2\left(\frac{\omega}{2}\right) = \left(\left(e^{i\omega/2} - e^{-i\omega/2}\right)/2i\right)^2$, so $(1 - y)^{\ell}P(y)$ is a polynomial in $e^{-i\omega/2}$ with terms running from degree $-2(2\ell + \tilde{\ell})$ through $2(2\ell + \tilde{\ell})$. Thus, by (8.74),

$$H(\omega) = e^{-i\omega/2}\cos\left(\frac{\omega}{2}\right)(1 - y)^{\ell}P(y) = \frac{1 + e^{i\omega}}{2}(1 - y)^{\ell}P(y)$$

and when viewed as a polynomial in $e^{-i\omega/2}$, $H(\omega)$ will have terms ranging from degree $-2(2\ell + \tilde{\ell})$ through $2(2\ell + \tilde{\ell}) + 2$. Equivalently, as a polynomial in $e^{-i\omega}$, $H(\omega)$ will have terms ranging from degree $-(2\ell + \tilde{\ell})$ through $\left(2\ell + \tilde{\ell}\right) + 1$. Using an argument similar to that from Theorem 5.4, we see that

$$\overline{\text{supp}(\phi)} = [-2\ell - \tilde{\ell}, 2\ell + \tilde{\ell} + 1]$$

and the number of nonzero coefficients in the filter is $(2\ell + \tilde{\ell}) + ((2\ell + \tilde{\ell}) + 1)$ plus one more for h_0. This sum simplifies to $4\ell + 2\tilde{\ell} + 2$ nonzero coefficients.

The proof in the odd-length case is very similar and is left as Problem 8.38.   ∎

Notation for Biorthogonal Spline Filter Pairs

Biorthogonal filter pairs are often referenced in books and software by the filter lengths of **h** (first) and the spline filter $\tilde{\mathbf{h}}$ (second). For example, the pair in Example 8.1 is called the $(9,3)$ biorthogonal filter pair or the $(9,3)$ biorthogonal spline filter pair since $\tilde{\mathbf{h}}$ is the spline filter for $n = 1$ and **h** was constructed using Algorithm 8.2. The filter pair in Example 8.10 is called the $(8,4)$ biorthogonal spline filter pair. In Problem 8.39 you will show that for N and $\tilde{N}$ of the same parity, if the length of $\tilde{\mathbf{h}}$ is $\tilde{N} + 1$, then the length of **h** is $2N + \tilde{N} - 1$. From these values of $\tilde{N}, N$, we can determine $\tilde{\ell}$ and ℓ.

We have then the following definition, which summarizes our "length" notation for the biorthogonal spline filter pairs.

Definition 8.13 (Notation for Biorthogonal Spline Filter Pairs) *Assume that N and $\tilde{N}$ have the same parity. Any biorthogonal spline filter pair built from Theorem 8.5 will be called the $(2N + \tilde{N} - 1, \tilde{N} + 1)$ biorthogonal spline filter pair.*   ∎

For example, the $(8,4)$ biorthogonal spline filter pair developed in Example 8.10 with $\tilde{\ell} = 1, \ell = 1$ has $4N + \tilde{N} - 1 = 8$ and $\tilde{N} + 1 = 4$, so that $\tilde{N} = 3$ and $N = 3$.

The CDF Filter Pairs

For some applications it is desirable to have biorthogonal filters $\mathbf{h}$ and $\tilde{\mathbf{h}}$ of nearly equal length, and to have nearly equal smoothness in the corresponding scaling functions. This is a weakness for most spline filters. Taking the $(8,4)$ biorthogonal spline pair as an example, we have similar smoothness with $\ell = \tilde{\ell} = 1$ but very different lengths. On the other hand, the $(9,7)$ biorthogonal filter pair have similar lengths, but $\tilde{\ell} = 3$ and $\ell = 1$. In Problem 8.36 you found the $(9,7)$ biorthogonal filter pair. We can use these filters to form $H(\omega)$ and $\tilde{H}(\omega)$. The modulus of each symbol is plotted in Figure 8.5.

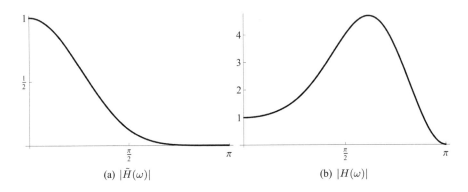

(a) $|\tilde{H}(\omega)|$ (b) $|H(\omega)|$

Figure 8.5   The modulus of $\tilde{H}(\omega)$ and $H(\omega)$ for the $(9,7)$ biorthogonal spline filter pair.

You can see from Figure 8.5 the effects in the disparity of the values of $\tilde{\ell}$ and ℓ. It can be shown that $\tilde{H}^{(k)}(\pi) = 0$ for $k = 0, \ldots, 5$, but $H^{(k)}(\pi) = 0$ only for $k = 0, 1$. Also, the shape of $|H(\omega)|$ is not ideal if $\mathbf{h}$ is supposed to be a lowpass filter (see the discussion preceding Figure 5.1 in Section 5.1).

Cohen, Daubechies and Feauveau wrote a 1992 paper [13] in which they developed a family of biorthogonal filter pairs $\mathbf{h}$, $\tilde{\mathbf{h}}$ satisfying three key properties:

1. Both $\mathbf{h}$ and $\tilde{\mathbf{h}}$ symmetric

2. Balanced (nearly equal) smoothness parameters ℓ and $\tilde{\ell}$

3. Balanced (nearly equal) lengths for the filters

In the remainder of this section we outline their development and explain the most important filter pair in detail. This is the CDF97 biorthogonal filter pair, used for the lossy version in the JPEG2000 compression standard and in the FBI fingerprint compression standard (see Section 9.4).

The CDF97 Biorthogonal Filter Pair

To ensure that $\tilde{\mathbf{h}}$ and $\mathbf{h}$ are symmetric, we write the symbols $H(\omega)$, $\tilde{H}(\omega)$ corresponding to $\mathbf{h}$, $\tilde{\mathbf{h}}$, respectively, as polynomials in $y = \sin^2\left(\frac{\omega}{2}\right)$, as developed in the discussion culminating in (8.63).

To obtain equal smoothness parameters for the two filters, we simply set $\ell = \tilde{\ell}$ and require $(1 - y)^\ell$ to be a factor of each symbol as discussed in the derivation of biorthogonal spline filter pairs. To summarize, the symbols corresponding to $\mathbf{h}$ and $\tilde{\mathbf{h}}$ must be of the form

$$H(\omega) = H_1(y) = (1 - y)^\ell q(y) \tag{8.83}$$
$$\tilde{H}(\omega) = \tilde{H}_1(y) = (1 - y)^\ell \tilde{q}(y)$$

for some polynomials $q(y)$ and $\tilde{q}(y)$, respectively. Of course these must be biorthogonal filters satisfying the fundamental biorthogonality condition (8.49), so we put $H_1(y)$ and $\tilde{H}_1(y)$ into this equation to obtain the requirement

$$(1 - y)^{2\ell} \tilde{q}(y)q(y) + y^{2\ell}\tilde{q}(1 - y)q(1 - y) = 1$$

We know from Theorem 6.1 that we can ensure a solution for $q(y)$ and $\tilde{q}(y)$ if they are chosen so that

$$\tilde{q}(y)q(y) = P(y) = \sum_{k=0}^{2\ell-1} \binom{4\ell - 1}{k} y^k (1 - y)^{2\ell-1-k} \tag{8.84}$$

In other words, we just need to factor $P(y)$ into $\tilde{q}(y)q(y)$ and at the same time require that $q(y)$ and $\tilde{q}(y)$ have real coefficients so that the filters coefficients are real, and we want to choose $q(y)$ and $\tilde{q}(y)$ to have (nearly) equal degrees. This will ensure (nearly) equal filter lengths, our third key property.

Recall that the linear factors of $P(y)$ come from its roots. Table 8.1 lists the roots of $P(y)$ built from (8.84) for the first few positive integers.

Table 8.1   The roots of the first few $P(y)$, where $P(y)$ is given by (8.84).

ℓ	$P(y)$	Roots of $P(y)$
1	$1 + 2y$	$-\frac{1}{2}$
2	$1 + 4y + 10y^2 + 20y^3$	-0.342384
		$-0.078808 \pm 0.373931i$
3	$1 + 6y + 21y^2 + 56y^3 + 126y^4 + 252y^5$	-0.297269
		$-0.19454 \pm 0.243998i$
		$0.0931748 \pm 0.358329i$

The choice of $\ell = 1$ results in a known biorthogonal spline filter pair (see Problem 8.45). The choice $\ell = 3$ yields a biorthogonal filter pair, and the derivation of the coefficients is left as Problem 8.46. The choice of $\ell = 2$ leads to the CDF97 filter pair, as we now show.

Remember that $q(y)$ and $\tilde{q}(y)$ must have real coefficients, so the root pair $r_1 \approx -0.0788 + 0.3739i, r_2 \approx -0.0788 - 0.3739i$ must stay together. Cohen, Daubechies, and Feauveau chose $q(y)$ to be composed of the linear factors $(y - r_1)$ and $(y - r_2)$ and $(y - r_3)$, with $r_3 \approx -0.3424$ as the sole linear factor of $\tilde{q}(y)$. Thus the polynomials must be of the form

$$q(y) = a\,(y - r_1)\,(y - r_2) \qquad \text{and} \qquad \tilde{q}(y) = b\,(y - r_3)$$

for some constants a and b. Since we require that $H(0) = \tilde{H}(0) = 1$, we insert $y = \sin^2(0) = 0$ into $q(y), \tilde{q}(y)$ above to obtain

$$1 = a\,(r_1 r_2) \qquad \text{and} \qquad 1 = -br_3$$

We can easily solve these equations to find that $a \approx 6.84768$ and $b \approx 2.9207$. Having $q(y), \tilde{q}(y)$, we use (8.83) to write down the symbols of the two filters in terms of y:

$$H_1(y) = a(1 - y)^2(y - r_1)(y - r_2)$$
$$\tilde{H}_1(y) = b(1 - y)^2(y - r_3)$$

Next we expand $H_1(y)$ with the substitution $y = \sin^2(\omega/2) = \left(\dfrac{e^{i\omega/2} - e^{-i\omega/2}}{2i}\right)^2$ and use a CAS to obtain

$$H(\omega) = \frac{1}{\sqrt{2}} \sum_{k=-4}^{4} h_k e^{-ik\omega}$$

with the nonzero filter coefficient values (rounded to four decimal digits)

$$\begin{aligned}
\mathbf{h} &= (h_{-4}, h_{-3}, h_{-2}, h_{-1}, h_0, h_1, h_2, h_3, h_4) \\
&= (0.0378, -0.0238, -0.1106, 0.3774, 0.8527, \\
&\qquad 0.3774, -0.1106, -0.0238, 0.0378)
\end{aligned} \qquad (8.85)$$

Similarly, for filter $\tilde{\mathbf{h}}$ we find the symbol

$$\tilde{H}(\omega) = \frac{1}{\sqrt{2}} \sum_{k=-3}^{3} \tilde{h}_k e^{-ik\omega}$$

with nonzero filter coefficient values (rounded to four decimal digits)

$$\begin{aligned}
\tilde{\mathbf{h}} &= (\tilde{h}_{-3}, \tilde{h}_{-2}, \tilde{h}_{-1}, \tilde{h}_0, \tilde{h}_1, \tilde{h}_2, \tilde{h}_3) \\
&= (-0.0645, -0.0407, 0.4181, 0.7885, 0.4181, -0.0407, -0.0645)
\end{aligned} \qquad (8.86)$$

So both $\mathbf{h}$ and $\tilde{\mathbf{h}}$ are symmetric with filters lengths 9 and 7, respectively. We record this important filter pair in Definition 8.14.

Definition 8.14 (The CDF97 Biorthogonal Filter Pair) *The* CDF97 biorthogonal filter pair $(\mathbf{h}, \tilde{\mathbf{h}})$ *is given by (8.85) and (8.86).* ∎

It is worthwhile to compare the CDF97 biorthogonal filter pair with the $(9,7)$ biorthogonal spline filter pair derived in Problem 8.36. $|H(\omega)|$ and $\tilde{H}(\omega)|$ for the spline filter pair are plotted in Figure 8.5. The graphs of $|H(\omega)|$ and $|\tilde{H}(\omega)|$ are plotted in Figure 8.6. As you can see, the balancing constraint $\ell = \tilde{\ell}$ makes both filters better suited to performing lowpass (or averages) filtering although still not as efficient as, say, the length 7 spline filter whose modulus is plotted in Figure 8.5(a). But both filters are improvements over the length 9 dual filter plotted in Figure 8.5(b).

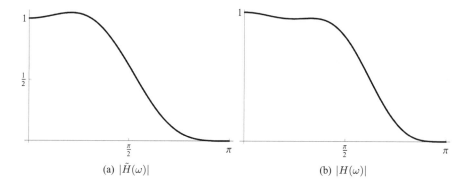

(a) $|\tilde{H}(\omega)|$ (b) $|H(\omega)|$

Figure 8.6   The modulus of $\tilde{H}(\omega)$ and $H(\omega)$ for the CDF97 biorthogonal filter pair.

Even-Length CDF Filters

The construction for odd-length CDF biorthogonal filter pairs can be modified to create even-length filter pairs with balanced smoothness. Now the symmetry for $\phi(t)$ is about $t = 1/2$ and the filter coefficients will have symmetry $h_k = h_{1-k}$. This shifting results in an extra factor of $e^{-i\omega/2}$ in the symbols. And just as in the spline filter construction, the even-length filter symbols will need an extra $\cos\left(\frac{\omega}{2}\right)$ factor. Consequently, the symbols corresponding to $\mathbf{h}$ and $\tilde{\mathbf{h}}$ must be of the form

$$H(\omega) = H_1(y) = e^{-i\omega/2} \cos\left(\tfrac{\omega}{2}\right) (1-y)^\ell q(y)$$
$$\tilde{H}(\omega) = \tilde{H}_1(y) = e^{-i\omega/2} \cos\left(\tfrac{\omega}{2}\right) (1-y)^\ell \tilde{q}(y)$$

$$\text{(8.87)}$$

for some polynomials $q(y)$ and $\tilde{q}(y)$, respectively. Here $y = \sin^2\left(\frac{\omega}{2}\right)$. As you will show in Problem 8.49, biorthogonality requires that $q(y)$ and $\tilde{q}(y)$ satisfy the condition

$$(1-y)^{2\ell+1}\tilde{q}(y)q(y) + y^{2\ell+1}\tilde{q}(1-y)q(1-y) = 1 \qquad \text{(8.88)}$$

If we choose $q(y)$ and $\tilde{q}(y)$ so that

$$\tilde{q}(y)q(y) = P(y) = \sum_{k=0}^{2\ell} \binom{4\ell+1}{k} y^k (1-y)^{2\ell-k} \tag{8.89}$$

then (8.88) is satisfied.

Table 8.2 records the the roots of $P(y)$ built from (8.89) for the integers $\ell = 1, 2, 3$.

Table 8.2   The roots of the first few $P(y)$ where $P(y)$ is given by (8.89).

ℓ	$\mathbf{P(y)}$	Roots of $\mathbf{P(y)}$
1	$1 + 3y + 6y^2$	$-0.25 \pm 0.322749i$
2	$1 + 5y + 15y^2 + 35y^3 + 70y^4$	$-0.275034 \pm 0.164277i$
		$0.025034 \pm 0.372249i$
3	$1 + 7y + 28y^2 + 84y^3 + 210y^4 + 462y^5 + 924y^6$	$-0.266501 \pm -0.107337i$
		$-0.124643 \pm 0.283191i$
		0.141144 ± 0.342103

For the case $\ell = 1$, there is no way to balance the roots since they are both complex. For the $\ell = 2$ case, the roots can be nicely balanced, as you will see in Problem 8.50. The filters have the same smoothness parameter value as the CDF97 pair, but with 10 nonzero filter coefficients each, they are longer. This extra length is generally a disadvantage in practice, which explains in part why the JPEG2000 committee and the FBI favored the CDF97 biorthogonal filter pair for their standards.

PROBLEMS

Note: *A computer algebra system and the software on the course Web site will be useful for some of these problems. See the Preface for more details.*

8.28   Use the translation rule from Section 2.2 to verify the symbol formula (8.51) for the B-spline $B_n(t)$ translated $\frac{n+1}{2}$ units left. Here n is odd. (*Hint:* See the proof of Theorem 5.3.)

8.29   Use the translation rule from Section 2.2 to verify the symbol formula (8.54) for the B-spline translated $\frac{n}{2}$ units left. Here n is even.

8.30   Mimic the argument leading to (8.56) to verify the formula (8.57) in Definition 8.12 in the case where n is even.

8.31   Use the symmetry of the binomial coefficients and (8.57) to show that $\tilde{h}_k = \tilde{h}_{-k}$ when n is odd and $\tilde{h}_k = \tilde{h}_{1-k}$ when n is even.

8.32   Find the spline filter $\tilde{\mathbf{h}}$ satisfying the following conditions:

(a) $\tilde{\mathbf{h}}$ has length 6.

(b) The symbol $\tilde{H}(\omega)$ is generated by the shifted quartic $B_4(t)$.

(c) n is even and the first nonzero coefficient is $\tilde{h}_{-3}$.

8.33   Which B-spline has a symmetric filter of length 7? Find this filter.

8.34   Use (8.59), the cascade algorithm from Section 6.2, and a CAS to create a plot of $B_3(t+2)$.

8.35   Use Algorithm 8.1 to systematically build the $(5,3)$ filter of Example 8.8.

8.36   Use Algorithm 8.1 to show that the $(9,7)$ biorthogonal spline filter pair is

$$\mathbf{h} = (h_{-4}, h_{-3}, h_{-2}, h_{-1}, h_0, h_1, h_2, h_3, h_4)$$
$$= \left(-\frac{5\sqrt{2}}{64}, \frac{15\sqrt{2}}{32}, -\frac{7\sqrt{2}}{8}, -\frac{7\sqrt{2}}{32}, \frac{77\sqrt{2}}{32}, -\frac{7\sqrt{2}}{32}, -\frac{7\sqrt{2}}{8}, \frac{15\sqrt{2}}{32}, -\frac{5\sqrt{2}}{64} \right)$$

and

$$\tilde{\mathbf{h}} = (\tilde{h}_{-3}, \tilde{h}_{-2}, \tilde{h}_{-1}, \tilde{h}_0, \tilde{h}_1, \tilde{h}_2, \tilde{h}_3)$$
$$= \left(\frac{\sqrt{2}}{64}, \frac{3\sqrt{2}}{64}, \frac{15\sqrt{2}}{64}, \frac{5\sqrt{2}}{8}, \frac{15\sqrt{2}}{64}, \frac{3\sqrt{2}}{64}, \frac{\sqrt{2}}{64} \right)$$

8.37   For $H(\omega)$ in Theorem 8.5, show that $H(0) = 1$ and $H(\pi) = 0$.

8.38   Prove Theorem 8.5 when n is odd.

8.39   In this problem you will characterize the lengths of the biorthogonal spline filter pairs. From Theorem 8.5 we know that the filter lengths must have the same parity (i.e., both even or both odd).

(a) For a given n we know that the length of $\tilde{\mathbf{h}}$ is $n+2$. Set $\tilde{N} = n+1$ (the power of $\cos\left(\frac{\omega}{2}\right)$) and write the length of $\tilde{\mathbf{h}}$ in terms of $\tilde{N}$. Explain why $\tilde{N}$ is even when n is odd and $\tilde{N}$ is odd when n is even.

(b) Suppose that $n = 2\tilde{\ell} - 1$ is odd. Since $\tilde{N}$ is even, set $N = 2\ell$. Show that the length of $\mathbf{h}$ is odd and equal to $2N + \tilde{N} - 1$.

(b) Suppose that $n = 2\tilde{\ell}$ is even. Then $\tilde{N}$ is odd, so we set $N = 2\ell + 1$. Show that the length of $\mathbf{h}$ is even and equal to $2N + \tilde{N} - 1$.

8.40   Find the N and $\tilde{N}$ values for the $(5,3)$ and $(9,7)$ biorthogonal spline filter pairs.

8.41   Suppose that we take $\tilde{\ell} = 3$ and $\ell = 4$ and wish to construct even-length filters $\tilde{\mathbf{h}}$ and $\mathbf{h}$. Use Problem 8.39 to find $\tilde{N}$ and N in this case and the lengths of each filter.

8.42   Suppose that we take $\tilde{\ell} = 2$ and $\ell = 4$ and wish to construct odd-length filters $\tilde{\mathbf{h}}$ and $\mathbf{h}$. Use Problem 8.39 to find $\tilde{N}$ and N in this case and the lengths of each filter.

8.43 Suppose that we take $\tilde{\mathbf{h}}$ to be the Haar filter $(\tilde{h}_0, \tilde{h}_1) = \left(\frac{\sqrt{2}}{2}, \frac{\sqrt{2}}{2}\right)$. What are the possible lengths of the dual filter $\mathbf{h}$?

8.44 Suppose that we take $\tilde{\mathbf{h}}$ to be the spline filter associated with $B_1(t+1)$. That is, $(\tilde{h}_{-1}, \tilde{h}_0, \tilde{h}_1) = \left(\frac{\sqrt{2}}{4}, \frac{\sqrt{2}}{2}, \frac{\sqrt{2}}{4}\right)$. What are the possible lengths of the dual filter $\mathbf{h}$?

8.45 Explain why $\ell = 1$ in (8.84) leads to a biorthogonal spline filter pair. Which pair is it?

8.46 Find the filter coefficients for the odd-length $\mathbf{h}$ and $\tilde{\mathbf{h}}$ biorthogonal filter pair when $\ell = 3$. Remember that there are two different choices for creating the polynomials $q(y)$ and $\tilde{q}(y)$, so you should get two different solutions.

8.47 Create plots of the $|\tilde{H}(\omega)|$ and $|H(\omega)|$ for the $\ell = 3$ filter pairs you found in Problem 8.46. Which seem closer to ideal?

8.48 Use the cascade algorithm to plot $\phi(t)$ and $\tilde{\phi}(t)$ for the CDF97 biorthogonal filter pair.

8.49 Verify that the symbols in (8.87) lead to (8.88) and (8.89).

8.50 Use Table 8.2 to find the biorthogonal filter pair coefficients for the even-length $\mathbf{h}$ and $\tilde{\mathbf{h}}$ filter when $\ell = 2$.

8.51 Discuss the $\ell = 3$ case for even-length CDF filters, in particular balancing lengths and smoothness.

8.52 Find the wavelet filter pair $\mathbf{g}$ and $\tilde{\mathbf{g}}$ for the CDF97 biorthogonal filter pair.

8.4   DECOMPOSITION AND RECONSTRUCTION

In this short section we develop multiscale wavelet decomposition and reconstruction for biorthogonal scaling functions and wavelets. This will allow us to move between finer and coarser scales of resolution within a dual MRA and naturally lead to the discrete biorthogonal wavelet transformation discussed in Section 8.5.

Decomposition at Different Scales

The ideas for decomposition using biorthogonal scaling functions and wavelets were introduced in Example 8.1. We now proceed in a more systematic way, beginning with a generalization of the dilation equations to link the spaces V_{j+1} and $\tilde{V}_{j+1}$ with the spaces V_j, W_j and $\tilde{V}_j$, $\tilde{W}_j$, respectively.

Proposition 8.7 (Dilation Equations at Different Scales) *Let $\phi(t)$ and $\tilde{\phi}(t)$ be scaling functions that generate a dual multiresolution analyses with associated biorthogonal wavelet functions $\psi(t)$ and $\tilde{\psi}(t)$. Then for any $j, k, \ell \in \mathbb{Z}$ the following dilation* *equations hold:*

$$\phi_{j,k}(t) = \sum_{\ell \in \mathbb{Z}} h_{\ell-2k}\phi_{j+1,\ell}(t)$$

$$\tilde{\phi}_{j,k}(t) = \sum_{\ell \in \mathbb{Z}} \tilde{h}_{\ell-2k}\tilde{\phi}_{j+1,\ell}(t)$$

$$\psi_{j,k}(t) = \sum_{n \in \mathbb{Z}} g_{n-2k}\phi_{j+1,n}(t)$$

$$\tilde{\psi}_{j,k}(t) = \sum_{n \in \mathbb{Z}} \tilde{g}_{n-2k}\tilde{\phi}_{j+1,n}(t)$$

where the coefficients h_k, $\tilde{h}_k$ and g_k, $\tilde{g}_k$ are defined by the dilation equations (8.11) and (8.36).   ∎

Proof: These dilation equations are generalizations of those given in (8.11) and (8.36). The proofs are very similar to the proof of Proposition 5.3 in Section 5.1 and are left as Problem 8.53.   ∎

With these dilation equations in hand, we can now find the projection coefficients in a dual multiresolution analyses.

Proposition 8.8 (Projections in a Dual Multiresolution Analyses) *Let $\phi(t)$, $\tilde{\phi}(t)$ be scaling functions that generate dual multiresolution analyses $\{V_j\}_{j \in \mathbb{Z}}$, $\{\tilde{V}_j\}_{j \in \mathbb{Z}}$ with associated biorthogonal wavelet functions $\psi(t)$ and $\tilde{\psi}(t)$. Suppose that $f_{j+1}(t) \in V_{j+1}$ and $\tilde{f}_{j+1}(t) \in \tilde{V}_{j+1}$ have the form*

$$f_{j+1}(t) = \sum_{k \in \mathbb{Z}} a_{j+1,k}\phi_{j+1,k}(t), \qquad \tilde{f}_{j+1}(t) = \sum_{k \in \mathbb{Z}} \tilde{a}_{j+1,k}\tilde{\phi}_{j+1,k}(t) \qquad (8.90)$$

Then the projections of $f_{j+1}(t)$ and $\tilde{f}_{j+1}(t)$ into V_j, $\tilde{V}_j$, respectively, are

$$f_j(t) = \sum_{k \in \mathbb{Z}} a_{j,k}\phi_{j,k}(t), \qquad \tilde{f}_j(t) = \sum_{k \in \mathbb{Z}} \tilde{a}_{j,k}\tilde{\phi}_{j,k}(t)$$

and the projections of $f_{j+1}(t)$ and $\tilde{f}_{j+1}(t)$ into W_j, $\tilde{W}_j$ respectively, are

$$g_j(t) = \sum_{k \in \mathbb{Z}} c_{j,k}\psi_{j,k}(t), \qquad \tilde{g}_j(t) = \sum_{k \in \mathbb{Z}} \tilde{c}_{j,k}\tilde{\psi}_{j,k}(t)$$

where

$$a_{j,k} = \sum_{\ell} \tilde{h}_{\ell-2k}a_{j+1,\ell}, \qquad \tilde{a}_{j,k} = \sum_{\ell} h_{\ell-2k}\tilde{a}_{j+1,\ell}$$

$$c_{j,k} = \sum_{\ell} \tilde{g}_{\ell-2k}a_{j+1,\ell}, \qquad \tilde{c}_{j,k} = \sum_{\ell} g_{\ell-2k}\tilde{a}_{j+1,\ell}$$

∎

Proof: These four projections have very similar proofs to that of Proposition 5.6 in the multiresolution analysis setting. We verify the formula for $\tilde{c}_{j,k}$ and leave the others to Problem 8.54. The projection of $\tilde{f}_{j+1}(t)$ into $\tilde{W}_j$ is

$$\tilde{g}_j(t) = \sum_{k \in \mathbb{Z}} \left\langle \tilde{f}_{j+1}(t), \psi_{j,k}(t) \right\rangle \tilde{\psi}_{j,k}(t)$$

Using (8.90), we expand $\tilde{f}_{j+1}(t)$ to obtain

$$\tilde{g}_j(t) = \sum_{k \in \mathbb{Z}} \sum_{\ell \in \mathbb{Z}} \tilde{a}_{j+1,\ell} \left\langle \tilde{\phi}_{j+1,\ell}(t), \psi_{j,k}(t) \right\rangle \tilde{\psi}_{j,k}(t)$$

Using the dilation equation for $\psi_{j,k}(t)$ from Proposition 8.7, the identity above can be rewritten as

$$\tilde{g}_j(t) = \sum_{k \in \mathbb{Z}} \sum_{\ell \in \mathbb{Z}} \tilde{a}_{j+1,\ell} \sum_{n \in \mathbb{Z}} g_{n-2k} \left\langle \tilde{\phi}_{j+1,\ell}(t), \phi_{j+1,n}(t) \right\rangle \tilde{\psi}_{j,k}(t)$$

By the duality of $\phi(t)$ and $\tilde{\phi}(t)$ we know that $\left\langle \tilde{\phi}_{j+1,\ell}(t), \phi_{j+1,n}(t) \right\rangle = \delta_{\ell,n}$ so

$$\tilde{g}_j(t) = \sum_{k \in \mathbb{Z}} \left(\sum_{\ell \in \mathbb{Z}} \tilde{a}_{j+1,\ell} g_{\ell-2k} \right) \tilde{\psi}_{j,k}(t)$$

and thus

$$\tilde{c}_{j,k} = \left\langle \tilde{f}_{j+1}(t), \psi_{j,k}(t) \right\rangle = \sum_{\ell \in \mathbb{Z}} \tilde{a}_{j+1,\ell} g_{\ell-2k}$$

as desired.   ∎

These equations for the projection coefficients $a_{j,k}$ and $c_{j,k}$ in Proposition 8.8 are the key to decomposition of functions. We now present a demonstration of these equations at work with spline filters.

Example 8.11 (Decomposition in Dual Multiresolution Analyses) *Suppose that*

$$f(t) = \sum_{k \in \mathbb{Z}} a_{4,k} \phi_{4,k}(t) \in V_4$$

for the dual multiresolution analysis generated by the (5,3) biorthogonal spline filter pair given in (8.50). For the projection of $f(t)$ into W_3, write the coefficients $c_{3,k}$ as a finite linear combination of $a_{4,k}$ values.
Solution
Using the decomposition formula for $c_{j,k}$ in Proposition 8.8, we write

$$c_{3,k} = \sum_{\ell \in \mathbb{Z}} \tilde{g}_{\ell-2k} a_{4,\ell} = \sum_{\ell \in \mathbb{Z}} (-1)^{\ell-2k} h_{1-(\ell-2k)} a_{4,\ell}$$

Now $\mathbf{h} = (h_{-2}, h_{-1}, h_0, h_1, h_2) = \sqrt{2}\left(-\frac{1}{8}, \frac{1}{4}, \frac{3}{4}, \frac{1}{4}, -\frac{1}{8}\right)$, *so* $h_{1-(\ell-2k)} = 0$ *except when* $1 - (\ell - 2k) = 0, \pm 1, \pm 2$. *We have*

$$c_{3,k} = (-1)^{-1}h_2 a_{4,2k-1} + (-1)^0 h_1 a_{4,2k} + (-1)^1 h_0 a_{4,2k+1}$$
$$+ (-1)^2 h_{-1} a_{4,2k+2} + (-1)^3 h_{-2} a_{4,2k+3}$$
$$= \sqrt{2}\left(\frac{1}{8}a_{4,2k-1} + \frac{1}{4}a_{4,2k} - \frac{3}{4}a_{4,2k+1} + \frac{1}{4}a_{4,2k+2} + \frac{1}{8}a_{4,2k+3}\right) \quad (8.91)$$

In Problem 8.56 you will find a similar formula for the $a_{3,k}$ *coefficients.* ∎

We can gain some insight into formula (8.91) by noticing that the computation can be viewed as the dot product of two vectors:

$$c_{3,k} = \sqrt{2}\left(\frac{1}{8}, \frac{1}{4}, -\frac{3}{4}, \frac{1}{4}, \frac{1}{8}\right) \cdot (a_{4,2k-1}, a_{4,2k}, a_{4,2k+1}, a_{4,2k+2}, a_{4,2k+3})$$
$$= \tilde{\mathbf{g}} \cdot (a_{4,2k-1}, a_{4,2k}, a_{4,2k+1}, a_{4,2k+2}, a_{4,2k+3}) \quad (8.92)$$

As in the orthogonal setting, we can view the calculation of projection coefficients as infinite matrix products. The matrix representation for obtaining the $c_{3,k}$, $k \in \mathbb{Z}$ in (8.92) is given in (8.93). The boldface elements in $\tilde{\mathcal{G}}$ lie along the main diagonal, with the row zero containing the boldface $\tilde{\mathbf{g}}_0$. Multiplying by the row k of $\tilde{\mathcal{G}}$ corresponds to (8.92). If we let $\mathbf{c}^3$ and $\mathbf{a}^4$ denote the bi-infinite sequences with entries $c_{3,k}$, $a_{4,k}$, respectively, we have $\mathbf{c}^3 = \tilde{\mathcal{G}}\mathbf{a}^4$, where

$$\tilde{\mathcal{G}} = \begin{bmatrix} \ddots & \ddots & & & & & & & & \\ \cdots & \tilde{g}_1 & \mathbf{\tilde{g}_2} & \tilde{g}_3 & 0 & 0 & 0 & 0 & \cdots & \cdots \\ \cdots & \tilde{g}_{-1} & \tilde{g}_0 & \mathbf{\tilde{g}_1} & \tilde{g}_2 & \tilde{g}_3 & 0 & 0 & \cdots & \cdots \\ \cdots & 0 & 0 & \tilde{g}_{-1} & \mathbf{\tilde{g}_0} & \tilde{g}_1 & \tilde{g}_2 & \tilde{g}_3 & 0 & 0 & \cdots \\ \cdots & & & 0 & 0 & \mathbf{\tilde{g}_{-1}} & \tilde{g}_0 & \tilde{g}_1 & \tilde{g}_2 & \tilde{g}_3 & \cdots \\ \cdots & & & & 0 & 0 & \mathbf{\tilde{g}_{-2}} & \tilde{g}_{-1} & \tilde{g}_0 & \tilde{g}_1 & \cdots \\ & & & & & & \ddots & & & \ddots \end{bmatrix} \quad (8.93)$$

We can form a similar infinite matrix $\tilde{\mathcal{H}}$ that can be multiplied with $\mathbf{a}^4$ to produce the coefficients $\mathbf{a}^3$ for projecting $f_4(t)$ into V_3. These infinite matrix products will yield the exact projection coefficients. If we truncate the matrices to make them finite, we arrive at nearly the discrete version of the projection given in Definition 8.16 in Section 8.5. Of course, the finite matrix products will miscalculate some of the projection coefficients, due to the truncation and row wrapping, just as in the orthogonal case discussed in Section 7.2. We do not pursue the nuances of this issue here, but the principles are the same as those given in the lengthy treatment for the discrete Daubechies transform.

It is natural to want to zoom in or out in the examination of functions or signals, which requires iterated applications of infinite matrices $\mathcal{H}$, $\tilde{\mathcal{H}}$, $\mathcal{G}$, and $\tilde{\mathcal{G}}$ to project, say, $\mathbf{a}^J = \{a_{J,k}\}_{k\in\mathbb{Z}}$ into coarser spaces V_j, W_j, with coefficient sequences $\mathbf{a}^j =$

$\{a_{j,k}\}_{k\in\mathbb{Z}}$, $\mathbf{c}^j = \{c_{j,k}\}_{k\in\mathbb{Z}}$, respectively, where $j < J$. The entire decomposition process is often called the *biorthogonal pyramid algorithm* and is represented in Figure 8.7.

$$
\begin{array}{ccccccccc}
 & & \text{Iteration 1} & & \text{Iteration 2} & & \text{Iteration 3} & \cdots \\
\mathbf{a}^J & \xrightarrow{\tilde{\mathcal{H}}} & \mathbf{a}^{J-1} & \xrightarrow{\tilde{\mathcal{H}}} & \mathbf{a}^{J-2} & \xrightarrow{\tilde{\mathcal{H}}} & \mathbf{a}^{J-3} & \cdots \\
 & \tilde{\mathcal{G}} \searrow & & \tilde{\mathcal{G}} \searrow & & \tilde{\mathcal{G}} \searrow & & \\
 & & \mathbf{c}^{J-1} & & \mathbf{c}^{J-2} & & \mathbf{c}^{J-3} & \cdots
\end{array}
$$

Figure 8.7 The biorthogonal pyramid algorithm.

This is very similar to the pyramid algorithm in Figure 7.10 for orthogonal wavelets discussed earlier, except that the dual filters $\tilde{\mathbf{h}}$ and $\tilde{\mathbf{g}}$ are used for the decomposition with coefficient equations from Proposition 8.8. This is in agreement with the discrete matrix methods discussed in Section 8.5. Naturally, the other filter pair $\mathbf{h}$, $\mathbf{g}$ should be used in the reconstruction process.

Reconstruction

Now suppose that we have a function $f(t) \in V_{j+1}$ whose projection coefficients $\mathbf{a}^j = \{a_{j,k}\}$ in V_j and $\mathbf{c}^j = \{c_{j,k}\}$ in W_j are known. How can we recover the coefficients $\{a_{j+1,m}\}$ of $f(t)$ in V_{j+1}? To solve this problem, we need to mimic the reconstruction algorithms from the orthogonal setting using dilation equations from Proposition 8.7:

$$f(t) = \sum_{k\in\mathbb{Z}} a_{j,k}\phi_{j,k}(t) + \sum_{k\in\mathbb{Z}} c_{j,k}\psi_{j,k}(t)$$

Multiply both sides of this equation by $\tilde{\phi}_{j+1,m}$ so that we can take advantage of duality, and integrate over $\mathbb{R}$ to obtain

$$
\begin{aligned}
\left\langle f(t), \tilde{\phi}_{j+1,m}(t)\right\rangle &= \sum_{k\in\mathbb{Z}} a_{j,k}\left\langle \phi_{j,k}(t), \tilde{\phi}_{j+1,m}(t)\right\rangle + \sum_{k\in\mathbb{Z}} c_{j,k}\left\langle \psi_{j,k}(t), \tilde{\phi}_{j+1,m}(t)\right\rangle \\
&= \sum_{k\in\mathbb{Z}} a_{j,k}\sum_{\ell\in\mathbb{Z}} h_{\ell-2k}\left\langle \phi_{j+1,\ell}(t), \tilde{\phi}_{j+1,m}\right\rangle \\
&\quad + \sum_{k\in\mathbb{Z}} c_{j,k}\sum_{\ell\in\mathbb{Z}} g_{\ell-2k}\left\langle \phi_{j+1,\ell}(t), \tilde{\phi}_{j+1,m}\right\rangle
\end{aligned}
$$

where the last step uses dilation equations from Proposition 8.7. The duality principle (8.1) tells us the inner products in both double sums above is $\delta_{\ell,m}$. We can thus simplify the inner product and write

$$\langle f(t), \phi_{j+1,m}(t)\rangle = \sum_{k\in\mathbb{Z}} a_{j,k}h_{m-2k} + \sum_{k\in\mathbb{Z}} c_{j,k}g_{m-2k}$$

But $a_{j+1,m} = \langle f(t), \phi_{j+1,m}(t)\rangle$, so we have the desired formula for reconstruction:

$$a_{j+1,m} = \sum_{k\in\mathbb{Z}} a_{j,k}h_{m-2k} + \sum_{k\in\mathbb{Z}} c_{j,k}g_{m-2k} \tag{8.94}$$

Iterating (8.94) produces the inverse pyramid algorithm given in Figure 8.8. Note in particular that the filters $\mathbf{h}$ and $\mathbf{g}$ are used for this reconstruction process.

$$\begin{array}{ccccccccc}
\mathbf{a}^{J+1} & \xleftarrow{\mathbf{h}} & \mathbf{a}^{J} & \xleftarrow{\mathbf{h}} & \mathbf{a}^{J-1} & \xleftarrow{\mathbf{h}} & \mathbf{a}^{J-2} & \cdots \\
 & \overset{\mathbf{g}}{\nwarrow} & & \overset{\mathbf{g}}{\nwarrow} & & \overset{\mathbf{g}}{\nwarrow} & & \\
 & & \mathbf{c}^{J} & & \mathbf{c}^{J-1} & & \mathbf{c}^{J-2} & \cdots
\end{array}$$

Figure 8.8 The inverse biorthogonal pyramid algorithm.

PROBLEMS

Note: *A computer algebra system and the software on the course Web site will be useful for some of these problems. See the Preface for more details.*

8.53 Prove Proposition 8.7.

8.54 Complete the proof of Proposition 8.8.

8.55 Write the formula for the coefficients $a_{3,k}$ in Example 8.11 as an infinite matrix product analogous to (8.93).

8.56 Find a formula for the coefficients $a_{3,k}$ in Example 8.11.

8.57 Consider the dual multiresolution analysis generated by the $(5,3)$ biorthogonal spline filter pair given in (8.50). Using the ideas from Example 8.11, write $c_{2,k}$ as a finite linear combination of $a_{4,k}$ values. Note that two iterations are needed.

8.58 Consider the dual multiresolution analysis generated by the $(5,3)$ biorthogonal spline filter pair given in (8.50). Using the ideas from Example 8.11 and an inverse pyramid algorithm, write $a_{2,j}$ as a finite linear combination of $a_{0,k}$, $c_{0,k}$, and $c_{1,k}$ values.

8.5 THE DISCRETE BIORTHOGONAL WAVELET TRANSFORM

Now that we have two classes of biorthogonal filter pairs available to us and an understanding of the pyramid algorithm in the biorthogonal setting, we can develop a discrete biorthogonal wavelet transformation. The construction is not unlike that performed for the discrete wavelet transformation, except that now there is a different filter pair for inverting the transformation. We will also see how to exploit the symmetry of the biorthogonal filter pairs we develop to modify the the discrete transform and eliminate the need for wrapping rows.

Constructing the Transformation Matrix

To construct the discrete biorthogonal transform matrix, we proceed in the exact same we used to build the wavelet transform matrix (7.14). As is the case in most of the literature, we use the filters $\tilde{\mathbf{h}}$ and $\tilde{\mathbf{g}}$ and build the transform matrix $\tilde{W}_N$. Unlike the orthogonal filter case, the biorthogonal filter pair coefficients have negative indices and this will slightly alter how we define the transform matrix.

Definition 8.15 (The Truncated Projection Matrices) *Suppose that $\tilde{\mathbf{h}}$ is a symmetric scaling filter from a biorthogonal filter pair with $\tilde{\mathbf{h}} = (\tilde{h}_{\tilde{\ell}}, \ldots, \tilde{h}_0, \ldots, \tilde{h}_{\tilde{L}})$, with $\tilde{\ell} \leq 0$, $\tilde{L} > 0$, and suppose that $\tilde{\mathbf{g}}$ is the associated wavelet filter with $\tilde{\mathbf{g}} = (\tilde{g}_{\tilde{m}}, \ldots, \tilde{g}_0, \ldots, \tilde{g}_{\tilde{M}})$, with $\tilde{m} \leq 0$, $\tilde{M} > 0$. Assume that N is an even positive integer with $N > \tilde{L}$ and $N > \tilde{M}$. Then we define the $\frac{N}{2} \times N$ biorthogonal truncated projection matrices $\tilde{H}_{N/2}$ and $\tilde{G}_{N/2}$ as*

$$
\tilde{H}_{N/2} = \begin{bmatrix}
\tilde{h}_0 & \tilde{h}_1 & \cdots & \tilde{h}_{\tilde{L}-1} & \tilde{h}_{\tilde{L}} & 0 & \cdots & 0 & \tilde{h}_{\tilde{\ell}} & \cdots & \tilde{h}_{-1} \\
\tilde{h}_{-2} & \tilde{h}_{-1} & \cdots & \tilde{h}_{\tilde{L}-3} & \tilde{h}_{\tilde{L}-2} & \tilde{h}_{\tilde{L}-1} & \cdots & 0 & \tilde{h}_{\tilde{\ell}-2} & \cdots & \tilde{h}_{-3} \\
\tilde{h}_{-4} & \tilde{h}_{-3} & \cdots & \tilde{h}_{\tilde{L}-5} & \tilde{h}_{\tilde{L}-4} & \tilde{h}_{\tilde{L}-3} & \cdots & & \tilde{h}_{\tilde{\ell}-4} & \cdots & \tilde{h}_{-5} \\
& & \vdots & & & & & & \vdots & & \\
\tilde{h}_4 & \tilde{h}_5 & & & & & \cdots & \tilde{h}_0 & \tilde{h}_1 & \tilde{h}_2 & \tilde{h}_3 \\
\tilde{h}_2 & \tilde{h}_3 & \cdots & & & & \cdots & \tilde{h}_{-2} & \tilde{h}_{-1} & \tilde{h}_0 & \tilde{h}_1
\end{bmatrix}
$$

and

$$
\tilde{G}_{N/2} = \begin{bmatrix}
\tilde{g}_0 & \tilde{g}_1 & \cdots & \tilde{g}_{\tilde{M}-1} & \tilde{g}_{\tilde{M}} & 0 & \cdots & 0 & \tilde{g}_{\tilde{m}} & \cdots & \tilde{g}_{-1} \\
\tilde{g}_{-2} & \tilde{g}_{-1} & \cdots & \tilde{g}_{\tilde{M}-3} & \tilde{g}_{\tilde{M}-2} & \tilde{g}_{\tilde{M}-1} & \cdots & 0 & \tilde{g}_{\tilde{m}-2} & \cdots & \tilde{g}_{-3} \\
\tilde{g}_{-4} & \tilde{g}_{-3} & \cdots & \tilde{g}_{\tilde{M}-5} & \tilde{g}_{\tilde{M}-4} & \tilde{g}_{\tilde{M}-3} & \cdots & & \tilde{g}_{\tilde{m}-4} & \cdots & \tilde{g}_{-5} \\
& & \vdots & & & & & & \vdots & & \\
\tilde{g}_4 & \tilde{g}_5 & & & & & \cdots & \tilde{g}_0 & \tilde{g}_1 & \tilde{g}_2 & \tilde{g}_3 \\
\tilde{g}_2 & \tilde{g}_3 & \cdots & & & & \cdots & \tilde{g}_{-2} & \tilde{g}_{-1} & \tilde{g}_0 & \tilde{g}_1
\end{bmatrix}
$$

∎

Using Definition 8.15, we can define the discrete biorthogonal transform matrix $\tilde{W}_N$.

Definition 8.16 (Discrete Biorthogonal Transform Matrix) *If $\tilde{H}_{N/2}$ and $\tilde{G}_{N/2}$ are given by Definition 8.15, then the* discrete biorthogonal transformation matrix $\tilde{W}_N$ *is*

$$
\tilde{W}_N = \begin{bmatrix} \tilde{H}_{N/2} \\ \tilde{G}_{N/2} \end{bmatrix}
\tag{8.95}
$$

∎

Let's look at an example illustrating the previous two definitions.

Example 8.12 (Biorthogonal Wavelet Transform Matrices) *Let $N = 10$. Construct $\tilde{W}_{10}$ for the $(5,3)$ biorthogonal spline filter pair*

$$\tilde{\mathbf{h}} = (\tilde{h}_{-1}, \tilde{h}_0, \tilde{h}_1) = \left(\frac{\sqrt{2}}{4}, \frac{\sqrt{2}}{2}, \frac{\sqrt{2}}{4} \right)$$

$$\mathbf{h} = (h_{-2}, h_{-1}, h_0, h_1, h_2) = \left(-\frac{\sqrt{2}}{8}, \frac{\sqrt{2}}{4}, \frac{3\sqrt{2}}{4}, \frac{\sqrt{2}}{4}, -\frac{\sqrt{2}}{8} \right)$$

Solution

Using formula (8.29), we can find the nonzero coefficients for the wavelet filter. We have $\tilde{\mathbf{g}} = (\tilde{g}_{-1}, \tilde{g}_0, \tilde{g}_1, \tilde{g}_2, \tilde{g}_3) = (\frac{\sqrt{2}}{8}, \frac{\sqrt{2}}{4}, -\frac{3\sqrt{2}}{2}, \frac{\sqrt{2}}{4}, \frac{\sqrt{2}}{8})$. Using Definitions 8.15 and 8.16, we write

$$\tilde{W}_{10} = \begin{bmatrix} \tilde{h}_0 & \tilde{h}_1 & 0 & 0 & 0 & 0 & 0 & 0 & 0 & \tilde{h}_{-1} \\ 0 & \tilde{h}_{-1} & \tilde{h}_0 & \tilde{h}_1 & 0 & 0 & 0 & 0 & 0 & 0 \\ 0 & 0 & 0 & \tilde{h}_{-1} & \tilde{h}_0 & \tilde{h}_1 & 0 & 0 & 0 & 0 \\ 0 & 0 & 0 & 0 & 0 & \tilde{h}_{-1} & \tilde{h}_0 & \tilde{h}_1 & 0 & 0 \\ 0 & 0 & 0 & 0 & 0 & 0 & 0 & \tilde{h}_{-1} & \tilde{h}_0 & \tilde{h}_1 \\ \tilde{g}_0 & \tilde{g}_1 & \tilde{g}_2 & \tilde{g}_3 & 0 & 0 & 0 & 0 & 0 & \tilde{g}_{-1} \\ 0 & \tilde{g}_{-1} & \tilde{g}_0 & \tilde{g}_1 & \tilde{g}_2 & \tilde{g}_3 & 0 & 0 & 0 & 0 \\ 0 & 0 & 0 & \tilde{g}_{-1} & \tilde{g}_0 & \tilde{g}_1 & \tilde{g}_2 & \tilde{g}_3 & 0 & 0 \\ 0 & 0 & 0 & 0 & 0 & \tilde{g}_{-1} & \tilde{g}_0 & \tilde{g}_1 & \tilde{g}_2 & \tilde{g}_3 \\ \tilde{g}_2 & \tilde{g}_3 & 0 & 0 & 0 & 0 & 0 & \tilde{g}_{-1} & \tilde{g}_0 & \tilde{g}_1 \end{bmatrix}$$

$$= \begin{bmatrix} \frac{\sqrt{2}}{2} & \frac{\sqrt{2}}{4} & 0 & 0 & 0 & 0 & 0 & 0 & 0 & \frac{\sqrt{2}}{4} \\ 0 & \frac{\sqrt{2}}{4} & \frac{\sqrt{2}}{2} & \frac{\sqrt{2}}{4} & 0 & 0 & 0 & 0 & 0 & 0 \\ 0 & 0 & 0 & \frac{\sqrt{2}}{4} & \frac{\sqrt{2}}{2} & \frac{\sqrt{2}}{4} & 0 & 0 & 0 & 0 \\ 0 & 0 & 0 & 0 & 0 & \frac{\sqrt{2}}{4} & \frac{\sqrt{2}}{2} & \frac{\sqrt{2}}{4} & 0 & 0 \\ 0 & 0 & 0 & 0 & 0 & 0 & 0 & \frac{\sqrt{2}}{4} & \frac{\sqrt{2}}{2} & \frac{\sqrt{2}}{4} \\ \frac{\sqrt{2}}{4} & -\frac{3\sqrt{2}}{4} & \frac{\sqrt{2}}{4} & \frac{\sqrt{2}}{8} & 0 & 0 & 0 & 0 & 0 & \frac{\sqrt{2}}{8} \\ 0 & \frac{\sqrt{2}}{8} & \frac{\sqrt{2}}{4} & -\frac{3\sqrt{2}}{4} & \frac{\sqrt{2}}{4} & \frac{\sqrt{2}}{8} & 0 & 0 & 0 & 0 \\ 0 & 0 & 0 & \frac{\sqrt{2}}{8} & \frac{\sqrt{2}}{4} & -\frac{3\sqrt{2}}{4} & \frac{\sqrt{2}}{4} & \frac{\sqrt{2}}{8} & 0 & 0 \\ 0 & 0 & 0 & 0 & 0 & \frac{\sqrt{2}}{8} & \frac{\sqrt{2}}{4} & -\frac{3\sqrt{2}}{4} & \frac{\sqrt{2}}{4} & \frac{\sqrt{2}}{8} \\ \frac{\sqrt{2}}{4} & \frac{\sqrt{2}}{8} & 0 & 0 & 0 & 0 & 0 & \frac{\sqrt{2}}{8} & \frac{\sqrt{2}}{4} & -\frac{3\sqrt{2}}{4} \end{bmatrix}$$

In Problem 8.59 you will show that

$$W_{10} = \left[\begin{array}{cccccccccc}
\frac{3\sqrt{2}}{4} & \frac{\sqrt{2}}{4} & -\frac{\sqrt{2}}{8} & 0 & 0 & 0 & 0 & 0 & -\frac{\sqrt{2}}{8} & \frac{\sqrt{2}}{4} \\
-\frac{\sqrt{2}}{8} & \frac{\sqrt{2}}{4} & \frac{3\sqrt{2}}{4} & \frac{\sqrt{2}}{4} & -\frac{\sqrt{2}}{8} & 0 & 0 & 0 & 0 & 0 \\
0 & 0 & -\frac{\sqrt{2}}{8} & \frac{\sqrt{2}}{4} & \frac{3\sqrt{2}}{4} & \frac{\sqrt{2}}{4} & -\frac{\sqrt{2}}{8} & 0 & 0 & 0 \\
0 & 0 & 0 & 0 & -\frac{\sqrt{2}}{8} & \frac{\sqrt{2}}{4} & \frac{3\sqrt{2}}{4} & \frac{\sqrt{2}}{4} & -\frac{\sqrt{2}}{8} & 0 \\
-\frac{\sqrt{2}}{8} & 0 & 0 & 0 & 0 & 0 & -\frac{\sqrt{2}}{8} & \frac{\sqrt{2}}{4} & \frac{3\sqrt{2}}{4} & \frac{\sqrt{2}}{4} \\
\hline
\frac{\sqrt{2}}{4} & -\frac{\sqrt{2}}{2} & \frac{\sqrt{2}}{4} & 0 & 0 & 0 & 0 & 0 & 0 & 0 \\
0 & 0 & \frac{\sqrt{2}}{4} & -\frac{\sqrt{2}}{2} & \frac{\sqrt{2}}{4} & 0 & 0 & 0 & 0 & 0 \\
0 & 0 & 0 & 0 & \frac{\sqrt{2}}{4} & -\frac{\sqrt{2}}{2} & \frac{\sqrt{2}}{4} & 0 & 0 & 0 \\
0 & 0 & 0 & 0 & 0 & 0 & \frac{\sqrt{2}}{4} & -\frac{\sqrt{2}}{2} & \frac{\sqrt{2}}{4} & 0 \\
\frac{\sqrt{2}}{4} & 0 & 0 & 0 & 0 & 0 & 0 & 0 & \frac{\sqrt{2}}{4} & -\frac{\sqrt{2}}{2}
\end{array}\right] \tag{8.96}$$

Note that the shift in the wavelet filter indices (8.29) requires us to be careful when writing down $\tilde{W}_N$. We use the following rules when constructing $\tilde{W}_N$:

1. $\tilde{\mathbf{h}}$ goes in the first row of $\tilde{W}_N$, with $\tilde{h}_0$ in the $(1,1)$ position and coefficients with positive indices following sequentially to the right. Coefficients with negative indices are wrapped to the end of the row with $\tilde{h}_{-1}$ in the $(1,N)$ position of the first row.

2. A new row is formed by shifting the row above it two units to the right.

3. Repeat the first two steps for $\tilde{\mathbf{g}}$.

In Problem 8.60, you will write $\tilde{W}_{12}$ and W_{12} for the $(8,4)$ biorthogonal spline filter. Let's look at an example.

Example 8.13 (The Discrete Biorthogonal Transform) *Let* $\mathbf{v} \in \mathbb{R}^{24}$ *with the elements* $v_k = 2k$, $k = 1, \ldots, 24$. *Using the* $(5,3)$ *biorthogonal spline filter pair, apply* $\tilde{W}_{24}$ *to* $\mathbf{v}$ *and plot the result.*
Solution
We use Example 8.12 as a guide to form the $\tilde{W}_{24}$. *The computation* $\mathbf{y} = \tilde{W}_{24}\mathbf{v}$ *yields an averages portion* $\mathbf{y}^{1,a}$ *and a details portion* $\mathbf{y}^{1,d}$. *The exact values are*

$$\mathbf{y}^{1,a} = [14\sqrt{2}, 6\sqrt{2}, 10\sqrt{2}, 14\sqrt{2}, 18\sqrt{2}, 22\sqrt{2},$$
$$26\sqrt{2}, 30\sqrt{2}, 34\sqrt{2}, 38\sqrt{2}, 42\sqrt{2}, 46\sqrt{2}]^T$$

and
$$\mathbf{y}^{1,d} = [6\sqrt{2}, 0, 0, 0, 0, 0, 0, 0, 0, 0, 0, -18\sqrt{2}]^T$$

These vectors are plotted in Figure 8.9.
As with the D4 scaling filter, we see some effects of row wrapping in the last entries for both $\mathbf{y}^{1,a}$ *and* $\mathbf{y}^{1,d}$, *but the nonlinearities are not nearly as severe with the biorthogonal filter pair. Because of the top row wrapping for* $\tilde{H}_{12}$ *and* $\tilde{G}_{12}$, *there is also some mild nonlinearity in the first entry for both* $\mathbf{y}^{1,a}$ *and* $\mathbf{y}^{1,d}$. *You are asked to explain the symmetry of the* $\mathbf{y}^{1,d}$ *entries in Problem 8.67.* ∎

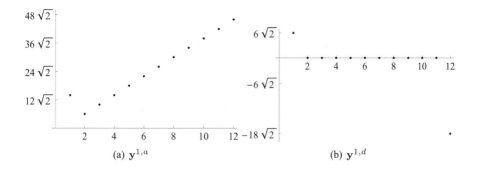

Figure 8.9 The output of the biorthogonal wavelet transformation applied to **v**.

Inverting the Transformation

If we use a CAS and multiply $\tilde{W}_{10}$ from Example 8.12 and W_{10}^T from (8.96), we find that

$$\tilde{W}_{10}W_{10} = I_{10}$$

so that $\tilde{W}_{10}$ is nonsingular with $\tilde{W}_{10}^{-1} = \tilde{W}_{10}^T$. In terms of blocks we have

$$\tilde{W}_{10}W_{10}^T = \left[\begin{array}{c} \tilde{H}_5 \\ \tilde{G}_5 \end{array}\right] \cdot \left[\begin{array}{c} H_5 \\ G_5 \end{array}\right]^T = \left[\begin{array}{c} \tilde{H}_5 \\ \tilde{G}_5 \end{array}\right] \cdot \left[\begin{array}{c|c} H_5^T & G_5^T \end{array}\right]$$

$$= \left[\begin{array}{c|c} \tilde{H}_5 H_5^T & \tilde{H}_5 G_5^T \\ \hline \tilde{G}_5 H_5^T & \tilde{G}_5 G_5^T \end{array}\right] = \left[\begin{array}{c|c} I_5 & 0_5 \\ \hline 0_5 & I_5 \end{array}\right]$$

The fact that $\tilde{W}_{10}^{-1} = W_{10}^T$ is not coincidental: The matrix products $\tilde{H}_5 H_5^T = I_5$, $\tilde{H}_5 G_5^T = 0_5$, $\tilde{G}_5 H_5^T = 0_5$, and $\tilde{G}_5 G_5^T = I_5$ all follow from formulas (8.20), (8.31)–(8.32) and the 2-cyclic structure of the rows of $\tilde{W}_{10}$ and W_{10}. We can generalize our findings with the following proposition.

Theorem 8.6 ($\tilde{W}_N$ and W_N^T Are Inverses) *Let $\tilde{H}_{N/2}$, $H_{N/2}$, $\tilde{G}_{N/2}$, and $G_{N/2}$ be given by Definition 8.15 and then use Definition 8.16 to construct $\tilde{W}_N$ and W_N. Then*

$$\begin{aligned} \tilde{H}_{N/2}H_{N/2}^T = \tilde{G}_{N/2}G_{N/2}^T = I_{N/2} \\ \tilde{H}_{N/2}G_{N/2}^T = \tilde{G}_{N/2}H_{N/2}^T = 0_{N/2} \end{aligned} \tag{8.97}$$

and

$$\tilde{W}_N^{-1} = W_N^T \tag{8.98}$$

∎

Proof: The proofs of the identities in (8.97) are very similar to the proof given in Proposition 7.2, with special care needed for the wrapped rows. These proofs are left as Problems 8.65 and 8.66. Using (8.97), we have in block matrix form

$$
\tilde{W}_N W_N^T = \begin{bmatrix} \tilde{H}_{N/2} \\ \tilde{G}_{N/2} \end{bmatrix} \cdot \begin{bmatrix} H_{N/2} \\ G_{N/2} \end{bmatrix}^T = \begin{bmatrix} \tilde{H}_{N/2} \\ \tilde{G}_{N/2} \end{bmatrix} \cdot \left[H_{N/2}^T \,\middle|\, G_{N/2}^T \right]
$$

$$
= \begin{bmatrix} \tilde{H}_{N/2} H_{N/2}^T & \tilde{H}_{N/2} G_{N/2}^T \\ \tilde{G}_{N/2} H_{N/2}^T & \tilde{G}_{N/2} G_{N/2}^T \end{bmatrix} = \begin{bmatrix} I_{N/2} & 0_{N/2} \\ 0_{N/2} & I_{N/2} \end{bmatrix}
$$

∎

In the next section we discuss iterating the discrete biorthogonal wavelet transform and how the iteration relates to the pyramid algorithm for biorthogonal wavelets.

Iterating the Transform

We can iterate the discrete biorthogonal wavelet transform in exactly the same way as was done for the discrete Haar wavelet transform and the discrete wavelet transform. We define $\tilde{W}_{N,0} = \tilde{W}_N$, and for $j \geq 1$,

$$
\tilde{W}_{N,j} = \mathrm{diag}\left[\tilde{W}_{N/2^j}, I_{N/2^j}, I_{N/2^{j-1}}, \ldots, I_{N/2}\right]
$$

Then we can produce j iterations of the discrete biorthogonal wavelet transformation via the matrix product

$$
\mathbf{y}^j = \tilde{W}_{N,j-1} \cdots \tilde{W}_{N,0} \mathbf{v}
$$

The inverse transformation can be iterated as well. We need to use W_N in this case. We can recover $\mathbf{v}$ via the formula

$$
\mathbf{v} = W_{N,0}^T W_{N,1}^T \cdots W_{N,j-1}^T \mathbf{y}^j
$$

The following example illustrates the iterated discrete biorthogonal transformation.

Example 8.14 (Iterating the Biorthogonal Wavelet Transformation) *Let $\mathbf{v}$ be the vector defined in Example 8.13. Use the $(5,3)$ biorthogonal filter pair and the iterated biorthogonal transformation to compute $\mathbf{y}^3$. Plot the result.*
Solution
We compute

$$
\mathbf{y}^3 = \tilde{W}_{24,2} \tilde{W}_{24,1} \tilde{W}_{24,0} \mathbf{v}
$$

The different portions of the $\mathbf{y}^3$ are

$$
\mathbf{y}^{3,a} = [46\sqrt{2}, 36\sqrt{2}, 68\sqrt{2}]
$$
$$
\mathbf{y}^{3,d} = [21\sqrt{2}, 0, -27\sqrt{2}]
$$
$$
\mathbf{y}^{2,d} = [18, 0, 0, 0, 0, -30]
$$
$$
\mathbf{y}^{1,d} = [6\sqrt{2}, 0, 0, 0, 0, 0, 0, 0, 0, 0, 0, 0, -18\sqrt{2}]
$$

These vectors are plotted in Figure 8.10.   ■

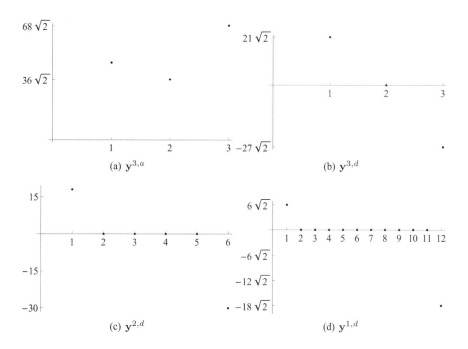

Figure 8.10   Three iterations of the discrete biorthogonal wavelet transform applied to **v**.

Minimizing Boundary Effects with Symmetric Filters

In Examples 7.4 and 8.13 we used the D4 orthogonal scaling filter and the $(5,3)$ biorthogonal spline filter, respectively, to transform the data. In both examples the data are linear, so it is natural, in particular, to expect the values in the averages portion of each transformation to be linear as well. Unfortunately, the blocks $H_{N/2}$ in the orthogonal transform and $\tilde{H}_{N/2}$ in the biorthogonal transform each contain a wrapping row (the last row in $H_{N/2}$ and the first row in $\tilde{H}_{N/2}$). In the case of the orthogonal transform, the last value in $H_{N/2}\mathbf{v}$ is computed using $v_1 = 2$, $v_2 = 4$, $v_{23} = 46$, and $v_{24} = 48$. The computation produces a weighted average $(h_1, h_2, h_3, h_4) \cdot (v_{23}, v_{24}, v_1, v_2)$, but these particular four values of **v** are not linear. A similar problem exists in the first value of $\tilde{H}_{N/2}\mathbf{v}$, where the computation is $(\tilde{h}_{-1}, \tilde{h}_0, \tilde{h}_1) \cdot (v_{24}, v_1, v_2)$. For the details portion of these transforms, there are also edge problems. You are asked to consider these in Problem 8.70.

We now illustrate how a discrete biorthogonal wavelet transform constructed with a symmetric biorthogonal filter pair can be modified to greatly soften these boundary or edge effects in the transformed signal.

An Example Using the Symmetry of the Filter Pair

Consider $\tilde{H}_4$ constructed from the biorthogonal spline filter $\tilde{\mathbf{h}} = (\tilde{h}_{-1}, \tilde{h}_0, \tilde{h}_1) = (\tilde{h}_1, \tilde{h}_0, \tilde{h}_1) = \left(\frac{\sqrt{2}}{4}, \frac{\sqrt{2}}{2}, \frac{\sqrt{2}}{4}\right)$. We write $\tilde{H}_4\mathbf{v}$ using symbolic notation for the filter:

$$\tilde{H}_4\mathbf{v} = \begin{bmatrix} \tilde{h}_0 & \tilde{h}_1 & 0 & 0 & 0 & 0 & 0 & \tilde{h}_1 \\ 0 & \tilde{h}_1 & \tilde{h}_0 & \tilde{h}_1 & 0 & 0 & 0 & 0 \\ 0 & 0 & 0 & \tilde{h}_1 & \tilde{h}_0 & \tilde{h}_1 & 0 & 0 \\ 0 & 0 & 0 & 0 & 0 & \tilde{h}_1 & \tilde{h}_0 & \tilde{h}_1 \end{bmatrix} \cdot \begin{bmatrix} v_1 \\ v_2 \\ v_3 \\ v_4 \\ v_5 \\ v_6 \\ v_7 \\ v_8 \end{bmatrix}$$

$$= \begin{bmatrix} \tilde{h}_1 v_8 + \tilde{h}_0 v_1 + \tilde{h}_1 v_2 \\ \tilde{h}_1 v_2 + \tilde{h}_0 v_3 + \tilde{h}_1 v_4 \\ \tilde{h}_1 v_4 + \tilde{h}_0 v_5 + \tilde{h}_1 v_6 \\ \tilde{h}_1 v_6 + \tilde{h}_0 v_7 + \tilde{h}_1 v_8 \end{bmatrix} = \begin{bmatrix} y_1^{1,a} \\ y_2^{1,a} \\ y_3^{1,a} \\ y_4^{1,a} \end{bmatrix}$$

As you can see, the problem with creating weighted averages of the v_k lies in the first component $y_1^{1,a}$. But notice that unlike the orthogonal filter case the symmetry of $\tilde{\mathbf{h}}$ ensures that each value $y_k^{1,a}$ will be computed by dotting a 3-vector of values of $\mathbf{v}$ with the vector $(\tilde{h}_1, \tilde{h}_0, \tilde{h}_1)$. If the original vector $\mathbf{v}$ were periodic, the symmetry of the filter would produce better correlated weighted averages. The solution? Simply alter $\mathbf{v}$ so that it is periodic!

Instead of applying $\tilde{H}_4$ to $\mathbf{v}$, let's apply $\tilde{H}_7$ to

$$\mathbf{v}^s = [v_1, v_2, \ldots, v_7, v_8, v_7, v_6, \ldots, v_3, v_2]^T \in \mathbb{R}^{14}$$

Now the last value of $\mathbf{v}^s$ is v_2, and this is much better correlated with the first two values of $\mathbf{v}$. We have

$$\mathbf{z} = \tilde{H}_7\mathbf{v}^s = \begin{bmatrix} \tilde{h}_1 v_2 + \tilde{h}_0 v_1 + \tilde{h}_1 v_2 \\ \tilde{h}_1 v_2 + \tilde{h}_0 v_3 + \tilde{h}_1 v_4 \\ \tilde{h}_1 v_4 + \tilde{h}_0 v_5 + \tilde{h}_1 v_6 \\ \tilde{h}_1 v_6 + \tilde{h}_0 v_7 + \tilde{h}_1 v_8 \\ \tilde{h}_1 v_8 + \tilde{h}_0 v_7 + \tilde{h}_1 v_6 \\ \tilde{h}_1 v_6 + \tilde{h}_0 v_5 + \tilde{h}_1 v_4 \\ \tilde{h}_1 v_4 + \tilde{h}_0 v_3 + \tilde{h}_1 v_2 \end{bmatrix}$$

Note that the last three values of $\tilde{H}_7\mathbf{v}^s$ are the same as the three values before them and since we only want four values for our averages portion of the transform, we take as our *modified averages portion*

$$\mathbf{y}^{1,Ma} = [z_1, z_2, z_3, z_4]^T \tag{8.99}$$

and observe that the last three values of $\mathbf{y}^{1,Ma}$ are the same as the last three values of $\mathbf{y}^{1,a}$!

Two questions arise immediately. Does this periodization $\mathbf{v}^s$ of $\mathbf{v}$ take care of wrapping row values in $\tilde{G}_{N/2}\mathbf{v}$? For general N, it is computationally inefficient to take a vector in $\mathbf{v} \in \mathbf{R}^N$, periodize it to obtain $\mathbf{v}^s \in \mathbb{R}^{2N-2}$, compute $\tilde{W}_{2N-2}\mathbf{v}^s$, and retain only the first N values of this product. In general, can we write an $N \times N$ matrix $\tilde{W}_N^p$ that produces the vector $\mathbf{y}^{Ma} = [\mathbf{y}^{1,Ma} \mid \mathbf{y}^{1,Md}]$, and is it invertible?

To answer the first question, consider the wavelet filter associated with $\tilde{h}$. Recall that this filter is built from symmetric filter $\mathbf{h}$:

$$
\begin{aligned}
\tilde{g} &= [\tilde{g}_{-1}, \tilde{g}_0, \tilde{g}_1, \tilde{g}_2, \tilde{g}_3]^T \\
&= [-h_2, h_1, -h_0, h_{-1}, -h_{-2}] \\
&= [-h_2, h_1, -h_0, h_1, -h_2] \\
&= \left[\frac{\sqrt{2}}{8}, \frac{\sqrt{2}}{4}, -\frac{3\sqrt{2}}{4}, \frac{\sqrt{2}}{4}, \frac{\sqrt{2}}{8}\right]^T
\end{aligned}
$$

This filter also displays some symmetry! We compute $\tilde{G}_4\mathbf{v}$

$$
\tilde{G}_4\mathbf{v} =
\begin{bmatrix}
h_1 & -h_0 & h_1 & -h_2 & 0 & 0 & 0 & -h_2 \\
0 & -h_2 & h_1 & -h_0 & h_1 & -h_2 & 0 & 0 \\
0 & 0 & -h_2 & h_1 & -h_0 & h_1 & -h_2 & 0 \\
0 & -h_2 & h_1 & -h_0 & h_1 & -h_2 & 0 & 0
\end{bmatrix}
\cdot
\begin{bmatrix}
v_1 \\ v_2 \\ v_3 \\ v_4 \\ v_5 \\ v_6 \\ v_7 \\ v_8
\end{bmatrix}
$$

$$
=
\begin{bmatrix}
-h_2\mathbf{v_8} + h_1\mathbf{v_1} - h_0\mathbf{v_2} + h_1\mathbf{v_3} - h_2\mathbf{v_4} \\
-h_2 v_2 + h_1 v_3 - h_0 v_4 + h_1 v_5 - h_2 v_6 \\
-h_2 v_4 + h_1 v_5 - h_0 v_6 + h_1 v_7 - h_2 v_8 \\
-h_2\mathbf{v_6} + h_1\mathbf{v_7} - h_0\mathbf{v_8} + h_1\mathbf{v_1} - h_2\mathbf{v_2}
\end{bmatrix}
=
\begin{bmatrix}
y_1^{1,d} \\ y_2^{1,d} \\ y_3^{1,d} \\ y_4^{1,d}
\end{bmatrix}
$$

We see that there are two wrapping rows. Let's see what happens when we compute $\tilde{G}_7\mathbf{v}^s$. We have

$$
\mathbf{w} = \tilde{G}_7\mathbf{v}^s =
\begin{bmatrix}
-h_2 v_2 + h_1 v_1 - h_0 v_2 + h_1 v_3 - h_2 v_4 \\
-h_2 v_2 + h_1 v_3 - h_0 v_4 + h_1 v_5 - h_2 v_6 \\
-h_2 v_4 + h_1 v_5 - h_0 v_6 + h_1 v_7 - h_2 v_8 \\
-h_2 v_6 + h_1 v_7 - h_0 v_8 + h_1 v_7 - h_2 v_6 \\
-h_2 v_8 + h_1 v_7 - h_0 v_6 + h_1 v_5 - h_2 v_4 \\
-h_2 v_6 + h_1 v_5 - h_0 v_4 + h_1 v_3 - h_2 v_2 \\
-h_2 v_4 + h_1 v_3 - h_0 v_2 + h_1 v_1 - h_2 v_2
\end{bmatrix}
$$

Notice now that the last three values of $\mathbf{z}$ are repeats of the first three values of $\mathbf{z}$, and the elements where the wrapping occurs, w_1 and w_4, are computed using consecutive values of $\mathbf{v}$. We take as our details portion of the transformation

$$\mathbf{y}^{1,Md} = [w_1, w_2, w_3, w_4]^T \tag{8.100}$$

so our *modified biorthogonal wavelet transformation* is

$$\mathbf{y}^{1,M} = \left[\mathbf{y}^{1,Ma} \mid \mathbf{y}^{1,Md}\right] \tag{8.101}$$

where $\mathbf{y}^{1,Ma}$ and $\mathbf{y}^{1,Md}$ are given by (8.99) and (8.100), respectively.

In Problem 8.71 you will create a matrix $\tilde{W}_8^M$ such that $\mathbf{y}^{1,M} = \tilde{W}_8^M \mathbf{v}$ and find its inverse.

This process will work for general even N. For the $(5,3)$ biorthogonal filter pair, there will always be one wrapping row in $\tilde{H}_{N/2}$ and two wrapping rows in $\tilde{G}_{N/2}$. Let's return to Example 8.13, but instead, use our modified transform.

Example 8.15 (The Modified Biorthogonal Wavelet Transform) *Apply the modified biorthogonal wavelet transformation (8.101) to the vector* $\mathbf{v} \in \mathbf{R}^{24}$ *whose components are* $v_k = 2k,\ k = 1, 2, \ldots, 24$.
Solution
We create the periodized vector $\mathbf{v}^s \in \mathbf{R}^{46}$:

$$\mathbf{v}^s = [2, 4, 6, \ldots, 44, 46, 48 \mid 46, 44, \ldots, 6, 4]^T$$

and compute $\tilde{H}_{23}\mathbf{v}^s$. *We keep only the first 12 entries of this product to form* $\mathbf{y}^{1,Ma}$:

$$\begin{aligned}
\mathbf{y}^{1,Ma} = [&3\sqrt{2}, 6\sqrt{2}, 10\sqrt{2}, 14\sqrt{2}, 18\sqrt{2}, 22\sqrt{2}\\
&26\sqrt{2}, 30\sqrt{2}, 34\sqrt{2}, 38\sqrt{2}, 42\sqrt{2}, 46\sqrt{2}]^T
\end{aligned}$$

We next compute $\tilde{G}_{23}\mathbf{v}^s$ *and retain the first 12 values to form the details portion of the modified transform*

$$\mathbf{y}^{1,Md} = \left[\frac{\sqrt{2}}{2}, 0, 0, 0, 0, 0, 0, 0, 0, 0, 0, -2\sqrt{2}\right]^T$$

The results are plotted in Figure 8.11.

While the averages portion is not exactly linear, it is a much better approximation of $\mathbf{v}$ *than* $\mathbf{y}^{1,a}$ *from Figure 8.9(a). The details portion* $\mathbf{y}^{1,Md}$ *also compares favorably to that obtained via the biorthogonal tranform plotted in Figure 8.9(b).* ∎

An Algorithm for the Modified Biorthogonal Transform Using Odd-Length Filters

It turns out that this periodization trick works for *any* odd-length symmetric biorthogonal filter pair. The following algorithm shows how to compute the modified biorthogonal wavelet transform using any symmetric odd-length biorthogonal filter pair.

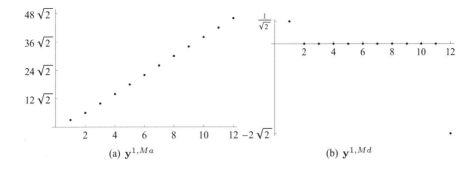

(a) $\mathbf{y}^{1,Ma}$ (b) $\mathbf{y}^{1,Md}$

Figure 8.11   The output of the modified biorthogonal wavelet transformation applied to $\mathbf{v}$.

Algorithm 8.3 (The Modified Biorthogonal Wavelet Transform) *Given a vector* $\mathbf{v} \in \mathbb{R}^N$, *with N an even positive integer, and an odd-length symmetric biorthogonal filter pair, this algorithm computes the modified biorthogonal transform of* $\mathbf{v}$.

1. Form $\mathbf{v}^s \in \mathbb{R}^{2N-2}$ *using the rule*

$$\mathbf{v}^s = \left[v_1, v_2, \ldots, v_{N-1}, v_N \mid v_{N-1}, \ldots, v_3, v_2\right]^T$$

2. Compute $\mathbf{z} = \tilde{H}_{N-1}\mathbf{v}^s$ *and* $\mathbf{w} = \tilde{G}_{N-1}\mathbf{v}^s$.

3. Take

$$\mathbf{y}^{1,Ma} = \left[z_1, \ldots, z_{N/2}\right]^T \qquad and \qquad \mathbf{y}^{1,Md} = \left[w_1, \ldots, w_{N/2}\right]^T$$

4. Return $\mathbf{y}^{1,M} = \left[\mathbf{y}^{1,Ma} \mid \mathbf{y}^{1,Md}\right]^T$. ■

The process is invertible: Given $\mathbf{y}^{Ma} = \left[\mathbf{y}^{1,Ma} \mid \mathbf{y}^{1,Md}\right]^T$, we can suitably periodize $\mathbf{y}^{1,Ma}$ and $\mathbf{y}^{1,Md}$ to create a vector of $\mathbf{z}$ of length $2N-2$:

$$\mathbf{z} = \left[y_1^{1,Ma}, \ldots, y_N^{1,Ma}, y_N^{1,Ma}, y_{N-1}^{1,Ma}, \ldots, y_2^{1,Ma} \mid \right.$$

$$\left. y_1^{1,Md}, \ldots, y_N^{1,Md}, y_N^{1,Md}, y_{N-1}^{1,Md}, \ldots, y_2^{1,Md}\right]^T$$

and compute $\tilde{W}_{2N-2}^{-1}\mathbf{z} = W_{2N-2}^T\mathbf{z}$ to obtain $\mathbf{v}^s$. We can extract $\mathbf{v}$ from $\mathbf{v}^s$.

Algorithm 8.3 and the inverse procedure just described are clearly not the most efficient way to compute the modified transform and its inverse. For a particular filter pair, it would be more efficient to determine $\tilde{W}_N^M$ and its inverse (see, e. g., Problem 8.71) and then exploit the structure of these two matrices to create fast algorithms for computing the modified transform and its inverse.

Of course, we can iterate the transform as we have done with the regular biorthogonal wavelet transform and the wavelet transforms from Chapters 4 and 7. In Problem 8.72 you will write an algorithm for the modified biorthogonal wavelet transformation using even-length symmetric biorthogonal filter pairs.

The Two–Dimensional Biorthogonal Wavelet Transformation

The process for computing the two-dimensional wavelet transform is almost the same as the two-dimensional transforms in Chapters 4 and 7, but in this case we use the matrix $\tilde{W}_N$ generated from filters $\tilde{\mathbf{h}}$ and $\tilde{\mathbf{g}}$. We have

$$B = \tilde{W}_N A \tilde{W}_N^T \tag{8.102}$$

Recall from (8.98) in Theorem 8.6 that we have $\tilde{W}_N^{-1} = W_N^T$, where W_N is the matrix constructed from filters $\mathbf{h}$ and $\mathbf{g}$. In Problem 8.74 you will show that $W_N^{-1} = \tilde{W}^T$, so (8.102) can be rewritten as

$$B = \tilde{W}_N A \tilde{W}_N^T = \tilde{W}_N A W_N^{-1}$$

To invert the transform, we multiply both sides of (8.102) on the left by W_N^T and on the right by W_N and then use (8.98) and Problem 8.74 to write

$$W_N^T B W_N = \left(W_N^T \tilde{W}_N \right) A \left(\tilde{W}_N^T W_N \right) I_N A I_N = A$$

We can also use the modified biorthogonal transform described in Algorithm 8.3 in the two-dimensional setting — the product $\tilde{W}_N A$ is simply an application of $\tilde{W}_N$ to each column of A and we can use Algorithm 8.3 for that computation.

We conclude this section by illustrating an application of the $(5,3)$ biorthogonal spline filter pair.

Example 8.16 (Lossless Compression and JPEG2000) *The JPEG2000 image compression standard [59] uses a variation of the $(5,3)$ biorthogonal spline filter pair to perform lossless image compression. Recall that in lossless compression, no quantization step is performed. In the simplest version of lossless compression, we simply subtract 127 from each element in image matrix A (to center the values around zero), apply the biorthogonal wavelet transformation, and encode the results. In this example we describe the transformation, compute the entropy of the transformed image, and perform Huffman coding.*

Consider the image matrix A from Example 7.9. We will perform lossless compression on this image. The JPEG2000 standard uses the LeGall filter pair $\tilde{\mathbf{h}}^L$, $\mathbf{h}^L$ to perform this task. The LeGall filter pair is easily obtained from the $(5,3)$ biorthogonal spline filter pair $\tilde{\mathbf{h}}$, $\mathbf{h}$. We take

$$\tilde{\mathbf{h}}^L = \frac{1}{\sqrt{2}}\mathbf{h} = \left(-\frac{1}{8}, \frac{1}{4}, \frac{3}{4}, \frac{1}{4}, -\frac{1}{8} \right) \tag{8.103}$$

$$\mathbf{h}^L = \sqrt{2}\tilde{\mathbf{h}} = \left(\frac{1}{2}, 1, \frac{1}{2} \right) \tag{8.104}$$

Note that the filters are switched and have been modified so that they do not include the $\sqrt{2}$. This modification helps with the coding step.

Let $\tilde{A}$ be the matrix obtained by subtracting 127 from each element of A. We form the matrix $\tilde{W}_{512}^L$ associated with the LeGall filter pair and use the biorthogonal transformation to compute $B = \tilde{W}_{512}^L \tilde{A} \left(\tilde{W}_{512}^L\right)^T$. The result is displayed in Figure 8.12.

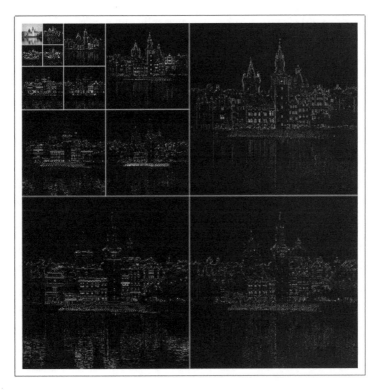

Figure 8.12   The two-dimensional discrete biorthogonal wavelet transformation matrix $B = W_{512}^L \tilde{A} \left(W_{512}^L\right)^T$.

If we Huffman-code the transformation, the bits per pixel is 5.09, a modest savings over the original 8bpp. The coding process is much more sophisticated and efficient than basic Huffman coding. An empirical study by Santa–Cruz and Ebrahimi [49], an empirical study comparing several lossless compression methods showed JPEG2000 averaging about 2.52bpp.[24]

Inverting the transformation is straightforward. In Problem 8.76, you will show that $\tilde{W}_N^{-1} = W_N^T$ just as is the case for the $(5,3)$ biorthogonal spline filter pair. ∎

---

[24]An interesting aside is that $\tilde{W}_N^L$ and W_N^L can be modified so that they map integers to integers. This is highly desirable in coding applications. The key ingredient in the process is called *lifting* and was introduced by Sweldens [58].

PROBLEMS

Note: *A computer algebra system and the software on the course Web site will be useful for some of these problems. See the Preface for more details.*

8.59   Verify (8.96) for the $(5,3)$ biorthogonal spline filter pair.

8.60   Find $\tilde{W}_{12}$ and W_{12} for the $(8,4)$ biorthogonal spline filter pair. Use Example 8.12 as a reference.

8.61   Let $\mathbf{v} \in \mathbb{R}^{24}$ with elements defined by $v_k = 2k$, $k = 1, \ldots, 24$. Decompose $\mathbf{v}$ using the CDF97 biorthogonal filter pair and the wavelet coefficients you found in Problem 8.52. Compare the results to Examples 7.4 and 8.13. Then recover $\mathbf{v}$.

8.62   Find the numerical values for W_{14} and $\tilde{W}_{14}$ for the $(8,4)$ spline filter pair in Example 8.10, and then use a CAS to confirm that $W_{14}^T \tilde{W}_{14} = I$.

8.63   Verify that $\left(\tilde{G} H^T \right)_{i,j} = 0$ for $i, j = 2, 3, 4$ in Example 8.12.

8.64   Verify that $\left(\tilde{G} H^T \right)_{i,j} = 0$ when i or j are 1 or 5 in Example 8.12.

8.65   Under the hypotheses of Theorem 8.6, prove the identities in (8.97) when none of the relevant rows of $\tilde{H}_{N/2}$, $H_{N/2}$, $\tilde{G}_{N/2}$, and $G_{N/2}$ involve row wrapping.

8.66   We saw in the proof given in Proposition 7.2 that special care was needed for the wrapped rows. In this problem you will address one of the cases for the proof of Theorem 8.6 in detail. You will show that entry $\left(\tilde{H}_{N/2} H_{N/2}^T \right)_{m,1} = \delta_{m,1}$, which is required for $W_N^T \tilde{W}_N = I_N$. You will examine the case where

$$2m < N + 1 - \tilde{L} - L \tag{8.105}$$

(a) Write down row 1 of $H_{N/2}$ and row m of $\tilde{H}_{N/2}$. In which columns do h_{-L} and $\tilde{h}_{\tilde{L}}$ appear in these rows?

(b) Show that (8.105) can be rewritten as $1 + \tilde{L} + 2(m-1) < N - L$. Then use part (a) to show that

$$\left(\tilde{H}_{N/2} H_{N/2}^T \right)_{m,1} = \sum_{k=0}^{L} h_k \tilde{h}_{k-2(m-1)} + \sum_{j=-1}^{-L} h_j \tilde{h}_{j-2(m-1)}$$

(c) Now use part (b) and Theorem 8.2 to show that $\left(\tilde{H}_{N/2} H_{N/2}^T \right)_{m,1} = \delta_{m,1}$.

8.67   Rigorously explain the symmetry and all the zero entries of $\mathbf{y}^{1,d}$ in Example 8.13.

8.68   Let $\mathbf{v} \in \mathbb{R}^{20}$ with $v_k = k/5$, $k = 1, \ldots, 20$.

(a) Use the $(5,3)$ biorthogonal spline filter pair to find $\mathbf{y}^2$, the second iteration of the discrete biorthogonal wavelet transform.

(b) Recover $\mathbf{v}$ from your answer in part (a).

8.69   Verify that $\tilde{W}_{N,j}^{-1} = W_{N,j}^T$ for $j \geq 1$.

8.70   Consider the computations $G_{N/2}\mathbf{v}$ and $\tilde{G}_{N/2}$ in Examples 7.4 and 8.13, respectively. Which values of these products are constructed from wrapping rows? Write these values as inner products of the filter coefficients and a portion of $\mathbf{v}$.

8.71   Let $\mathbf{v} \in \mathbf{R}^8$ and $\tilde{\mathbf{h}}, \mathbf{h}$ be the $(5,3)$ biorthogonal spline filter pair.

(a) Find a 4×8 matrix $\tilde{H}_4^M$ so that $\tilde{H}_4^M \mathbf{v}$ results in $\mathbf{y}^{1,Ma}$ given in (8.99).

(b) Find a 4×8 matrix $\tilde{G}_4^M$ so that $\tilde{G}_4^M \mathbf{v}$ results in $\mathbf{y}^{1,Md}$ given in (8.100).

(c) Show that the matrix $\tilde{W}_8^M = \begin{bmatrix} \tilde{H}_4^M \\ \tilde{G}_4^M \end{bmatrix}$ is invertible.

8.72   In this problem you will write an algorithm similar to Algorithm 8.3 using even-length symmetric biorthogonal filter pairs. The following steps will help you organize your work.

(a) Suppose that $\mathbf{v} \in \mathbb{R}^{12}$. Compute $\mathbf{y}^{1,a} = \tilde{H}_6\mathbf{v}$ using the $(8,4)$ biorthogonal spline filter pair.

(b) Now form the vector

$$\mathbf{v}^s = [v_1, \ldots, v_{12} \mid v_{12}, v_{11}, \ldots, v_2, v_1]^T$$

Compute $\mathbf{z} = \tilde{H}_{12}\mathbf{v}^s$ and compare the values of $\mathbf{y}^{1,a}$ to $\mathbf{z}$. We take $\mathbf{y}^{1,Ma} = [z_1, \ldots, z_6]^T$.

(c) Repeat parts (a) and (b) for the products $\mathbf{y}^{1,d} = \tilde{G}_6\mathbf{v}$ and $\mathbf{w} = \tilde{G}_{12}\mathbf{v}^s$. We take $\mathbf{y}^{1,Md} = [w_1, \ldots, w_6]^T$.

(d) Use parts (a)–(c) to write an algorithm for computing the modified biorthogonal wavelet transformation of $\mathbf{v} \in \mathbb{R}^N$ using an even-length symmetric biorthogonal filter pair.

8.73   Let $\mathbf{v} \in \mathbb{R}^{24}$ with $v_k = \left(\frac{k-1}{2}\right)^2$, $k = 1, 2, \ldots, 24$.

(a) Find the average portion $\mathbf{y}^{1,Ma}$ of the modified biorthogonal transform using Problem 8.72 and the $(8,4)$ biorthogonal spline filter pair.

(b) Find the average portion $y^{1,a}$ of the regular biorthogonal transform using the $(8,4)$ biorthogonal spline filter pair.

(c) Compare the edge effects from the vectors in parts (a) and (b).

8.74   Use (8.98) to show that $W_N^{-1} = \tilde{W}_N^T$.

8.75   Let $\tilde{\mathbf{g}}$ and $\mathbf{g}$ be the wavelet filters associated with the $(5,3)$ biorthogonal spline filter pair. Show that the wavelet filters associated with the LeGall filter pair (8.103) and (8.104) can be expressed as $\tilde{\mathbf{g}}^L = \sqrt{2}\tilde{\mathbf{g}}$ and $\mathbf{g}^L = \frac{1}{\sqrt{2}}\mathbf{g}$.

8.76   Let $\tilde{\mathbf{h}}$ and $\mathbf{h}$ be the $(5,3)$ biorthogonal spline filter pair with associated wavelet matrices $\tilde{W}_N$ and W_N. In this problem you will show that the matrix $\tilde{W}_N^L$ constructed from the LeGall filter pair (8.103), (8.104) has inverse $\left(W_N^L\right)^T$. The following steps will help you organize your work.

(a) Using (8.103) and Problem 8.75, write $\tilde{H}_{N/2}^L$, $\tilde{G}_{N/2}^L$ in terms of $H_{N/2}$ and $G_{N/2}$ and $H_{N/2}^L$, $G_{N/2}^L$ in terms of $\tilde{H}_{N/2}$ and $\tilde{G}_{N/2}$.

(b) Use part (a) to write $\tilde{W}_N^L$ and W_N^L in block form. Use (8.97) from Theorem 8.6 to perform the block matrix multiplication $\tilde{W}_N^L \left(W_N^L\right)^T$.

8.6   RIESZ BASIS THEORY

The idea of a multiresolution analysis can effectively be generalized to that of a dual multiresolution analysis under certain conditions on the dual generators $\phi(t)$ and $\tilde{\phi}(t)$. As mentioned in Section 8.1, we need the sets $\{\phi(t-j)\}_{j\in\mathbb{Z}}$ and $\left\{\tilde{\phi}(t-j)\right\}_{j\in\mathbb{Z}}$ to each be "nearly" orthogonal. In this section we formalize this notion with the definition of Riesz bases. Before giving this definition, we need a little terminology and some facts.·

Preliminary Notions

Definition 8.17 (Square–Summable Sequence) *A bi-infinite sequence* $\mathbf{c} = (\dots, c_{-1}, c_0, c_1, \dots)$ *is* square-summable *if*

$$\sum_{k\in\mathbb{Z}} c_k^2 < \infty$$

∎

In this section we work with the spaces $L^2(\mathbb{R})$ and $L^2(I)$, where $I = [a,b]$ is a finite interval. To reduce notational confusion, we label their norms as

$$\|f(t)\|_{\mathbb{R}} = \int_{\mathbb{R}} |f(t)|^2 \, dt \qquad \text{and} \qquad \|f(t)\|_I = \int_a^b |f(t)|^2 \, dt$$

In order to reduce notational clutter we will often write $\phi(t-n)$ as $\phi_n(t)$. Here are some interesting facts about orthogonal bases and square-summable sequences that are useful in our study of Riesz bases.

Proposition 8.9 (Orthogonal Bases and Square–Summable Sequences) *Suppose that* $\mathbf{c} = (\dots, c_{-1}, c_0, c_1, \dots)$ *is a square-summable sequence and* $\{\phi_n(t)\}$ *is an orthogonal basis of* V_0, *where* $\phi(t)$ *generates an orthogonal multiresolution analysis of* $L^2(\mathbb{R})$. *Then*

$$f(t) = \sum_{n\in\mathbb{Z}} c_n e^{in\omega} \in L^2\left([0, 2\pi)\right) \qquad and \qquad g(t) = \sum_{n\in\mathbb{Z}} c_n \phi(t-n) \in L^2(\mathbb{R})$$

with

$$\|f(t)\|_I^2 = \left\|\sum_{n\in\mathbb{Z}} c_n e^{in\omega}\right\|_I^2 = 2\pi \sum_{n\in\mathbb{Z}} c_n^2 \tag{8.106}$$

$$\|g(t)\|_{\mathbb{R}}^2 = \left\|\sum_{n\in\mathbb{Z}} c_n \phi_n(t)\right\|_{\mathbb{R}}^2 = \sum_{n\in\mathbb{Z}} c_n^2 \tag{8.107}$$

∎

Proof: We will show the second result, and leave the first to Problem 8.78. Since $\{\phi_n(t)\}_{n\in\mathbb{Z}}$ is an orthogonal set, we have

$$\left\|\sum_{n\in\mathbb{Z}} c_n \phi_n(t)\right\|_{\mathbb{R}}^2 = \left\langle \sum_{n\in\mathbb{Z}} c_n \phi_n(t), \sum_{k\in\mathbb{Z}} c_k \phi_k(t) \right\rangle$$

$$= \sum_{n\in\mathbb{Z}} \sum_{k\in\mathbb{Z}} c_n c_k \langle \phi_n(t), \phi_k(t) \rangle$$

$$= \sum_{n\in\mathbb{Z}} c_n^2 \langle \phi_n(t), \phi_n(t) \rangle$$

$$= \sum_{n\in\mathbb{Z}} c_n^2$$

since we know that $\|\phi_n(t)\|_{\mathbb{R}} = 1$ by Proposition 5.1. The fact that $\mathbf{c}$ is a square-summable sequence means that the sum above is finite, so $g(t) \in L^2(\mathbb{R})$ and (8.107) is proved. ∎

Riesz Basis Defined

For $\phi(t)$ to be one of the generators of a dual multiresolution analysis, it turns out that the strict equality in (8.107) can be relaxed to a pair of inequalities making the set $\{\phi_n(t)\}$ "nearly" orthogonal. Here is the formal definition making these ideas precise.

Definition 8.18 (Riesz Basis) *Suppose that $\phi(t) \in L^2(\mathbb{R})$ and $0 < A < B$ are constants such that*[25]

$$A \sum_{n \in \mathbb{Z}} c_n^2 \leq \left\| \sum_{n \in \mathbb{Z}} c_n \phi_n(t) \right\|_{\mathbb{R}}^2 \leq B \sum_{n \in \mathbb{Z}} c_n^2 \tag{8.108}$$

for any square-summable sequence **c**. *Then we say that the translates* $\{\phi_n(t)\}_{n \in \mathbb{Z}}$ *form a* Riesz basis *of* $V_0 = \text{span}\{\phi_n(t)\}_{n \in \mathbb{Z}}$. ∎

Of course, all orthogonal bases are Riesz bases, with $A = B = 1$. All of the biorthogonal sets $\{\phi_n(t)\}_{n \in \mathbb{Z}}$ and $\left\{\tilde{\phi}_n(t)\right\}_{n \in \mathbb{Z}}$ built from dual generators in this book are actually Riesz bases, and as mentioned in Definition 8.9 they can be used to create a dual multiresolution analysis. We will show that the B-splines generate Riesz bases later in this section. But first we present a contrary function that satisfies a dilation equation and looks much like the Haar scaling function, but does not even generate a Riesz basis! This example should give a little more insight into Definition 8.18.

Example 8.17 (A Function Satisfying a Dilation Equation But Not a Riesz Basis)
Define $\phi(t)$ *by*

$$\phi(t) = \begin{cases} \frac{1}{3}, & 0 \leq t < 3 \\ 0, & otherwise \end{cases} = \sqcap\left(\frac{t}{3}\right) \tag{8.109}$$

This function looks like a stretched Haar scaling function, and satisfies a similar dilation equation, as you will show in Problem 8.79. The translates of the Haar scaling function form an orthogonal basis, but for this $\phi(t)$, it is not necessarily the case that $\langle \phi(t - \ell), \phi(t - k) \rangle = \delta_{\ell,k}$. We use Definition 8.18 to show that $\phi(t)$ does not generate a Riesz basis for the span of its integer translates. For positive integer m, define the bi-infinite sequences $\mathbf{c}^m$ *component-wise by*

$$c_n^m = \frac{1}{\sqrt{6m}} \begin{cases} 0, & n \leq 0, n > 3m \\ 1, & n = 1 + 3k, k = 0, \ldots, m - 1 \\ 1, & n = 2 + 3k, k = 0, \ldots, m - 1 \\ -2, & n = 3k, k = 1, \ldots, m \end{cases} \tag{8.110}$$

That is, $\mathbf{c}^m$ *is*

$$\mathbf{c}^m = \left(\ldots, c_{-1}^m, c_0^m, c_1^m, c_2^m, c_3^m, c_4^m, c_5^m, c_6^m, \ldots, c_{3m-2}^m, c_{3m-1}^m, c_{3m}^m, c_{3m+1}^m, \ldots \right)$$

$$= \frac{1}{\sqrt{6m}}(\ldots, 0, 0, 1, 1, -2, 1, 1, -2, \ldots, 1, 1, -2, 0, \ldots)$$

[25]The inequalities in (8.108) are often called a *frame* in wavelet literature. See Mallat [42] for more details.

This sequence has been chosen carefully so that it is square-summable to 1 for every positive integer m (see Problem 8.80). Now define the functions $f_m(t) \in V_0$ by

$$f_m(t) = \sum_{n \in \mathbb{Z}} c_n^m \phi(t - n) \tag{8.111}$$

For example,

$$f_6(t) = \frac{1}{6} \left(\phi(t-1) + \phi(t-2) - 2\phi(t-3) \right.$$
$$+ \cdots + \phi(t-16) + \phi(t-17) - 2\phi(t-18))$$

For this choice of $m = 6$, it is easy to see that

$$\sum_{n \in \mathbb{Z}} \left(c_n^6 \right)^2 = \sum_{n=1}^{18} \left(c_n^6 \right)^2 = \frac{1}{36} \sum_{n=1}^{6} \left(1^2 + 1^2 + (-2)^2 \right) = \frac{1}{36} \cdot 36 = 1$$

A plot of $f_6(t)$ is shown in Figure 8.13. There you can see how the overlap of the $\phi_n(t)$ and choice of c_n^m values force $f_6(t) = 0$ for $t \in (-\infty, 0) \cup [3, 19) \cup [21, \infty)$. We can also compute $\|f_6(t)\|_{\mathbb{R}} = \frac{\sqrt{10}}{18}$. In Problem 8.82 you will show that, in general, $f_m(t) = 0$ for $t \in (-\infty, 1) \cup [3, 3m + 1) \cup [3(m + 1), \infty)$ and

$$\|f_m(t)\|_{\mathbb{R}} = \sqrt{\frac{5}{27m}} \tag{8.112}$$

and this norm converges to 0 as $m \to \infty$.

   Now the bi-infinite sequence $\mathbf{c}$ is square-summable to 1, so the condition (8.108) becomes

$$A < \|f_m(t)\|_{\mathbb{R}} < B$$

But $\|f_m(t)\|_{\mathbb{R}} \to 0$ as $m \to \infty$ so we can never find $A > 0$ to satisfy this inequality. Thus the set $\{\phi_n(t)\}_{n \in \mathbb{Z}}$ cannot be a Riesz basis for its span.   ∎

Riesz Bases and the Stability Function

We saw a nice characterization of orthonormal bases in the frequency domain in Theorem 5.5 that used the stability function $\mathcal{A}(\omega)$. Recall that $\{\phi(t - n)\}_{n \in \mathbb{Z}}$ is an orthonormal set if and only if the stability function $\mathcal{A}(\omega)$ is constant. In particular, we have

$$\mathcal{A}(\omega) = \sum_{\ell \in \mathbb{Z}} \left| \hat{\phi}(\omega + 2\pi\ell) \right|^2 = \frac{1}{2\pi}$$

for all $\omega \in \mathbb{R}$. This condition can be relaxed to a nice analog for Riesz bases.

Theorem 8.7 (The Stability Function and Riesz Bases) *Suppose that $\phi(t) \in L^2(\mathbb{R})$ and let $\mathcal{A}(\omega)$ be the stability function given by (5.40). Then $\{\phi_n(t)\}_{n \in \mathbb{Z}}$ is a Riesz basis of its span V_0 if and only if there are constants $0 < \alpha < \beta$ such that*

$$\alpha < \mathcal{A}(\omega) < \beta \tag{8.113}$$

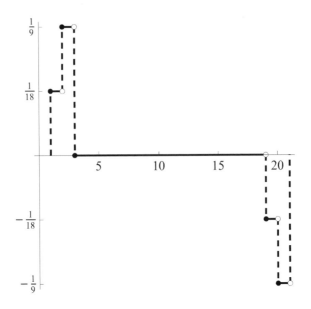

Figure 8.13   The function $f_6(t)$.

■

Proof: Let **c** be a square-summable sequence. From Plancherel's identity, Corollary 2.1, and the translation rule (Proposition 2.6), for Fourier transforms, we know that

$$\left\| \sum_{n \in \mathbb{Z}} c_n \phi_n(t) \right\|_{\mathbb{R}}^2 = \left\| \widehat{\sum_{n \in \mathbb{Z}} c_n \phi_n(t)} \right\|_{\mathbb{R}}^2$$

$$= \left\| \sum_{n \in \mathbb{Z}} c_n e^{-in\omega} \widehat{\phi}(\omega) \right\|_{\mathbb{R}}^2$$

$$= \int_{\mathbb{R}} \left| \sum_{n \in \mathbb{Z}} c_n e^{-in\omega} \widehat{\phi}(\omega) \right|^2 \, d\omega \tag{8.114}$$

Using the same argument as in Theorem 5.5, we partition $\mathbb{R}$ into intervals of length 2π to obtain

$$\int_{\mathbb{R}} \left| \sum_{n \in \mathbb{Z}} c_n e^{-in\omega} \widehat{\phi}(\omega) \right|^2 \, d\omega = \int_0^{2\pi} \left| \sum_{n \in \mathbb{Z}} c_n e^{-in\omega} \right|^2 \sum_{\ell \in \mathbb{Z}} \left| \widehat{\phi}(\omega + 2\pi\ell) \right|^2 \, d\omega$$

Putting these equations together and defining

$$F(\omega) = \sum_{n \in \mathbb{Z}} c_n e^{-in\omega}$$

we have

$$\left\|\sum_{n\in\mathbb{Z}} c_n\phi_n(t)\right\|_{\mathbb{R}}^2 = \int_0^{2\pi} |F(\omega)|^2 \, \mathcal{A}(\omega)\, d\omega \qquad (8.115)$$

Now assume that (8.113) is true. We need to show that (8.108) is valid. We first multiply (8.113) by $|F(\omega)|^2$ and integrate to obtain

$$\alpha \int_0^{2\pi} |F(\omega)|^2 \, d\omega < \int_0^{2\pi} \mathcal{A}(\omega)\, |F(\omega)|^2 \, d\omega < \beta \int_0^{2\pi} |F(\omega)|^2 \, d\omega \qquad (8.116)$$

By Proposition 8.9 we know that

$$\int_0^{2\pi} |F(\omega)|^2 \, d\omega = \left\|\sum_{n\in\mathbb{Z}} c_n e^{-in\omega}\right\|_I^2 = 2\pi \sum_{n\in\mathbb{Z}} c_n^2$$

so applying this and (8.115) to (8.116) yields

$$2\pi\alpha \sum_{n\in\mathbb{Z}} c_n^2 < \left\|\sum_{n\in\mathbb{Z}} c_n\phi_n(t)\right\|_{\mathbb{R}}^2 < 2\pi\beta \sum_{n\in\mathbb{Z}} c_n^2$$

which proves that the Riesz basis condition (8.108) holds with $A = 2\pi\alpha$ and $B = 2\pi\beta$.

We will give a proof adapted from Hong, Wang and Gardner [36] in the other direction under the assumption that $\mathcal{A}(\omega)$ is continuous.[26] Our proof uses the contrapositive, so we assume that (8.113) is false. There are two possibilities — either α or β do not exist. We investigate the case where α does not exist and leave the easier case involving β as Problem 8.83.

Since $\mathcal{A}(\omega)$ is continuous and 2π-periodic, it has a minimum. But since $\mathcal{A}(\omega) \geq 0$ for all $\omega \in \mathbb{R}$ and no $\alpha > 0$ exists to satisfy (8.108), this minimum must be zero, attained at some point $\omega_0 \in [0,2\pi)$ with $\mathcal{A}(\omega_0) = 0$. Now fix a positive integer ℓ. By the continuity of $\mathcal{A}(w)$ there is an interval J about ω_0 for which $\mathcal{A}(\omega) \leq \frac{1}{\ell}$. Let δ denote the width of interval J and define a function $Q(\omega)$ by

$$Q(\omega) = \begin{cases} \frac{1}{\sqrt{\delta}} & \omega \in J \\ 0 & \text{otherwise} \end{cases}$$

Now, clearly, $Q(\omega)$ is in $L^2([0,2\pi))$ with $\|Q(\omega)\|_I = 1$. Since $Q(\omega) \in L^2(\mathbb{R})$, there is a square-summable sequence $\mathbf{c}^\ell$ for which $Q(\omega)$ has the Fourier series $Q(\omega) = \sum_{n\in\mathbb{Z}} c_n^\ell e^{-in\omega}$. By Proposition 8.9 we know that $\sum_{n\in\mathbb{Z}} \left(c_n^\ell\right)^2 = \frac{1}{2\pi}$. Now define another function $f_\ell(t) \in V_0$ by

$$f_\ell(t) = \sum_{n\in\mathbb{Z}} c_n^\ell \phi(t - n)$$

[26] A proof without this assumption is very similar to the one that we present but requires some technical definitions that we wish to avoid.

Then by the same argument as in (8.114) and (8.115), we have

$$\|f_\ell(t)\|_{\mathbb{R}}^2 = \int_0^{2\pi} |Q(\omega)|^2 \, \mathcal{A}(\omega) \, d\omega$$

By the definition of $Q(\omega)$,

$$\int_0^{2\pi} |Q(\omega)|^2 \, \mathcal{A}(\omega) \, d\omega \leq \int_J |Q(\omega)|^2 \, \frac{1}{\ell} \, d\omega = \delta \cdot \frac{1}{\delta} \cdot \frac{1}{\ell} = \frac{1}{\ell}$$

so $\|f_\ell\|_{\mathbb{R}}^2 \leq \frac{1}{\ell}$.

Now we repeat this computation for each integer $\ell > 0$ and in so doing, find a sequence $\mathbf{c}^\ell$ and a function $f_\ell(t)$ of the form $f_\ell(t) = \sum_{n \in \mathbb{Z}} c_n^\ell \phi(t - n)$ with

$$\left\| \sum_{n \in \mathbb{Z}} c_n^\ell \phi(t - n) \right\|_{\mathbb{R}}^2 \leq \frac{1}{\ell} \quad \text{and} \quad \sum_{n \in \mathbb{Z}} \left(c_n^\ell \right)^2 = \frac{1}{2\pi} \tag{8.117}$$

Hence there can be no constant $A > 0$ for which

$$A \sum_{n \in \mathbb{Z}} c_n^2 \leq \left\| \sum_{n \in \mathbb{Z}} c_n \phi_n(t) \right\|^2 \leq \frac{1}{\ell}$$

for all integers $\ell > 0$ and square-summable sequences $\mathbf{c}$, for otherwise (8.117) would imply that

$$0 < A \frac{1}{2\pi} \leq \frac{1}{\ell}$$

for all $\ell > 0$. Then by Definition 8.18 the set $\{\phi_n(t)\}_{n \in \mathbb{Z}}$ is not a Riesz basis of its span V_0, which is what we needed to show to prove the contrapositive. ∎

B–Splines Generate Riesz Bases

We can use Theorem 8.7 to show that the B-splines generate Riesz bases.

Proposition 8.10 (Riesz Bases from B–Splines) *Let $B_n(t)$ be a B-spline. Then the set $\{B_n(t - k)\}_{k \in \mathbb{Z}}$ forms a Riesz basis of $V_0 = \text{span}\{B_n(t - k)\}_{k \in \mathbb{Z}}$.* ∎

Proof: We will show that $\alpha < \mathcal{A}(\omega)$ in (8.113) is valid for $\alpha = \frac{1}{2\pi} \left(\frac{2}{\pi} \right)^{2n}$ and leave other inequality with $\beta = \frac{1}{2\pi}$ to Problem 8.84. We know from Proposition 2.17 that

$$\widehat{B_{n-1}}(\omega) = (2\pi)^{-1/2} e^{-i\omega n/2} \text{sinc}^n \left(\frac{\omega}{2} \right)$$

Recall that $\mathcal{A}(\omega)$ is a 2π-periodic function so it suffices to show that $\alpha < \mathcal{A}(\omega)$ for $\omega \in [0, 2\pi)$. Using some elementary calculus (see Problem 8.85), we see that

$$\frac{\sin(\omega/2)}{\omega/2} \geq \frac{2}{\pi}, \quad 0 < \omega \leq \pi \tag{8.118}$$

$$\frac{\sin(\omega/2)}{\pi - \omega/2} \geq \frac{2}{\pi}, \quad \pi \leq \omega < 2\pi \tag{8.119}$$

For $\omega \in [0, 2\pi)$ and B-spline $B_{n-1}(t)$,

$$\mathcal{A}(\omega) = \sum_{\ell \in \mathbb{Z}} \left| \hat{\phi}(\omega + 2\pi\ell) \right|^2$$

$$= \frac{1}{2\pi} \sum_{\ell \in \mathbb{Z}} \left| \mathrm{sinc}^n \left(\frac{\omega}{2} + \pi\ell \right) \right|^2 \qquad (8.120)$$

$$\geq \frac{1}{2\pi} \left(\left| \mathrm{sinc}^n \left(\frac{\omega}{2} \right) \right|^2 + \left| \mathrm{sinc}^n \left(\frac{\omega}{2} - \pi \right) \right|^2 \right)$$

$$= \frac{1}{2\pi} \left(\left| \frac{\sin(\omega/2)}{\omega/2} \right|^{2n} + \left| \frac{\sin(\omega/2)}{\omega/2 - \pi} \right|^{2n} \right)$$

Then by (8.118), whether $\omega \in [0, \pi]$ or $\omega \in [\pi, 2\pi)$ we know that

$$\frac{1}{2\pi} \left(\left| \frac{\sin(\omega/2)}{\omega/2} \right|^{2n} + \left| \frac{\sin(\omega/2)}{\omega/2 - \pi} \right|^{2n} \right) \geq \frac{1}{2\pi} \left| \frac{2}{\pi} \right|^{2n}$$

Putting this identity together with (8.120) yields

$$\mathcal{A}(\omega) \geq \frac{1}{2\pi} \left(\frac{2}{\pi} \right)^{2n}$$

which finishes the proof of the inequality (8.113) for $\alpha = \frac{1}{2\pi} \left(\frac{2}{\pi} \right)^{2n}$.   ∎

Although it goes beyond the scope of this book to do so, it can be shown that the duals $\tilde{\phi}(t)$ for the B-splines constructed in Section 8.3 also generate Riesz bases, and together with the B-splines generate a dual multiresolution analysis. For more details on this theory, the interested reader can consult Walnut [62] or Hong, Wang and Gardner [36].

PROBLEMS

Note: *A computer algebra system and the software on the course Web site will be useful for some of these problems. See the Preface for more details.*

8.77   Show that the bi-infinite sequence **c** with $c_n = 3^{|n|/2}$, $n \in \mathbb{Z}$ is a square-summable sequence. Find $\sum_{n \in \mathbb{Z}} c_n^2$.

8.78   Prove (8.106). (*Hint:* Review (2.14).)

8.79   Show that $\phi(t)$ defined by (8.109) satisfies a dilation equation. (*Hint:* Sketch a graph of $\phi(t)$ and $\phi(2t - k)$ for different values of k and try something similar to the Haar dilation equation.)

8.80   Show that the sequences defined by (8.110) all satisfy $\sum_{k \in \mathbb{Z}} (c_k^n)^2 = 1$.

8.81   Sketch $f_{10}(t)$ as defined by (8.111). Find $\|f_{10}(t)\|_{\mathbb{R}}$ and show that $f_m(t) = 0$ for $t \in (-\infty, 1) \cup [3,31) \cup [33, \infty)$.

8.82   Show that $f_m(t) = 0$ for $t \in (-\infty, 1) \cup [3, 3m + 1) \cup [3(m + 1), \infty)$ and verify the norm value given in (8.112). Use (8.112) to show that $\lim\limits_{m \to \infty} \|f_m(t)\|_{\mathbb{R}} = 0$.

8.83   Under the assumption that $\mathcal{A}(\omega)$ is continuous, show that there must be a β in (8.113) for the proof of Theorem 8.7.

8.84   In this problem you will finish the proof of Proposition 8.10.

(a) Use the calculus fact $\sin\left(\frac{\omega}{2}\right) \leq \frac{\omega}{2}$ to show that

$$\left| \operatorname{sinc}\left(\frac{\omega}{2} + \pi\ell\right) \right|^{2n} \leq \left| \operatorname{sinc}\left(\frac{\omega}{2} + \pi\ell\right) \right|^{2}$$

(b) Use part (a), Theorem 5.5, and the stability function $\mathcal{A}(\omega)$ for the Haar scaling function to show that

$$\sum_{\ell \in \mathbb{Z}} \left| \operatorname{sinc}^n\left(\frac{\omega}{2} + \pi\ell\right) \right|^{2} \leq 1$$

(c) Use part (b) to show that (8.113) is valid with $\beta = \frac{1}{2\pi}$ for every B-spline $B_n(t)$.

8.85   Use ideas from calculus to establish (8.118) and (8.119).

CHAPTER 9

WAVELET PACKETS

In this chapter we consider an alternative decomposition of $L^2(\mathbb{R})$. For the orthogonal multiresolution analysis $\{V_j\}_{j\in\mathbb{Z}}$, we can use the fact that

$$V_J = V_0 \oplus W_0 \oplus \cdots \oplus W_{J-1}$$

and decompose function $f_J(t) \in V_J$ into component functions $f_0(t) \in V_0$ and $g_k(t) \in W_k, k = 0, \ldots, J-1$. Coifman, Meyer, and Wickerhauser [14, 15] proposed a further decomposition of the wavelet spaces W_k. Their idea was to "split" these spaces into two orthogonal subspaces using the same rule for splitting $V_{j+1} = V_j \oplus W_j$. They called the resulting basis functions (constructed from $\phi(t)$ and $\psi(t)$) for the resulting spaces *wavelet packet functions*. Taken as a set, all the wavelet packet functions form an orthonormal basis for $L^2(\mathbb{R})$. In Section 9.1 we present the basic ideas of their construction. In Section 9.2 we consider the spaces formed by wavelet packet functions and learn how to represent $L^2(\mathbb{R})$ using them. In addition, we see how to form redundant representations of $f(t) \in L^2(\mathbb{R})$ using these spaces. We end this section with a short description of biorthogonal wavelet packets and an example.

The wavelet packet decomposition naturally leads to a discrete algorithm for computing projection coefficients in the wavelet packet spaces. We describe both the

Wavelet Theory: An Elementary Approach with Applications. By D. K. Ruch and P. J. Van Fleet

one- and two-dimensional versions of the discrete wavelet packet transformation in Section 9.3.

Unlike the pyramid algorithm for orthogonal and biorthogonal wavelets and scaling functions, the discrete wavelet packet transform leads to a redundant representation of the input data. Coifman and Wickerhauser in [16] proposed a *best basis algorithm* for determining the best packet decomposition of the discrete input. We describe this algorithm in Section 9.3.

The discrete wavelet packet transform is perhaps best-known as the transformation tool selected by Bradley, Brislawn and Hopper at Los Alamos National Laboratory and the Federal Bureau of Investigation in the early 1990s for use in the *FBI Fingerprint Compression Specification* [8]. This application is discussed in the final section of this chapter.

9.1   CONSTRUCTING WAVELET PACKET FUNCTIONS

We motivate the construction of wavelet packet functions by recalling the problem of decomposing a function $f_3(t) \in V_3$ into its component functions $f_0(t) \in V_0$ and $g_k \in W_k, k = 0,1,2$. If the V_k spaces are generated by an orthogonal scaling function $\phi(t)$, then we can use the operators $\mathcal{H}$ (5.16) and $\mathcal{G}$ (5.27) to perform the task. We have

$$f_3(t) = \sum_{k \in \mathbb{Z}} a_{3,k} \phi_{3,k}(t)$$

The coefficients $a_{j,k}$, $c_{j,k}$ are the projection coefficients into V_j, W_j, $j = 0, 1, 2$, respectively. A schematic of the decomposition appears in Figure 9.1.

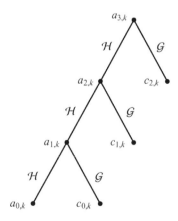

Figure 9.1   The pyramid algorithm for computing the coefficients of the component functions of $f_3(t) \in V_3$.

The nodes of Figure 9.1 are labeled with the component function coefficients while the branches are labeled with the part of the wavelet transformation that was used to create the coefficients.

We can view the process illustrated by Figure 9.1 as a *partial tree decomposition* of the initial coefficients $a_{3,k}$. If we so desired, we could create a *full tree decomposition* of the coefficients by applying $\mathcal{H}, \mathcal{G}$ to $\{c_{2,k}\}$ and then $\mathcal{H}, \mathcal{G}$ to the resulting coefficients as well as to $\{c_{1,k}\}$. The full tree decomposition is plotted in Figure 9.2.

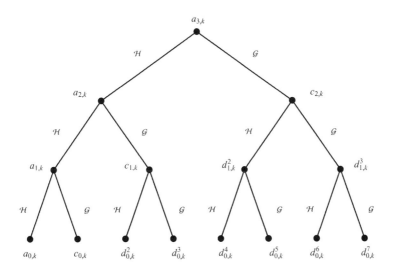

Figure 9.2   Three iterations of the full tree decomposition.

Although we can certainly apply the transform to create the full tree decomposition, it is natural to ask if this decomposition can be represented with some combination of scaling functions or wavelet functions, and if so, what are the spaces represented by these new nodes? Let's try one particular example. Consider the coefficients $d_{1,k}^{3}$. These coefficients are obtained by applying $\mathcal{G}$ to the bi-infinite sequence $\{c_{2,k}\}$:

$$d_{1,k}^{3} = \sum_{\ell \in \mathbb{Z}} g_{\ell - 2k} c_{2,\ell} \tag{9.1}$$

Recall that the $c_{2,k}$ are obtained by projecting $f_3(t)$ into the space W_2. Since this space is spanned by the orthogonal set $\{\psi_{2,\ell}\}_{\ell \in \mathbb{Z}}$, we have $c_{2,\ell} = \langle f_3(t), \psi_{2,\ell}(t) \rangle$. Inserting this identity into (9.1) and using properties of the inner product and the substitution $\ell \mapsto \ell - 2k$ gives

$$
\begin{aligned}
d_{1,k}^3 &= \sum_{\ell \in \mathbb{Z}} g_{\ell-2k} c_{2,\ell} = \sum_{\ell \in \mathbb{Z}} g_{\ell-2k} \big\langle f_3(t), \psi_{2,\ell}(t) \big\rangle \\
&= \Big\langle f_3(t), \sum_{\ell \in \mathbb{Z}} g_{\ell-2k} \psi_{2,\ell}(t) \Big\rangle = \Big\langle f_3(t), \sum_{\ell \in \mathbb{Z}} g_\ell \psi_{2,\ell+2k}(t) \Big\rangle \\
&= \Big\langle f_3(t), \sum_{\ell \in \mathbb{Z}} g_\ell \cdot 2\,\psi(4t - \ell - 2k) \Big\rangle \\
&= \Big\langle f_3(t), \sum_{\ell \in \mathbb{Z}} g_\ell \cdot 2\,\psi\big(2(2t - k) - \ell\big) \Big\rangle \\
&= \Big\langle f_3(t), \sqrt{2}\,w^3(2t - k) \Big\rangle
\end{aligned}
$$

where

$$
w^3(t) = \sqrt{2} \sum_{\ell \in \mathbb{Z}} g_\ell \psi(2t - \ell)
$$

We can replace g_k by h_k in the argument above and easily see that

$$
d_{1,k}^2 = \sum_{\ell \in \mathbb{Z}} h_{\ell-2k} c_{2,\ell} = \big\langle f_3(t), \sqrt{2}\,w^2(2t - k) \big\rangle \tag{9.2}
$$

where

$$
w^2(t) = \sqrt{2} \sum_{\ell \in \mathbb{Z}} h_\ell \psi(2t - \ell) \tag{9.3}
$$

The coefficients $d_{1,k}^2$ and $d_{1,k}^3$ are projection coefficients of functions in spaces spanned by translates and two-dilates of the *wavelet packet functions* $w^2(t)$ and $w^3(t)$, respectively. We learn in Section 9.2 that the span of the translates and dilates of these functions form the *wavelet packet spaces* $\mathcal{W}_{2,1}$ and $\mathcal{W}_{3,1}$. In particular, we have

$$
\mathcal{W}_{2,1} = \mathrm{span}\big\{w^2(2t - k)\big\}_{k \in \mathbb{Z}} \qquad \text{and} \qquad \mathcal{W}_{3,1} = \mathrm{span}\big\{w^3(2t - k)\big\}_{k \in \mathbb{Z}}
$$

It stands to reason that these translates and two-dilates are orthogonal to each other since they are constructed from the translates and two-dilates of the orthogonal wavelet function $\psi(t)$. Indeed, in Section 9.2 we will learn that

$$
\mathcal{W}_2 = \mathcal{W}_{2,1} \oplus \mathcal{W}_{3,1}
$$

Now let's consider the coefficients $d_{0,k}^4$. These values are obtained by applying $\mathcal{H}$ to the coefficients $d_{1,k}^2$. We have

$$
d_{0,k}^4 = \sum_{\ell \in \mathbb{Z}} h_{\ell-2k} d_{1,\ell}^2 \tag{9.4}
$$

We know from (9.2) that the coefficients $d_{1,k}^2$ are projection coefficients of $f_3(t)$ into $\mathcal{W}_{2,1}$. But recall that $g_2(t)$ is the projection of $f_3(t)$ into $\mathcal{W}_2$, so if we project $g_2(t)$

into the space $\mathcal{W}_{2,1}$, we should obtain the same projection coefficients $d_{1,k}^2$. Then we can write

$$d_{1,\ell}^2 = \langle g_2(t), \sqrt{2}\, w^2(2t - \ell) \rangle \tag{9.5}$$

Inserting (9.5) into (9.4) and using simplification similar to that used to find $d_{1,k}^2$ gives

$$
\begin{aligned}
d_{0,k}^4 &= \sum_{\ell \in \mathbb{Z}} h_{\ell - 2k} \langle g_2(t), \sqrt{2}\, w^2(2t - \ell) \rangle \\
&= \Big\langle g_2(t), \sqrt{2} \sum_{\ell \in \mathbb{Z}} h_{\ell - 2k} w^2(2t - \ell) \Big\rangle \\
&= \Big\langle g_2(t), \sqrt{2} \sum_{\ell \in \mathbb{Z}} h_\ell w^2(2t - 2k - \ell) \Big\rangle \\
&= \Big\langle g_2(t), \sqrt{2} \sum_{\ell \in \mathbb{Z}} h_\ell w^2\big(2(t - k) - \ell\big) \Big\rangle \\
&= \langle g_2(t), w^4(t - k) \rangle
\end{aligned}
$$

where

$$w^4(t) = \sqrt{2} \sum_{\ell \in \mathbb{Z}} h_\ell w^2(2t - \ell) \tag{9.6}$$

In a similar manner, we can show that the coefficients $d_{0,k}^5$ can be expressed as

$$d_{0,k}^5 = \langle g_2(t), w^5(t - k) \rangle$$

where

$$w^5(t) = \sqrt{2} \sum_{\ell \in \mathbb{Z}} g_\ell w^2(2t - \ell) \tag{9.7}$$

The wavelet packet functions $w^4(t)$ and $w^5(t)$ are linear combinations of the wavelet packet functions $w^2(2t - \ell)$ and are thus elements of $\mathcal{W}_{2,1}$. In Section 9.2 we show that $\mathcal{W}_{2,1}$ can be decomposed as

$$\mathcal{W}_{2,1} = \mathcal{W}_{4,0} \oplus \mathcal{W}_{5,0}$$

where $\{w^4(t - k)\}_{k \in \mathbb{Z}}$, $\{w^5(t - k)\}_{k \in \mathbb{Z}}$ are orthonormal bases for the spaces $\mathcal{W}_{4,0}$ and $\mathcal{W}_{5,0}$, respectively. In Problem 9.1 you will show that

$$w^6(t) = \sqrt{2} \sum_{\ell \in \mathbb{Z}} h_\ell w^3(2t - \ell) \quad \text{and} \quad w^7(t) = \sqrt{2} \sum_{\ell \in \mathbb{Z}} g_\ell w^3(2t - \ell)$$

can be used to generate the coefficients $d_{0,k}^6$ and $d_{0,k}^7$, respectively.

If we let $w^0(t) = \phi(t)$, $w^1(t) = \psi(t)$, and in general,

$$w_{j,k}^n = 2^{j/2} w^n(2^j t - k)$$

then we can again plot our full tree decomposition, but this time we label the nodes with the functions that produce the coefficients. This tree is plotted in Figure 9.3.

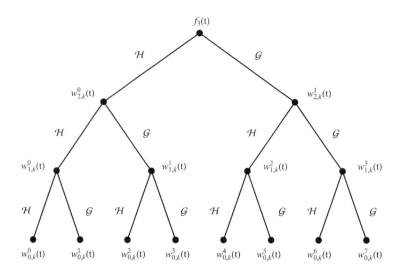

Figure 9.3 Three iterations of the full tree decomposition. Here the nodes are labeled with the functions used to generate the projection coefficients.

Wavelet Packet Functions Defined

Note that once we have established $w^0(t) = \phi(t)$ and $w^1(t) = \psi(t)$, the remaining functions can be generated from them. Those functions with odd superscripts $2j + 1$ are generated from $w^j(t)$ using the wavelet filter **g**, while functions with even superscripts $2j$ are generated from $w^j(t)$ using the scaling filter **h**. We are ready for the following definition:

Definition 9.1 (Wavelet Packet Functions) *Suppose that $\phi(t)$ generates an orthogonal multiresolution analysis $\{V_j\}_{j \in \mathbb{Z}}$ with associated wavelet function $\psi(t)$. The wavelet packet functions are defined by $w^0(t) = \phi(t)$, $w^1(t) = \psi(t)$ and for $n = 2, 3, \ldots$*

$$w^{2n}(t) = \sqrt{2} \sum_{k \in \mathbb{Z}} h_k w^n(2t - k)$$

$$w^{2n+1}(t) = \sqrt{2} \sum_{k \in \mathbb{Z}} g_k w^n(2t - k)$$

(9.8)

∎

Let's look at an example.

Example 9.1 (Haar Wavelet Packet Functions) *Find formulas for the wavelet packet functions $w^k(t)$, $k = 0, \ldots, 7$ for the Haar scaling function $\phi(t) = \sqcap(t)$. Plot these wavelet packet functions.*

Solution

Recall that $\psi(t) = w^1(t) = \sqcap(2t) - \sqcap(2t - 1)$. We use (9.8) to write

$$w^2(t) = \sqrt{2} \sum_{k=0}^{1} h_k w^1(2t - k) = \sqrt{2} \left(\frac{\sqrt{2}}{2} w^1(2t) + \frac{\sqrt{2}}{2} w^1(2t - 1) \right)$$

$$= \sqcap(4t) - \sqcap(4t - 1) + \sqcap(4t - 2) - \sqcap(4t - 3) \qquad (9.9)$$

and

$$w^3(t) = \sqrt{2} \sum_{k=0}^{1} g_k w^1(2t - k) = \sqrt{2} \left(\frac{\sqrt{2}}{2} w^1(2t) - \frac{\sqrt{2}}{2} w^1(2t - 1) \right) \qquad (9.10)$$

$$= \sqcap(4t) - \sqcap(4t - 1) - \sqcap(4t - 2) + \sqcap(4t - 3) \qquad (9.11)$$

These packet functions are plotted in Figure 9.4.

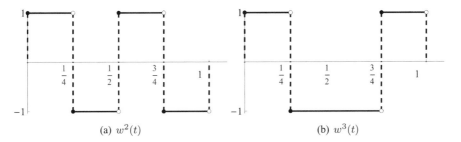

(a) $w^2(t)$ (b) $w^3(t)$

Figure 9.4 The Haar wavelet packet functions $w^2(t)$ and $w^3(t)$.

For $w^4(t)$ we use (9.8) to write

$$w^4(t) = \sqrt{2} \sum_{k=0}^{1} h_k w^2(2t - k) = w^2(2t) + w^2(2t - 1)$$

and (9.9) to obtain

$$\begin{aligned} w^4(t) &= w^2(2t) + w^2(2t - 1) \\ &= \sqcap(8t) - \sqcap(8t - 1) + \sqcap(8t - 2) - \sqcap(8t - 3) \\ &\quad + \sqcap(8t - 4) - \sqcap(8t - 5) + \sqcap(8t - 6) - \sqcap(8t - 7) \end{aligned}$$

Using (9.8) and (9.9) we have

$$\begin{aligned} w^5(t) &= w^2(2t) - w^2(2t - 1) \\ &= \sqcap(8t) - \sqcap(8t - 1) + \sqcap(8t - 2) - \sqcap(8t - 3) \\ &\quad - \sqcap(8t - 4) + \sqcap(8t - 5) - \sqcap(8t - 6) + \sqcap(8t - 7) \end{aligned}$$

In Problem 9.2 you will verify that the formulas for $w^6(t)$ and $w^7(t)$ are

$$w^6(t) = \sqcap(8t) - \sqcap(8t - 1) - \sqcap(8t - 2) + \sqcap(8t - 3)$$
$$+ \sqcap(8t - 4) + \sqcap(8t - 5) + \sqcap(8t - 6) - \sqcap(8t - 7) \qquad (9.12)$$

and

$$w^7(t) = \sqcap(8t) - \sqcap(8t - 1) - \sqcap(8t - 2) + \sqcap(8t - 3)$$
$$- \sqcap(8t - 4) + \sqcap(8t - 5) + \sqcap(8t - 6) - \sqcap(8t - 7) \qquad (9.13)$$

The Haar wavelet packet functions $w^4(t), \ldots, w^7(t)$ are plotted in Figure 9.5.

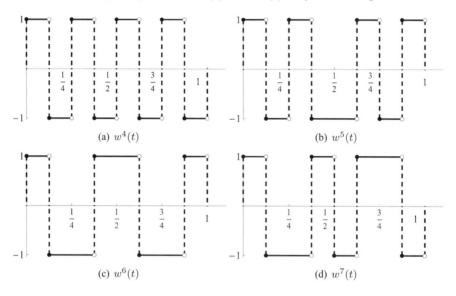

(a) $w^4(t)$ (b) $w^5(t)$

(c) $w^6(t)$ (d) $w^7(t)$

Figure 9.5   The Haar wavelet packet functions $w^4(t)$, $w^5(t)$, $w^6(t)$, and $w^7(t)$.

■

Support and Inner Product Properties of Wavelet Packet Functions

We are now ready to state and prove some properties obeyed by wavelet packet functions. The first gives information about the support of wavelet packet functions.

Proposition 9.1 (Support Properties for Wavelet Packet Functions) *Suppose that the wavelet packet functions $w^n(t)$ given in (9.8) have been constructed using orthogonal scaling function $\phi(t) = w^0(t)$ with nonzero scaling filter values $h_0, \ldots, h_L$. Further suppose that $\overline{\mathrm{supp}}(\phi) = [0, L]$. If*

$$\psi(t) = \sqrt{2} \sum_{k=0}^{L} g_k \phi(2t - k)$$

where $g_k = (-1)^k h_{L-k}$, then for each $n = 1, 2, \ldots, \overline{\text{supp}}\,(w^n) = [0, L]$.   ∎

Proof: The proof of this proposition uses induction on n. See Problem 9.3.   ∎

Let's look at an example that illustrates the support properties given in Proposition 9.1.

Example 9.2 (Wavelet Packets from the D4 Scaling Function) *Suppose that $\phi(t)$ is the Daubechies scaling function generated from the length four scaling filter (6.39). Plot $w^n(t)$ for $n = 0, 1, 2, 3$.*

Solution

We take $g_k = (-1)^k h_{3-k}$, $k = 0, \ldots, 3$ as our wavelet filter. From Problem 7.9 in Section 7.1, we know that $\text{supp}(\psi) = [0, 3]$. We use (9.8) to generate the linear combinations of $\psi(2t - k)$ needed for $w^2(t)$ and $w^3(t)$. All four packet functions are plotted in Figure 9.6, and each is supported on the interval $[0,3]$.   ∎

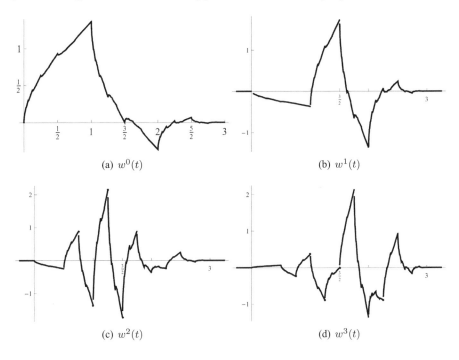

(a) $w^0(t)$ (b) $w^1(t)$

(c) $w^2(t)$ (d) $w^3(t)$

Figure 9.6   The Daubechies wavelet packet functions $w^0(t)$, $w^1(t)$, $w^2(t)$, and $w^3(t)$.

Since wavelet packet functions are constructed from a scaling function $\phi(t)$ and wavelet function $\psi(t)$ that generate orthonormal sets $\{\phi(t - k)\}_{k \in \mathbb{Z}}$ and $\{\psi(t - k)\}_{k \in \mathbb{Z}}$, it stands to reason that packet functions should form an orthonormal set as well. The next few propositions establish orthogonality relations obeyed by wavelet packet functions. These propositions culminate in Theorem 9.1, which establishes the fact that $\{w^n(t-k)\}$ where, $j, n \in \mathbb{Z}$ with $n \geq 0$, is an orthonormal set of functions.

Wavelet Packet Functions and Orthogonality

The next result establishes the fact that integer translates of a wavelet packet function are orthogonal to each other.

Proposition 9.2 (Orthogonality of Translates) *Suppose that $w^n(t)$ is given by (9.8). If j and k are integers, then*

$$\langle w^n(t-j), w^n(t-k)\rangle = \delta_{j,k} \tag{9.14}$$

∎

Proof: Using (1.20) from Proposition 1.7, it is enough to show that

$$\langle w^n(t), w^n(t-k)\rangle = \delta_{0,k} \tag{9.15}$$

We proceed by induction on n. Consider the case where $n = 0$. Then we have

$$\langle w^0(t), w^0(t-k)\rangle = \langle \phi(t), \phi(t-k)\rangle = \delta_{0,k}$$

since $\{\phi(t-k)\}_{k\in\mathbb{Z}}$ is an orthonormal set of functions. Now assume that

$$\langle w^\ell(t), w^\ell(t-k)\rangle = \delta_{0,k}$$

for $0 \leq \ell < n$. Then if $n > 0$ is even, we use the substitution $p = 2k + j$ to write

$$\langle w^n(t), w^n(t-k)\rangle = \left\langle \sqrt{2}\sum_{i\in\mathbb{Z}} h_i w^{n/2}(2t-i), \sqrt{2}\sum_{j\in\mathbb{Z}} h_j w^{n/2}(2(t-k)-j)\right\rangle$$

$$= 2\sum_{i\in\mathbb{Z}}\sum_{j\in\mathbb{Z}} h_i h_j \left\langle w^{n/2}(2t-i), w^{n/2}(2t-(2k+j))\right\rangle$$

$$= 2\sum_{i\in\mathbb{Z}}\sum_{p\in\mathbb{Z}} h_i h_{p-2k}\left\langle w^{n/2}(2t-i), w^{n/2}(2t-p)\right\rangle$$

Applying (1.21) from Proposition 1.7 (with $m = 1$) to the last identity gives

$$\langle w^n(t), w^n(t-k)\rangle = \sum_{i\in\mathbb{Z}}\sum_{p\in\mathbb{Z}} h_i h_{p-2k}\left\langle w^{n/2}(t), w^{n/2}(t-(p-i))\right\rangle$$

But the inner product on the right-hand side of this identity is $\delta_{0,p-i} = \delta_{p,i}$ by the induction hypothesis and we can write

$$\langle w^n(t), w^n(t-k)\rangle = \sum_{i\in\mathbb{Z}}\sum_{p\in\mathbb{Z}} h_i h_{p-2k}\delta_{p,i} = \sum_{i\in\mathbb{Z}} h_i h_{i-2k}$$

By Proposition 5.5(ii), the last sum above reduces to $\delta_{0,k}$ and the result is established for even n. The case of n odd is similar and is left as Problem 9.4. ∎

We are now ready to establish the orthogonality relationship between $w^n(t - j)$ and $w^0(t - k) = \phi(t - k)$.

Proposition 9.3 (Orthogonality at Level Zero) *Suppose that $\phi(t)$ is a scaling function that along with its integer translates, forms an orthonormal basis for $V_0 \in L^2(\mathbb{R})$. Assume that the wavelet packet functions $w^m(t)$ are given by (9.8). Then*

$$\langle w^m(t - j), w^0(t - k) \rangle = \delta_{m,0} \delta_{j,k} \tag{9.16}$$

for integers j and k and nonnegative integer m. ∎

Proof: Using (1.20) from Proposition 1.7, it is enough to prove that

$$\langle w^m(t), w^0(t - k) \rangle = \delta_{m,0} \delta_{0,k} \tag{9.17}$$

for all $k \in \mathbb{Z}$.

The proof to establish (9.17) is by mathematical induction. The base step, when $m = 0$, is established in Proposition 9.2. Now suppose $m > 0$ is even and that

$$\langle w^\ell(t), w^0(t - k) \rangle = 0 \tag{9.18}$$

holds for $\ell = 0, \ldots, m - 1$. We have

$$\langle w^m(t), w^0(t - k) \rangle = \left\langle \sqrt{2} \sum_{i \in \mathbb{Z}} h_i w^{m/2}(2t - i), \phi(t - k) \right\rangle$$

$$= \left\langle \sqrt{2} \sum_{i \in \mathbb{Z}} h_i w^{m/2}(2t - i), \sqrt{2} \sum_{j \in \mathbb{Z}} h_j \phi(2(t - k) - j) \right\rangle$$

$$= 2 \sum_{i \in \mathbb{Z}} \sum_{j \in \mathbb{Z}} h_i h_j \langle w^{m/2}(2t - i), \phi(2t - (2k + j)) \rangle$$

$$= 2 \sum_{i \in \mathbb{Z}} \sum_{p \in \mathbb{Z}} h_i h_{p-2k} \langle w^{m/2}(2t - i), \phi(2t - p) \rangle$$

$$= \sum_{i \in \mathbb{Z}} \sum_{p \in \mathbb{Z}} h_i h_{p-2k} \langle w^{m/2}(t), \phi(t - (p - i)) \rangle$$

The last identity was obtained using (1.21) from Proposition 1.7 (with $m = 1$). By the induction hypothesis (9.18), the inner products on the right-hand side of the identity above are zero and the result is established. The case where m is odd is similar and is left as Problem 9.5. ∎

We are now ready to state and prove a main result concerning orthogonality and wavelet packet functions.

Theorem 9.1 (Orthogonality and Wavelet Packet Functions) *Suppose that $\phi(t)$ is a scaling function that generates the orthonormal set $\{\phi(t-k)\}_{k\in\mathbb{Z}}$, and the associated packet functions $w^n(t)$ are given by (9.8). Then for integers j, k and nonnegative integers m, n, we have*

$$\langle w^m(t-j), w^n(t-k)\rangle = \delta_{m,n}\delta_{j,k} \tag{9.19}$$

∎

Proof: According to (1.20) from Proposition 1.7, we need only show

$$\langle w^m(t), w^n(t-k)\rangle = \delta_{m,n}\delta_{0,k} \tag{9.20}$$

for all $k \in \mathbb{Z}$.

The proof is induction on n. The base step when $n = 0$ is established in Proposition 9.3. The induction hypothesis says that for all nonnegative integers m and integers $0 \le \ell < n$, we have

$$\langle w^m(t), w^\ell(t-k)\rangle = 0 \tag{9.21}$$

We have four cases: m, n both even, m, n both odd, m even and n odd, and m odd and n even. Let's consider the case when both m and n are even.

If $m = n$, then $\frac{m}{2} = \frac{n}{2}$ and the result follows from Proposition 9.2. Now if $m \ne n$, we write

$$
\begin{aligned}
\langle w^m(t), w^n(t-k)\rangle &= \Big\langle \sqrt{2}\sum_{i\in\mathbb{Z}} h_i w^{m/2}(2t-i), \sqrt{2}\sum_{j\in\mathbb{Z}} h_j w^{n/2}(2(t-k)-j)\Big\rangle \\
&= 2\sum_{i\in\mathbb{Z}}\sum_{j\in\mathbb{Z}} h_i h_j \big\langle w^{m/2}(2t-i), w^{n/2}(2t-(2k+j))\big\rangle \\
&= 2\sum_{i\in\mathbb{Z}}\sum_{p\in\mathbb{Z}} h_i h_{p-2k} \big\langle w^{m/2}(2t-i), w^{n/2}(2t-p)\big\rangle \\
&= 2\sum_{i\in\mathbb{Z}}\sum_{p\in\mathbb{Z}} h_i h_{p-2k} \big\langle w^{m/2}(2t-i), w^{n/2}(2t-p)\big\rangle \\
&= \sum_{i\in\mathbb{Z}}\sum_{p\in\mathbb{Z}} h_i h_{p-2k} \big\langle w^{m/2}(t), w^{n/2}(t-(p-i))\big\rangle \tag{9.22}
\end{aligned}
$$

The next-to-last identity follows from the substitution $p = 2k + j$, while the last identity follows from (1.21) from Proposition 1.7 (with $m = 1$). By the induction hypothesis (9.21),

$$\big\langle w^{m/2}(t), w^{n/2}(t-(p-i))\big\rangle = 0$$

and (9.20) is established in this case. The case where both m, n are odd is similar and is left as Problem 9.7.

Now suppose that m is even and n is odd. If $m \ne n - 1$, then $\frac{m}{2} \ne \frac{n-1}{2}$ and we can write

$$\langle w^m(t), w^n(t-k)\rangle = \Big\langle \sqrt{2}\sum_{i\in\mathbb{Z}} h_i w^{m/2}(2t-i), \sqrt{2}\sum_{j\in\mathbb{Z}} g_j w^{(n-1)/2}(2(t-k)-j)\Big\rangle$$

An argument similar to (9.22) gives

$$\langle w^m(t), w^n(t-k)\rangle = \sum_{i\in\mathbb{Z}}\sum_{p\in\mathbb{Z}} h_i g_{p-2k}\langle w^{m/2}(t), w^{(n-1)/2}(t-(p-i))\rangle$$

and we know that the inner products on the right-hand side of this identity are zero by the induction hypothesis. Now if $m = n - 1$, we use simplifications similar to those above and write

$$\langle w^m(t), w^n(t-k)\rangle = \langle w^{n-1}(t), w^n(t-k)\rangle$$
$$= \sum_{i\in\mathbb{Z}}\sum_{p\in\mathbb{Z}} h_i g_{p-2k}\langle w^{(n-1)/2}(t), w^{(n-1)/2}(t-(p-i))\rangle$$
$$= \sum_{i\in\mathbb{Z}}\sum_{p\in\mathbb{Z}} h_i g_{p-2k}\delta_{i,p}$$
$$= \sum_{i\in\mathbb{Z}} h_i g_{i-2k}$$

where the next-to-last identity follows from Proposition 9.2. The last sum above is zero by (5.58) from Problem 5.44 in Section 5.2. Thus $\langle w^m(t), w^n(t-k)\rangle = 0$, which establishes the result for this case.

The last case, m odd and n even, is very similar except that we consider the cases $n = m - 1$ and $n \neq m - 1$. See Problem 9.8. ∎

The following corollary is immediate.

Corollary 9.1 (Wavelet Packet Functions are Linearly Independent) *The wavelet packet functions* $\{w^n(t-k)\}_{k,n\in\mathbb{Z}, n\geq 0}$ *given in Theorem 9.1 is a linearly independent set of functions.* ∎

Proof: The proof is left as Problem 9.9. ∎

Biorthogonal Wavelet Packet Functions

We conclude this section with a short discussion on wavelet packet functions that are constructed from biorthogonal scaling functions. For biorthogonal scaling functions $\tilde{\phi}(t)$, $\phi(t)$ and their associated wavelet functions $\tilde{\psi}(t)$, $\psi(t)$, we can construct the wavelet packet functions $w^n(t)$ as described in (9.8) and the dual wavelet packet functions $\tilde{w}^0(t) = \tilde{\phi}(t)$, $\tilde{w}^1(t) = \tilde{\psi}(t)$ and for $n = 2, 3, \ldots$

$$\tilde{w}^{2n}(t) = \sqrt{2}\sum_{k\in\mathbb{Z}} \tilde{h}_k \tilde{w}^n(2t-k)$$

$$\tilde{w}^{2n+1}(t) = \sqrt{2}\sum_{k\in\mathbb{Z}} \tilde{g}_k \tilde{w}^n(2t-k) \tag{9.23}$$

We state the following theorem and leave the proof as Problem 9.10.

Proposition 9.4 (Dual Wavelet Packets Are Orthogonal) *If $w^n(t)$ and $\tilde{w}^m(t)$ are given by (9.8) and (9.23), respectively, then*

$$\langle w^n(t-j), \tilde{w}^m(t-k) \rangle = \delta_{n,m}\delta_{j,k} \tag{9.24}$$

∎

Proof: The proof of this proposition is left as Problem 9.10. ∎

Let's look at an example.

Example 9.3 (Linear B–Spline Wavelet Packets) *Let $\tilde{\phi}(t)$ and $\phi(t)$ be the scaling functions associated with the $(5,3)$ biorthogonal spline filter pair. In particular, $\tilde{\phi}(t)$ is the symmetric linear B-spline $\tilde{\phi}(t) = B_1(t+1) = \wedge(t-1)$. Find formulas for $\tilde{w}^p(t)$, $p = 0, 1, 2, 3$, and plot the results.*
Solution
   We have $\tilde{w}^0(t) = \tilde{\phi}(t)$. Recall that the scaling filter $\tilde{\mathbf{h}}$ for $\tilde{\phi}(t)$ is

$$\tilde{\mathbf{h}} = \left(\tilde{h}_{-1}, \tilde{h}_0, \tilde{h}_1 \right) = \left(\frac{\sqrt{2}}{4}, \frac{\sqrt{2}}{2}, \frac{\sqrt{2}}{4} \right) \tag{9.25}$$

and the dual filter is $\mathbf{h} = (h_{-2}, h_{-1}, h_0, h_1, h_2) = \left(-\frac{\sqrt{2}}{8}, \frac{\sqrt{2}}{4}, \frac{3\sqrt{2}}{4}, \frac{\sqrt{2}}{4}, -\frac{\sqrt{2}}{8} \right)$ and we can use $\mathbf{h}$ to construct the wavelet filter $\tilde{\mathbf{g}}$. Using the formula $\tilde{g}_k = (-1)^k h_{1-k}$, we have

$$\tilde{\mathbf{g}} = (\tilde{g}_{-1}, \tilde{g}_0, \tilde{g}_1, \tilde{g}_2, \tilde{g}_3) = \left(\frac{\sqrt{2}}{8}, \frac{\sqrt{2}}{4}, -\frac{3\sqrt{2}}{4}, \frac{\sqrt{2}}{4}, \frac{\sqrt{2}}{8} \right) \tag{9.26}$$

We can use (9.25) and (9.26) to write

$$\tilde{w}^1(t) = \tilde{\psi}(t) = \sqrt{2} \sum_{k=-1}^{3} \tilde{g}_k \tilde{\phi}(2t-k)$$

$$\tilde{w}^2(t) = \sqrt{2} \sum_{k=-1}^{1} \tilde{h}_k \tilde{w}^1(2t-k)$$

$$\tilde{w}^3(t) = \sqrt{2} \sum_{k=-1}^{3} \tilde{g}_k \tilde{w}^1(2t-k)$$

The four biorthogonal wavelet packet functions are plotted in Figure 9.7. ∎

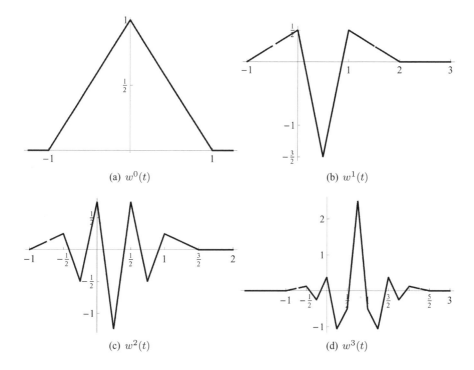

Figure 9.7   The symmetric linear B-spline wavelet packet functions $\tilde{w}^0(t)$, $\tilde{w}^1(t)$, $\tilde{w}^2(t)$, and $\tilde{w}^3(t)$.

PROBLEMS

Note: *A computer algebra system and the software on the course Web site will be useful for some of these problems. See the Preface for more details.*

9.1   Use arguments similar to the one preceding (9.6) to show that

$$d^6_{0,k} = \langle h_2(t), w^6(t-k) \rangle \quad \text{and} \quad d^7_{0,k} = \langle g_2(t), w^7(t-k) \rangle$$

where

$$w^6(t) = \sqrt{2} \sum_{\ell \in \mathbb{Z}} h_\ell w^3(t-\ell) \quad \text{and} \quad w^7(t) = \sqrt{2} \sum_{\ell \in \mathbb{Z}} g_\ell w^3(t-\ell)$$

9.2   Verify formulas (9.12) and (9.13) from Example 9.1.

9.3   Use an induction argument to prove Proposition 9.1.

9.4   In this problem you will complete the proof of Proposition 9.2 by considering n odd in (9.15).

(a) Let $n = 1$. Show that

$$\langle w^1(t), w^1(t-k) \rangle = \sum_{p \in \mathbb{Z}} g_p g_{p-2k}$$

and then use Problem 5.11 in Section 5.1 to complete the proof.

(b) For $n > 1$ odd, use an induction argument similar to the one used for even n in the proof of Proposition 9.2 to establish the result.

9.5   In this problem you will complete the induction proof of Proposition 9.3 for m odd.

(a) Suppose that $m = 1$. Show that

$$\langle w^1(t), w^0(t-k) \rangle = \sum_{p \in \mathbb{Z}} g_p h_{p-2k}$$

and then use Problem 5.44 in Section 5.2 to show that the inner product is zero.

(b) Now consider $m > 1$ odd. Mimic the induction argument for m even to complete the proof.

9.6   Establish (9.19) for $m > 1$ odd in the proof of Theorem 9.1.

9.7   Perform the induction proof for Theorem 9.1 for m, n both odd.

9.8   Perform the induction proof for Theorem 9.1 for m odd and n even. Consider the cases $n = m - 1$ and $n \neq m - 1$.

9.9   Prove Corollary 9.1.

9.10   Prove Proposition 9.4.

9.11   Using a CAS, construct and plot $\tilde{w}^4(t), \ldots, \tilde{w}^7(t)$ for the symmetric linear B-spline scaling function of Example 9.3.

9.12   Using a CAS, construct and plot $\tilde{w}^n(t)$, where $\tilde{\phi}(t)$ is the symmetric quadratic B-spline function. That is, $\tilde{\phi}(t) = B_2(t+2)$, where $B_2(t)$ is given by (2.60) in Section 2.4.

9.2   WAVELET PACKET SPACES

The integer translates of the wavelet packet functions $w^0(t) = \phi(t)$, $w^1(t) = \psi(t)$, $w^2(t), \ldots, w^n(t)$ are used to create the projection coefficients at level n of a full tree decomposition (refer to Figure 9.3). Theorem 9.1 shows these functions and their integer translates form an orthonormal set. In this section we will learn about the subspaces of $L^2(\mathbb{R})$ formed by these orthonormal sets and how $L^2(\mathbb{R})$ can be constructed from them. We will also learn how these spaces *split* and provide redundant representations of functions in $L^2(\mathbb{R})$. We begin with the following definition.

Definition 9.2 (The Wavelet Packet Space W_n) *We define the* wavelet packet space *W_n as the linear span of the integer translates of the wavelet packet function $w^n(t)$ intersected with $L^2(\mathbb{R})$. That is,*

$$W_n = \mathrm{span}\{\ldots, w^n(t+1), w^n(t), w^n(t-1), \ldots\} \cap L^2(\mathbb{R})$$

$$= \mathrm{span}\{w^n(t-k)\}_{k \in \mathbb{Z}} \cap L^2(\mathbb{R})$$

(9.27)

∎

Wavelet Packet Spaces and $L^2(\mathbb{R})$

Note that $W_0 = V_0$ and $W_1 = W_0$ where V_0 is the space in the multiresolution analysis generated by the integer translates of scaling function $\phi(t)$ and W_1 is the wavelet space generated by the integer translates of $\psi(t)$. The next proposition follows immediately from Theorem 9.1.

Proposition 9.5 (Wavelet Packet Spaces Are Perpendicular) *Suppose that W_n are the wavelet packet spaces given in Definition 9.2. If $n \neq m$, then $W_n \perp W_m$.* ∎

Proof: Let $f(t) \in W_n$ and $g(t) \in W_m$. Then

$$f(t) = \sum_{j \in \mathbb{Z}} d_j^n w^n(t-j) \quad \text{and} \quad g(t) = \sum_{k \in \mathbb{Z}} d_k^m w^m(t-k)$$

and

$$\langle f(t), g(t) \rangle = \left\langle \sum_{j \in \mathbb{Z}} d_j^n w^n(t-j), \sum_{k \in \mathbb{Z}} d_k^m w^m(t-k) \right\rangle$$

$$= \sum_{j \in \mathbb{Z}} \sum_{k \in \mathbb{Z}} d_j^n d_k^m \langle w^n(t-j), w^m(t-k) \rangle$$

But the inner products on the right-hand side of the identity above are zero by Theorem 9.1 and the proof is complete. ∎

The following result characterizes $L^2(\mathbb{R})$ in terms of wavelet packet functions and wavelet packet spaces.

Theorem 9.2 (Wavelet Packet Spaces and $\mathbf{L^2(\mathbb{R})}$) *The set of wavelet packet functions $\{w^n(t-k)\}_{k,n \in \mathbb{Z}, n \geq 0}$ is an orthonormal basis for $L^2(\mathbb{R})$. Moreover,*

$$L^2(\mathbb{R}) = W_0 \oplus W_1 \oplus \cdots \oplus \cdots W_n \oplus \cdots$$

$$= V_0 \oplus W_0 \oplus W_2 \oplus \cdots \oplus \cdots W_n \oplus \cdots$$

∎

Proof: We have shown that $\{w^n(t-k)\}_{k,n\in\mathbb{Z},n\geq 0}$ is an orthonormal set of functions. The spaces $\mathcal{W}_n$, $n \geq 0$ are perpendicular to each other by Proposition 9.5. What remains to be shown is that the set $\{w^n(t - k)\}_{k,n\in\mathbb{Z},n\geq 0}$ is *complete*. That is, we must show that any function $f(t) \in L^2(\mathbb{R})$ can be written as a linear combination of the basis functions. This proof is quite technical and thus is omitted. The interested reader is referred to Walnut [62] or Wickerhauser [63].   ∎

Wavelet Packet Functions at Different Scales

The preceding results dealt with wavelet packet functions at the base level of the full tree decomposition. One of the advantages of wavelet packet functions is the choice of decompositions they allow us. To understand how we can create different decompositions of a function $f(t) \in L^2(\mathbb{R})$, we need to investigate the wavelet packet functions and associated spaces that occur at higher levels in the full tree decomposition. We start with the following definition.

Definition 9.3 (Translates and Dilates of Wavelet Packet Functions) *Let $w^n(t)$ be a wavelet packet function associated with scaling function $\phi(t)$. Here n is a nonnegative integer. For integers j and k we define*

$$\boxed{w_{j,k}^n(t) = 2^{j/2}w^n(2^j t - k)} \tag{9.28}$$

∎

As was the case with level zero wavelet packet functions, for fixed nonnegative integer n and integer j, $\{w_{j,k}^n(t)\}_{k\in\mathbb{Z}}$ is an orthonormal set of functions.

Proposition 9.6 (Level j Wavelet Packet Functions Are Orthonormal) *Let j, k, ℓ, m, and n be integers with $m, n \geq 0$. Then*

$$\langle w_{j,k}^n(t), w_{j,\ell}^n(t)\rangle = \delta_{k,\ell} \tag{9.29}$$

and

$$\langle w_{j,k}^n(t), w_{j,\ell}^m(t)\rangle = \delta_{m,n}\delta_{k,\ell} \tag{9.30}$$

∎

Proof: The proof of Proposition 9.6 is straightforward and is left as Problems 9.16 and 9.17.   ∎

There is a natural relation that connects $w_{j-1,k}^{2n}$, $w_{j-1,k}^{2n+1}(t)$ and $w_{j,k}^n(t)$.

Proposition 9.7 (Relationship Between Wavelet Packet Functions) *Let n, j, and k be integers with $n \geq 0$. Then the following identities hold for wavelet packet* *functions:*

$$w_{j-1,k}^{2n}(t) = \sum_{p \in \mathbb{Z}} h_{p-2k} w_{j,p}^{n}(t) \tag{9.31}$$

$$w_{j-1,k}^{2n+1}(t) = \sum_{p \in \mathbb{Z}} g_{p-2k} w_{j,p}^{n}(t) \tag{9.32}$$

∎

Proof: We prove (9.31) — the proof for (9.32) is nearly identical save for using the wavelet filter **g** instead of the scaling filter **h**. We use the defining relation (9.28) to write

$$w_{j-1,k}^{2n}(t) = 2^{(j-1)/2} w^{2n}(2^{j-1}t - k)$$

and then use (9.8) to expand the right-hand side of the identity above:

$$
\begin{aligned}
w_{j-1,k}^{2n}(t) &= 2^{(j-1)/2} w^{2n}(2^{j-1}t - k) \\
&= 2^{(j-1)/2} \sqrt{2} \sum_{m \in \mathbb{Z}} h_m w^n \left(2\left((2^{j-1}t - k) - m\right) \right) \\
&= 2^{j/2} \sum_{m \in \mathbb{Z}} h_m w^n \left(2^j t - (2k + m) \right) \\
&= 2^{j/2} \sum_{p \in \mathbb{Z}} h_{p-2k} w^n \left(2^j t - p \right)
\end{aligned}
$$

where the last identity is obtained via the substitution $p = 2k + m$ in the summation. We use (9.28) again to complete the proof:

$$w_{j-1,k}^{2n}(t) = \sum_{p \in \mathbb{Z}} h_{p-2k} 2^{j/2} w^n \left(2^j t - p \right) = \sum_{p \in \mathbb{Z}} h_{p-2k} w_{j,p}^{n}(t)$$

∎

Splitting Wavelet Packet Spaces

We can now define wavelet packet spaces at different scales.

Definition 9.4 (Wavelet Packet Spaces at Level j) *For fixed nonnegative integer n and $j \in \mathbb{Z}$, we define the* wavelet packet space at level j *to be*

$$
\begin{aligned}
\mathcal{W}_{n,j} &= \text{span}\{2^{j/2} w^n(2^j t - k)\}_{k \in \mathbb{Z}} \cap L^2(\mathbb{R}) \\
&= \text{span}\{w_{j,k}^{n}(t)\}_{k \in \mathbb{Z}} \cap L^2(\mathbb{R})
\end{aligned} \tag{9.33}
$$

∎

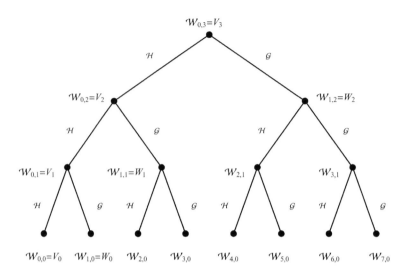

Figure 9.8 Three iterations of the full tree decomposition with nodes labeled by wavelet packet spaces.

We note that

$$\mathcal{W}_{n,0} = \mathcal{W}_n, \quad \mathcal{W}_{0,j} = V_j, \quad \mathcal{W}_{1,j} = W_j$$

We can again refer to a full tree decomposition, plotted in Figure 9.8 to better understand how these packet spaces relate to each other.

We have the following proposition.

Proposition 9.8 (Splitting Spaces) *Suppose that n is a nonnegative integer. Then the wavelet packet spaces from Definition 9.4 satisfy*

$$\boxed{\mathcal{W}_{n,j} = \mathcal{W}_{2n,j-1} \oplus \mathcal{W}_{2n+1,j-1}} \tag{9.34}$$

∎

Proof: Let $f(t) \in \mathcal{W}_{2n,j-1}$ and $g(t) \in \mathcal{W}_{2n+1,j-1}$. We need to show that $f(t) + g(t) \in \mathcal{W}_{n,j}$. Since $\{w_{j-1,k}^{2n}\}_{k\in\mathbb{Z}}$ is an orthonormal basis for $\mathcal{W}_{2n,j-1}$, we can write

$$f(t) = \sum_{k\in\mathbb{Z}} a_k w_{j-1,k}^{2n}(t)$$

We now use Proposition 9.7 to write

$$f(t) = \sum_{k \in \mathbb{Z}} a_k w_{j-1,k}^{2n}(t)$$

$$= \sum_{k \in \mathbb{Z}} a_k \sum_{p \in \mathbb{Z}} h_{p-2k} w_{j,p}^{n}(t)$$

$$= \sum_{p \in \mathbb{Z}} \left(\sum_{k \in \mathbb{Z}} a_k h_{p-2k} \right) w_{j,p}^{n}(t)$$

The last identity shows that $f(t) \in \mathcal{W}_{n,j}$. In a similar manner, we can show that $g(t) \in \mathcal{W}_{n,j}$ so that $\mathcal{W}_{2n,j-1} \oplus \mathcal{W}_{2n+1,j-1} \subseteq \mathcal{W}_{n,j}$. To show inclusion in the other direction, we show that any basis function $w_{j,\ell}^{n}(t) \in \mathcal{W}_{n,j}$ can be written as a linear combination of functions $w_{j-1,k}^{2n}(t)$ and $w_{j-1,m}^{2n+1}(t)$. Let $\ell \in \mathbb{Z}$ and consider the sum

$$\frac{1}{2} \sum_{k \in \mathbb{Z}} h_{\ell - 2k} w_{j-1,k}^{2n}(t)$$

We use (9.31) Proposition 9.7 to write

$$\frac{1}{2} \sum_{k \in \mathbb{Z}} h_{\ell - 2k} w_{j-1,k}^{2n}(t) = \frac{1}{2} \sum_{k \in \mathbb{Z}} h_{\ell - 2k} \sum_{p \in \mathbb{Z}} h_{p-2k} w_{j,p}^{n}(t)$$

$$= \frac{1}{2} \sum_{p \in \mathbb{Z}} \left(\sum_{k \in \mathbb{Z}} h_{\ell - 2k} h_{p-2k} \right) w_{j,p}^{n}(t)$$

From Problem 5.6 in Section 5.1 we know that the factor inside the parentheses is zero unless $\ell = p$, and in this case the value is 1. Thus we have

$$\frac{1}{2} \sum_{k \in \mathbb{Z}} h_{\ell - 2k} w_{j-1,k}^{2n}(t) = \frac{1}{2} w_{j,\ell}^{n}(t) \tag{9.35}$$

In an analogous manner (see Problem 9.18), we can show that

$$\frac{1}{2} \sum_{k \in \mathbb{Z}} g_{\ell - 2k} w_{j-1,k}^{2n+1}(t) = \frac{1}{2} w_{j,\ell}^{n}(t) \tag{9.36}$$

Combining (9.35) and (9.36) gives

$$w_{j,\ell}^{n}(t) = \frac{1}{2} \sum_{k \in \mathbb{Z}} h_{\ell - 2k} w_{j-1,k}^{2n}(t) + \frac{1}{2} \sum_{k \in \mathbb{Z}} g_{\ell - 2k} w_{j-1,k}^{2n+1}(t)$$

and in this way we show that any basis function $w_{j,\ell}^{n}(t) \in \mathcal{W}_{n,j}$ can be expressed as a linear combination of basis functions $w_{j-1,k}^{2n}(t) \in \mathcal{W}_{2n,j-1}$ and $w_{j-1,k}^{2n+1}(t) \in \mathcal{W}_{2n+1,j-1}$. Thus $\mathcal{W}_{n,j} \subset \mathcal{W}_{2n,j-1} \oplus \mathcal{W}_{2n+1,j-1}$ and the proof is complete. ∎

Redundant Bases Representations

We conclude this section with an example that illustrates one of the advantages of wavelet packet decompositions. Unlike the wavelet decomposition, the full tree decomposition allows for different representations of the original function. This redundancy is useful in applications, as it allows the user to select the "best" representation with which to solve or model a problem.

Example 9.4 (Decomposition into Different Packet Spaces) *Consider the function $f(t) = \max\{0, 1 - t^2\} \in L^2(t)$. For this example we use the Haar scaling function $\phi(t) = \sqcap(t)$. We first project $f(t)$ into V_2. Recall that V_2 is spanned by the functions $2\sqcap(4t - k)$, $k \in \mathbb{Z}$. We have*

$$f_2(t) = 2 \sum_{k=-4}^{3} a_{2,k} \sqcap (4t - k), \quad \text{with} \quad a_{2,k} = \langle f_2(t), \phi_{2,k}(t) \rangle$$

where

$$a_{2,k} = 2 \int_{\mathbb{R}} f(t) \sqcap (4t - k) \, dt = 2 \int_{k/4}^{(k+1)/4} 1 - t^2 \, dt$$

The projections of $f_2(t) \in V_2$ and $f(t)$ are plotted in Figure 9.9.

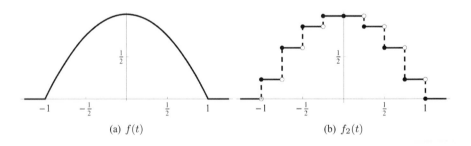

(a) $f(t)$ (b) $f_2(t)$

Figure 9.9   The functions $f(t)$ and $f_2(t)$.

We project $f_2(t)$ into wavelet packet spaces $W_{n,j}$, $n = 0, 1, 2$ and $j = 0, 1$ and then describe the different representations of $f_2(t)$ in these spaces. The full tree decomposition is plotted in Figure 9.10, and the functions that generate the wavelet packet spaces are listed in Table 9.1.

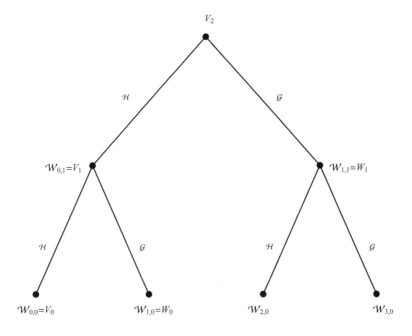

Figure 9.10 The tree of wavelet packet spaces for Example 9.4.

Table 9.1 Basis functions for the wavelet packet decomposition. The functions $w^2(t)$ and $w^3(t)$ are plotted in Figure 9.4.

Space	Generator	Formula
$\mathcal{W}_{0,1}$	$w_{0,1}^1(t)$	$\sqrt{2}\,\Pi\,(2t)$
$\mathcal{W}_{1,1}$	$w_{1,0}^1(t)$	$\sqrt{2}\,(\Pi(4t) - \Pi(4t - 1))$
$\mathcal{W}_{0,0}$	$w^0(t)$	$\phi(t) = \Pi(t)$
$\mathcal{W}_{0,1}$	$w^1(t)$	$\psi(t) = \Pi(2t) - \Pi(2t - 1)$
$\mathcal{W}_{0,2}$	$w^2(t)$	$\Pi(4t) - \Pi(4t - 1) + \Pi(4t - 2) - \Pi(4t - 3)$
$\mathcal{W}_{0,3}$	$w^3(t)$	$\Pi(4t) - \Pi(4t - 1) - \Pi(4t - 2) + \Pi(4t - 3)$

We first project $f_2(t)$ into the wavelet packet spaces $\mathcal{W}_{0,1} = V_1$ and $\mathcal{W}_{1,1} = W_1$ using the projection formula (1.31). In these cases we obtain

$$f_1(t) = \sum_{k=-2}^{1} \langle f_2(t), \phi_{1,k}(t)\rangle \phi_{1,k}(t)$$

$$= \frac{5\sqrt{2}}{24}\phi(2t + 2) + \frac{11\sqrt{2}}{24}\phi(2t + 1) + \frac{11\sqrt{2}}{24}\phi(2t) + \frac{5\sqrt{2}}{24}\phi(2t - 1)$$

$$g_1(t) = \sum_{k=-2}^{1} \langle f_2(t), \psi_{1,k}(t)\rangle \psi_{1,k}(t)$$

$$3\sqrt{2} \qquad\qquad \sqrt{2} \qquad\qquad \sqrt{2} \qquad\qquad 3\sqrt{2}$$

These functions are plotted in Figure 9.11.

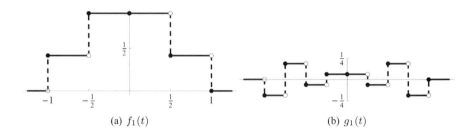

(a) $f_1(t)$ (b) $g_1(t)$

Figure 9.11   The functions $f_1(t)$ and $g_1(t)$. The breakpoints for $g_1(t)$ are at the quarter-integers.

We next project $f_1(t)$ into $\mathcal{W}_{0,0} = V_0$ and $\mathcal{W}_{1,0} = W_1$. We have

$$f_0(t) = \sum_{k=-1}^{0} \langle f_1(t), \phi(t-k) \rangle \phi(t-k) = \frac{2}{3}\phi(t+1) + \frac{2}{3}\phi(t)$$

$$g_0(t) = \sum_{k=-1}^{0} \langle f_1(t), \psi(t-k) \rangle \psi(t-k) = -\frac{1}{4}\psi(t+1) + \frac{1}{4}\psi(t)$$

These functions are plotted in Figure 9.12.

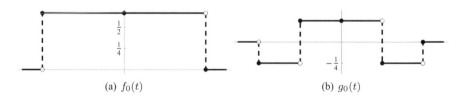

(a) $f_0(t)$ (b) $g_0(t)$

Figure 9.12   The functions $f_0(t)$ and $g_0(t)$. The breakpoints for $g_0(t)$ are at the half-integers.

Our final step is to project $g_1(t)$ into $\mathcal{W}_{2,0}$ and $\mathcal{W}_{3,0}$. We denote these projections $p_2(t)$ and $p_3(t)$, where

$$p_2(t) = \sum_{k=-1}^{0} \langle g_1(t), w^2(t-k) \rangle w^2(t-k)$$

$$= -\frac{1}{8}w^2(t+1) + \frac{1}{8}w^2(t)$$

$$p_3(t) = \sum_{k=-1}^{0} \langle g_1(t), w^3(t-k) \rangle w^3(t-k)$$

$$= -\frac{1}{16}w^3(t+1) - \frac{1}{16}w^3(t)$$

The functions $p_2(t)$ and $p_3(t)$ are plotted in Figure 9.13.

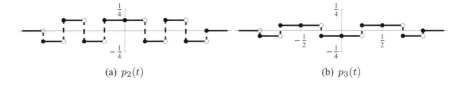

(a) $p_2(t)$ (b) $p_3(t)$

Figure 9.13   The functions $p_2(t)$ and $p_3(t)$. The breakpoints for $p_2(t)$ are at the quarter-integers.

We can represent $f_2(t)$ in several different ways. The classical wavelet decomposition is

$$f_2(t) = f_0(t) + g_0(t) + g_1(t)$$

But there are other possibilities. Using Figure 9.10 we can write

$$f_2(t) = f_1(t) + p_2(t) + p_3(t)$$
$$= f_1(t) + g_1(t)$$
$$= f_0(t) + g_0(t) + p_2(t) + p_3(t)$$

∎

In the next section we examine the discrete version of these multiple decompositions.

PROBLEMS

Note: *A computer algebra system and the software on the course Web site will be useful for some of these problems. See the Preface for more details.*

9.13   Show that $\|\overline{w_{j,k}^n(t)}\| = 1$.

9.14   What is $\operatorname{supp}\left(\overline{w_{j,k}^n}\right)$? (*Hint:* Use Proposition 9.1.)

9.15   Use Definition 9.3 and Proposition 9.1 to find the support of $w_{j,k}^n(t)$, where $w^n(t)$ is constructed from the Daubechies $2L$ scaling function $\phi(t)$.

9.16   Prove (9.29) from Proposition 9.6. (*Hint:* Write down the inner product, make the substitution $u = 2^j t$, and then use Proposition 9.3.)

9.17   Prove (9.30) from Proposition 9.6. (*Hint:* Write down the inner product, make the substitution $u = 2^j t$, and then use Theorem 9.1.)

9.18   In the proof of Proposition 9.7, show that

$$\frac{1}{2} \sum_{k \in \mathbb{Z}} g_{\ell - 2k} w_{j-1,k}^{2n+1}(t) = \frac{1}{2} w_{j,\ell}^n(t)$$

9.19   Suppose that we started Example 9.4 by projecting $f(t)$ into V_3 so that our full tree decomposition consisted of three levels instead of two. How many possible reconstruction formulas exist for $f_3(t)$?

9.20   Repeat Example 9.4 using the biorthogonal scaling functions $\tilde{\phi}(t)$ and $\phi(t)$, where $\tilde{\phi}(t) = B_1(t+1) = \wedge(t-1)$ from Example 9.3.

9.3   THE DISCRETE PACKET TRANSFORM AND BEST BASIS ALGORITHM

In this section we construct the one- and two-dimensional discrete wavelet packet transformation. We will see that this transformation is very similar to those we constructed in Chapters 4, 7, and 8. The main difference between the discrete wavelet packet transformation and other wavelet transformations developed previously is that the discrete wavelet packet transformation gives multiple decompositions of the data. This allows the user to pick the representation that is best-suited for a particular application. The question of how to select the best representation was first addressed by Coifman and Wickerhauser [16]. Their *best basis algorithm* is an effective, flexible method that allows an application-driven choice of the best representation of the transformed data. We describe the best basis algorithm in the second half of the section.

Motivation for the Discrete Packet Transformation

Suppose that $\phi(t)$ is a Daubechies scaling function from Chapter 6, and let $\mathcal{W}_{n,j}$ be the wavelet packet space constructed from it. Next assume that $f(t) \in \mathcal{W}_{n,j}$ with

$$f(t) = \sum_{k \in \mathbb{Z}} a_{j,k}^n w_{j,k}^n(t)$$

where

$$a_{j,k}^n = \langle f(t), w_{j,k}^n(t) \rangle$$

Then, by Proposition 9.8, we know that

$$f(t) \in \mathcal{W}_{n,j} = \mathcal{W}_{2n,j-1} \oplus \mathcal{W}_{2n+1,j-1}$$

Since $\{w_{j-1,k}^{2n}\}$ forms an orthonormal basis for $\mathcal{W}_{2n,j-1}$, we can create the projection $f^{2n}(t)$ of $f(t)$ into $\mathcal{W}_{2n,j-1}$ using the formula

$$f^{2n}(t) = \sum_{k \in \mathbb{Z}} a_{j-1,k}^{2n} w_{j-1,k}^{2n}(t)$$

where

$$a_{j-1,k}^{2n} = \langle f(t), w_{j-1,k}^{2n}(t) \rangle$$

We can use Proposition 9.7 to write

$$
\begin{aligned}
a_{j-1,k}^{2n} &= \langle f(t), w_{j-1,k}^{2n}(t) \rangle \\
&= \left\langle f(t), \sum_{p \in \mathbb{Z}} h_{p-2k} w_{j,p}^n(t) \right\rangle \\
&= \sum_{p \in \mathbb{Z}} h_{p-2k} \langle f(t), w_{j,p}^n(t) \rangle \\
&= \sum_{p \in \mathbb{Z}} h_{p-2k} a_{j,k}^n
\end{aligned}
\tag{9.37}
$$

In a similar way we see that the projection $f_{2n+1}(t)$ of $f(t)$ into $\mathcal{W}_{2n+1,j-1}$ can be written as

$$f^{2n+1}(t) = \sum_{k \in \mathbb{Z}} b_{j-1,k}^{2n+1} w_{j-1,k}^{2n+1}(t)$$

where

$$b_{j-1,k}^{2n+1} = \sum_{p \in \mathbb{Z}} g_{p-2k} a_{j,k}^n \tag{9.38}$$

Equations (9.37) and (9.38) tell us that the projection coefficients $a_{j-1,k}^{2n}$ and $b_{j-1,k}^{2n+1}$ can be obtained using the projection matrices $\mathcal{H}$ and $\mathcal{G}$ given in (5.16) and (5.27), respectively! The same analysis holds true if we were to use biorthogonal scaling functions $\phi(t)$ and $\tilde{\phi}(t)$ to construct the packet spaces.

The One–Dimensional Wavelet Packet Transformation

The only difference in performing i iterations of the discrete wavelet packet transform and i iterations of the discrete wavelet transform is that we apply a wavelet transformation matrix to both the averages and details portion at each packet iteration step. We also retain *all* averages and details portions of the discrete packet transformation. So for, say, $i = 3$ iterations, our discrete wavelet packet transformation would decompose the input data $\mathbf{w}^3$ as shown in Figure 9.14. Here $w_{j-1}^{2n} = \mathcal{H}w_j^n$ and $w_{j-1}^{2n+1} = \mathcal{G}w_j^n$, where $\mathcal{H}$ and $\mathcal{G}$ are given in (5.16) and (5.27), respectively.

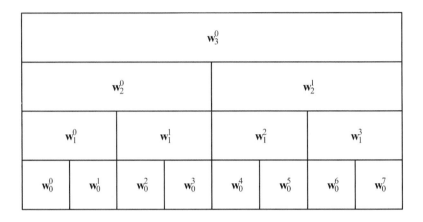

Figure 9.14   Three iterations of the discrete wavelet packet transformation.

Of course, if we use the discrete wavelet packet decomposition to obtain projection coefficients in different wavelet packet spaces, then we would have to pad the coefficient vector with zeros as described in Section 7.2. For applications such as signal and image processing, we typically do not pad input with zeros. Let's look at an example.

Example 9.5 (One-Dimensional Discrete Wavelet Packet Transform) *Consider the vector* $\mathbf{v} \in \mathbb{R}^{2048}$ *whose elements are defined by* $v_k = d\left(\frac{k-1}{2048}\right)$, $k = 1, \ldots, 2048$, *where*

$$d(t) = \sqrt{t(1-t)} \sin\left(\frac{2\pi \cdot 1.05}{t + 0.05}\right)$$

is the Doppler function from the test set suggested by Donoho and Johnstone [24]. Using the Daubechies D4 scaling filter, apply three iterations of the discrete wavelet packet transformation to $\mathbf{v}$.

Solution

We apply the discrete wavelet transformation to each average or detail block of the preceding iteration. The results are shown in Figure 9.15. The top row is the vector $\mathbf{v}$. ∎

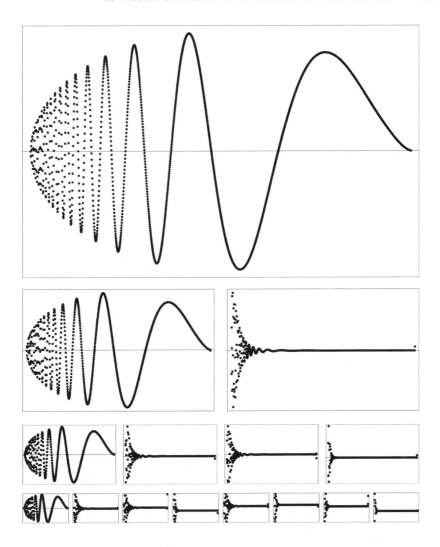

Figure 9.15   Three iterations of the discrete wavelet packet transformation using the D4 filter.

The Two-Dimensional Wavelet Packet Transformation

It is also very easy as well to extend the ideas above to create a two-dimensional wavelet transformation. Given matrix $A \in \mathbb{R}^{M \times N}$, where both M and N are divisible by 2^i, we apply the discrete two-dimensional (bi)orthogonal wavelet transform to each average and detail portion of the previous result. The process is a bit more complex in this case because one application of a discrete two-dimensional wavelet transformation results in four different blocks of output (recall (4.23) from Section 4.2).

The following example shows how to compute the two-dimensional wavelet packet transformation.

Example 9.6 (Two-Dimensional Discrete Wavelet Packet Transform) *Compute three iterations of the two-dimensional discrete wavelet transformation to the* 512×512 *digital image from Example 7.9 in Section 7.2. In this case we use the Daubechies D6 scaling filter.*

Solution

   We start by applying the two-dimensional wavelet transformation to the image matrix A. The result is four distinct blocks of output. We next apply the two-dimensional wavelet transformation to each of these blocks and iteratively continue one more time to complete the computation. The various iterations are plotted in Figure 9.16. ■

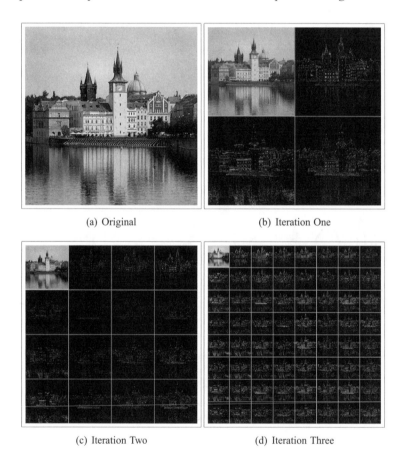

(a) Original   (b) Iteration One

(c) Iteration Two   (d) Iteration Three

Figure 9.16   The various levels of three iterations of the two-dimensional discrete wavelet packet transformation. Note that the elements in the last iteration have been scaled to facilitate viewing.

Motivating the Best Basis Algorithm

One of the biggest advantages of the discrete wavelet packet transformation is the fact that it returns redundant representations of the input data. In applications, this flexibility allows the user to choose the representation that best models the problem or solution. Let's look at an example that illustrates the various basis representations returned to us via the discrete wavelet packet transformation.

Example 9.7 (Redundant Representations) *Suppose that we apply two iterations of the discrete wavelet packet transformation to* $\mathbf{v} \in \mathbb{R}^N$, *where* N *is divisible by* 4. *How many different representations of the original data are available to us via the transformation?*

Solution

We illustrate the possibilities graphically in Figure 9.17. We see that counting the original input $\mathbf{v}$ *itself, there are five possible ways to represent* $\mathbf{v}$. *Note that Figure 9.17(d) is the wavelet transformation representation of* $\mathbf{v}$. ∎

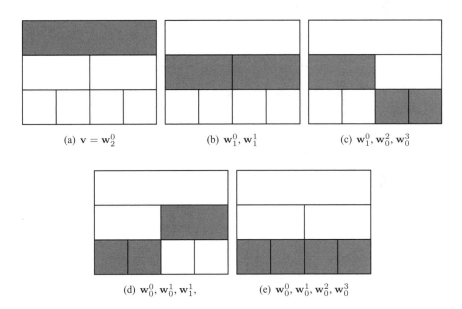

(a) $\mathbf{v} = \mathbf{w}_2^0$ (b) $\mathbf{w}_1^0, \mathbf{w}_1^1$ (c) $\mathbf{w}_1^0, \mathbf{w}_0^2, \mathbf{w}_0^3$

(d) $\mathbf{w}_0^0, \mathbf{w}_0^1, \mathbf{w}_1^1,$ (e) $\mathbf{w}_0^0, \mathbf{w}_0^1, \mathbf{w}_0^2, \mathbf{w}_0^3$

Figure 9.17   Various ways that we can represent $\mathbf{v}$ available to us via two iterations of the wavelet packet transformation.

Number of Bases

As you might guess from the preceding example, the number of possible bases can grow quite large as the number of iterations increases. Clearly, if no transformation is performed, then there is the trivial representation of **v** by **v** itself. If one iteration of the transformation is performed, then there are two ways (**v** and the transformed vector) to represent **v**. The following proposition gives us a recursive method to compute the number of different ways we can represent **v** via the wavelet packet transformation.

Proposition 9.9 (Number of Bases) *Let $c(i)$ represent the number of possible basis representation of* **v** *at level i of the discrete wavelet packet transformation. Then $c(0) = 1$, $c(1) = 2$, and for $i > 1$ we have*

$$c(i) = c(i - 1)^2 + 1 \qquad (9.39)$$

■

Proof: We have already verified that $c(0) = 1$ and $c(1) = 2$. We now proceed by induction. Assume that (9.39) holds for $0, \ldots, i$ and consider a wavelet packet decomposition with $i + 1$ levels.

Suppose that we perform an $(i+1)$-level wavelet packet decomposition of **v**. Then the first level is simply the wavelet transform **y** with average and detail blocks $\mathbf{y}^{1,a}$ and $\mathbf{y}^{1,d}$ (see Figure 9.18).

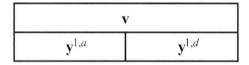

Figure 9.18   The first step in a wavelet packet transformation.

The decomposition of $\mathbf{y}^{1,a}$ is nothing more than a level i wavelet packet transformation, and by the induction hypothesis, we know that there are $c(i)$ possible representations of $\mathbf{y}^{1,a}$. In a similar way, we see that there are $c(i)$ wavelet packet representations of $\mathbf{y}^{1,d}$. Since for every wavelet packet representation of $\mathbf{y}^{1,a}$, there are $c(i)$ packet representations for $\mathbf{y}^{1,d}$, there are a total of $c(i)^2$ wavelet packet representations of the vector $\left[\mathbf{y}^{1,a}, \mathbf{y}^{1,d}\right]^T$. Add one more to this number to reflect that **v** itself is a possible representation and we see there are $c(i + 1) = c(i)^2 + 1$ wavelet packet representations of the $(i + 1)$-level wavelet packet decomposition of **v** and the proof is complete. ■

The Cost Function

As we have seen from Proposition 9.9, the number of possible packet representations for the given vector $\mathbf{v} \in \mathbb{R}^N$ grows quite rapidly. We thus have the opportunity to pick the "best" representation for use in applications. Each of these representations can be viewed as a linear combination of basis vectors $\{\mathbf{u}^k \mid k = 1, \ldots, N\}$ in $\mathbb{R}^N$.

So if

$$\mathbf{v} = c_1 \mathbf{u}^1 + \cdots + c_N \mathbf{u}^N \tag{9.40}$$

and we have several basis candidates, it is natural to choose the basis that requires the fewest coefficients c_k. To perform this task, Coifman and Wickerhauser [16] employed a *cost function*. This nonnegative real-valued function is typically used to compare two representations of the same data and assign a smaller number to the one that is deemed more efficient. The advantage here is the fact that the cost function can be application dependent — we can determine what comprises an efficient representation based on a particular application and design our cost function accordingly.

In applications such as image compression, we typically use cost functions that consider efficient representations those that use relatively few coefficients. For the purposes of fast computation in the best–basis algorithm, we impose an additivity condition on the cost function. We have the following definition:

Definition 9.5 (Cost Function) *Suppose that* $\mathbf{v} \in \mathbb{R}^M$. *Then a* cost function $\mathcal{C}(\mathbf{v})$ *is any nonnegative real-valued function that satisfies* $\mathcal{C}(\mathbf{0}) = 0$, *where* $\mathbf{0}$ *is the length* M *zero vector, and for any partition*

$$\mathbf{v} = [\mathbf{v}_\ell \mid \mathbf{v}_r]^T = [v_1, \ldots, v_m \mid v_{m+1}, \ldots, v_M]^T$$

where $1 \le m \le M$, *we have*

$$\mathcal{C}(\mathbf{v}) = \mathcal{C}\left([\mathbf{v}_\ell \mid \mathbf{v}_r]^T\right) = \mathcal{C}(\mathbf{v}_\ell) + \mathcal{C}(\mathbf{v}_r) \tag{9.41}$$

∎

There are several different cost functions described in the literature. We provide four in the example that follows.

Example 9.8 (Examples of Cost Functions) *In this example we list and describe four cost functions. The proofs that each are cost functions are left as exercises.*

Shannon Entropy. *This function is closely related to the entropy function (4.9) given in Definition 4.3. For* $\mathbf{v} \in \mathbb{R}^M$ *with* $\|\mathbf{v}\| = 1$, *we have*

$$\mathcal{C}_s(\mathbf{v}) = -\sum_{k=1}^{M} v_k^2 \ln\left(v_k^2\right) \tag{9.42}$$

where we define $v_k^2 \ln\left(v_k^2\right) = 0$ *in the case that* $v_k = 0$ *(see Problem 9.24). It is clear that* $\mathcal{C}_s(\mathbf{0}) = 0$, *and in Problem 9.25 you will show that* $\mathcal{C}_s(\mathbf{v})$ *satisfies Definition 9.5. Note that if* v_k^2 *is close to zero or one, the value of* $v_k^2 \ln\left(v_k^2\right)$ *will be small, so this function rewards representations that come "close" to using some of the basis vectors.*

Number Above Threshold. *For* $\mathbf{v} \in \mathbb{R}^M$ *and* $t > 0$, *we define*

$$\mathcal{C}_t(\mathbf{v}) = \#\{ v_k : |v_k| \ge t\} \tag{9.43}$$

For example, if $\mathbf{v} = [3, -4, -1, 0, 3, 1, 5]^T$, *then* $\mathcal{C}_2(\mathbf{v}) = 4$. *This function counts the number of "essential" coefficients in a given representation.*

Number of Nonzero Elements. *Let* $\mathbf{v} \in \mathbb{R}^M$ *and define*

$$\mathcal{C}_z(\mathbf{v}) = \#\{ v_k : v_k \neq 0 \} \tag{9.44}$$

For example, if $\mathbf{v} = [2, 0, -13, 0, 0, 9, 0, 0, 1, 5]^T$, *then* $\mathcal{C}_z(\mathbf{v}) = 5$. *This function counts the number of basis vectors (9.40) needed for a given representation and does not take into account the magnitude of the coefficients.*

Sum of Powers. *For* $\mathbf{v} \in \mathbb{R}^M$ *and* $p > 0$, *we define*

$$\mathcal{C}_p(\mathbf{v}) = \sum_{k=1}^{M} |v_k|^p \tag{9.45}$$

Suppose that $p > 1$. *Then* $|v_k|^p < |v_k|$ *whenever* $|v_k| < 1$. *In such cases this function "rewards" those representations that use coefficients that are small in magnitude.*

■

Let's look at an example that will serve as good motivation for how the cost function is used in the best basis algorithm.

Example 9.9 (The Cost Function and the Wavelet Packet Transform) *Consider the wavelet packet transformation from Example 9.5. Using* $\mathcal{C}_s$ *and the number above threshold cost function* $\mathcal{C}_t$ *with* $t = 0.05$, *compute the cost of each component of the wavelet packet transform plotted in Figure 9.15.*
Solution
The results of the computations are plotted in Figure 9.19.   ■

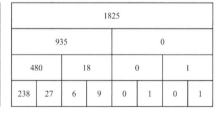

(a) Shannon Entropy   (b) Number Above Threshold $t = 0.05$

Figure 9.19   The cost function values for each component of the wavelet packet transformation of Example 9.5.

The Best Basis Algorithm

The idea of the best basis algorithm is quite straightforward. We use the fact that $\mathcal{W}_{n,j} = \mathcal{W}_{2n,j-1} \oplus \mathcal{W}_{2n+1,j-1}$ and the additivity of the cost function to make determinations about which of the portions of the packet transformation we should use.

As an example, consider Figure 9.14, Figure 9.19(b), and the cost function values 238 and 27 associated with packet components $\mathbf{w}_0^0$ and $\mathbf{w}_0^1$, respectively, in Example 9.9. It is natural to compare the cost associated with the vector $\mathbf{y} = \left[\mathbf{w}_0^0 \mid \mathbf{w}_0^1\right]^T$ with the component $\mathbf{w}_1^0$ directly above it. Using the additivity of the cost function, we have $\mathcal{C}_{0.05}(\mathbf{y}) = 238 + 27 = 265$. We compare this value to $\mathcal{C}_{0.05}\left(\mathbf{w}_1^0\right) = 480$ and since the former value is smaller, we will choose $\mathbf{w}_1^0$ over $\mathbf{y}$ as we seek the best basis. This process is repeated for the remaining components in rows three and four of Figure 9.19(b) and then iteratively for the remaining rows. We complete the algorithm in the following example.

Example 9.10 (Best Basis Algorithm) *Find the best basis for the wavelet packet decomposition from Example 9.5 using the cost function $\mathcal{C}_{0.05}$.*
Solution
We start by creating a four-level best basis tree with the bottom nodes labeled with "1" and all other nodes marked with "0" (see Figure 9.20(a)). The nodes labeled with "1" at the completion of the algorithm represent the components of the packet transform that constitute the best basis. The initial cost values are plotted in Figure 9.20(b). Figure 9.14 will be of great assistance as we proceed through this example.

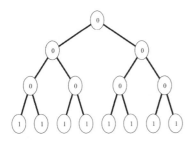

(a) Initial Best Basis Tree

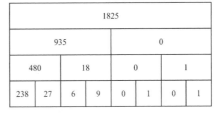

(b) Initial Cost Values

Figure 9.20 The initial best basis tree and cost function values for Example 9.10.

We start by adding the numbers pairwise on the bottom row of Figure 9.20(b). The results are 265, 15, 1, 1. We compare each of these values to the corresponding four values 480, 18, 0, 1 in the third row of Figure 9.20(b). If a value from row four is smaller than the corresponding value from row three (in this case $265 < 480$ and $15 < 18$), we replace the corresponding cost function value in row three and leave *our basis tree unchanged. If a value from row four is greater than or equal to a corresponding value in row three, we leave the cost functions unchanged but mark the row three node with a "1" and the two nodes below it with "0" in the best basis tree. The results of this step are plotted in Figure 9.21.*

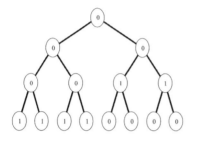

(a) Updated Best Basis Tree

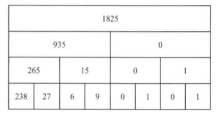

(b) Updated Cost Values

Figure 9.21   The updated best basis tree and cost function values for Example 9.10.

We now add pairwise the cost function values in the third row and compare them with the corresponding values in row two of Figure 9.21(b). In this case $265 + 15 = 290 < 935$ and $0 + 1 > 0$, so we update the first value in the second row of Figure 9.21(b) and replace the second "0" in row two of Figure 9.21(a) with a "1" and relabel the two nodes below it with "0." The results of the second step are shown in Figure 9.22.

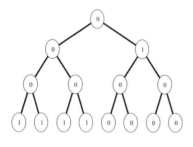

(a) Updated Best Basis Tree

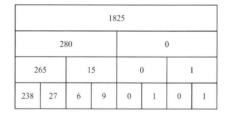

(b) Updated Cost Values

Figure 9.22   The updated best basis tree and cost function values for Example 9.10.

The final step in the algorithm is to add the two values in row two of Figure 9.22(b) and compare them to the cost value in the top row of Figure 9.22. Since $280 + 0 < 1825$, nothing changes with regard to our best basis tree and the algorithm is complete. The best basis positions are shown in Figure 9.23.

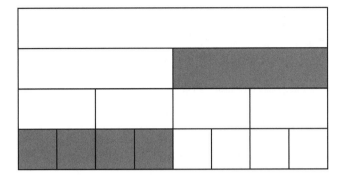

Figure 9.23   Best basis positions for Example 9.10.

■

We next state the best basis algorithm of Coifman and Wickerhauser [16].

Algorithm 9.1 (Best Basis Algorithm) *Suppose that we have performed an* $(i+1)$-*level (or* i *iterations) wavelet packet decomposition of vector* $\mathbf{v} = \mathbf{w}_i^0$ *and cost function* $\mathcal{C}$. *The following algorithm returns the* best basis *representation of* $\mathbf{v}$ *relative to the cost function* $\mathcal{C}$.

Inputs:
   $(i+1)$-*level wavelet packet decomposition* $\{\mathbf{w}_j^k\}$, $j = 0, \ldots, i$, $k = 0, \ldots, 2^{i-j} - 1$.
   Cost function $\mathcal{C}$.
Output:
   Best basis representation of $\mathbf{v} = \mathbf{w}_i^0$.

// Create arrays **c** *and* **b** *to hold cost function and basis tree information, respectively.*
// The first i *rows of* **b** *are set to zero while the bottom row consists of ones.*
For $j = 0$, $j \leq i$, $j{+}{+}$,
   For $k = 0$, $k < 2^{i-j}$, $k{+}{+}$,
      $c_{j,k} = \mathcal{C}\left(\mathbf{w}_j^k\right)$
      If $(j = 0)$
         $b_{j,k} = 1$
      Else
         $b_{j,k} = 0$
      EndIf
   EndLoop
EndLoop

// Do the best basis search.

For $j = 0$, $j \leq i$, $j {+}{+}$,

    For $k = 0$, $k < 2^{i-j-1}$, $k {+}{+}$,

        $t = c_{j,2k} + c_{j,2k+1}$

        If $(t < c_{j+1,k})$ *// Keep these basis components*

            $c_{j+1,k} = t$

        Else *// Keep the basis component in the row above.*

            $b_{j+1,k} = 1$

            // Zero out $b_{j,k}$ below.

            For $m = j$, $m \geq 0$, $m {-}{-}$,

                For $n = 2^{j-m+1}k$, $n < 2^{j-m+1}(k+1)$, $n {+}{+}$,

                    $b_{m,n} = 0$

                EndLoop

            EndLoop

        EndIf

    EndLoop

EndLoop

Return: *Basis tree array* **b** *and those* $\mathbf{w}_j^k$ *for which* $b_{j,k} = 1$.

■

A couple of comments are in order regarding Algorithm 9.1. If the algorithm does not return the basis tree array **b**, then we have no hope of inverting the wavelet packet transformation. The algorithm is not as efficient as it could be — the innermost for loops (on m and n) convert *all* values b_{mn} to zero that lie below $b_{j+1,k}$. In fact, many of these values are already possibly zero, so these assignments are redundant. We have chosen readability of the pseudocode over efficiency. The interested reader is encouraged to perform the steps necessary to make the code more efficient.

Best Basis and Image Compression

We can easily extend the best basis algorithm to two-dimensional wavelet packet transformations. The only changes in the code would be the fact that the array **c** that holds the cost function values would need three indices (to account for the four components in each transform step instead of the two in the one-dimensional case) and the computation of t in Algorithm 9.1 would sum four cost function values instead of two.

We conclude this example by using the best basis algorithm in conjunction with the two-dimensional wavelet packet transformation to perform image compression.

Example 9.11 (Wavelet Packets and Image Compression) *We return to the compression problem considered in Examples 4.10 and 7.9. We use the Daubechies D4 scaling filter and perform four iterations of the wavelet packet transformation. We use the best basis algorithm with the cost function $C_{25}(\mathbf{v})$ given by (9.43). The result is plotted in Figure 9.24(a).*

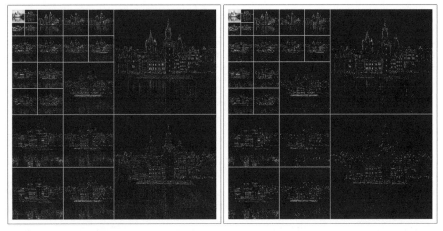

(a) Packet Transformation   (b) Quantized Packet Transformation

Figure 9.24   Four iterations of the wavelet packet transformation and the quantized transformation.

We again use the cumulative energy to quantize the transformation. We compute the cumulative energy vector for the packet transformation and plot it in Figure 9.25.

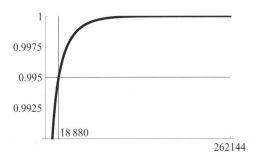

Figure 9.25   The cumulative energy vector for the packet transformation.

As with the previous examples, we retain 99.5% *of the energy of the transformation. For this example, the largest (in absolute value)* $18{,}880$ *terms constitute* 99.5% *of the energy of the transformation. Table 9.2 shows how this value compares to those obtained in Examples 4.10 and 7.9.*

We next perform Huffman coding (see Appendix A) on the quantized transformation. Table 9.3 shows that the packet D4 transformation slightly outperforms the transformations from previous examples with regard to bpp and is comparable to the D4 and D6 transformations with regard to PSNR.

Table 9.2 Cumulative energy information for the Haar, D4, D6, and D4 packet transforms.

Transform	Nonzero Terms	% Zero
Haar	72996	72.15
D4	19523	92.56
D6	19708	92.48
D4 packet	18880	92.80

Table 9.3 Coding information for the Haar, D4, D6, and D4 packet transformations.

Transform	Bitstream Length	bpp	Entropy	PSNR
Haar	705536	2.69	2.53	43.90
D4	426240	1.63	1.01	27.04
D6	427309	1.63	1.01	27.04
D4 packet	421669	1.61	0.98	27.04

We apply the inverse wavelet packet transformation to the quantized packet transformation to view the compressed image. This image and the original are plotted in Figure 9.26. ■

(a) Original Image   (b) Compressed Image

Figure 9.26 Four iterations of the wavelet packet transformation and the quantized transformation.

There are still a few improvements that we can make on the compression process. The coding step can be more efficient if the wavelet (packet) transformation is adjusted to map integers to integers. We do not address that topic in this book but the interested reader can consult Van Fleet [60]. The use of cumulative energy as a quantization tool is not optimal; a better quantization method is discussed in Section 9.4.

PROBLEMS

Note: *A computer algebra system and the software on the course Web site will be useful for some of these problems. See the Preface for more details.*

9.21 Let $v = [0, 3, 4, 3, 0, -5, -6, -5, -2, 1, 2, 1, 0, -3, -4, -3]^T$. Using the Haar filter, compute by hand three iterations of the discrete wavelet packet transform of $\mathbf{v}$.

9.22 Using (9.39) and a CAS, compute $c(i)$ for $i = 3, \ldots, 10$.

9.23 Use mathematical induction to show that $c(i) > 2^{2^i}$ whenever $i \geq 2$.

9.24 Use L'Hôpital's rule to prove that $\lim_{t \to 0} t \ln(t) = 0$. In this way we justify the assignment $v_k^2 \ln \left(v_k^2 \right) = 0$ when $v_k = 0$ in (9.42).

9.25 Show that $C_s(\mathbf{v})$ defined in (9.42) is a nonnegative real-valued function with $C_s(\mathbf{0}) = 0$ and also satisfies (9.41).

9.26 We know that $C_s(\mathbf{0}) = 0$. Describe all other vectors $\mathbf{v}$ for which $C_s(\mathbf{v}) = 0$.

9.27 Find $\mathbf{C}_s(\mathbf{v})$ for each of the following vectors.

(a) $\mathbf{v} \in \mathbf{R}^M$ where $v_k = \frac{1}{\sqrt{M}}$, for $k = 1, \ldots, M$.

(b) $\mathbf{v} \in \mathbf{R}^M$ where M is an even integer. Half of the $v_k = 0$ and the other half take the value $\pm\sqrt{2/M}$.

9.28 Show that the function $C_t(\mathbf{v})$ given by (9.43) satisfies Definition 9.5.

9.29 Let $C_t(\mathbf{v})$ be the cost function defined by (9.43). Show that $\mathbf{C}_t(\mathbf{v}) = 0$ if and only if $\mathbf{v}$ is the zero vector.

9.30 Show that $C_p(\mathbf{v})$ given by (9.45) satisfies Definition 9.5.

9.31 For the vector $\mathbf{v}$ given in Problem 9.21, compute $C_s(\mathbf{v})$, $C_2(\mathbf{v})$, and $C_z(\mathbf{v})$. Note that for $C_s(\mathbf{v})$ you will have to normalize $\mathbf{v}$ so that it has unit length.

9.32 Let $\mathbf{v} \in \mathbb{R}^M$ and consider the function $f(\mathbf{v}) = \max\{|v_k| : 1 \leq k \leq M\}$. Is $f(\mathbf{v})$ a cost function?

9.33 Repeat Example 9.10 using

(a) $C_z(\mathbf{v})$ given by (9.44).

(b) $C_p(\mathbf{v})$ given by (9.45). Use $p = 3$.

(c) $\mathcal{C}_{.4}(\mathbf{v})$ given by (9.43).

9.34   Let $\mathbf{v} \in \mathbf{R}^M$ with $\|\mathbf{v}\|^2 = 1$ and consider $\mathcal{C}(\mathbf{v}) = -\sum_{k=1}^{M} f\left(v_k^2\right)$ where

$$f(t) = \begin{cases} \ln(t), & t \neq 0 \\ 0, & t = 0 \end{cases}$$

Show that $\mathcal{C}(\mathbf{v})$ is a cost function.

9.35   Let $\mathbf{v} \in \mathbb{R}^M$ and consider the *nonnormalized Shannon entropy function*

$$\mathcal{C}(\mathbf{v}) = \sum_{k=1}^{M} s(v_k)\, v_k^2 \ln\left(v_k^2\right)$$

with the convention that $0 \ln(0) = 0$. Here $s(t)\colon [0, \infty) \to \{-1, 1\}$ is given by

$$s(t) = \begin{cases} -1, & 0 \le t < 1 \\ 1, & t \ge 1 \end{cases}$$

Show that $\mathcal{C}(\mathbf{v})$ is a cost function.

9.36   Is the entropy function (4.9) from Definition 4.3 a cost function? Either prove it is or provide a counterexample to show that it is not.

9.37   Write a best basis algorithm (use Algorithm 9.1 as a guide) for matrix input.

9.4   THE FBI FINGERPRINT COMPRESSION STANDARD

According to its Web site [26], the Federal Bureau of Investigation (FBI) started collecting and using fingerprints in 1902, and the Identification Division of the FBI was established in 1921. Bradley, Brislawn, and Hopper [8] reported that the FBI had about 810,000 fingerprint cards in 1924. Each card consists of 14 prints. Each finger is printed, flat impressions are taken of both thumbs, and each hand is printed simultaneously. In 1992, the Identification Division was renamed the Criminal Justice Information Services (CJIS) Division. In 1996, the CJIS fingerprint collection consisted of over 200 million print cards (see Bradley, Brislawn, Onyshczak, and Hopper [9]) and now totals over 250 million cards. The CJIS Web site [26] reports that about 80 million of these cards are digitized and that they add or digitally convert about 7000 new cards daily to their electronic database.

Each digitized print (an example is plotted in Figure 9.27) from a fingerprint card is scanned at 500 dots per inch (dpi) and each print is of varying size (e. g., a flat thumbprint measuring 0.875×1.875 inches is stored in a matrix of dimensions 455×975 pixels). According to Bradley, Brislawn, and Hopper [7], about 10.7 megabytes is necessary to store each fingerprint card. Thus over 800 terabytes of space would be needed to store the digitized cards as raw data. Given the daily

Figure 9.27   A digital fingerprint of size 832×832 pixels.

growth rate of the digitized card database, it is clear why the CJIS Division decided to use image compression to store the fingerprint cards.

The FBI first considered the JPEG image compression standard [46] but ultimately decided not to use it for fingerprint compression. The JPEG standard starts by partitioning the image into 8×8 blocks and then applying the *discrete cosine transform* to each block. Quantization is applied individually to each block, which is then encoded using a version of Huffman coding. The compressed image is recovered by unencoding the image data and then applying the discrete cosine transform to each 8×8 block. The method is highly effective, but the decoupling of the image into 8×8 blocks often gives the compressed image a "blocky" look. The image in Figure 9.27 was compressed using JPEG, and the center 96×96 portion is enlarged in Figure 9.28. The blocky structure of the compression method is evident in this figure.

In 1993 the FBI adopted the *Wavelet/Scalar Quantization Standard* (WSQ) [7]. This standard was developed by Thomas Hopper of the FBI in conjunction with researchers Jonathan Bradley and Christopher Brislawn at Los Alamos National Laboratory [8, 9, 10].

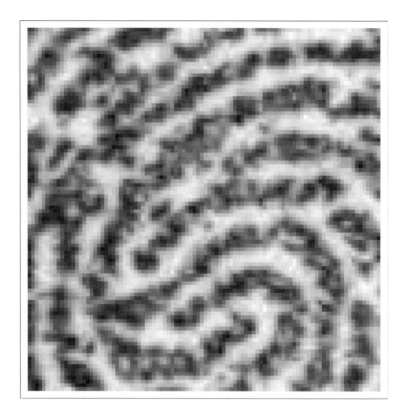

Figure 9.28   The enlarged 96×96 center of the JPEG-compressed fingerprint from Figure 9.27. The decoupling of 8×8 blocks is evident in this image.

Our interest in this standard is the fact that it uses the CDF97 biorthogonal filter pair developed in Section 8.3 and the discrete wavelet packet transform from Section 9.3. Unlike the image compression schemes described earlier in the book, this standard allows the user to set the compression rate *before* the quantization step. Although we can set the compression rate to whatever value we desire, we must be aware of the fact that there is an objective test for the effectiveness of this standard. In a US court of law, experts use 10 points on a fingerprint to uniquely identify the person whose finger produced the print (see Hawthorne [33]). If the compressed version of a digitized fingerprint does not preserve these 10 points accurately, then the compression scheme is useless in a court of law. Remarkably, a typical compression rate used by the FBI is 0.75 bit per pixel, or a 10.7 : 1 ratio if we consider 8 bits per pixel a 1 : 1 compression ratio.

The Basic Wavelet/Scalar Quantization Standard Algorithm

We now present the basic algorithm for implementing the FBI WSQ standard.

Algorithm 9.2 (FBI Wavelet/Scalar Quantization Standard) *This algorithm gives a basic description of the FBI WSQ standard developed by Bradley, Brislawn and Hopper [7, 8, 9, 10] for compressing digitized fingerprints. Suppose that matrix A holds a grayscale version of a digitized fingerprint image.*

1. *Normalize A.*

2. *Apply the discrete wavelet packet transform using the CDF97 biorthogonal filter pair. At each iteration of the packet transform, use the modified transformation (see Algorithm 8.3) to better handle boundary effects. Rather than using a best basis algorithm, use a prescribed basis for the packet transform.*

3. *Perform quantization on each portion of the transform.*

4. *Apply adaptive Huffman coding.*

■

Note: The material that follows is based on the work [7, 8, 9, 10] of the designers Bradley, Brislawn and Hopper of the WSQ standard.

We will look at each step of Algorithm 9.2 in more detail using the fingerprint plotted in Figure 9.27 as a running example. Each step contains some computations that we will not attempt to justify — to do so would be beyond the scope of this book. The standard designers have done much empirical work and analysis in developing the first three steps of the algorithm. For more details the interested reader is encouraged to see the references cited in the note above.

Normalizing the Image

The first step in Algorithm 9.2 is to normalize the digitized fingerprint image A. If μ is the mean value of the elements of A and m_1 and M_1 are the minimum and maximum elements of A, then we normalize A to obtain $\tilde{A}$, where the elements of $\tilde{A}$ are

$$\tilde{A}_{i,j} = \frac{A_{i,j} - \mu}{R}$$

Here

$$R = \frac{1}{128}\max\{M_1 - \mu, \mu - m_1\} \tag{9.46}$$

Bradley, Brislawn, and Hopper [7] remark that this normalization produces a mean of approximately zero in the approximation portion of the iterated packet transformation (region 0 in Figure 9.29) that is computed in the second step of the algorithm. In this way the distribution of values in this portion of the transformation is similar to the distribution of other portions of the transformation, making the Huffman coding in the final step more efficient. For the fingerprint plotted in Figure 9.27, we have $\mu = 193.074$, $m_1 = 22$, and $M_1 = 253$. The mean of $\tilde{A}$ is 1.22171×10^{-15}.

Applying the Discrete Wavelet Packet Transform

The second step in Algorithm 9.2 is the application of a discrete wavelet packet transformation to $\tilde{A}$. This packet transform is constructed from the CDF97 biorthogonal filter pair from Section 8.3 and is modified via Algorithm 8.3 to better handle the boundary effects produced by the transform matrix. The similar lengths of each filter in the pair and the balanced smoothness of the corresponding scaling functions were desirable features for the designers of the standard.

The number of possible representations of the two-dimensional transform grows quickly as the number of iterations increase,[27] and it is thus interesting to learn how the standard designers choose the "best" packet representation. We introduced the best basis algorithm in Section 9.3 and it would seem natural to expect it to be used in conjunction with some cost function that measures the effectiveness of each portion of the transformation with regard to image compression. But the designers of the standard did not use the best basis algorithm — undoubtedly, the cost of computing a best basis for each fingerprint weighed in their decision not to use it. Unlike the compression of generic images, the fingerprint compression considers a class of images with many similar traits and tendencies. The authors of the standard [7] used empirical studies and analysis of the *power spectral density*[28] of fingerprint images to determine a static packet representation that could be used for *all* digitized fingerprints. The power spectral density analysis gives information about the frequencies at which values in the packet transform occur. This information is useful when designing a quantization method that will be most effective with the static wavelet packet representation selected by the standard designers.

For fingerprint compression, the static representation used in the WSQ standard is displayed in Figure 9.29. It is interesting to note that the schematic in Figure 9.29 implies that five iterations of the discrete wavelet packet transform are computed for the standard. The 64 portions of this discrete wavelet packet transform representation are numbered and referred to as the *bands* of the transform. Note that no portions of the first and third iterations are used, and only four bands of the fifth iteration are computed. We recognize bands 0 through 3 as the fifth iteration of the modified biorthogonal wavelet transform from Algorithm 8.3 applied to $\tilde{A}$. The transform of the fingerprint image from Figure 9.27 is plotted in Figure 9.30.

---

[27]Walnut remarks [62] that if A is $M \times M$ with $M = 2^i$, then there are more than $2^{M^2/2}$ possible wavelet packet representations.

[28]The power spectral density of a continuous signal can be viewed as the norm squared of its Fourier transform. If the signal is discrete, then the power spectral density is in terms of the norm squared of the Fourier series built from the signal.

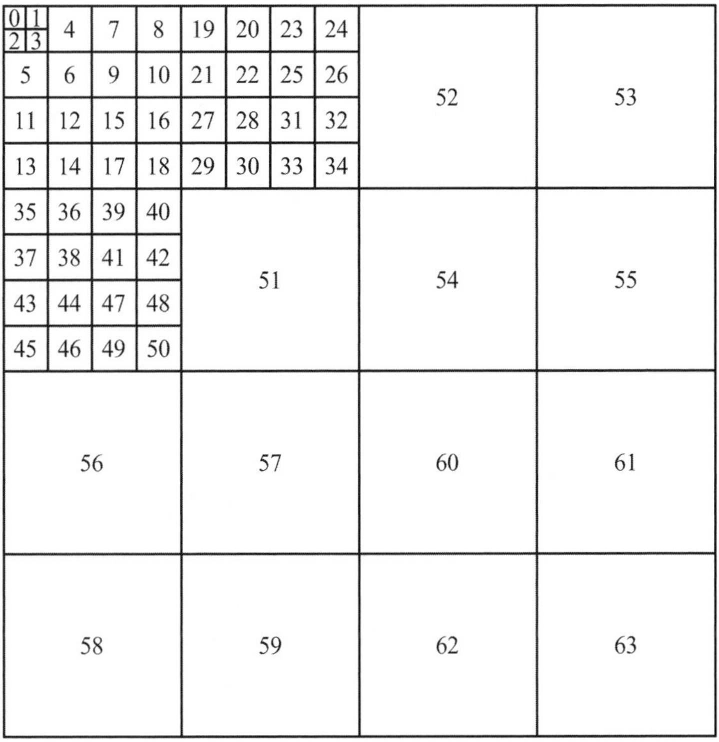

0 1 2 3	4	7	8	19	20	23	24		
5	6	9	10	21	22	25	26	52	53
11	12	15	16	27	28	31	32		
13	14	17	18	29	30	33	34		
35	36	39	40						
37	38	41	42	51		54		55	
43	44	47	48						
45	46	49	50						
56			57			60		61	
58			59			62		63	

Figure 9.29   The wavelet packet representation used for the second step of Algorithm 9.2.

Performing Quantization

Perhaps the most important step in the WSQ standard algorithm is quantization. Remarkably, bands 60 through 63 of the transform are discarded. Mathematically, we can think of the elements in these bands as quantized to zero. Quantization is performed on bands 0 through 59 using the piecewise constant function

$$f(t, Q, Z) = \begin{cases} \left\lfloor \dfrac{t - Z/2}{Q} \right\rfloor + 1, & t > Z/2 \\ 0, & -Z/2 \le t \le Z/2 \\ \left\lceil \dfrac{t + Z/2}{Q} \right\rceil - 1, & t < -Z/2 \end{cases} \tag{9.47}$$

Here Z and Q are positive numbers. The breakpoints of this function are $\pm \left(\frac{Z}{2} + kQ \right)$, $k \in \mathbb{Z}$. The value Z is called the *zero bin width* since it determines the range of values quantized to zero. The value Q is called the *bin width*. The quantization function is plotted in Figure 9.31.

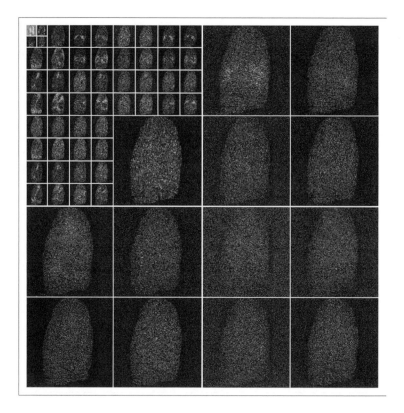

Figure 9.30   The wavelet packet transformation of the fingerprint from Figure 9.27.

For $k = 0, \ldots, 59$ we compute the bin widths Q_k and the zero bin width Z_k for band k and use these values to compute the quantized transform elements for each band. Indeed, if $w_{i,j}^k$ are the elements in band k, then we compute the quantized values

$$\hat{w}_{i,j}^k = f\left(w_{i,j}^k, Q_k, Z_k\right)$$

The zero bin width Z_k is assigned the value

$$Z_k = 1.2 Q_k \tag{9.48}$$

for $k = 0, \ldots, 59$ and Q_k is computed using

$$Q_k = \begin{cases} \dfrac{1}{q}, & 0 \le k \le 3 \\[3mm] \dfrac{10}{q A_k \ln\left(\sigma_k^2\right)}, & 4 \le k \le 59 \end{cases} \tag{9.49}$$

The values Q_k depend on values q, A_k, and σ_k. The constants $A_k \approx 1$ chosen by the FBI are listed in Table 9.4.

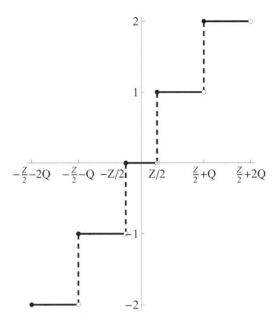

Figure 9.31   The quantization function for the FBI WSQ algorithm.

Table 9.4   The weights A_k used to determine the bin width (9.49).

Band	A_k
$4, \ldots, 51$	1.00
$52, 56$	1.32
$53, 55, 58, 59$	1.08
$54, 57$	1.42

The value σ_k^2 is the variance of a modified version of band k. In particular, we compute σ_k^2 from a subregion of band k. Suppose that the elements of band k are $\hat{w}_{i,j}^k$, where $0 \le i < M_k$ and $0 \le j < N_k$. Instead of using the entire $M_k \times N_k$ matrix, we instead set

$$\tilde{M}_k = \left\lfloor \frac{3M_k}{4} \right\rfloor \qquad \tilde{N}_k = \left\lfloor \frac{7N_k}{16} \right\rfloor \tag{9.50}$$

and consider the subregion with indices

$$i_{0,k} = \left\lfloor \frac{M_k}{8} \right\rfloor \le i \le i_{0,k} + \tilde{M}_k - 1 = i_{1,k}$$

$$j_{0,k} = \left\lfloor \frac{9N_k}{32} \right\rfloor \le j \le j_{0,k} + \tilde{N}_k - 1 = j_{1,k}$$

We compute σ_k^2 using the formula

$$\sigma_k^2 = \frac{1}{\tilde{M}_k \tilde{N}_k - 1} \sum_{i=i_{0,k}}^{i_{1,k}} \sum_{j=j_{0,k}}^{j_{1,k}} \left(\hat{w}_{i,j}^k - \mu_k \right)^2 \tag{9.51}$$

where μ_k is the mean of band k.

For example, band 39 is a square matrix with $M_{39} = N_{39} = 52$. Using (9.50) we see that $\tilde{M}_{39} = 39$ and $\tilde{N}_{39} = 22$. We compute the subregion boundary indices to be

$$i_{0,39} = \left\lfloor \frac{52}{8} \right\rfloor = 6, \quad i_{1,39} = 6 + 39 - 1 = 44$$

and

$$j_{0,39} = \left\lfloor \frac{9 \cdot 52}{32} \right\rfloor = 14, \quad j_{1,39} = 14 + 22 - 1 = 35$$

Using (9.51) we compute $\sigma_{39}^2 = 244.707$. If the variance $\sigma_k^2 < 1.01$ for any band k, then that band has all elements set to 0.

The final value that is used to compute the bin widths Q_k in (9.49) is the parameter q. This important value depends on the rate r selected for the compression process. A typical value of r is $r = 0.75$bpp and that is what we use for our example.

To compute q, we need some more values for each band. For $k = 0, \ldots, 59$, let

$$P_k = qQ_k = \begin{cases} 1, & 0 \leq k \leq 3 \\ \dfrac{10}{A_k \ln \left(\sigma_k^2 \right)}, & 4 \leq k \leq 59 \end{cases}$$

If the dimensions of A are $M \times N$, then we set

$$m_k = \frac{MN}{M_k N_k}$$

where the dimensions of the matrix in band k are $M_k \times N_k$. In engineering terms we can think of m_k as the *downsampling* rate for band k. For example, the dimensions of the fingerprint in Figure 9.27 are 832×832 and the dimensions of band 39 are $M_{39} \times N_{39} = 52 \times 52$. Thus $m_k = \frac{832^2}{52^2} = 256$. The values of m_k for the fingerprint in Figure 9.27 are given in Table 9.5.

Table 9.5   The values m_k for the fingerprint in Figure 9.27.

Band	m_k
$0, \ldots, 3$	1024
$4, \ldots, 50$	256
$51, \ldots 59$	16

For $S = \sum_{k=0}^{59} \frac{1}{m_k}$, Bradley and Brislawn [5, 6] showed that if q is defined by

$$q = (0.4) \cdot 2^{r/S-1} \left(\prod_{k=0}^{59} \left(\frac{\sigma_k}{P_k} \right)^{1/m_k} \right)^{-1/S} \tag{9.52}$$

then the quantization scheme will produce a bit rate of r bits per pixel when the quantized data are encoded. Using Table 9.5 we find that $S = \frac{3}{4}$. Plugging these values and the values for σ_k, P_k, and $r = 0.75$ into (9.52) gives $q = 0.0375$. So that we can get a sense of the size of the bin widths Q_k, we have listed them in Table 9.6 for the fingerprint in Figure 9.27.

Table 9.6   The bin widths Q_k from (9.49) for the fingerprint in Figure 9.27.

Band	Q_k	Band	Q_k	Band	Q_k	Band	Q_k
0	26.702	15	3.513	30	3.869	45	2.844
1	26.702	16	2.966	31	3.474	46	2.956
2	26.702	17	3.007	32	3.723	47	3.453
3	26.702	18	2.475	33	3.090	48	2.989
4	2.779	19	4.569	34	3.173	49	3.703
5	2.480	20	3.939	35	4.224	50	3.126
6	2.594	21	4.592	36	4.244	51	4.196
7	2.735	22	3.880	37	3.641	52	4.978
8	2.639	23	2.820	38	3.741	53	4.294
9	2.858	24	3.101	39	4.855	54	5.871
10	2.521	25	2.909	40	4.361	55	5.015
11	2.647	26	3.100	41	4.496	56	4.486
12	2.796	27	5.103	42	3.775	57	5.341
13	2.355	28	4.553	43	2.731	58	4.052
14	2.397	29	4.552	44	2.863	59	4.830

Using the values of Q_k from Table 9.6 and $Z_k = 1.2 Q_k$ in the quantization function (9.47), we quantize the first 60 bands of the discrete packet transform and replace the elements in the remaining four bands with zeros. The quantized packet transformation is plotted in Figure 9.32.

Encoding the Quantized Transform

The final step in Algorithm 9.2 is to encode the values in the quantized transformation. The WSQ standard uses an adaptive Huffman coding method that is much like the one used by the JPEG group (see Pennebaker and Mitchell [46]) for its compression standard for this task. This coding scheme is a bit more sophisticated than the basic Huffman coding method detailed in Appendix A. The interested reader is referred to Gersho and Gray [28]. Recall that the last four bands are not even included in the modified transform. The image is coded at 0.75 bit per pixel or at a compression ratio of 10.7 : 1.

Recovering the Compressed Image

To recover the compressed image, we perform the following steps:

1. Decode the Huffman codes.

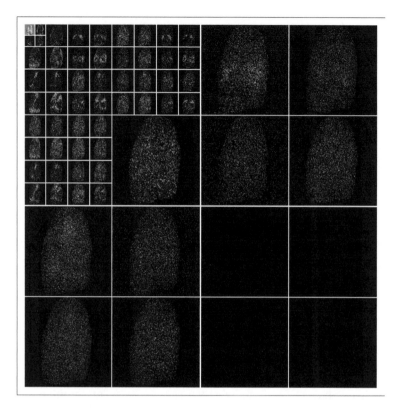

Figure 9.32   The quantized wavelet packet transformation of the fingerprint from Figure 9.27.

2. Apply a *dequantization function*.

3. Compute the inverse wavelet packet transform to obtain the normalized approximation $\tilde{B}$ to $\tilde{A}$.

4. For μ the mean of A and R given by (9.46), we compute

$$B_{i,j} = R\tilde{B}_{i,j} + \mu \qquad (9.53)$$

The dequantization function is interesting and we describe it now. The quantization function $y = f(t, Q, Z)$ (9.47) maps values larger than $Z/2$ to positive numbers, values between and including $-Z/2$ and $Z/2$ to zero, and values smaller than $-Z/2$ to negative numbers. Then it makes sense that the dequantization function $d(y, Q, Z)$ should be piecewise defined with different formulas for positive, negative, and zero values. The easiest case is $y = 0$. Here we define $d(0, Q, Z) = 0$, since we have no way of knowing which $t \in [-Z/2, Z/2]$ yields $f(t, Q, Z) = 0$. If $t > Z/2$, we have

$$y = f(t, Q, Z) = \left\lfloor \frac{t - Z/2}{Q} \right\rfloor + 1 \qquad \text{or} \qquad y - 1 = \left\lfloor \frac{t - Z/2}{Q} \right\rfloor$$

Exact inversion is impossible due to the floor function. Instead of subtracting 1 from y, the designers of the standard subtracted a value $0 < C < 1$ instead to account for the floor function. So for $y > 0$, we define

$$d(y, Q, Z) = (y - C)Q + Z/2$$

We can add the same "fudge factor" in the case when $y < 0$ and thus define our dequantization function as

$$d(y, Q, Z) = \begin{cases} (y - C)Q + Z/2, & y > 0 \\ 0, & y = 0 \\ (y + C)Q - Z/2, & y < 0 \end{cases} \qquad (9.54)$$

We apply $f(y, Q_k, Z_k)$ to each element in band k, $k = 0, \ldots, 59$. A typical value for C used by the FBI is $C = 0.44$. We have applied $d(y, Q_k, Z_k)$ to the packet transformation in Figure 9.32. The result is plotted in Figure 9.33. Note that we have added zero matrices in the place of bands 60 through 63.

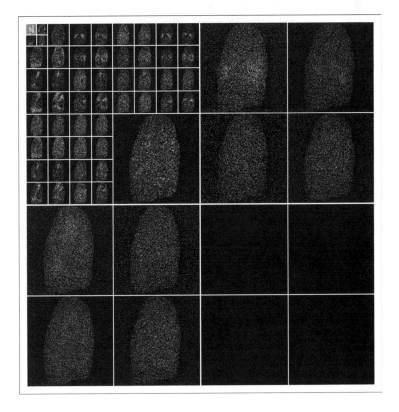

Figure 9.33   The dequantized wavelet packet transformation constructed by applying $d(y, Q_k, Z_k)$ to the matrix in Figure 9.32.

The last step in recovering the compressed fingerprint image is to apply (9.53) to each element in the image matrix in Figure 9.33. The result is plotted in Figure 9.34.

Figure 9.34   The fingerprint from Figure 9.27 compressed at 0.75 bit per pixel or at a ratio of 10.7 : 1.

The images in Figures 9.27 and 9.34 are difficult to distinguish. Indeed, the PSNR of the two images is 38.5462. In Figures 9.35 and 9.36 we have plotted the fingerprint of Figure 9.27 compressed at 0.40 and 0.25 bits per pixel, respectively. Table 9.7 summarizes the results.

Table 9.7   Different compression results for the fingerprint in Figure 9.27.

Bits per Pixel	Compression Ratio	q	PSNR
0.75	10.7 : 1	0.0375	38.5462
0.40	20 : 1	0.0271	37.3248
0.25	32 : 1	0.0236	36.7256

Figure 9.35 The fingerprint from Figure 9.27 compressed at 0.40 bit per pixel or at a ratio of 20 : 1.

Figure 9.36 The fingerprint from Figure 9.27 compressed at 0.25 bit per pixel or at a ratio of $32 : 1$.

Appendix A

Huffman Coding

The material that follows is adapted from Section 3.4 of Discrete Wavelet Transformations: An Elementary Approach with Applications*, by Patrick J. Van Fleet [60]
with permission from John Wiley & Sons, Inc.*

In this appendix we discuss a simple method for reducing the number of bits needed
to represent a signal or digital image.

Huffman coding, introduced by David Huffman [37], is an example of lossless
compression. The routine can be applied to integer-valued data and the basic idea is
quite simple. The method exploits the fact that signals and digital images often contain
elements that occur with a much higher frequency than other elements (consider, for
example, the digital image and its corresponding histogram in Figures 7.20 and 7.21).
Recall that intensity values from grayscale images are integers that range from 0 to
255, and each integer can be represented with an 8-bit ASCII code (see, e. g., Oualline

*

[45]). Thus, if the dimensions of an image are $N \times M$, then we need $8 \cdot NM$ bits to store it. Instead of insisting that each intensity be represented by 8 bits, Huffman suggested that we could use a variable number of bits for each intensity. That is, intensities that occurred most would be represented by a low number of bits, and intensities occurring infrequently would be represented by a larger number of bits.

Generating Huffman Codes

Let's illustrate how to create Huffman codes via an example.

Example A.1 *Consider the 5×5 digital grayscale image given in Figure A.1. The bitstream for this image is created by writing each character in binary form and then listing them consecutively. Here is the bit stream:*

> 00100001010100000101000001010000001000010101000000111000
> 01101000001110000101000000111000010100001111110001010000
> 00111000010100000011100001101000001110000101000000100001
> 01010000010100000101000000100001

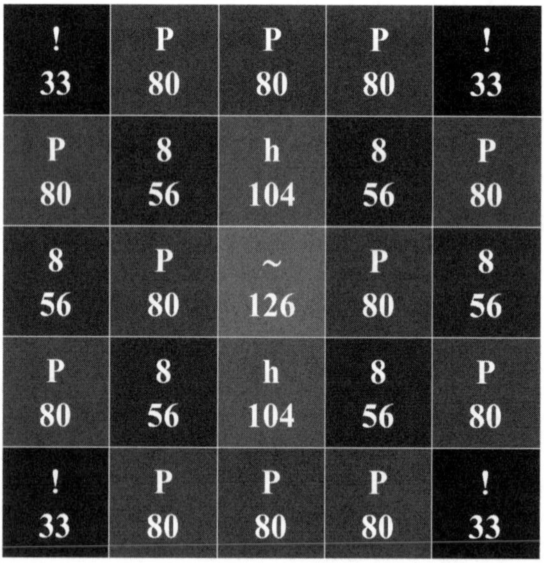

Figure A.1 A 5×5 grayscale image with its intensity levels superimposed.

Table A.1 lists the binary representations, ASCII codes, frequencies, and relative frequencies for each intensity.

To assign new codes to the intensities, we use a tree diagram. As is evident in Figure A.2, the first tree is simply one line of nodes that lists the intensities and their

Table A.1   The ASCII codes, binary representation, frequencies, and relative frequencies for the intensities that appear in Figure A.1.

Intensity	ASCII Code	Binary Rep.	Frequency	Rel. Frequency
33	!	00100001_2	4	.16
56	8	00111000_2	6	.24
80	P	01010000_2	12	.48
104	h	01101000_2	2	.08
126	~	01111110_2	1	.04

relative frequencies in nondecreasing order. Note that for this example, all the relative frequencies were distinct. When some characters have the same relative frequency, the order in which you list these characters does not matter — you need only ensure that all relative frequencies are listed in nondecreasing order.

Figure A.2   The first tree in the Huffman coding scheme. The characters and their associated relative frequencies are arranged so that the relative frequencies are in nondecreasing order.

The next step in the process is to create a new node from the two leftmost nodes in the tree. The probability of this new node is the sum of the probabilities of the two leftmost nodes. In our case, the new node is assigned the relative frequency $0.12 = 0.04 + 0.08$. Thus, the first level of our tree now has four nodes (again arranged in nondecreasing order), and the leftmost node spawns two new nodes on a second level as illustrated in Figure A.3.

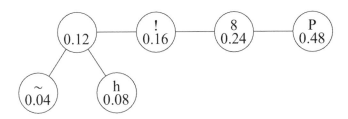

Figure A.3   The second tree in the Huffman coding scheme.

We repeat the process and again take the two leftmost nodes and replace them with a single node. The relative frequency of this new node is the sum of the relative frequencies of these two leftmost nodes. In our example, the new node is assigned the relative frequency 0.28 = 0.12 + 0.16. We again arrange the first level so that the relative frequencies are nondecreasing. Note that this new node is larger than the node with frequency 0.24. Our tree now consists of three levels, as illustrated in Figure A.4(a). We repeat this process for the two remaining characters at the top level. Figure A.4(b) and (c) illustrate the last two steps in the process and the final Huffman code tree.

Notice that on the final tree in Figure A.4(c), we label branches at each level using a 0 for left branches and a 1 for right branches. From the top of the tree, we can use these numbers to describe a path to a particular character. This description is nothing more than a binary number, and thus the Huffman code for the particular character. Table A.2 shows the Huffman codes for each character and the total bits needed to encode the image.

Table A.2 The Huffman code, frequency, and total bits required are given for each character.

Character	Huffman Code	Frequency	Total Bits
!	111	4	12
8	10	6	12
P	0	12	12
h	1101	2	8
~	1100	1	4
		Bits needed to encode the image:	**48**

Here is the new bitstream using Huffman codes:

111000111010110110010011000100101101100111000111

Note that the characters that appear most have the least number of bits assigned to their new code. Moreover, the savings is substantial. Using variable-length codes for the characters, the Huffman scheme needs 48 bits to represent the image. With no coding, each character requires 8 bits. Since our image consists of 25 pixels, we would need 200 bits to represent it. ∎

Algorithm for Generating Huffman Codes

Next, we review the general algorithm for creating Huffman codes for the elements in a set.

Algorithm A.1 (Huffman Code Generation) *Given set A, this algorithm generates the Huffman codes for each element in A. Let |A| denote the number of elements in A.*

Here are the basic steps:

1. For each distinct element in A, create node n_k with associated relative frequency $p(n_k)$.

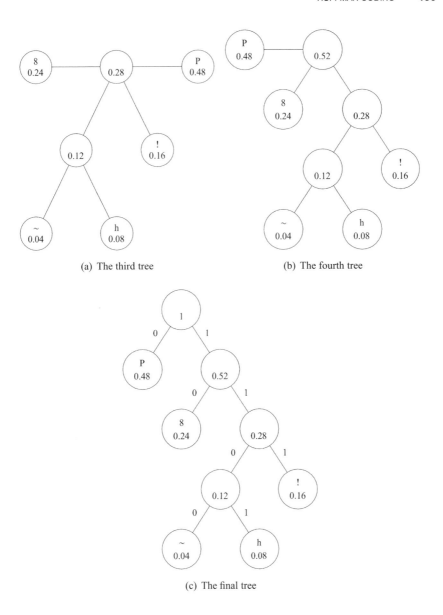

(a) The third tree   (b) The fourth tree

(c) The final tree

Figure A.4   The final steps in the Huffman coding algorithm.

2. *Order the nodes so that the relative frequencies form a nondecreasing sequence.*

3. *While $|A| > 1$:*

    (a) *Find nodes v_j and v_k with the smallest relative frequencies.*

(b) *Create a new node n_ℓ from nodes n_j and n_k with relative frequency $p(n_\ell) = p(n_j) + p(n_k)$.*

(c) *Add two descendant branches to node n_ℓ with descendant nodes n_j and n_k. Label the left branch with a 0 and the right branch with a 1.*

(d) *Drop nodes n_j and n_k from A and add node n_ℓ to A.*

(e) *Order the nodes so that the relative frequencies form a nondecreasing sequence.*

EndWhile

4. *For each distinct element in A, create a Huffman code by converting all the 0's and 1's on the descendant branches from the one remaining node in A to the node corresponding to the element into a binary number. Call the set of Huffman codes B.*

5. *Return B.*

■

It should be pointed out that simply generating the Huffman codes for signal is not the same as compressing it. Although Huffman codes do provide a way to reduce the number of bits needed to store the signal, we still must *encode* these data. The encoded version of the signal not only includes the new bit stream for the signal, but also some header information (essentially the tree generated by Algorithm A.1) and a pseudo end-of-file character. The encoding process is also addressed in

Not only is Huffman coding easy to implement, but as Gonzalez and Woods point out in [31], if we consider all the variable-bit-length coding schemes we could apply to a signal of fixed length, where the scheme is applied to one element of the signal at a time, then the Huffman scheme is optimal. There are several variations of Huffman coding as well as other types of coding. See Gonzalez and Wood [31] or Sayood [51] (or references therein) if you are interested in learning more about coding.

Decoding

Decoding Huffman codes is quite simple. You need only the code tree and the bitstream. Consider the following example.

Example A.2 *Consider the bitstream*

$$100111100110101111001100 \qquad (A.1)$$

that is built using the coding tree plotted in Figure A.5. The bitstream represents some text. Determine the text.

Solution

From the tree we note that $e = 0$, $w = 10$, $sp =$ " " $= 111$, $r = 1101$, and $t = 1100$. We begin the decoding process by examining the first few characters until

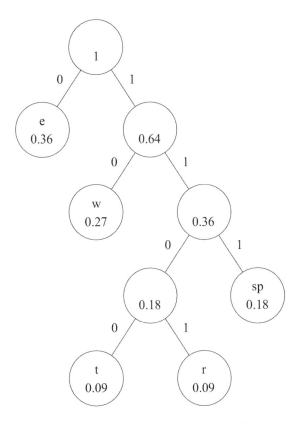

Figure A.5   The Huffman code tree for Example A.2. Here "sp" denotes a space.

we find a code for one of the letters above. Note that no character has 1 for a code, so we look at the first two characters, 01. This is the code for w, so we now have the first letter. We continue in this manner taking only as many bits as needed to represent a character in the tree. The next bit is 0, which is the code for e. The next code needs three bits — 111 is the space key. The next few characters are w (01), e (0), r (1101), and e (0). The last four characters are the space key (111), then w, e, and t, so that the coded string is "we were wet." ∎

Bits per Pixel

In this book we are interested primarily in using Huffman codes to assess the effectiveness of wavelet transformations in image compression. We usually measure this effectiveness by computing the bits per pixel (bpp) needed to store the image.

Definition A.1 (Bits per Pixel) *We define the* bits per pixel *of an image, denoted* bpp, *as the number of bits used to represent the image divided by the number of pixels.*

∎

For example, the 5×5 image in Figure A.1 is originally coded using 8 bpp. Since Huffman coding resulted in a bitstream of length 48, we say that we can code the image via the Huffman scheme using an average of $48/25 = 1.92$ bpp. As you will soon see, wavelets can drastically reduce this average!

PROBLEMS

A.1   Consider the string *mississippi*.

(a) Generate the Huffman code tree for the string.

(b) Look up the ASCII value for each character in *mississippi*, convert them to base 2, and then write the bitstream for *mississippi*.

(c) Write the bitstream for the string using the Huffman codes.

A.2   Repeat Problem A.1 for the string *terminal*.

A.3   Consider the 4×4 image whose intensity matrix is

$$
\begin{matrix}
100 & 100 & 120 & 100 \\
100 & 50 & 50 & 40 \\
100 & 40 & 40 & 50 \\
120 & 120 & 100 & 100
\end{matrix}
$$

(a) Generate the Huffman code tree for the image.

(b) Write the bitstream for the image using Huffman codes.

(c) Compute the bpp for this bitstream.

A.4   Suppose that a and b are two characters in a string. Is it possible for the Huffman code tree to represent a as 00 and b as 000? Why or what not? (*Hint:* If so, how would you interpret a segment of a bitstream that was represented by 00000?)

A.5   As a generalization of Problem A.4, explain why it is impossible for a Huffman code consisting of n bits to be the first n bits of a longer Huffman code. This is an important feature for decoding — since no code can serve as a *prefix* for another code, we can be assured of no ambiguities when decoding bitstreams.

A.6   Suppose that a 40×50 image consists of the four intensities 50, 100, 150, and 200 with relative frequencies $\frac{1}{8}, \frac{1}{8}, \frac{1}{4}$, and $\frac{1}{2}$, respectively.

(a) Create the Huffman tree for the image.

(b) What is the bpp for the Huffman bitstream representation for the image?

(c) What is the entropy of the image?

(d) Recall in the discussion preceding Definition 4.3 that Shannon [53] showed that the best bpp we could hope for when compressing a signal is the signal's entropy. Do your findings in parts (b) and (c) agree with Shannon's postulate?

A.7   Can you create an image so that the bpp for the Huffman bitstream is equal to the entropy of the image?

A.8   Consider the Huffman codes $e = 0$, $n = 10$, $s = 111$, and $t = 110$.

(a) Draw the Huffman code tree for these codes.

(b) Assume that the frequencies for e, n, s, and t are $4, 2, 2$, and 1, respectively. Find a word that uses all these letters.

(c) Are these the only frequencies that would result in the tree you plotted in part (a)? If not, find some other frequencies for the letters that would result in a tree with the same structure.

A.9   Given the Huffman codes $g = 10, n = 01, o = 00, space\, key = 110, e = 1110$, and $i = 1111$, draw the Huffman code tree and use it to decode the bitstream

$$10001111011011010001111011011010000011110$$

A.10   Suppose that you wish to draw the Huffman tree for a set of letters S. Let $s \in S$. Explain why the node for s cannot spawn new branches of the tree. Why is this fact important?

References

1. Robert G. Bartle and Donald R. Sherbert, *Introduction to Real Analysis*, 3rd ed., John Wiley, New York, 1999.

2. George Benke, Maribeth Bozek-Kuzmicki, David Colella, Garry M. Jacyna, and John J. Benedetto, *Wavelet-based analysis of electroencephalogram (EEG) signals for detection and localization of epileptic seizures*, Proceedings of SPIE: Wavelet Applications II (Harold H. Szu, ed.), vol. 2491, April 1995, pp. 760–769.

3. A. Bijaoui, E. Slezak, F. Rue, and E. Lega, *Wavelets and the study of the distant universe*, Proc. of the IEEE **84** (1996), no. 4, 670–679.

4. Albert Boggess and Francis J. Narcowich, *A First Course in Wavelets with Fourier Analysis*, Prentice Hall, Upper Saddle River, NJ, 2001.

5. J. N. Bradley and C. M. Brislawn, *Proposed first-generation WSQ bit allocation procedure*, Proceedings of the Symposium on Criminal Justice Information Services Technology (Gaithersburg, MD), Federal Bureau of Investigation, September 1993, pp. C11–C17.

6. _____, *The wavelet/scalar quantization compression standard for digital fingerprint images*, Proceedings of the International Symposium on Circuits and Systems (London), vol. 3, IEEE Circuits and Systems Society, June 1994, pp. 205–208.

7. J. N. Bradley, C. M. Brislawn, and T. Hopper, *WSQ gray–scale fingerprint image compression specification*, Tech. Report Revision 1, Federal Bureau of Investigation, Washington, DC, February 1992.

8. _____, *The FBI wavelet/scalar quantization standard for gray-scale fingerprint image compression*, Proceedings of SPIE: Visual Information Processing (Orlando, FL), vol. 1961, April 1993, pp. 293–304.

9. J. N. Bradley, C. M. Brislawn, R. J. Onyshczak, and T. Hopper, *The FBI compression standard for digitized fingerprint images*, Proceedings of SPIE: Applied Digital Image Processing XIX (Denver, CO), vol. 2847, August 1996, pp. 344–355.

10. C. M. Brislawn, *Fingerprints go digital*, Notices Amer. Math. Soc. **42** (1995), no. 11, 1278–1283.

11. A. Calderbank, I. Daubechies, W. Sweldens, and B.-L. Yeo, *Wavelet transforms that map integers to integers*, Appl. Comp. Harm. Anal. **5** (1998), no. 3, 332–369.

12. E. Ward Cheney and David R. Kincaid, *Numerical Analysis: Mathematics of Scientific Computing*, 3rd ed., Brooks/Cole, Belmont, CA, 2001.

13. A. Cohen, I. Daubechies, and J.-C. Feauveau, *Biorthogonal bases of compactly supported wavelets*, Comm. Pure Appl. Math. **45** (1992), 485–560.

14. R. Coifman, Y. Meyer, and M. V. Wickerhauser, *Size properties of wavelet packets*, Wavelets and Their Applications, Jones and Bartlett, Boston, MA, 1992, pp. 453–470.

15. _____, *Wavelet analysis and signal processing*, Wavelets and Their Applications, Jones and Bartlett, Boston, MA, 1992, pp. 153–178.

16. R. Coifman and M. V. Wickerhauser, *Entropy-based algorithms for best basis selection*, IEEE Trans. Inform. Theory **38** (1992), 713–718.

17. I. Daubechies, *Orthogonal bases of compactly supported wavelets*, Comm. Pure Appl. Math. **41** (1988), 909–996.

18. _____, *Orthonormal bases of compactly supported wavelets II. Variations on a theme*, SIAM J. Math. Anal. **24** (1993), no. 2, 499–519.

19. I. Daubechies and J. Lagarias, *Two-scale difference equations I. Existence and global regularity of solutions*, SIAM J. Math. Anal. **22** (1991), 1388–1410.

20. Ingrid Daubechies, *Ten Lectures on Wavelets*, Society for Industrial and Applied Mathematics, Philadelphia, 1992.

21. Carl de Boor, *A Practical Guide to Splines*, revised ed., Springer-Verlag, New York, 2001.

22. Morris H. DeGroot and Mark J. Schervish, *Probability and Statistics*, 3rd ed., Addison-Wesley, Reading, MA, 2002.

23. D. Donoho, *Wavelet shrinkage and W.V.D.: A 10-minute tour*, Progress in Wavelet Analysis and Applications (Toulouse, France) (Y. Meyer and S. Rogues, eds.), Editions Frontiers, 1992, pp. 109–128.

24. D. Donoho and I. Johnstone, *Ideal spatial adaptation via wavelet shrinkage*, Biometrika **81** (1994), 425–455.

25. _____, *Adapting to unknown smoothness via wavelet shrinkage*, J. Amer. Stat. Assoc. **90** (1995), no. 432, 1200–1224.

26. Federal Bureau of Investigation, *Criminal Justice Information Services (CJIS) Web site*, http://www.fbi.gov/hq/cjisd/, 2009.

27. Michael W. Frazier, *An Introduction to Wavelets Through Linear Algebra*, Undergraduate Texts in Mathematics, Springer-Verlag, New York, 1999.

28. Allen Gersho and Robert M. Gray, *Vector Quantization and Signal Compression*, Kluwer Academic, Norwell, MA, 1992.

29. J. Gibbs, *Fourier series*, Nature **59** (1898), 200.

30. _____ , *Fourier series*, Nature **59** (1899), 606.

31. Rafael C. Gonzalez and Richard E. Woods, *Digital Image Processing*, 2nd ed., Pearson Prentice Hall, Upper Saddle River, NJ, 2002.

32. F. Hampel, *The influence curve and its role in robust estimation*, J. Amer. Stat. Assoc. **69** (1974), no. 346, 383–393.

33. Mark R. Hawthorne, *Fingerprints: Analysis and Understanding*, CRC Press, Boca Raton, FL, 2008.

34. Stephen G. Henry, *Catch the (seismic) wavelet*, AAPG Explorer (1997), 36–38.

35. _____ , *Zero phase can aid interpretation*, AAPG Explorer (1997), 66–69.

36. Don Hong, Jianzhong Wang, and Robert Gardner, *Real Analysis with an Introduction to Wavelets and Applications*, Academic Press, San Diego, CA, 2004.

37. David A. Huffman, *A method for the construction of minimum-redundancy codes*, Proc. Inst. Radio Eng. **40** (1952), 1098–1101.

38. David W. Kammler, *A First Course in Fourier Analysis*, 2nd ed., Cambridge University Press, New York, 2008.

39. Fritz Keinert, *Wavelets and Multiwavelets*, Chapman & Hall/CRC, Boca Raton, FL, 2004.

40. Thomas W. Körner, *Fourier Analysis*, Cambridge University Press, New York, 1989.

41. Stéphane Mallat, *Multiresolution approximations and wavelet orthonormal bases of $l^2(\mathbb{R})$*, Trans. Amer. Math. Soc. **315** (1989), 69–87.

42. _____ , *A Wavelet Tour of Signal Processing*, 3rd ed., Academic Press, San Diego, 2008.

43. Carl D. Meyer, *Matrix Analysis and Applied Linear Algebra*, Society for Industrial and Applied Mathematics, Philadelphia, 2000.

44. Yves Meyer, *Wavelets and Operators*, Advanced Mathematics, Cambridge University Press, 1992.

45. Steve Oualline, *Practical C++ Programming*, O'Reilly, Sebastopol, CA, 2003.

46. William B. Pennebaker and Joan L. Mitchell, *JPEG Still Image Compression Standard*, Van Nostrand Reinhold, New York, 1992.

47. D. Ruch and P. J. Van Fleet, *Gibbs' phenomenon for nonnegative compactly supported scaling vectors*, J. Comp. Appl. Math. **304** (2004), 370–382.

48. Walter Rudin, *Principles of Mathematical Analysis*, 3rd ed., McGraw-Hill, New York, 1976.

49. Diego Santa-Cruz and Touradj Ebrahimi, *An analytical study of JPEG2000 functionalities*, Proceedings of the International Conference on Image Processing (Vancouver, CA), IEEE, September 2000.

50. Karen Saxe, *Beginning Functional Analysis*, Springer-Verlag, New York, 2002.

51. Khalid Sayood, *Introduction to Data Compression*, 2nd ed., Morgan Kaufmann, San Francisco, 2000.

52. Larry L. Schumaker, *Spline Functions: Basic Theory*, 3rd ed., Cambridge University Press, New York, 2007.

53. Claude E. Shannon, *A mathematical theory of communication*, Bell Syst. Tech. J. **27** (1948), no. 3, 379–423, Continued in 27(4): 623–656, October 1948.

54. _____, *Communication in the presence of noise*, Proc. Inst. Radio Eng. **37** (1949), no. 1, 10–21.

55. James Stewart, *Calculus: Early Transcendentals*, 6th ed., Brooks/Cole, Belmont, CA, 2003.

56. Gilbert Strang, *Introduction to Linear Algebra*, 2nd ed., Wellesley-Cambridge Press, Wellesley, MA, 1998.

57. Gilbert Strang and Truong Nguyen, *Wavelets and Filter Banks*, Wellesley-Cambridge Press, Wellesley, MA, 1996.

58. W. Sweldens, *The lifting scheme: A construction of second generation wavelets*, SIAM J. Math. Anal. **29** (1997), no. 2, 511–546.

59. David Taubman and Michael Marcellin, *JPEG2000: Image Compression Fundamentals, Standards and Practice*, The International Series in Engineering and Computer Science, Kluwer Academic, Norwell, MA, 2002.

60. Patrick J. Van Fleet, *Discrete Wavelet Transformations: An Elementary Approach with Applications*, John Wiley, Hoboken, NJ, 2008.

61. Brani Vidakovic, *Statistical Modeling by Wavelets*, John Wiley, New York, 1999.

62. David F. Walnut, *An Introduction to Wavelet Analysis*, Birkhäuser, Boston, 2002.

63. Mladen Victor Wickerhauser, *Adapted Wavelet Analysis from Theory to Software*, A. K. Peters, Natick, MA, 1994.

Topic Index

Author Index